DOLAUCOTHI-PUMSAINT

Survey and excavations at a Roman Gold-Mining complex 1987–1999

by

Barry and Helen Burnham

with contributions by

Alwyn Annels, R. N. E. Barton, Edward Besly, Dee Brennan, Astrid Caseldine, Hilary Cool, Peter Crew, John Crowther, Brenda Dickinson, Quentin Dresser, Gill Evans, Martin Henig, Jennifer Hillam, Robert Hopkins, Heather James, Su Johnson, David Jordan, Michael Walker and Peter Webster

UNIVERSITY OF WALES
LAMPETER

1·8·2·2

PRIFYSGOL CYMRU
LLANBEDR PONT STEFFAN

Cadw

Oxbow Books

YR YMDDIRIEDOLAETH GENEDLAETHOL
THE NATIONAL TRUST

Published by
Oxbow Books, Park End Place, Oxford OX1 1HN

ISBN 1 84217 112 7

A CIP record for this book is available from the British Library

This book is available direct from
Oxbow Books, Park End Place, Oxford OX1 1HN
(Phone: 01865-241249; Fax: 01865-794449)

and

The David Brown Book Company
PO Box 511, Oakville, CT 06779, USA
(Phone: 860-945-9329; Fax: 860-945-9468)

and

via our website
www. oxbowbooks.com

Front cover picture: the late burnt timber building within the fort at Pumsaint during excavation in 1989

Back cover, main picture: general view of the Gwenlais leat.
Top left: a cornelian gemstone from the 1989 excavations inside the fort at Pumsaint, depicting the goddess Nemesis.
Top right: the stone wall of the reduced fortlet at Pumsaint during the 1989 excavations.
Bottom left: the processing wastes from the 1992 excavations near the Carreg Pumsaint

Printed in Great Britain by
Antony Rowe, Chippenham

CONTENTS

LIST OF ILLUSTRATIONS

LIST OF FIGURES

LIST OF TABLES

ACKNOWLEDGEMENTS

Numerous bodies and individuals have contributed in various ways over the years to the work presented in this volume. While we have endeavoured to mention all the major players here, we are conscious that others too numerous to name individually have also helped in lesser, but nonetheless much appreciated, ways; our thanks are due to them all.

Chief among those without whom the work could never have proceeded is the National Trust, the owner of the Dolaucothi estate on which all the projects except the 1990 Gwenlais Leat section were carried out. The Trust's help has gone far beyond the mere granting of access permission, however. Their works teams provided practical assistance, particularly in the form of site preparation and fencing on all the major excavations, and undertook the demolition of the standing building prior to the 1989 work. In addition, Trust facilities both at the mine and, more recently, in the village were made available to field teams and were much appreciated. Substantial Trust funding was also crucial to the completion of most of the projects, in particular that of 1989 which it financed in full. Trust staff who have been particularly helpful over the years include (in alphabetical order): Emma Plunkett Dillon, Emrys Evans, Susan Lloyd Fern, Carl Jenkins, Alison Mellor, Peter Mitchell, Brian Parry and David Thackray. Access was also kindly provided to the vernacular report and photographs on the standing buildings, reproduced here as section 2.2.3(N).

Several of the projects were carried out on land farmed by the National Trust's tenant farmers, who have also been most helpful and co-operative. Particular thanks are due to Mr A Davies, Ynysau Uchaf (Annell Leat 1993 and Roman Road 1997), who put up with a fair amount of disruption and kindly provided a tow out of the mire when needed, while Mr E Williams, Pentwyn, Mr G Williams, Llwynceiliog and Mr N Williams, Dolaucothi Farm have all at one time or another allowed us to work on their land and/or use their access tracks.

We must also gratefully acknowledge the interest and various acts of kindness of the people of Pumsaint. Gerallt and Daphne Morgan of the Post Office in particular took a lively interest in the proceedings, lent occasional equipment, provided a constant supply of cold drinking water to a parched 1989 excavation crew and lent the photograph of the early 20th century village (Fig. 2.71). Successive tenants of the Dolaucothi Arms, the Forge (who forgot to send us the bill!), the Co-op and the local cricket team (who allowed the 1999 work in the middle of their playing season) all also contributed in their various ways. The late Tom Cleminson and his wife, Pearl, of Ogofau Lodge were particularly helpful to the 1991–3 excavation, allowing the use of their facilities and helping to make the site good afterwards, as well as helping to relocate some of the 'lost' millstones from the earlier investigations.

The Dyfed Archaeological Trust (now Archaeoleg Cambria Archaeology) was the formal contractor for a number of the projects, in particular the 1987 and 89 excavations, the 1995 and 96 watching briefs and the excavation element of the 1997 work. Without their input, much of this work would not have been possible. They have provided very considerable and much appreciated help in various forms over the years. Special thanks go to Don Benson and Heather James for helping to facilitate various parts of the work, to Ken Murphy and Andrew Manning for exceptional supervisory assistance, and to Dee Brennan for keeping track of the finds. The bulk of the illustrations for this volume were expertly drawn by Trust staff Ian Darke and Neil Ludlow, while Kevin Evans and Ian Clewes at Lampeter oversaw their photographic reproduction and scanning respectively. The final process of generating digitised copy lay with HBB.

Our heartfelt thanks are also due to the University of Wales, Lampeter, formerly St David's University College, our home institution, and to successive Heads of its Department of Archaeology. They encouraged the work and provided time, equipment, student workforce and technical services in various forms, as well as several grants from the Pantyfedwen Fund and from Departmental resources.

Cadw: Welsh Historic Monuments has taken a close interest in much of the work, a good deal of which was carried out within scheduled areas as the subject of consents. Crucially, Cadw also funded the post-excavation work on the 1989 site and the 1990 leat sections project, neither of which would otherwise have been possible. Particular thanks must go to the successive Regional Inspectors, Sian Rees and Rick Turner, for their support and encouragement.

A number of other sources of funding have also

provided generous and much appreciated support for the various projects over the years. In alphabetical order these are: the Board of Celtic Studies, the Carmarthenshire Antiquarian Society, the Mining Field Centre (University of Wales College, Cardiff) and the Society of Antiquaries of London. Generous contributions towards the cost of publication were provided by Cadw: Welsh Historic Monuments, The National Trust and the Pantytedwen Fund, University of Wales, Lampeter.

While the support of all those already mentioned has been crucial to the running of the various projects, we must not forget our debt to all the others who have worked with us. Many of those who willingly gave up considerable time and effort to offer us the benefits of their specialist knowledge are credited with their own contributions elsewhere in the volume. More generally, however, we should like to pay tribute here to the stimulus of the late Professor Barri Jones, whose work did so much to generate the framework for our research. Though not always agreeing with our conclusions, he always gave generously of his time and ideas when discussing the sites.

None of the work would have been possible, however, without our valiant workforce on the ground, who toiled in conditions varying from the heatwave of summer 1989 to the deluges of summers 1990 and 1997, endured the repetitive complexities of the 1991–3 site with good humour (and helped with the baling out), and some of whom even kept coming back for more. Though too numerous to mention individually, we extend our deepest thanks to them all.

Last, but not least, we would like to thank Jeff Davies who undertook the soul-destroying task of reading through and commenting on the finished text, prior to its publication. His comments, as usual, were helpful and much appreciated.

Key to all sections

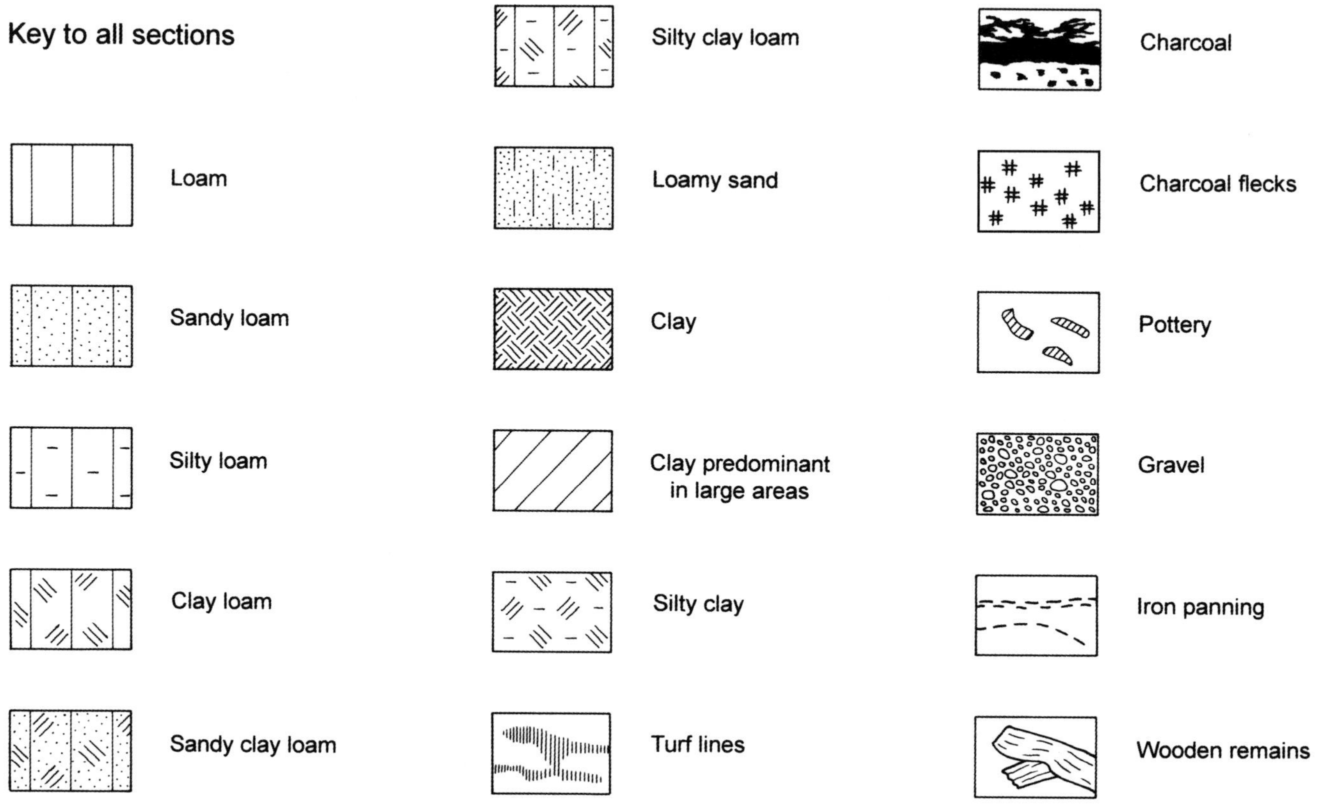

1 INTRODUCTION

The gold-mines at Dolaucothi have exercised a particular fascination in much recent literature, because they represent the only site in Britain where the Romans are certainly known to have mined for gold. They are situated in south-west Wales, centred at SN 664 402, just off the A482 road from Llanwrda to Lampeter, *c.* 1 km south-east of the modern village of Pumsaint, itself the site of a recorded Roman fort (Fig. 1.1). The main workings, which are thought to span various phases of exploitation from the pre-Roman through to the present, can be traced over a distance of more than a kilometre, extending north-east to south-west along a mountain spur forming the south-eastern side of the Cothi valley. They comprise a series of large opencast pits, trenches and underground galleries, several leats supplying water to a system of tanks and reservoirs, and the remains of various processing mills and their associated buildings (for a recent, general account, see Annels and Burnham 1995). Today these features form part of a much larger estate extending over *c.* 2500 acres, which was given to the National Trust in 1941–1943

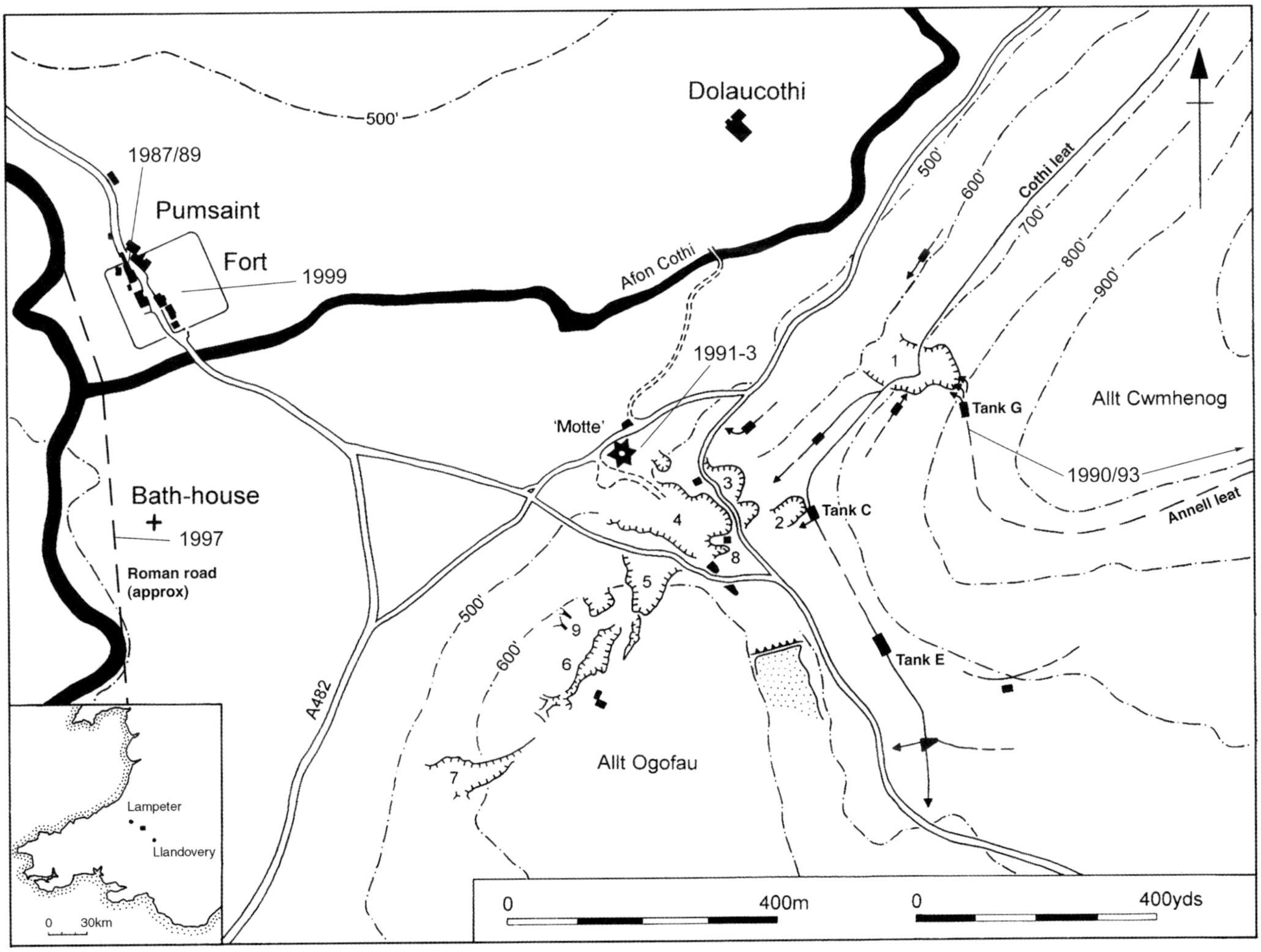

Fig. 1.1 Location plan of gold-mines, fort and bath-house

by Mr H. T. C. Lloyd-Johnes, whose family is known to have owned the land since the time of Henry VII. This volume seeks to report on a series of investigations carried out at this important complex over a period of 12 years from 1987 to 1999.

1.1 HISTORY OF EXCAVATION AND RESEARCH

The mine workings and associated features in the Dolaucothi-Pumsaint area have been a source of increasing speculation and interest for over three centuries, since their first brief mention in the 1695 version of Camden's *Britannia* (edited by Gibson). Somewhat more informative are the extracts from a journal by Sir Joseph Banks, dated 26th August 1767, conveniently collated by G. Eyre Evans (1921–2, 14–15), which clearly record a visit to one of the underground galleries on Allt Ogofau, most likely that now known to us as Upper Roman adit (Fig. 1.1; no. 9). Both he and slightly later antiquarian observers (*e.g.* Williams 1818) were also well aware of the existence of an old watercourse or leat, which could be traced over a distance of *c.* 11 km from the point where it tapped the river Cothi to near the top of the workings, and of the Carreg Pumsaint which, even then, was recognised as a probable mortar stone, despite its local association with the legend of the 'Five Saints'. Already by the early 19th century, local tradition was ascribing the workings to the Romans (*e.g.* Murchison 1839, 367–9), a theory which readily gained credence from the existence of a hoard of Roman gold jewellery, first discovered in 1796 or 1797, and subsequently added to in 1819 (Bosanquet 1917, 25–7), and from the haphazard 'excavation' of the remains of a possible bath-house in the 1830s south of the river and west of the A482 (*ibid.*, 30–1). The nature of the metal being mined was not apparently established for certain, however, until as late as 1844, when members of the Geological Society identified a speck of free gold in a fragment of the ubiquitous quartz found at the site (Smyth 1846). Subsequent mining operations, inevitably prompted by this important discovery, have often brought new information to light, but they have also added a new element of complexity to the surviving surface remains (see Hall 1993, 39–48).

The mine workings themselves were first systematically described in the early years of the 20th century by Haverfield (1909–10) and Bosanquet (1917, 27–30), at which time the chance was also taken to explore the line of the Cothi leat and one of the tanks along its course (Fig. 1.1; Tank E). Interest was further aroused during renewed mining operations in the 1930s, when tunnelling from the mine-shaft at the entrance to the Ogofau Pit allowed access to a series of large underground stopes extending up to 45m below the surface (Nelson 1944, 6–7, fig. 3); here the miners discovered a number of well-preserved wooden objects, including a possible panning cradle (or shovel?), a ladder and, most important of all, a fragment from the rim of a drainage wheel of well-known Roman type, now in the National Museum of Wales (Davies 1936; Boon and Williams 1966). These and earlier discoveries were conveniently brought together in a single summary by Nash Williams (1952). Further fieldwork in 1959 confirmed the detailed line of the Cothi leat, whilst also estimating that it could deliver between 2.5 and 3.5 million gallons per day (Jones *et al.* 1960). Such discoveries paved the way for a more exhaustive exploration of the surface remains in the Dolaucothi-Pumsaint area.

The most detailed investigations of the site were undertaken in the late 1960s and early 1970s, under the aegis of the Dolaucothi Research Committee of the National Museum of Wales, involving two separate, though inter-related, projects: the first combined surface survey and selective excavation to elucidate, in particular, the history of Roman exploitation at the site, directed primarily by G. D. B. Jones and P. R. Lewis; the second saw an exploration of the accessible underground galleries, most of them of 19th- and 20th-century date, by the then Department of Mineral Exploitation, University College, Cardiff. The archaeological dimension of this work included one of the first truly systematic descriptions of the principal surface workings, a further investigation of the Cothi leat, the discovery of the so-called Annell leat, running at a higher level above the workings, and the excavations within the recently discovered fort beneath the modern village of Pumsaint (Lewis and Jones 1969; Jones and Lewis 1971a; Jones and Little 1973; 1974; Lewis 1977). Parallel investigations in the gold-mining areas of Spain, especially in the vicinity of Las Medulas, also provided valuable comparative material, not least in terms of the importance of water to the mining operations (Bird 1972; 1984; Jones and Bird 1972; Lewis and Jones 1970; Jones 1980). Such work provided the framework for a simple developmental model of Roman mining at Dolaucothi (see section 1.3 below), the broad details of which continue to be rehearsed by recent commentators (*e.g.* Bird 2001).

Following this pioneering research, much of which was sadly never published beyond interim summaries, there was a lull in the pace of archaeological activity until renewed operations began in 1982 under the aegis of the then Archaeology Unit, now the Department of Archaeology, University of Wales, Lampeter. This led initially to the identification of a possible water-driven mill complex and its associated features in 1982 (Austin and Burnham 1984), the recording of the exposed archaeological section south of the fort in 1985 (Burnham and Burnham 1987a), and limited survey work on Upper and Lower Roman adits in 1986, prior to their being lit and opened to the public (Burnham and Burnham 1988); a student field class in 1983 also led to the identification and subsequent recording of 11 complete pillow mounds, scattered in five groups, on Allt Ogofau (Gerrard 1988). Such small-scale interventions paved the way for a more concerted programme of work over an extended period from 1987 to 1999, the detailed results of which are reported in this volume. For the location of the key sites see Fig. 1.1.

1987/89: In 1987, trial excavations were undertaken within the north-west part of the fort at Pumsaint, west of the A482, to assess the likely archaeological implications of a National Trust proposal to build houses within the modern village (Burnham and Burnham 1987b). This paved the way for a major 13–week excavation in 1989, carried out in conjunction with the Dyfed Archaeological Trust (site code PUM 89), which added significant new details about the chronology and internal layout of the site, including evidence relating to its two-phase defences, the *intervallum* area, and the successive phases of internal building (see section 2.2 below).

1990/93: The discovery of the Gwenlais leat in 1988 (see Figs 1.2 and 3.1; Bick 1989) refocused attention briefly on the water systems associated with the mines. As a consequence, in 1990, in response to a commission from Cadw: Welsh Historic Monuments, sections were cut at various points across the visible remains in the vicinity of the upper Annell and Gwenlais rivers, recovering in the process valuable evidence for the reconstruction of the local environment. This was followed, in 1993, by a survey of the earthworks associated with Tank G, at the postulated terminus of the so-called Annell leat, and by a section across the line of the latter in the field immediately to the south. Despite the new details which emerged much still remains uncertain (see chapter 3).

1991–93: More importantly, the identification of the site of a possible mill complex in 1982 raised several important research questions, which were initially addressed by small-scale work in 1984, in the area immediately north of the complex, adjacent to the possible 'motte' (Austin and Burnham 1985). The results proved sufficiently promising to warrant a larger-scale operation over three successive seasons between 1991 and 1993 (site code DOL 91–93), which revealed much new evidence about ore crushing and processing at the site, whilst at the same time advancing a probable Roman date for such activities and, potentially, also for the postulated mill (see chapter 4).

1997: Prior to 1997, little work had been directed at investigating areas of related settlement. In that year, at the request of the National Trust, trial excavations were carried out in field 5532, on a headland formed by a fossil river terrace south of the Cothi and west of the postulated bath-house (site code DOL 97). These were designed to assess the nature of a possible corridor-type villa, first reported from a resistivity survey by Geophysical Surveys of Bradford (Report 94/19, March 1994). In the event, the building proved to be no more than the line of the Roman road from Llandovery to Pumsaint, flanked by evidence of associated settlement along its eastern side (see chapter 5). These investigations were also accompanied by a campaign of geophysical survey across both the headland and the two fields to the south and south-east, which provided additional evidence about the nature and extent of the site.

1999: In the wake of the successful work south of the river, small-scale geophysical survey was also undertaken in late 1997 across the east side of the fort at Pumsaint. This produced sufficiently promising results to warrant a more extensive survey across the whole of the cricket field in 1999, which located the likely line of the east defences and an area of associated civilian settlement, including a possible bath-house (see section 2.6 below).

Such survey and excavation by the authors has now ceased temporarily, to facilitate the full publication of the results obtained over the period from 1987–99. Work has, however, continued at the mines, first in the form of a detailed surface survey of the workings in 1999, with an appropriate overlay of archaeological interpretation (Landmark Surveys and Cambria Archaeology 1999, supplemented by Cambria Archaeology 2002), and second in the form of a short season of survey on the main areas of surface and underground workings by a French team led by Béatrice Cauuet (Ancel *et al.* 2000). The results from this work will be considered in the overview at the end of this volume.

1.2 DESCRIPTION OF PRINCIPAL ARCHAEOLOGICAL FEATURES

In the absence of documentary sources, our knowledge of the earliest phases of mining and settlement in the Dolaucothi-Pumsaint area has been almost entirely dependent upon the surviving archaeological evidence. This is best described under three headings: the mine workings; the leat systems and related reservoirs; the ancillary processing and settlement areas (see Fig. 1.1).

1.2.1 Mine Workings

These have been shown to comprise a palimpsest of surface and underground features, focused on the saddle now occupied by the main Ogofau Pit (Fig. 1.1, no. 4), and on the slopes of Allt Cwmhenog to the north-east and Allt Ogofau to the south-west. The most prominent are the large opencast pits and trenches (nos 1 to 7), which attest to the successful exploitation of the ore bodies. They were initially worked by opencast methods, more than likely with the use of water, and were subsequently extended underground in places. By far the largest is the Ogofau Pit, which is currently some 24m deep from its upper rim and extends over 150m east-west by 100m north-south. Recent drilling within the pit, however, has recorded a rubble and tailings infill up to 12.50m thick in places, above what was originally a highly irregular floor. Similar, if somewhat smaller and less drastically altered, opencasts lie above this on the slopes of Allt Cwmhenog, including the so-called 'Roman Pit' (no. 3) and the Mitchell Pit (no. 2), the latter immediately below the line of the Cothi leat and one of its associated tanks (C). Both these pits still preserve much of their original profile, despite their original floor being obscured by recent mining debris.

To the north-east, at the extremity of the lode zone, lies another large opencast, the Allt Cwmhenog Pit (no. 1), now much obscured by a forestry plantation, but originally comprising two coalescing pits, the lower one cutting into the base of the upper. Several possible tanks (G and Q) have been postulated around its rim, apparently served by the feeder leats associated with the so-called Annell system. Similar opencast workings exist on Allt Ogofau to the south-west, including those now known as the Niagara Pit (no. 5), beyond which a series of trenches extend over half a kilometre as far as Cwrt-y-Cilion farm (nos 6 and 7), some of them apparently affected by later collapses into underground galleries. Besides these conspicuous surface workings, there are other far less prominent features associated with small-scale exploration and pitting on exposed outcrops. They tend to survive in areas of peripheral or unsuccessful exploitation, both on Allt Cwmhenog and Allt Ogofau, but in practice remain poorly known in the absence of any detailed survey. Particular interest attaches to one area on the southern slopes of Allt Ogofau, where several circular huts close to a quartz exposure have been claimed as evidence for pre-Roman activity (Jones 1969).

The early miners also pursued the ore-bearing lodes below ground. This is clearly demonstrated by various cavernous entrances and, more importantly, by the two 60m long adits on the northern slopes of Allt Ogofau (no. 9) and by the galleries or stopes beneath the Ogofau Pit (no. 4). The adits, known as Lower and Upper Roman, were both hand-driven, as the surviving pick and chisel marks clearly attest. Lower Roman, which is 2–3m high, has a very characteristic wedge-shaped profile, somewhat wider at the top than the bottom. Upper Roman, by contrast, is square-cut, *c.* 2m across. Both are said to have served as drainage and haulage passages for a series of underground stopes, all but one of which, at the end of Upper Roman, have now collapsed, though at least one commentator has suggested a link with hydraulic mining (Hall 1988). The galleries or stopes beneath the Ogofau Pit, which were discovered during tunnelling in 1935, clearly extended to a depth of *c.* 45m below the current ground surface. Though they are inaccessible today, their date seems assured by the discovery of part of a Roman drainage wheel which has produced a radiocarbon date of 90±70 BC (HAR 2809).

Typologically there is nothing specific about these mine workings to suggest a Roman date, though the drainage wheel fragment must have implications at least for the mining in and under the Ogofau Pit, since it is unlikely to have been moved far from its original location. Less certainty attaches to the suggestive fragment of Roman glass apparently found 'stratified' in the Mitchell Pit, since the full details have never been published. No comparable *in-situ* finds are known from other parts of the mine workings.

1.2.2 Leat systems and related reservoirs

The extensive provision and use of water at Dolaucothi has been seen as one of its major features. Lewis and Jones claimed to have identified some four separate systems (Fig. 1.2), apparently all focused on the mines, though only one, the Cothi leat, has ever been fully published (Jones *et al.* 1960). This tapped the river *c.* 11 km upstream and ran on a carefully-engineered line to the top of the opencast workings on Allt Cwmhenog, where it served at least two separate tanks or reservoirs (Fig. 1.1; C and E). Estimates based on excavation suggest that it was capable of supplying at least 2.5 million gallons daily to the mines. Of the reservoirs, that at C lies immediately above the rock-cut face of the Mitchell Pit (no. 2) and was arguably linked with its exploitation. Excavation has indicated that its outer bank was originally pierced at one corner by some sort of sluice gate, apparently serving a rock-cut overflow channel, but was later remodelled to supply a set of three stepped washing tables used in ore-processing. Beyond this, to the south, lies the largest of the reservoirs, Tank E, which was capable of holding an estimated quarter of a million gallons. Excavation has demonstrated that its rock-cut outflow channel was controlled by a timber sluice gate.

Traces of a second leat system were discovered in 1969, serving the workings at a higher level than the Cothi and extending along the southern flanks of Allt Cwmhenog. Only a few isolated sections were actually identified, however, because so much of its

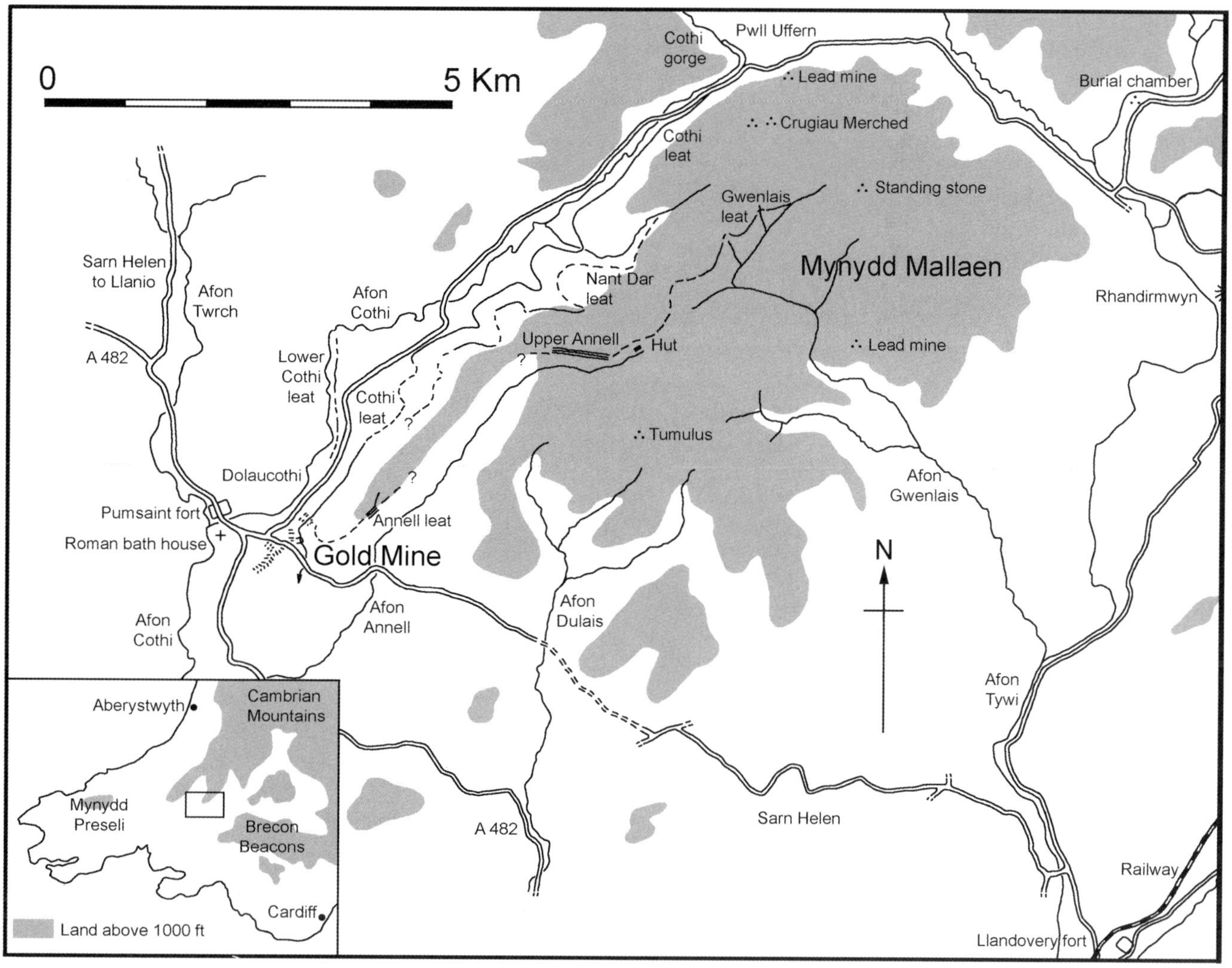

Fig. 1.2 Leat systems serving the mine workings (after Lewis 1977, fig. 1, with additions)

course had been destroyed by afforestation and land improvement, and even these have never been fully published. As reconstructed, it was thought to have tapped a marshy area near the headwaters of the Annell, *c.* 7 km upstream, before descending to the mines along a relatively steep gradient. Its ultimate destination is said to have been a pair of tanks (G and Q), perched above, or alongside, the Allt Cwmhenog Pit (no. 1). The importance of Tank G lies in the existence of a series of deep gullies which drop away steeply from its sluice gate and then sub-divide, before plunging into the opencast. These provide the clearest evidence at Dolaucothi for the use of water in the hushing process. The situation in the upper reaches of the Annell has been complicated, however, by the recent discovery of the Gwenlais leat, apparently extending the system up to 3 km beyond its postulated source, to tap the headwaters of the Gwenlais (Bick 1989). The problems associated with this have led some commentators to question the existence of the so-called 'Annell' leat along the flanks of Allt Cwmhenog (Bick and Boon 1993), which only further fieldwork can now resolve (for a more detailed discussion see chapter 3).

Two lesser water supplies have also been postulated, one tapping the Nant Dâr, a tributary of the Cothi, the other coming off the latter river lower down its course than the main Cothi leat. Unfortunately, the relevant details have remained unpublished and neither was actually traced into the mine area proper. A number of isolated tanks, among them I and J perched on the edge of the main Ogofau Pit, have also been claimed as evidence of exploitation by fire-setting and quenching, but their remains are less than convincing on the ground and their source of supply unknown.

Although leat systems are common on mining sites of all periods, the examples at Dolaucothi have traditionally been assigned to the Roman period, partly on the excavated evidence, partly by analogy with the evidence from the Roman mines in Spain, and partly because of what seems to be an eyewitness account of the large-scale use of water in Roman mining recorded by Pliny the Elder (*Natural History*

33, 70–74). More importantly, two radiocarbon dates have recently been published from a peat horizon infilling a section across the upper reaches of the so-called Annell leat (GRN 16553 and 16720), which suggest that it was already three-quarters full by *c.* AD 800 (Jones and Maude 1991). Though not without its difficulties, this has potentially far-reaching implications for our wider understanding of the leat systems as a whole.

1.2.3 Ancillary ore-processing and settlement areas

While considerable fieldwork has been directed at examining the mine workings and the leat systems, the related question of ore-processing has not always attracted the attention it deserves. Lewis and Jones were able to identify several possible sets of stepped washing tables, most notably in association with Tank C, apparently comprising a series of rock-cut steps in the local shale. These could have supported a set of inclined wooden washing tables or employed the broken character of the shale to trap the heavier gold particles during the washing process. It is doubtful, however, whether such washing tables could have coped with the quantity of material produced by the opencasts, which must surely have been more effectively processed downslope from the workings rather than uphill. Evidence in support of this has been noted in the fields to the north of the Ogofau Lodge, where abundant traces of crushed quartz have been noted during ploughing. This area has also produced fragments of several hand querns and one stone tentatively identified as an axle socket. Potentially far more important for our understanding of ore-processing was the discovery in 1982, on the sloping ground above the Carreg Pumsaint, of the remains of a possible water-driven mill complex (see Fig. 4.1). At the time, it was thought to date on technological grounds to the medieval or early modern period, which opened up the possibility of a hitherto unrecognised phase of mining at Dolaucothi; recent suggestions that it might be associated with the Roman processing wastes excavated nearby would, however, raise other possibilities and would, if proven, represent a significant advance in our knowledge of early exploitation (Burnham 1997; see also chapter 4).

As far as associated settlement areas are concerned, the recorded evidence is almost exclusively Roman. Particular importance obviously attaches to the discovery of an auxiliary fort under the village of Pumsaint, north of the river Cothi, which was clearly founded in the 70s AD, after the completion of the conquest of Wales. As with many other Welsh forts, it was apparently reduced in size, before being abandoned in the course of the 2nd century. South of the river, the remains of a possible bath-house were located in the 1830s, probably in an area marked today by disturbed ground and a scatter of brick and tile. Pottery associated with this has suggested that occupation here continued down to the end of the 2nd century at least. The occasional discovery of bricks and tiles in the surrounding fields has been taken to indicate the presence of a more extensive civilian settlement, a suggestion to which recent work has lent further credence (see chapter 5). The possibility of an interest in the mines early in the medieval period has also been suggested by the identification of two mounds as possible Norman mottes, one on Allt Ogofau above the 'Roman' adits (Bosanquet 1917, 29, fig. 33), the other at the entrance to the Ogofau Pit, adjacent to the Carreg Pumsaint (Nash-Williams 1952, fig. 3). Both were accepted by Hogg and King (1963, 93), though there remains the suspicion that some element of confusion has arisen between the sources; irrespective of this the mound at the entrance to the workings cannot now be interpreted as a motte in view of its identification as little more than a conical spoil heap (see section 4.3.3(E) below).

1.3 HISTORICAL RECONSTRUCTION: THE LEWIS AND JONES MODEL

The background to the work considered in this volume necessarily lies in the specific historical reconstruction proposed by Lewis and Jones in the 1970s. On the evidence available to them, they argued that the single most important phase in the development of the site occurred in the Roman period, when it was probably the only gold-mine in Britain and almost certainly one of the most technically advanced. Any pre-Roman activity at the site was thought to have been destroyed by the scale of the Roman operations, with the possible exception of the circular huts on the southern slopes of Allt Ogofau, well away from the main focus of the workings. A pre-Roman origin was not considered absolutely essential, however, as the Romans would have been well able to recognise the potential of the local geology and the placer deposits in the valley bottom, given the proximity of the fort at Pumsaint and the speed with which they tended to exploit the mineral resources of recently annexed territories.

As far as the Roman operations were concerned, Lewis and Jones devised a simple developmental model, based upon a logical progression from prospecting, through opencast development, to underground stoping, all closely associated with a system of water supply designed for a variety of uses, including hushing, fire-setting and quenching, hydraulicing and ore-processing (Jones and Lewis 1971a, 297–300). The key to the proposed sequence lay in the relationships between the various opencasts, themselves notoriously difficult to date on surface evidence alone, and the relative sequences of the leat systems and their associated features, the Roman dating of which seemed beyond doubt. Especially critical to this was the primacy of the Annell leat and the associated Allt Cwmhenog Pit, apparently guaranteed by the way in which the Cothi leat was subsequently contoured around the latter at a later date. The resulting model can be summarised simply (Fig. 1.3):

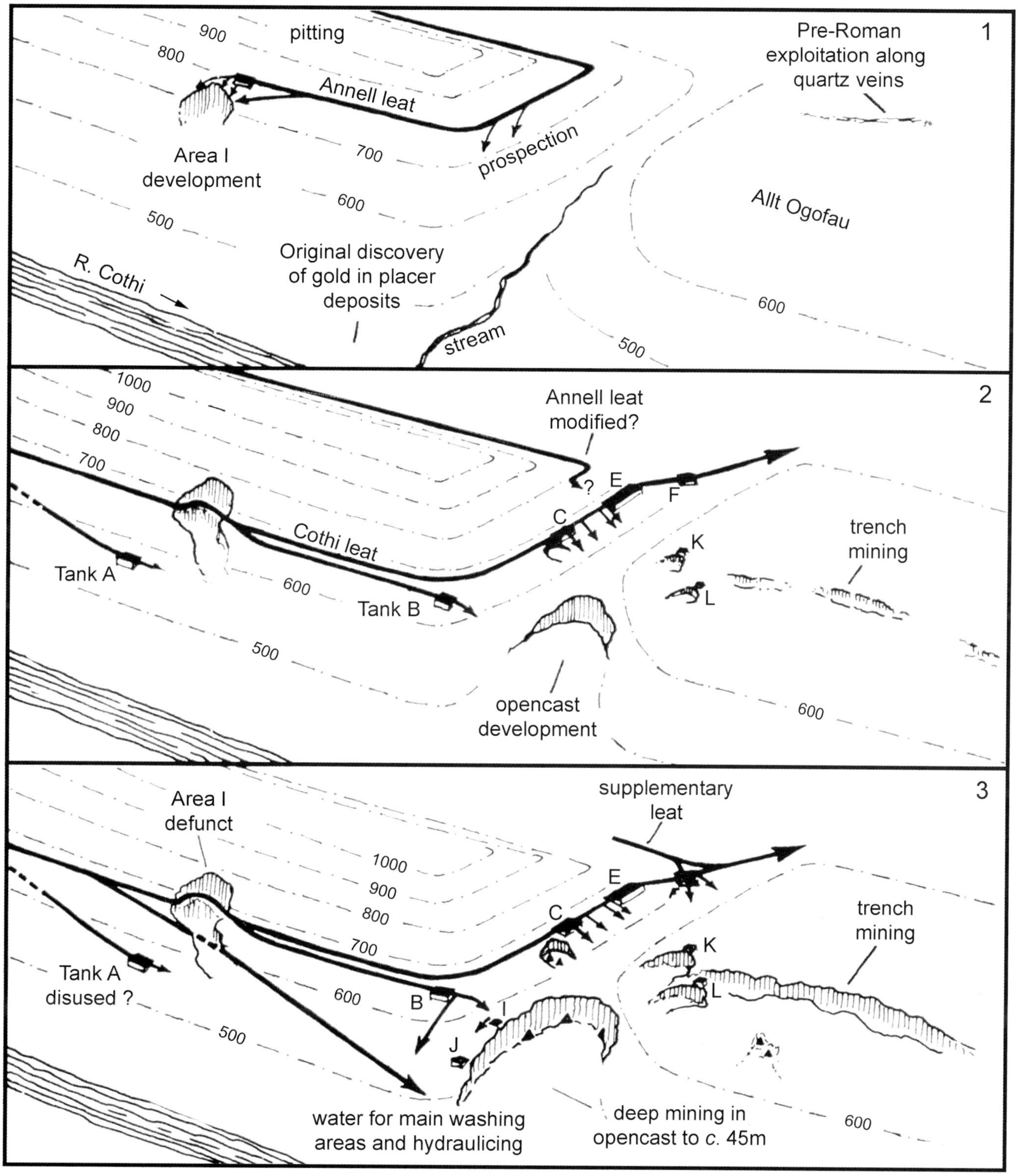

Fig. 1.3 Schema for mine development (after Jones and Lewis 1971a, fig. 9)

(a) limited pre-Roman exploitation;
(b) the development of the Annell leat and its associated tanks in connection with the exploitation of the Allt Cwmhenog Pit;
(c) the development of the larger and more efficient Cothi leat and its associated tanks in connection with the exploitation of the main Ogofau Pit and the other opencast pits on Allt Cwmhenog and, less certainly, Allt Ogofau;
(d) the exploitation of underground resources beneath the Ogofau Pit and the creation of at least two adits on Allt Ogofau in connection with stoping beneath the surface trenches.

The minor water supplies were obviously less easy to fit into this neat schema, while its applicability to the workings on Allt Ogofau also remained uncertain, given the rather inconclusive nature of the postulated tanks and the absence of a definitive link to the leat systems. Even so, Lewis and Jones felt confident to argue that the basic form of the mines had already taken shape in the Roman period, principally because most of the opencasts still preserved traces of the associated tanks at critical points around their edges.

Any refinement of the dating of this activity was necessarily a matter of speculation, depending for the most part upon material derived from the fort at Pumsaint, the bath-house and the modern dam at Melin-y-Milwyr, rather than from the workings *per se*. This suggested an initial phase of direct military control soon after the foundation of the fort in the 70s, with the army providing the necessary technical expertise for the development of the Annell and Cothi leats. The subsequent reduction in the size of the fort by the early 2nd century was thought, in part, to reflect garrison changes elsewhere in Wales, though it was also suggested that the mines became less important to the state, perhaps coinciding with the increased output from the newly developed gold-mines in Dacia. Though the reduced fort did not outlast the mid 2nd century, the pottery from the bath-house and Melin-y-Milwyr clearly attested occupation continuing throughout the rest of the Roman era. At this point it was considered likely that the mines were leased to civilian concerns, a practice well attested elsewhere in mining, though the details remained unclear.

While this developmental model proposed by Lewis and Jones has proved attractive, given the nature of the evidence and the growing body of datable material found in the vicinity of the mines, doubts have been expressed about the extent to which all the pre-modern surface remains should be ascribed to the Roman period. It is no longer safe, for instance, to assume that earlier mining was necessarily small-scale and, therefore, easily destroyed by later Roman operations, given recent work on sites of bronze-age date in Wales (see Crew and Crew 1990) and, more importantly, on the pre-Roman gold-mining sites in the Limousin area of central France (Cauuet 1999). Likewise, later exploitation of the mines cannot be dismissed, even if it has proved difficult to trace either on the ground or in the documentary sources, partly because basic mining techniques are such that relatively small-scale, intermittent usage could probably pass unnoticed. This might well have been the case early in the medieval period, though the reinterpretation of the so-called 'motte' at the foot of the workings as a conical spoil heap has obviously removed one line of evidence. Questions have also been raised as to whether such a well-worked site could have avoided the attentions of later miners, not least the numerous entrepreneurs active throughout west Wales in the 16th and 17th centuries (*e.g.* Boon 1981, for Cardiganshire). Some such activity could well account for the small amount of 17th-century pottery from Melin-y-Milwyr, but for the moment this must stand in isolation.

Given the self-evident importance of the site, both nationally and internationally, the publication of this volume provides a welcome opportunity not just to present the results of the most recent investigations within the Dolaucothi-Pumsaint area, undertaken between 1987 and 1999, but also to review how they have contributed to our wider understanding of the mine's development, the technologies involved, and the nature and extent of the associated settlements. Such issues will form the core of the overview at the end of this volume.

2 EXCAVATIONS AND SURVEY ON THE FORT AND VICUS AT PUMSAINT, 1987–1999

2.1 PREVIOUS EXCAVATIONS

The possibility that a Roman settlement or fort might underlie the modern village of Pumsaint was first suspected in the late 1960s, as part of the wider investigation of the surface workings and the associated leat systems (see section 1.1 above) at that time. The first evidence came in the form of Flavian-Trajanic pottery, discovered during the construction of the manse, and when erosion by the river Cothi south of the village revealed several clay-floored buildings in association with Flavian samian (Boon 1971, 500, fig. 30). The suggestion that this was an auxiliary fort was further enhanced by the distinctive setting of the site, located as it was on a fairly level terrace above the flood-plain, at the confluence of the rivers Cothi and Twrch, which ran along the south and west sides respectively. Its nature and extent, however, were not confirmed until the early 1970s (Fig. 2.1), when an opportunity was afforded for excavation by the creation of a car park in a former vegetable garden south of the hotel and by the realignment of the modern road through the village (Jones and Little 1973; 1974).

These excavations revealed an interesting sequence, commencing with the foundation in the 70s of a normal auxiliary fort occupying an area of *c.* 1.9ha (4.75 acres). Its defences were first encountered on the south and west in 1972 (Trenches 6 and 10), revealing in the process two distinct phases, the first a turf-clay rampart with two external ditches, the second a stone wall constructed on a foundation of heavy river cobbles and stiff grey clay infilling the original inner ditch. This sequence was subsequently confirmed on the northern side in 1973 (Trench 18), where traces of a gate were also recovered. On the east, however, the sequence proved to be somewhat different in 1972 (Trenches 11 and 12), there being only a single-phase ditch system on the predicted line, infilled with clay-rampart material and destruction debris. This anomaly was only explained by the discovery of the later stone wall and its associated ditches on an entirely different line inside the earlier fort, immediately east of the A482 (Trenches 5 and 7), effectively enclosing a small fortlet of *c.* 0.9ha (2.2 acres) on its western side. From the associated pottery it was argued that this reduction had occurred in the early years of the 2nd century, and that the fortlet had not outlived the middle of that century.

Within the area of the new car park, the 1972 excavations sampled a cross-section of the internal buildings on either side of one of the fort's principal north-south roads, which itself had evidence for eleven separate surfaces (Fig. 2.2). To its east lay a substantial stone-built and buttressed granary, measuring 17.00m north-south by at least 12.00m east-west, which had apparently replaced a timber predecessor and clearly formed part of the central range of buildings within the fort. It had apparently gone out of use when the fort was reduced in size, because its demolished remains were overlain at one point by the gravel road and a drain associated with the corner of the fortlet's defences.

West of the main road, a complex succession of timber buildings was identified, with evidence for a dramatic change of alignment and function over time. The earliest phase (1A) included a pair of probable barracks aligned east-west, each with a side drain flanking a narrow street or alley. The southern of the two, at least, preserved traces of partitions at intervals of 3.60m. Both were assigned to the Flavian period on the basis of the pottery found in their construction trenches. There was even a suggestion that they could have continued below the main north-south road and the granaries, though this would be difficult to accommodate within the fort's layout as it is currently conceived. This first phase was then followed by a major realignment of the site through 90°, evidenced by the construction of a further north-south street with some six surfaces across the top of the earlier buildings.

This new street served to separate a barrack at the western end of the site from a series of four successive buildings to the east. The lowest of these proved to be a barrack divided into two sections, 3.80 and 2.10m wide respectively, which was associated with late Flavian to Trajanic pottery. This structure had subsequently been demolished in the early years of the 2nd century to make way for three successive

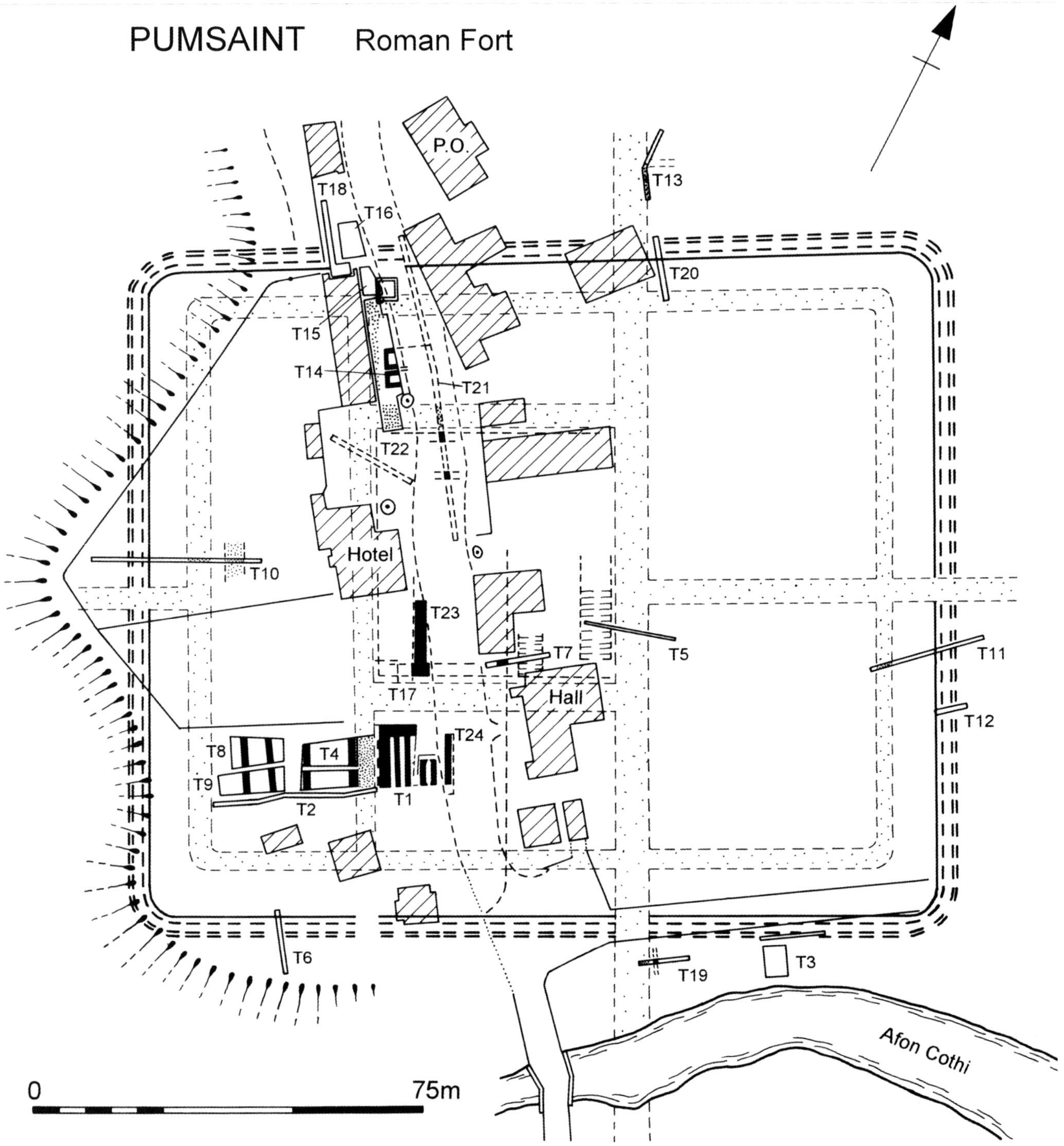

Fig. 2.1 Plan of fort and excavation trenches (1970s) in relation to village (after Jones and Little 1973, fig. 1)

phases of workshop, the first of which (1C) had been sub-divided into separate working rooms by butt-ended partitions. Its successor (1D) retained much the same plan, with evidence for no less than five furnaces, in one of which nails had clearly been reworked from a bent state. The third phase (1E) was less well preserved, but several burnt patches in its flooring confirm its continued industrial function. Above these workshops, there were the very fragmentary remains of even later structures, one of which was assigned to the mid 2nd century on the basis of the pottery in its construction trenches.

Elsewhere in the fort, rescue excavations were carried out in 1973 (Trenches 14 to 18), immediately to the east and north of the 1989 site (see Fig. 2.103). These confirmed the line of the northern defences, as well as revealing evidence for an original timber gateway, the eastern side of which was represented

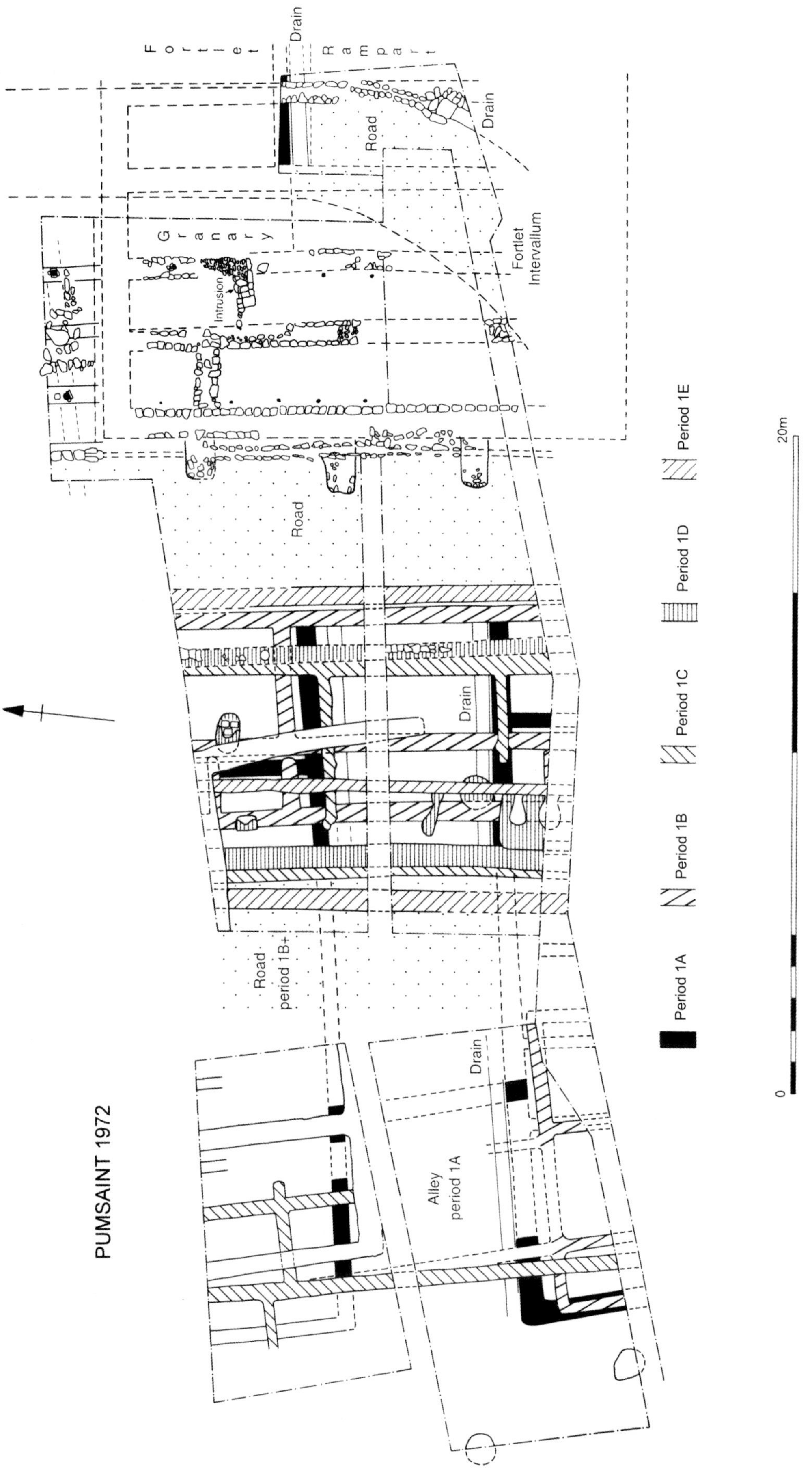

FIG. **2.2** Plan of 1972 excavations in car park (after Jones and Little 1973, fig. 4)

by a construction trench and several large post-pits. This had apparently been blocked when the stone wall was inserted, because its associated ditches could be shown to continue right across the original gap without interruption. The road itself proved to be little more than 3.00m wide, extending north-south along the eastern side of the 1989 site to connect up eventually with the main road in the former vegetable garden. To the east of this lay two phases of timber building, the earlier of which was identified as part of the original *praetorium*. This was later replaced by two separate shed-like structures, *c.* 4.40 and 4.50m wide respectively, each with a clay floor. Various other features were recorded in 1973, including details relating to the *principia*, which was said to be large by normal standards.

Prior to the work in 1987 and 1989 these excavations formed the basis of our understanding of the military sequence at Pumsaint, which makes it particularly unfortunate that the results have been published only in interim form. Some additional insights were gleaned, however, in 1985, when the chance was taken to clean and record the eroding river section south of the fort and the village (Burnham and Burnham 1987a). This revealed an extensive spread of material most plausibly interpreted, at the time, as demolition debris connected with the fort/fortlet transition. From the associated pottery, it was possible to argue that this had probably occurred in the early Hadrianic period, though this is clearly in need of revision in the light of more recent thinking on the dating of Black-burnished ware in Wales (see section 2.2.4(Cii) below). There was also evidence for several clay floors and a number of related pits, including one possible latrine shaft.

2.2 EXCAVATIONS IN THE NORTH-WEST PART OF THE FORT, 1987 and 1989

2.2.1 Introduction

Between July and September 1989, large-scale excavations were carried out at Pumsaint, by the Dyfed Archaeological Trust and St David's University College (now University of Wales, Lampeter) under the direction of Barry and Helen Burnham. The site lay immediately to the west of the main A482 through the centre of the modern village (Fig. 2.3), to the north of the Dolaucothi Arms Hotel, in an area occupied prior to the excavation by a range of 19th-century farm buildings and their associated yards (Fig. 2.4). This work was carried out in response to a detailed National Trust plan to lease the site to a Housing Association for the construction of several new cottages.

The primary interest of this site lay in the fact that it occupied an area known to include the northern defences, the east-west *intervallum* road and the associated structures of a two-phase Roman fort first examined by Jones and Little (1973; 1974) in the early 1970s. This fort lay on the main Roman road from Llandovery to Llanio, in close proximity to the gold-mining complex at Dolaucothi, with which it was thought to be linked in a supervisory capacity. The site also afforded a chance of exploring whether there had been any medieval activity on the site of the village, and of elucidating the history and development of the extant buildings.

Long before the formulation of any detailed plans for the 1989 site, the National Trust had been granted outline planning consent in 1987 for the development of housing across an extensive parcel of land north-west of the Dolaucothi Arms Hotel. This provided a welcome opportunity to undertake small-scale evaluation work at an early stage in the planning process, with the aim of assessing the likely threat to the surviving archaeology. The work was duly carried out by the authors in June 1987, working on behalf of the Dyfed Archaeological Trust, with the active collaboration of the National Trust (Burnham and Burnham 1987b). Given the limited time available and uncertainties over the National Trust's detailed plans, the main thrust of the excavations was to determine the depth at which undisturbed archaeology was reached. For the most part this was expected to be Roman, though along the east of the threatened area there was a possibility of encountering material from earlier phases of the modern village, as well as the remains of the recent farmyard levels associated with the standing building.

Accordingly, six trial holes (TH1–6) were excavated to the west of the existing 19th-century building, both to ensure as wide a spread as possible across the proposed development area and to allow the recovery of a section of the buried deposits along both its north-south and east-west axes (Fig. 2.5). Once the initial objectives had been achieved and undisturbed deposits had been reached, TH 1 and 2 and TH 4 and 5 were linked up to allow a more detailed assessment of the uppermost archaeological horizons in plan. Subsequently, the chance was also taken to put a small trial hole (TH7) inside the building itself, near to the east side of the site, to enable a closer inspection of the stratigraphic sequence beneath its floor and to cover against the possibility that development might involve the upgrading or demolition of the structure. The discovery of a defunct sewer trench in TH5 also allowed an extension westwards so as to remove its fill in its entirety; this had the advantage of providing a preview of the surviving Roman archaeology in section.

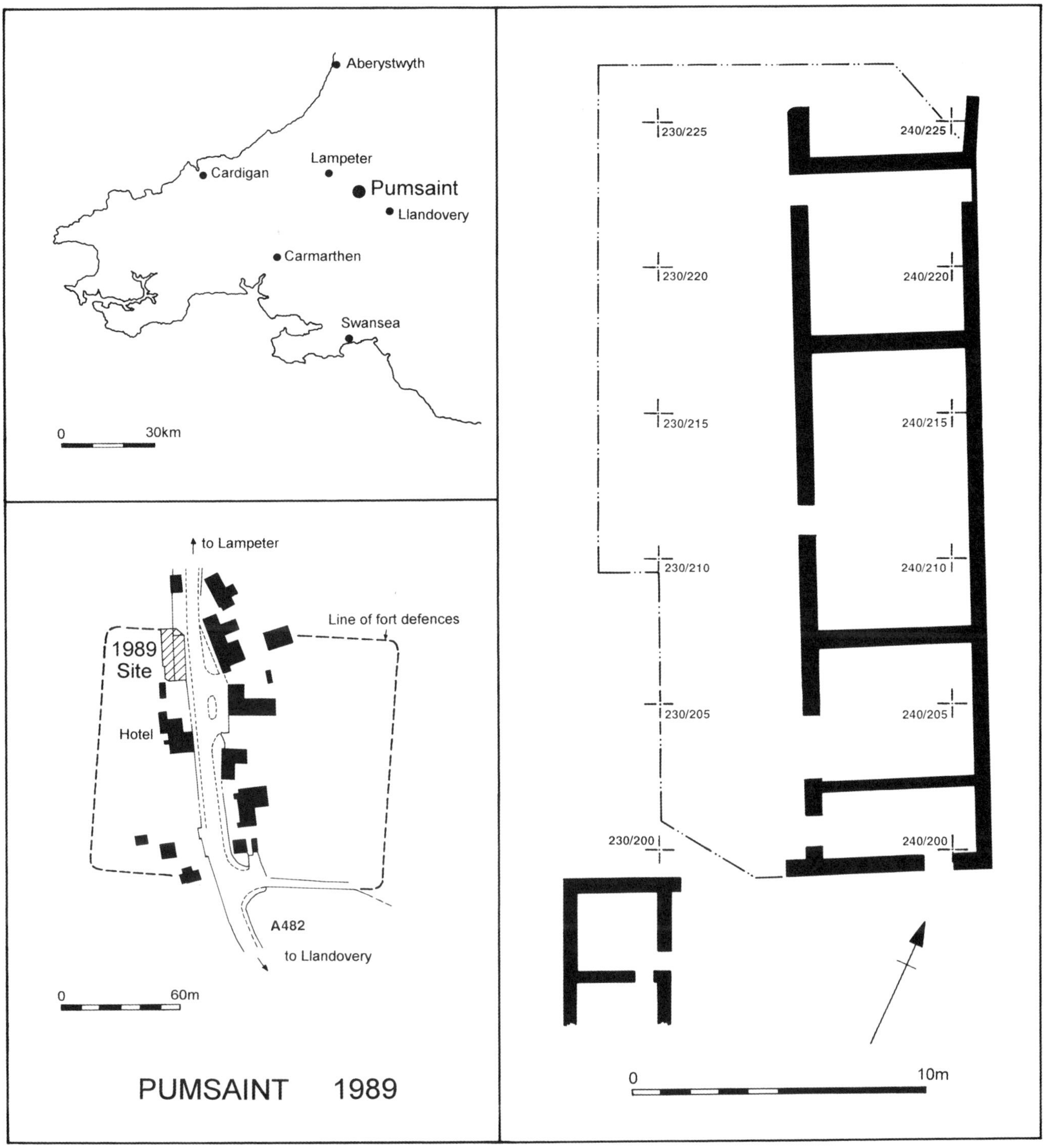

Fig. 2.3 Location plan of 1989 site

Fig. 2.4 19th-century cowshed/combination range occupying the 1989 site prior to excavation, looking north-east

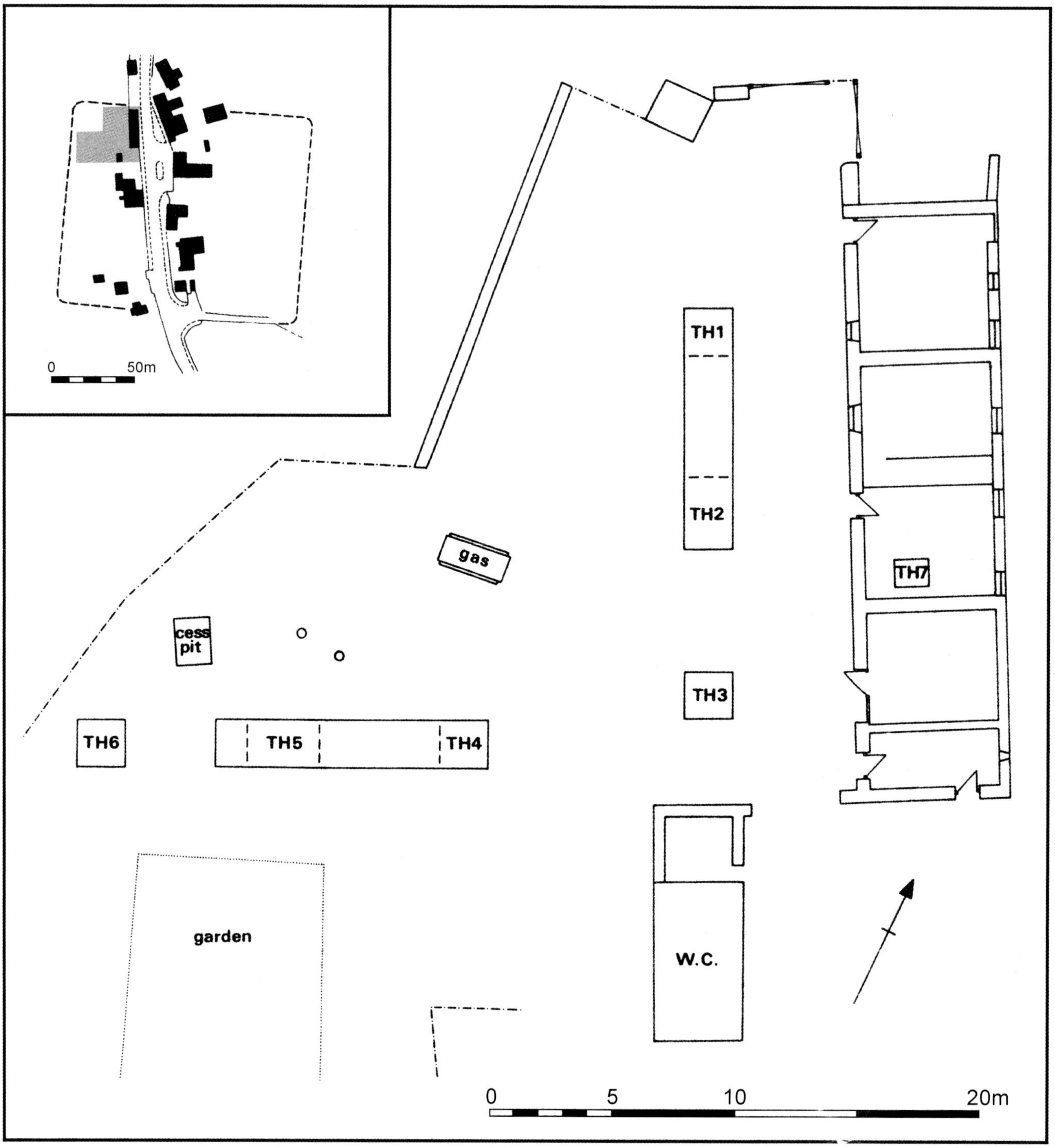

Fig. 2.5 Location of 1987 trial holes (1–7)

The surviving archaeology across the different parts of the site clearly reflected their very diverse histories. In Trench 1/2 and TH3, the excavations encountered the farmyard levels associated with the standing building along the frontage. These, in their turn, overlay an undisturbed horizon, taken to be the top of Roman archaeology, at a depth which trended from as little as 0.20m at the north end of the site to *c.* 0.50m at the south. In Trench 1/2, in particular, a characteristic orange clay spread was tentatively identified as the back of the fort defences, while a cobbled spread to the south was thought to represent the remains of the *via sagularis*. This hypothesis was duly proven, when this part of the 1987 site was re-examined in 1989.

The excavations along the east-west axis, in Trench 4/5 and TH6, revealed a completely different and much simpler sequence, in which an extensive turf, topsoil and disturbed subsoil horizon, up to 0.40m thick, overlay what were clearly undisturbed Roman levels. These latter included a possible gravelled street across the east end of Trench 4/5 and an extensive spread of clay loam containing charcoal flecks, burnt clay and several large-headed nails, which is best interpreted as the demolished remains of earlier timber buildings. The excavation of the defunct sewer trench at the west end of Trench 4/5 demonstrated that this latter horizon overlay an undisturbed stratigraphic sequence, up to 0.50m thick, including evidence of unmistakable yellow clay floors, charcoal lenses, and what appeared to be at least one beam slot on an alignment parallel with the road surface. The associated finds were exclusively Roman, confirming that preservation here is as good as that encountered by Jones and Little under the Hotel car park in 1972.

A distinctive sequence was also recorded in TH7 which investigated a small area beneath the central room (Room 2) of the existing building. This revealed an earlier cobbled floor, beneath which lay a lens of grey and white ashy material sealing a patch of heavily fired red clay of uncertain industrial or domestic origin. Though nothing was recovered to date this activity, the relative levels between the different trenches were sufficiently discrete as to suggest that the excavation had ceased above the Roman levels, with the implication that potentially interesting post-Roman deposits were to be expected just below the floor levels of the 19th-century building. This hypothesis was duly confirmed by the excavations of 1989.

When the National Trust's plans were finalised in 1989, it rapidly became clear that the development would be confined to the eastern part of the 1987 site, and that it would also involve the demolition of the existing 19th-century building. Given that Roman archaeology had been located just below the associated farmyard levels right across this part of the site, any such development would clearly involve significant disturbance particularly to the surviving evidence for the defences at the northern end of the site. All this increased the urgency of the case for excavation and, incidentally, opened up the opportunity of correlating the results with the 1973 work immediately to the north and east. As a result of considerable discussion between the interested parties, it was agreed that prior excavation was essential, with the National Trust providing the necessary funds and Cadw agreeing to cover the post-excavation costs. The preliminary results of this work were published in a series of interim reports and articles (Burnham and Burnham 1989; 1991a; 1991b).

2.2.2 The Roman Levels

(A) The principal elements of the 1989 site (Fig. 2.6)

For convenience the evidence will be discussed in sections working from north to south across the site, leaving aside the limited evidence for any pre-military activity prior to the fort's construction.

(i) The defences

Two phases of Roman defence were located running east-west across the northern end of the site, commencing with a considerable stretch of the original turf-clay and gravel rampart which terminated just short of a timber structure or tower on the east. This latter was represented by three of its original post-holes, no doubt connected in some way with the gateway postulated to the east by Jones and Little. To the north, the inner of the two first-phase ditches was also relocated and part sectioned. At a secondary stage, a stone wall had been constructed to the north of the original rampart, partly overlying the infilled line of the ditch, with the robbed-out foundations of a stone tower at its eastern end. Considerable modification was inevitably necessary to the existing structures.

(ii) The well and cistern

Immediately behind the two successive towers lay a very substantial feature associated throughout with the fort's water supply. It seems to have begun life as a well, with a stone lining towards the base at least, if not throughout, which was later converted into a clay-lined cistern. The associated infills proved to be particularly rich in finds, but their value was somewhat reduced by the fact that all the relevant stratigraphic relationships with neighbouring Roman features had been removed by the foundations of the much later clom and stone-built structures.

(iii) The east-west road and the associated buildings

To the south of the defences and the well lay the east-west *intervallum* road of the fort, with its associated drains, which had been cut into the back of the turf-clay rampart. This would originally have formed a junction with the north-south road, first identified along the eastern side of the 1989 site in 1973, but of this very little survived. In the junction between the two lay a series of foundation trenches representing the remains of the earliest timber buildings inside the fort, though none of their associated floor levels had survived later disturbances across this part of the site. Despite the problems of identifying different phases, three separate buildings seem to be represented. In the next stage of development in this part of the site, both the east-west road and the associated structures were overlain by two further phases of buildings, whose plans were much better preserved.

(iv) The burnt timber structure

The most surprising feature lay at the southern end of the excavated area, at the bottom of a large, vertically-sided pit, which had been cut down to a depth of

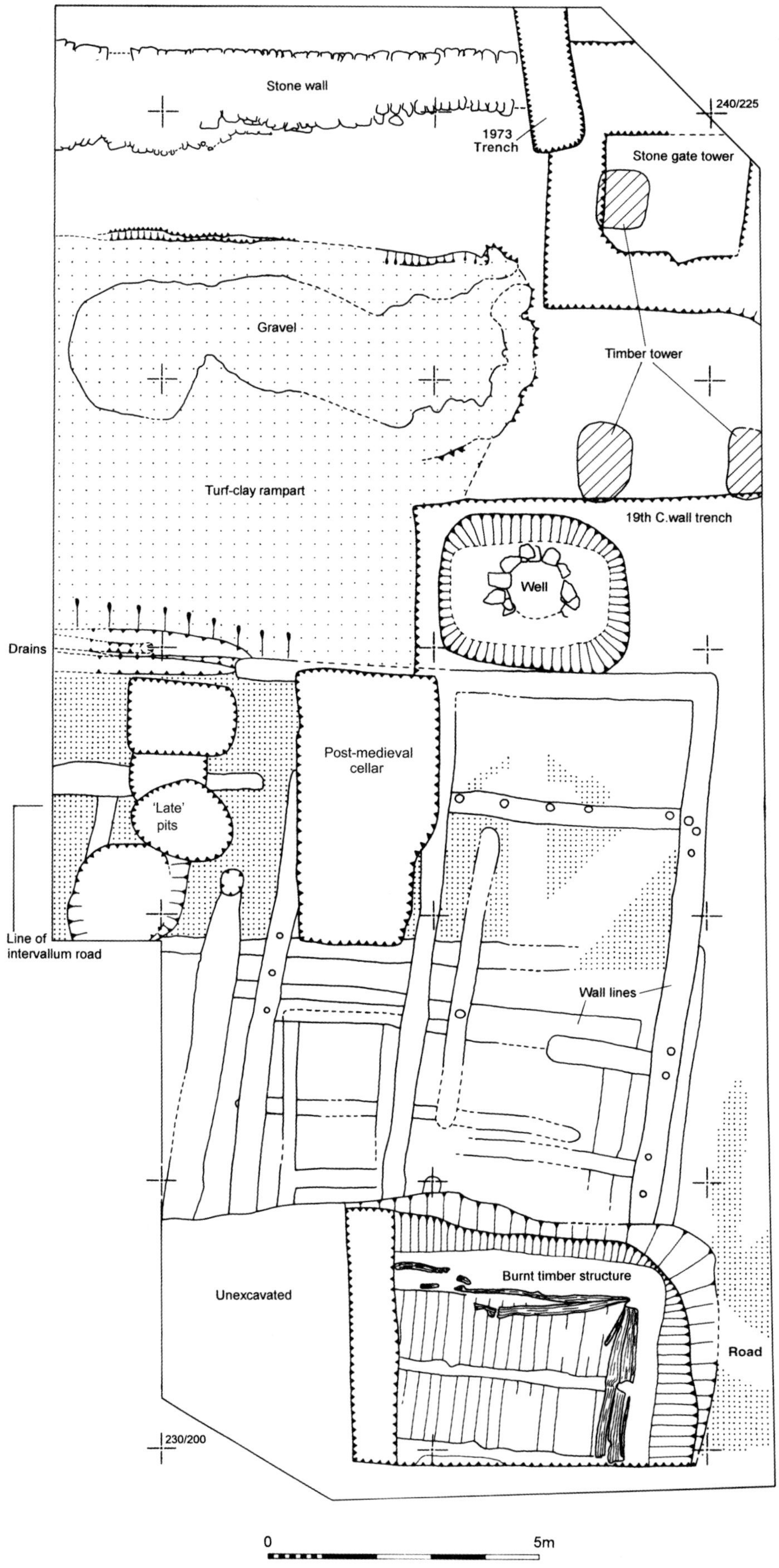

Fig. 2.6 Principal features of 1989 site

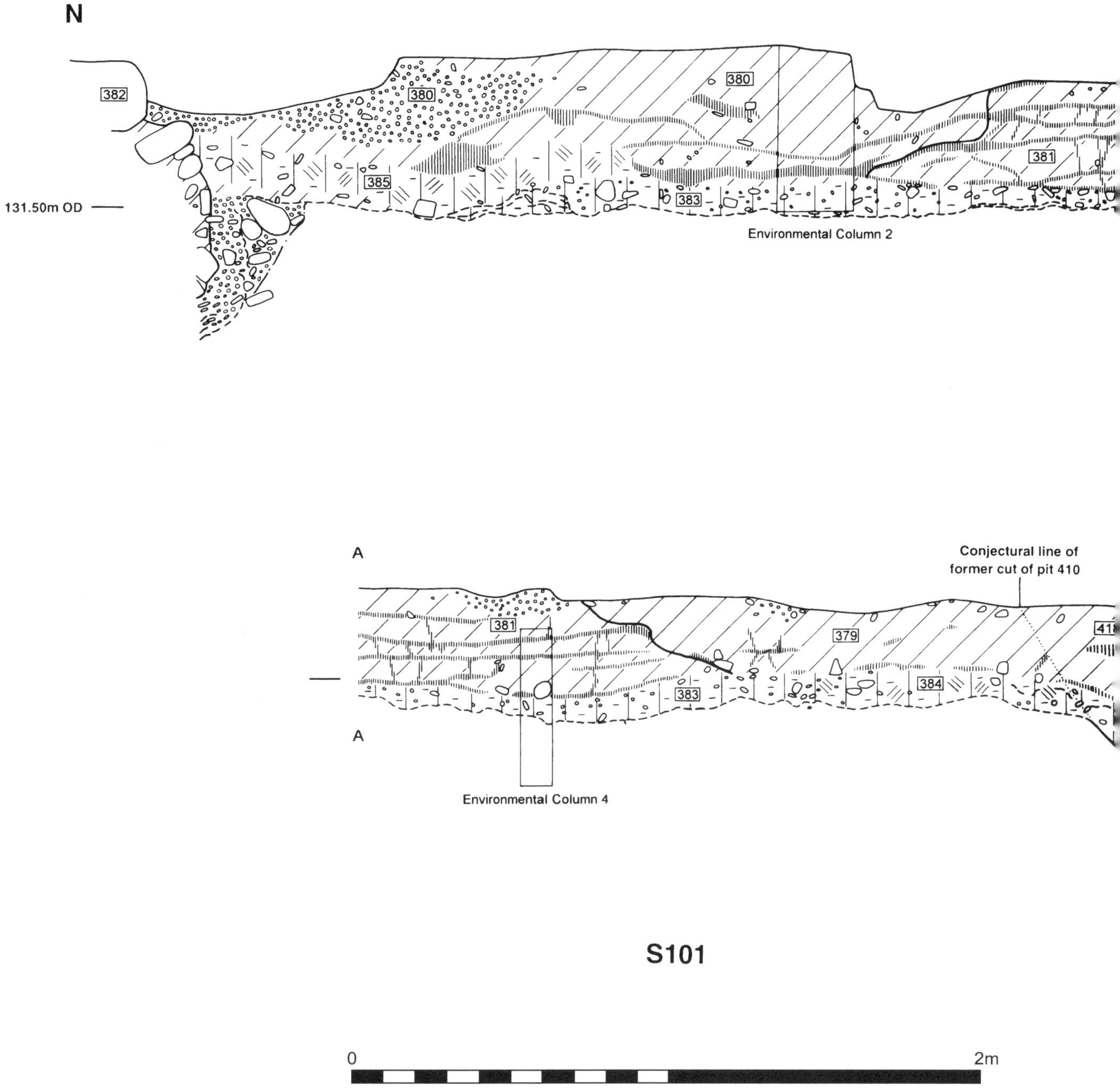

Fig. 2.8 East section of trial trench across rampart

A

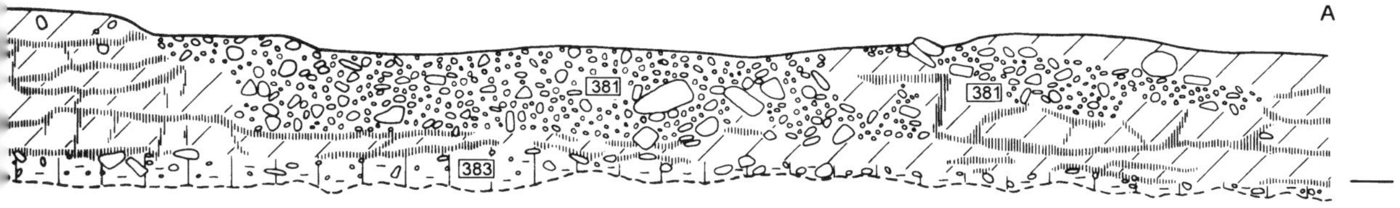

A

S

— 131.50m OD

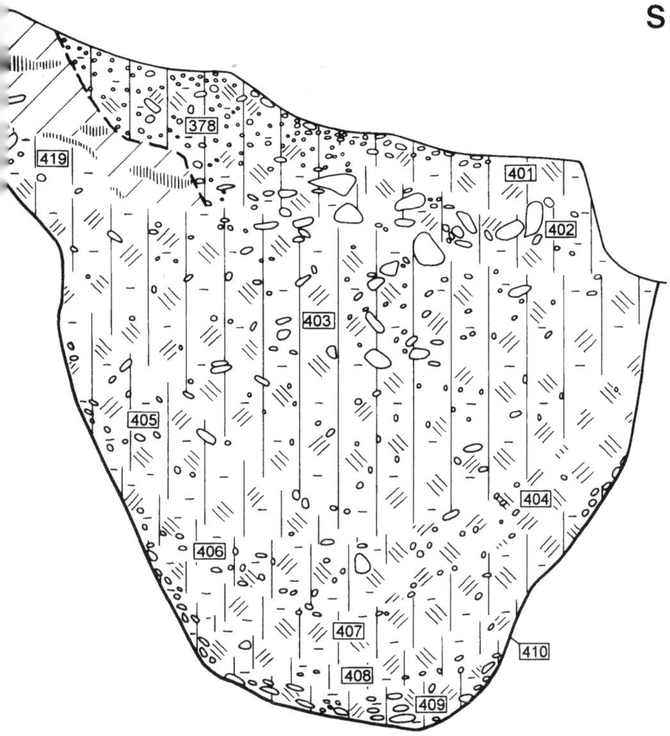

and small stones.
milar fills on the southern and
respectively, comprising a silty
ıdmixture of small stones; the
llowish brown (2.5 Y 5/4) to
8).
llowish silty clay loam (10 YR
ure of small and medium-sized
clay similar to that used in the
flecks of charcoal.

these fills and the absence of
the uppermost layer [403],
ess of natural silting into an
nstruction of the first-phase
ıtext [403] on this interpreta-
ıt a more rapid infilling (or
infillings) of the remaining
Roman period, which would
e of several clay fragments
oining sherd from a redware
pretation assumes that the
and 419] are identical with
the north, [419] being an
ed horizon [384] behind the
ıd [418] being part of the
n [379], and that both had
nto the top of the pit above

retation, which was advanced
as a later feature cut through
ıd would interpret contexts
g the result of considerable
owing the extensive weather-
riginal sides. This could well
with the sequence of inter-
8; 351], which had been dug
est road at the western side of
2.2.4(I) below). Whichever
d, the pit continued to slump,
velling in the form of contexts
these, only the lowest [378]
ic finds, in the form of a rim
e bowl (CP 10) and several
phora (not catalogued).
y features were identified on
ıg the general view that the
relatively open terrace for the
t. There is evidence, however,
l samples which were taken
d of the disturbed ground
the first-phase rampart, to
ıltivation in the immediate
section 2.2.4(R) below).

he gateway/tower
rt (Figs 2.7 to 2.9; 2.11 to 2.13)
he excavation, an extensive
ealed across the northern end

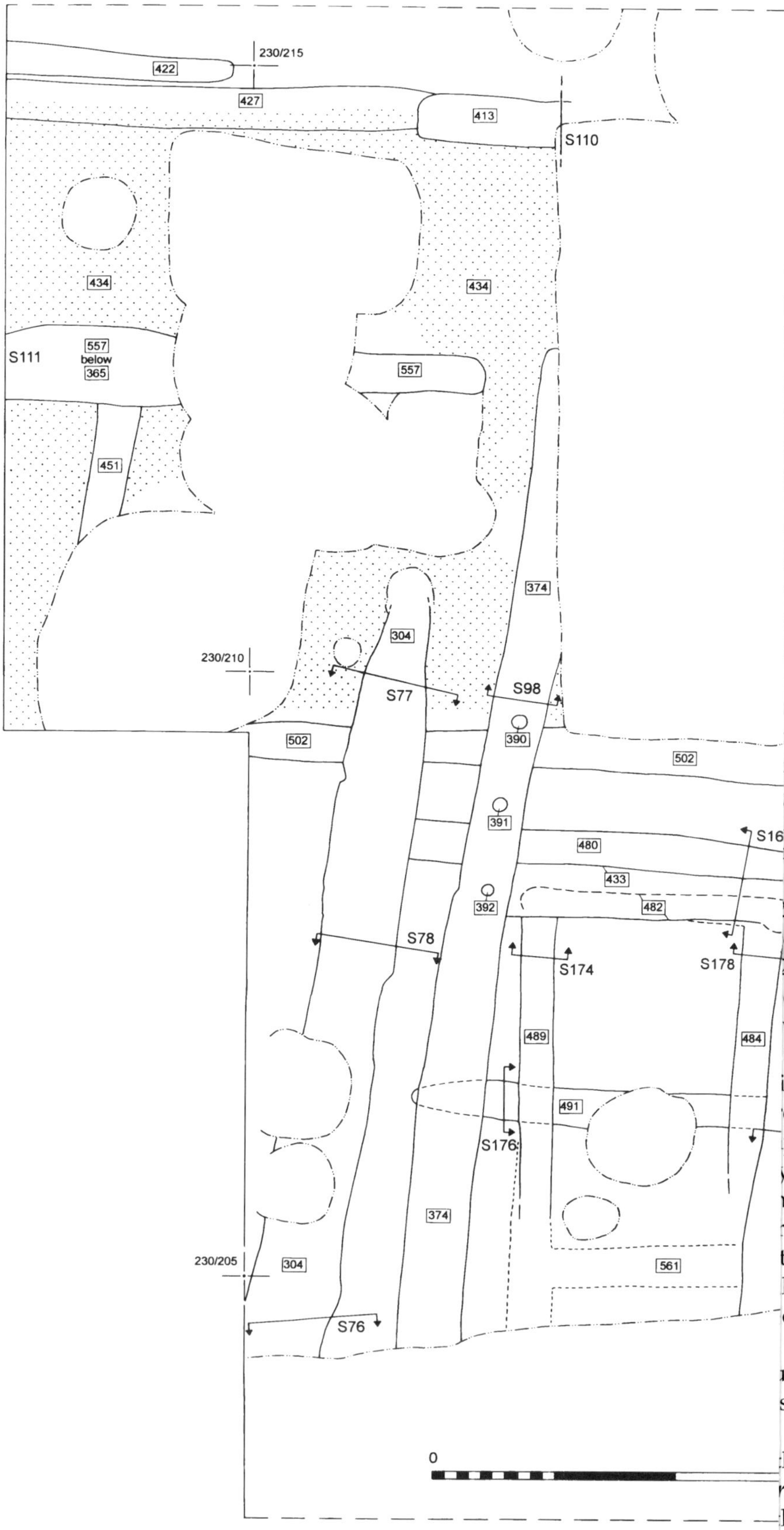

Fig. 2.10 Composite plan of east-west road and successive Roman buildings

Fig. 2.11 The two-phase defences, looking west

of the site, immediately to the south of the disturbed areas associated with the robbing, probably in the 17th or 18th century, of the stone wall and tower (robber trenches [292 and 331]). Towards its western end, this clay was seen to incorporate an irregular, sausage-shaped wedge of gravel, some 8.00m long, and to extend for at least 9.00m south of the wall, to the point at which it was overlain by the gravel makeup [434] of the Roman road. Further east, its southern edge had been cut away by one of the foundation trenches [74] of the 19th-century stone building, thereby destroying all the stratigraphic connections with the well and clay-lined cistern, which lay immediately to the south. Judging from the relatively flat appearance of its upper surface, this clay and gravel horizon would seem to have been levelled at some point in its history (Figs 2.9 and 2.11).

The overall dimensions of this clay and gravel spread, as first exposed, were clearly too wide for a normal fort rampart, but it was not possible at the outset to identify any clear-cut internal sub-divisions within its upper surface. For this reason, a trench *c.* 1.30m wide was initially excavated due south of the stone wall, with the intention of clarifying the overall stratigraphy. In the event, its eastern section (Fig. 2.8) rapidly demonstrated that the central core [381] of the clay and gravel spread formed part of an original rampart, roughly 4.60m wide and 0.35m high at this point, with later clay spreads to the north [380] and south [379]. This core was characterised by two dumps of compact clay, lying to either side of a central gravel and stone infill, which clearly corresponded with the sausage-shaped feature already described. The clay varied in colour from light grey (10 YR 8/1) to bright yellowish brown (10 YR 6/6), with characteristic dark brown (10 YR 3/4) bands at regular intervals representing the decayed turf horizon on the original sods.

Subsequent excavation in plan to the east of this section revealed that the southern clay cheek did not maintain its original alignment, but progressively curved northwards to enclose the eastern end of the gravel and stone infill. Unfortunately, this particular part of the rampart was not very well preserved, surviving to a height of only 0.10m above the surrounding ground surface, but even so it created an obvious, if somewhat ragged butt end, terminating *c.* 1.50m west of the post-holes associated with the timber tower described in the next section. This is best interpreted as the truncated remains of the original rampart terminal, which must have been cut back and modified when the stone tower was inserted immediately to the east.

In the western section of the trial trench, surprisingly, the story was somewhat different; here the rampart [381] proved to be much wider than expected and there was no obvious trace of any clay addition at the back to balance layer [379] to the east (Fig. 2.12). This difference was duly confirmed by the subsequent excavation of the rampart in plan to the west of the trial trench, and by a close examination of the main section along the western edge of the site (Fig. 2.7). Both clearly demonstrated that the original rampart must have been *c.* 8.00m wide at the base, with a narrow clay cheek at the front, bands of gravel and stone behind, and a much wider dump of clay to the rear. This discrepancy between the two sections of the trial trench can only be explained by assuming that the rampart had been deliberately widened at this point, perhaps to afford access or to accommodate some other feature, all trace of which has now been lost. Sadly, no trace of the presumed boundary between the rampart [381] and the later clay dump [379] was seen during the excavation of the trial trench; this was an obvious consequence of the excavation strategy at this point, exacerbated by the problems associated with the deep pit [410] already described in the preceding section, and by its neighbour to the west [396] which had clearly been cut through the rampart from a higher level.

The clay cheek along the northern face of the rampart was easily traced west of the trial trench, its near-vertical face rising progressively to a maximum height of 0.50m (Fig. 2.13). Unfortunately, all trace of it in the western section had been removed by a later post-hole [398]. The clay and gravel spread behind this extended in uninterrupted fashion for some 7.00m, beyond which it progressively sloped away downhill to the point where it became little more than a very thin spread beneath the makeup of the gravel road [434]. Along the line of this slope, three separate gullies [416; 422; 427] had been dug, presumably to carry excess water away to the west (see section 2.2.2(E) below).

Despite careful excavation in plan, too little of the actual rampart survived to see if it had originally been timber laced, nor was there any trace of associated uprights. Beneath its makeup, however, the very tenuous remains of a possible log corduroy or some form of timber strapping [525] were detected after careful trowelling, represented by several parallel dark brown stains (7.5 YR 3/4) running north-south across the underlying horizon [383]. These stains were only located towards the front end of the rampart, extending from the surviving vertical face beneath the narrow clay cheek and part of the gravel infill. Whether they were originally more extensive remains uncertain.

Beneath the rampart and its associated features was a mixed layer [383] of friable, dark greyish-yellow silty loam (2.5 Y 4/2), containing 20–30% coarse components and some charcoal flecks, the bottom of which was defined by a distinct iron pan

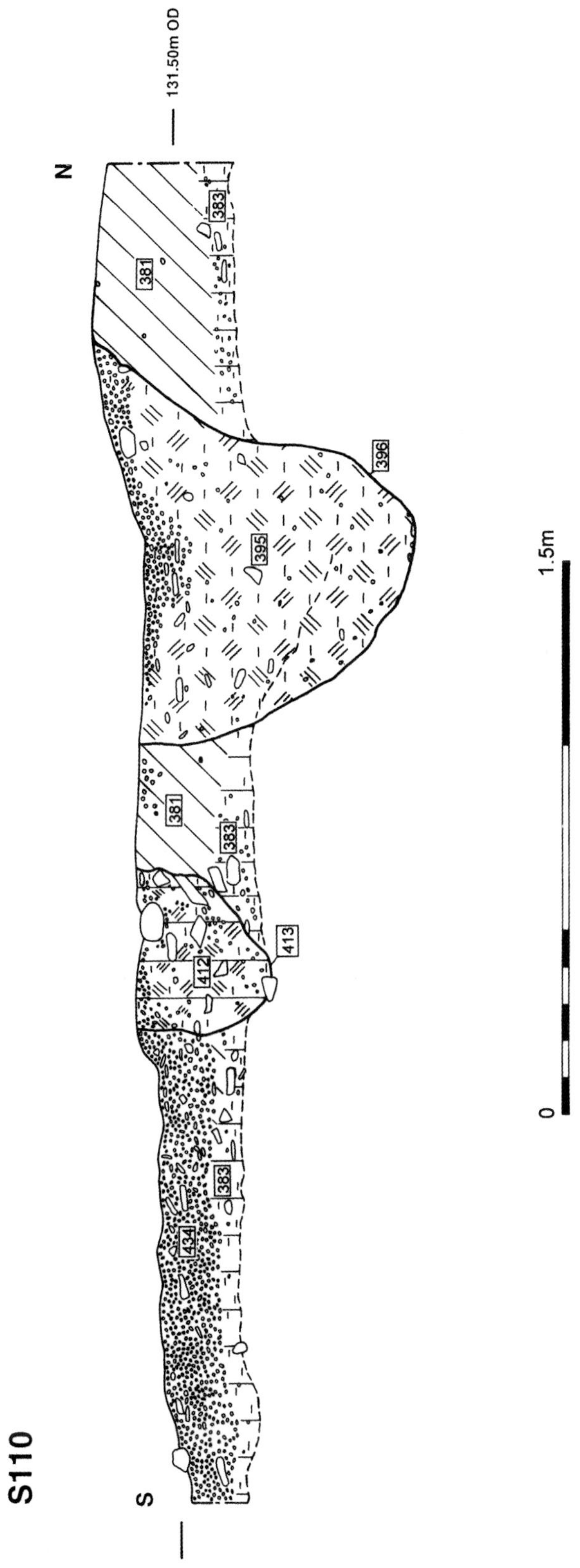

Fig. 2.12 Partial west section of trial trench across back of rampart

Fig. 2.13 Front face of clay rampart, looking south

overlying the natural soil horizons. This would seem to represent a trample layer created during the laying of the timber strapping and the construction of the rampart and, as such, it probably derives either from the pre-rampart soil or from prior turf clearance. Outside the confines of the rampart, this layer merged with similar horizons to the north [385] and to the east and south east [384], in areas which had been left open and, therefore, subject to later disturbance.

None of the layers associated with the turf-clay rampart produced any finds. This is hardly surprising, however, because the gravel infill must have been derived directly from the excavation of the fort's ditches to the north, while the turf-clay was no doubt brought in from somewhere on the nearby floodplain.

(ii) ***The timber tower*** (Figs 2.9, 2.14 to 2.15)
At the north-eastern corner of the site, east of the surviving butt end of the turf-clay rampart, lay three substantial post-holes [517; 538; 539], apparently forming the western and southern sides of a rectangular timber structure or tower (Fig. 2.14). Of these, only the pair on the west could be fully examined in 1989. The south-western post-hole [538] proved to be roughly rectangular in shape, *c.* 1.50m long by 1.00m wide, with almost vertical sides dug to a depth of at least 1.15m. Both in its base and in the half section (Fig. 2.15, S185), the impression of the

Fig. 2.14 Timber gate-posts and robber trench of stone tower, looking south

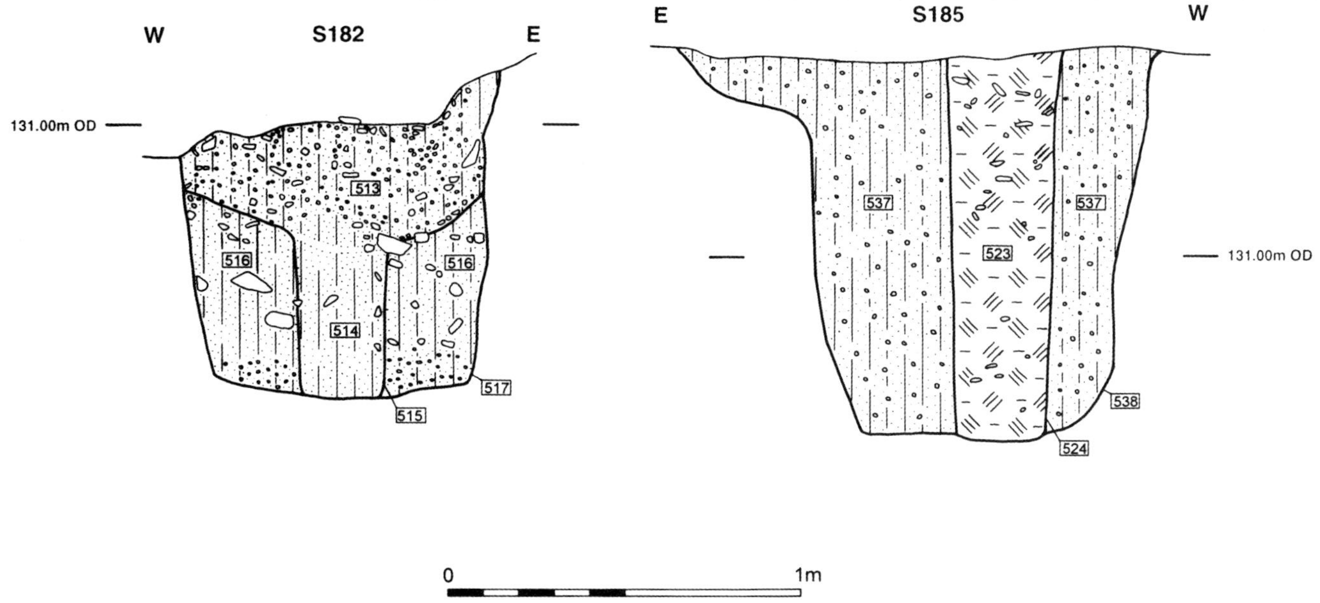

Fig. 2.15 Half-sections across north-west and south-west gate posts

original timber upright [524] was clearly visible, measuring *c.* 0.30m square, with a very distinctive fill [523] of brown silty clay (10 YR 4/6) containing only 10–15% coarse components. This latter extended right up the profile, strongly suggesting that, at the end of its life, the post had been sawn off at ground level and left to rot *in situ*. The original timber, once inserted, had been held in place by a backfill [537] which mainly comprised small to medium-sized gravels (80%), set within a matrix of greyish-yellow-brown loamy coarse sand (10 YR 5/2).

The corresponding north-western post-hole proved to be very similar, though its uppermost fill had been truncated on the west partly by the foundation trench [495] for the stone tower and partly by the later robber trench [292]. Its location was first recorded as a sinkage along the western side of the overlying gravel spread [526], the removal of which revealed a rectangular post-hole [517], measuring *c.* 1.10m by 0.90m. This had rounded corners and had been dug almost vertically into the underlying gravels to a depth of 1.05m. Again, the impression of the original timber upright [515], measuring *c.* 0.30m square, was clearly visible in the half section (Fig. 2.15, S182), surrounded by much the same sort of gravelly backfill [516] as its south-western counterpart.

On this occasion, however, neither the post-impression nor its fill [514] extended all the way up the profile, terminating instead *c.* 0.45m above the base of the pit. Above this, in the topmost part of the pit, was a further backfill [513] of small to medium-sized stones, set within a yellowish-brown loamy coarse sand (10 YR 5/8). This proved to be very much the same as the original backfill [516] within the post-hole and the fill [514] of the post-impression, which also comprised a high percentage (65%) of small to medium-sized stones, set within a loamy coarse sand of variable colour (10 YR 5/2 to 5/8). This would strongly suggest that the top of the original post-hole had been dug out to allow the post to be cut off at a much lower level, an action perhaps necessitated by the proximity of the foundations for the stone tower.

On the south-eastern side, only part of the original post-hole [539] lay within the 1989 site, but even this had been severely disturbed by a later intrusion [322], which had cut away the vital area where the timber upright would have lain. Like its western counterparts, it seems to have been rectangular in plan, measuring 1.30m by at least 0.60m, with rounded corners and near vertical sides. It had been dug to a depth of at least 0.85m, though its full extent could not be ascertained due to a shortage of time at the end of the excavation. It contained a single fill [535] of brown loamy coarse sand (10 YR 4/4), intermixed with very small to medium-sized gravels. Whether the later disturbance [322] had anything to do with the removal of the post or the later alterations in this part of the site must remain uncertain, because too little was available for excavation.

Unfortunately no trace of any north-eastern post-hole was located in 1989, to complete the four sides of the original structure, no doubt because it lay on the very edge of the area available for excavation, where the ground had been much disturbed by the foundation trench for the later stone tower. Even so, it is possible to restore the original dimensions from the surviving post-holes to suggest a rectangular structure *c.* 4.50 by 2.70m in size. This would seem to be far too narrow for a free-standing tower, indicating perhaps that the features identified in 1989 must have formed part of a more extensive structure, further elements of which were located by Jones and Little immediately to the east in 1973. The reconciliation of the evidence from the two sites is discussed in more detail in section 2.2.5(B) below.

(iii) The inner ditch section (Figs 2.9 and 2.16)
Of the two first-phase ditches recorded by Jones and Little (1974), only part of the inner one was actually re-sectioned in 1989, by partially re-opening and widening one of the original 1973 trenches. This approach was necessitated by the decision to leave the overlying stone wall *in situ* throughout its entire length. Just over half the original width of the ditch was recorded in this way, revealing a fairly typical V-shaped profile [508] with traces of an 'ankle-breaker' or clearing-out channel at its base. Assuming an even profile at this point, the ditch must have been over 3.00m wide and at least 1.40m deep.

Besides the profile, the excavation also provided a useful section through the successive deposits infilling the ditch, and thereby confirmed the basic sequence identified by Jones and Little (1974). The lowest fill [507], which had accumulated to a depth of *c.* 0.35m across the cleaning channel, proved to be composed almost entirely (95%) of small to medium-sized gravels, set within a coarse matrix of dull yellowish-brown sand (10 YR 5/4). This clearly represents a primary fill, derived almost exclusively from the sides of the ditch. Above this lay a more extensive infill [506] to a depth of 0.70m, which comprised a dull yellowish-brown sandy clay loam (10 YR 5/4) with a 60% admixture of small to medium-sized gravels. This layer had clearly accumulated over a longer period than [507], to judge from the characteristic traces of banding within it, which included alternating lenses of gravel and more silty material.

Above this the character of the infill changed dramatically, with the dumping of an extensive layer [505] of homogeneous yellowish-brown silty clay (2.5 YR 5/3 to 10 YR 5/4), which averaged 0.30m thick across much of its width. This in turn was overlain by a thin spread [504] of dull yellowish-brown sandy clay loam (10 YR 5/3) with a 60% admixture of small gravels, which appeared to be petering out towards its northern end. Both horizons are best interpreted as deliberate dumps, designed to level up the first-phase ditch prior to the construction of the stone rampart; they were both, no doubt, derived from the adjoining turf-clay rampart [381], which must have

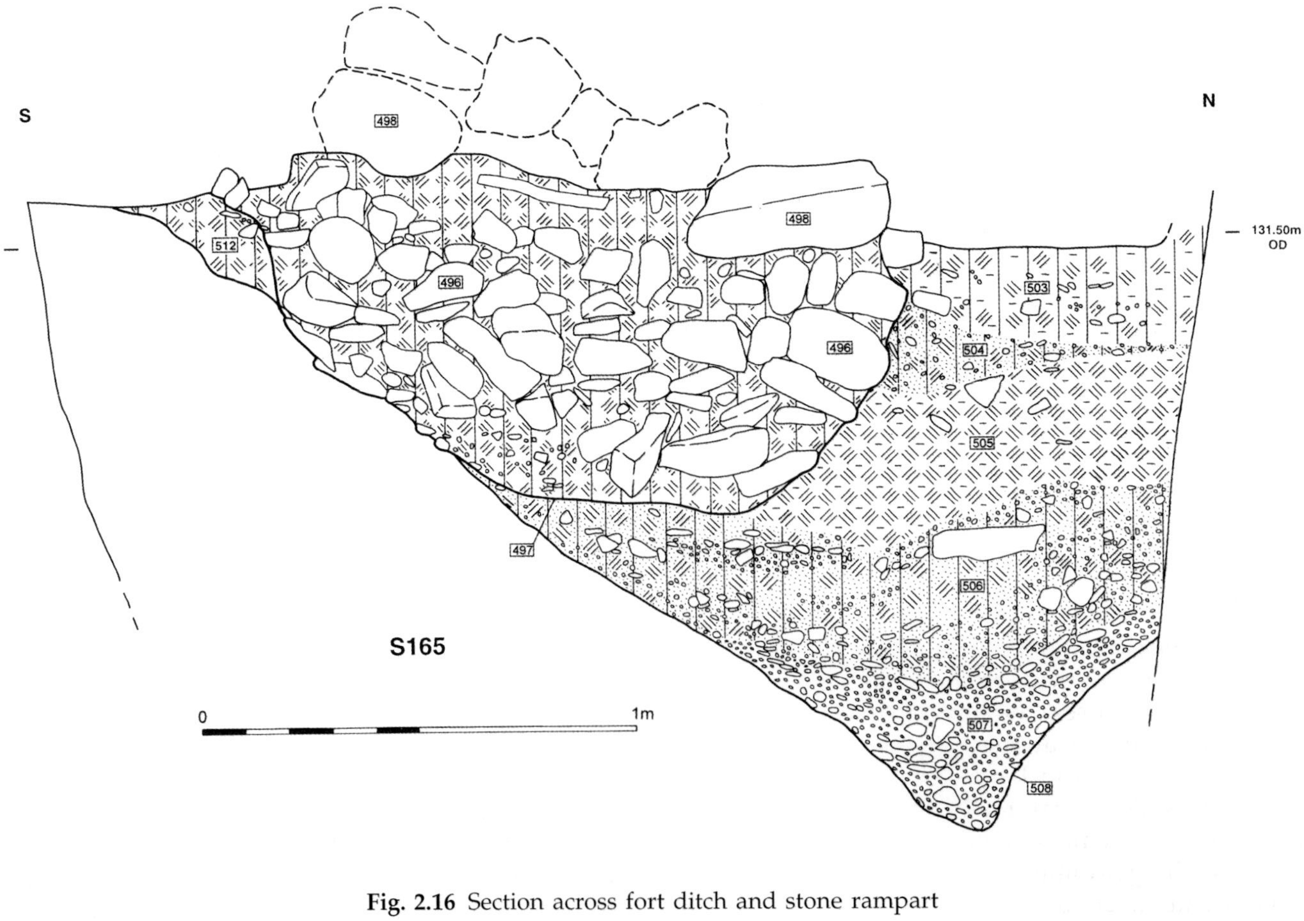

Fig. 2.16 Section across fort ditch and stone rampart

been modified in the process. The insertion of the stone wall and its associated foundation trench [497] is discussed in the next section; once this happened, a further 'fill' [503] of yellowish-brown silty clay loam (2.5 Y 5/4), with lenses of silty material and a 15% admixture of small gravels, seems to have accumulated on its northern side.

The silty clay horizon [505], described above, can be correlated with two other contexts of identical character in this part of the site (see Fig. 2.9); the first [512] was visible both in section and in plan to the west of the foundation cut [497], while the second [536] was recorded in plan to the east of the 1973 trench, where it had been much truncated by later activity. All three must be part of the same infilling operation, which makes it interesting to note that context [536] seems to be petering out towards its eastern end, as if respecting an underlying ditch terminal close to the site of the postulated gateway. Moreover, the oblique alignment of context [512] relative to the later stone wall, might be an indication that the ditch was curving inwards at this point, much as Jones and Little (1974) argued for the outer ditch.

None of the layers associated with the inner ditch produced any finds, with the exception of the silty clay horizon [536], which yielded an indeterminate body sherd from a redware vessel (not catalogued).

(iv) The stone rampart and tower (Figs 2.17 to 2.20)
At a secondary stage, a stone wall and tower had been constructed to the north of the first-phase rampart, partly overlying the deliberately infilled line of the original inner ditch. By comparison with earlier sightings, the stone wall was surprisingly well preserved, standing in places up to three courses high on a foundation of stiff clay and water-worn cobbles, whereas the tower had been heavily robbed, leaving only fragments of its original clay and cobble foundation *in situ*.

Just over 8.00m of the wall and its associated foundations were investigated in 1989, though this did not involve the removal of any of the structure itself, so as to ensure its preservation. Along its length, there was a marked variation in the width of the wall (Figs 2.17 and 2.18), even though the materials used and the method of construction appeared to be identical. The eastern segment [498], which extended for just over 2.50m west of the tower, proved to be relatively narrow, averaging 1.20m in width, whereas its longer, western counterpart [382] was actually wider, with an average width of 1.40m. Despite this, both segments were constructed above a one-course offset at the front and comprised a mixture of medium to large-sized quartz and mudstone blocks. Both were readily available locally, the quartz perhaps being a by product from the working of the nearby gold-mines. There seems to have been some attempt to use quartz as the predominant facing stone, with the mudstone largely confined to the core, perhaps suggesting a desire to create a deliberate effect.

The variation in the width of the wall was also noted at the foundation level, which progressively increased from *c.* 1.50m on the east to over 2.00m on the west, though the average in the central area seems to have been more like 1.65m. The foundations themselves were examined at two points, first in connection with the inner ditch section (Fig. 2.16) and second along their inner face (Fig. 2.19), just to the west of the junction between the narrow and broad walls. At the point where they overlay the inner ditch, the foundations [496] had been laid somewhat irregularly into a U-shaped foundation trench [497], *c.* 1.50m wide and 0.75m deep, which had been cut down into the underlying ditch fills [504 and 505] to accommodate them. They comprised a mass of large rounded stones and boulders set somewhat irregularly within a matrix of compact, light grey clay (2.5 Y 7/1). Further west, where the foundations [499] overlay the original berm, they had apparently been set into a vertically-sided trench [387] some 0.50m deep, although only its southern side was actually investigated. Here the stones and boulders had clearly been more systematically laid within the clay matrix, such that they comprised three or four layers of quasi 'herringbone' masonry.

At the eastern end of the stone wall lay the foundations [494] of a stone tower (Figs 2.14 and 2.17), much of which had been robbed by a post-Roman robber trench [292]. Enough survived, however, to demonstrate that they had been constructed in much the same way as the adjoining wall, with a mixture of rounded stones and boulders set in a compact matrix of light grey clay. These had been laid in a foundation trench [495], which proved to be 1.60m wide on its northern side, where it had been cut down into the underlying deposit [536], and only 1.10m wide on the other two sides. To judge from the surviving remains, the tower must originally have measured *c.* 5.00 by 4.50m externally. Unlike the wall, nothing survived of the stonework above foundation level. One further feature which ought to belong with this phase is a roughly rectangular layer of compacted gravel [526], *c.* 2.40 by 2.00m in extent, confined within what must have been the interior of the tower; it is best interpreted as a gravel floor, even though its relationship with the surrounding features had largely been destroyed by one of the 1973 excavation trenches on the north and by a robber trench [292] on the remaining sides.

Although certainty is impossible, it looks as if the tower was built first, probably with a short length of narrow curtain or wing wall, which was subsequently joined rather roughly by the wider curtain of the main stone rampart. This would also account for the progressive narrowing of the foundations as they

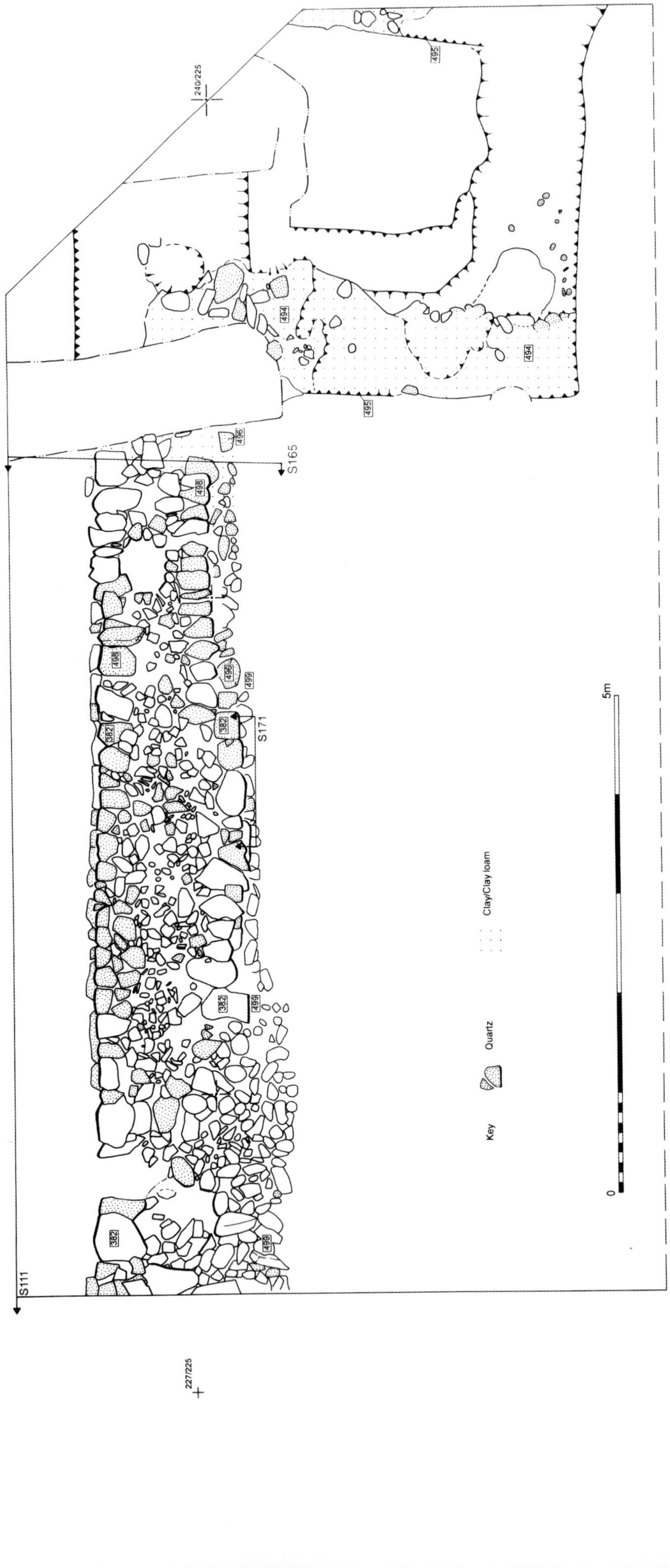

Fig. 2.17 Composite plan of stone wall and tower

Fig. 2.18 Stone wall and secondary rampart infill, looking east

approached the tower. Such a sequence would be logical, conforming with the pattern recognised on Hadrian's Wall, where the milecastles and turrets were also built first with wing walls ready to bond with the wall at a later stage. Unfortunately no datable material was found in association with either the wall or its foundations.

The decision to construct a stone wall necessarily had implications for the existing defences. The most obvious would have been the removal of the timber breastwork, the demolition of the timber structure or tower, and the infilling of the ditch system. The first of these has left little or no archaeological trace; the second saw the timber uprights cut off at or below ground level (see section 2.2.2(Cii) above); the third involved the dumping of two successive layers [504 and 505] above the existing fills in the ditch (see section 2.2.2(Ciii) above). It has also been suggested that the somewhat ragged end of the clay cheek at the eastern end of the turf-clay rampart is a strong indication that it had been cut back to make way for the insertion of the stone tower (see section 2.2.2(Ci) above). By contrast, there was no evidence to suggest any similar truncation along the northern face, which would otherwise have disturbed the continuous turf and trample horizon [385] across the extent of the berm.

Once the stone wall and tower had been constructed, or perhaps even at stages during their construction, the intervening space between them and the original rampart was infilled (Fig. 2.20). This was best preserved along the northern side of the turf-clay defences, where the berm was overlain by a very mixed layer [380] of compact silty clay which contained patches of pure clay and thin lenses of gravel and small stones and which varied in colour from dull yellowish brown (10 YR 5/3) to dull yellow (2.5 Y 6/3). It was also characterised by thin lenses of dark brown (10 YR 3/4) silty clay, much like the rampart core [381], though here they were neither as continuous nor so clearly defined.

This infill had sealed an interesting horizon across the top of the berm, which comprised a well-established turf horizon capping a layer [385] of dark brown silty clay loam (10 YR 3/3) containing 10% coarse components, charcoal flecks and thin lenses of clay (see Fig. 2.8). In section this layer merged with and overlay the trample horizon beneath the rampart core, suggesting that it was derived in part from the pre-rampart soil and in part from erosion off the rampart. The subsequent development of the turf horizon above this clearly represents a period of stability preceding the construction of the stone defences.

The corresponding infill [448] east of the turf-clay defences, behind the stone tower, was far less well preserved, being no more than 0.10m thick at its western end and progressively petering out towards the east, which made it impossible to define its relationship with the elongated intrusion [322] right at the eastern side of the site. Upon excavation this infill proved to be a very mixed horizon, containing quite a high clay content on the surface, but rapidly giving way to a dark brown silty loam (10 YR 3/4) containing 30% coarse components; this latter probably represents the very bottom of the layer, close to its junction with the underlying trample. Towards the south-west, however, its clay content became more pronounced, such that it progressively merged with the mixed layer [379] of compact clay which overlay the original rampart at this point. This clay varied in colour from light grey (10 YR 8/2) to yellowish brown (10 YR 5/8), with traces of characteristic banding at intervals, much like the rampart core, though they were nowhere as clearly defined.

The difficulty of defining any junction between these two spreads makes it highly likely that they

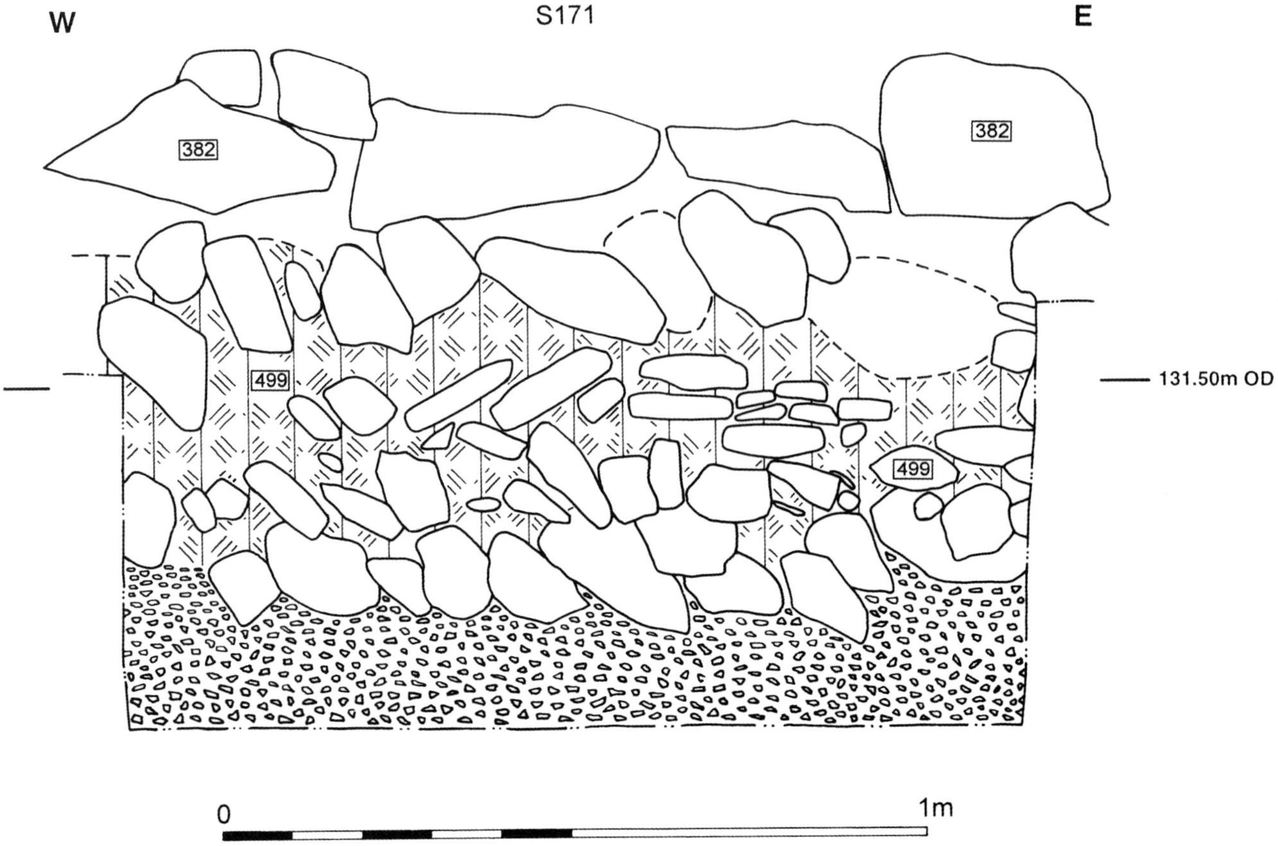

Fig. 2.19 Section of back face of wall

were part of the same infilling operation, east of the original rampart terminal. Both overlay a horizon [384] of dark brown silty clay loam (10 YR 3/3), containing about 20% coarse components, which was very similar in character to the trampled layer found beneath the rampart core. This would seem to represent the churned-up ground surface east and south east of the original defences, its slightly higher clay content resulting from erosion off the back of the rampart.

In the eastern section of the trial trench (see Fig. 2.8), both the clay dump [379] and the trample horizon [384] were seen to be slumping down into the northern side of a large pit [410]. At this point, they were deliberately given separate context numbers, [418] and [419] respectively, because it was thought that the pit was a post-rampart feature into which they had subsequently slumped following the extensive weathering and collapse of its original sides. For reasons already outlined, however, there are strong grounds for arguing that the pit was a pre-military feature, which would make the sub-division of the contexts unnecessary (see section 2.2.2(B) above).

Virtually nothing was found in association with these infills, with the exception of part of a samian base from [380] and a fragmentary iron object from [448] (see below). This is not surprising, perhaps, if we assume that the material used in the infill was largely derived from the modifications to the upper part of the original rampart, which have otherwise left no archaeological traces.

(v) Dating evidence and chronological discussion

Despite the fact that a substantial length of the defences was examined on the 1989 site, when compared with earlier sightings, only three contexts produced any finds, in each case a single item, all associated with the second phase of construction. Leaving aside the undiagnostic iron object from the rampart infill [448], this leaves only two fragments of pottery from other contexts:

380 rampart infill: samian base of form 18 or 18/31 (Table 2.2)

536 inner ditch fill: indeterminate body sherd, possibly from a redware flagon (not catalogued)

Neither of these, unfortunately, provides any help in clarifying the chronology of the defences as defined by Jones and Little in the early 1970s. The 1989 site did, however, confirm the two-phase character of the defences and added extra detail about the western side of the structure(s) identified to the east in 1973, even though the two sites are not easy to reconcile (see section 2.2.5(B) below). Some insight into the general chronology can obviously be gleaned from an

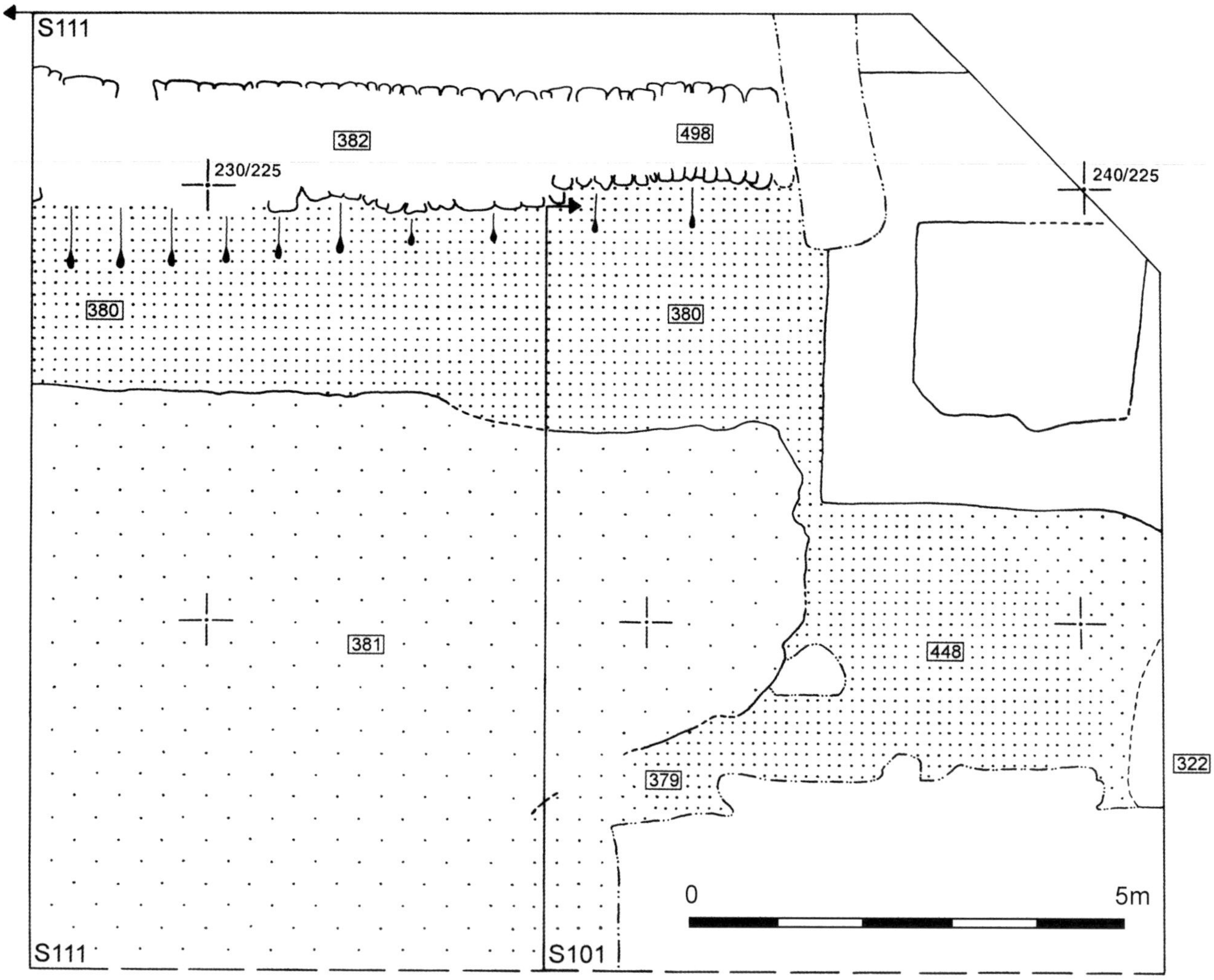

Fig. 2.20 Composite plan of turf infills between the two phases of defence

analysis of the abundant finds recovered from the successive fills of the well and cistern, which lay between the secondary infill [379 and 448] east of the rampart and the east-west gravel street [434] to the south (see section 2.2.2(Diii) below), but these cannot provide any refinement of the defensive sequence given the fact that all the critical stratigraphic relationships had been removed by one of the foundation trenches [74] of the 19th-century stone building.

(D) The well and the clay-lined cistern

(i) The main pit and stone-lined well (Figs 2.9, 2.21 to 2.25)

Immediately behind the two successive towers and the butt end of the first-phase rampart lay a substantial feature which eventually proved to be associated with the fort's water supply. When first encountered it took the form of an enigmatic clay edge extending east-west in the base of the foundation trench [74] which housed the north wall of the central room (Room 2) in the 19th-century stone building. This clay was subsequently seen to curve southwards at both ends, demonstrating that it formed the outer edge of a much larger feature, the southern extent of which remained somewhat ill-defined because of later intrusions. For this reason, excavation was initially directed at sectioning the obvious northern side, between the visible clay edges, revealing in the process a clay-lined feature [376] up to 3.00m wide and 1.90m deep, with evidence for some ten successive fills (Fig. 2.21).

As work progressed on the removal of these fills south of the recorded section, the full extent of the clay lining rapidly became clear, thereby demonstrating that the section line lay somewhat north of the original centre of the feature, which also proved to have a square sump at its base. At this point, a new half-section was established across the centre of the sump and the area between it and the original section was excavated through the clay lining into the surrounding deposits; this revealed the presence of a substantial pit [560] beneath the clay-lined feature, with a mixture of fills which extended for at least a further 1.50m down

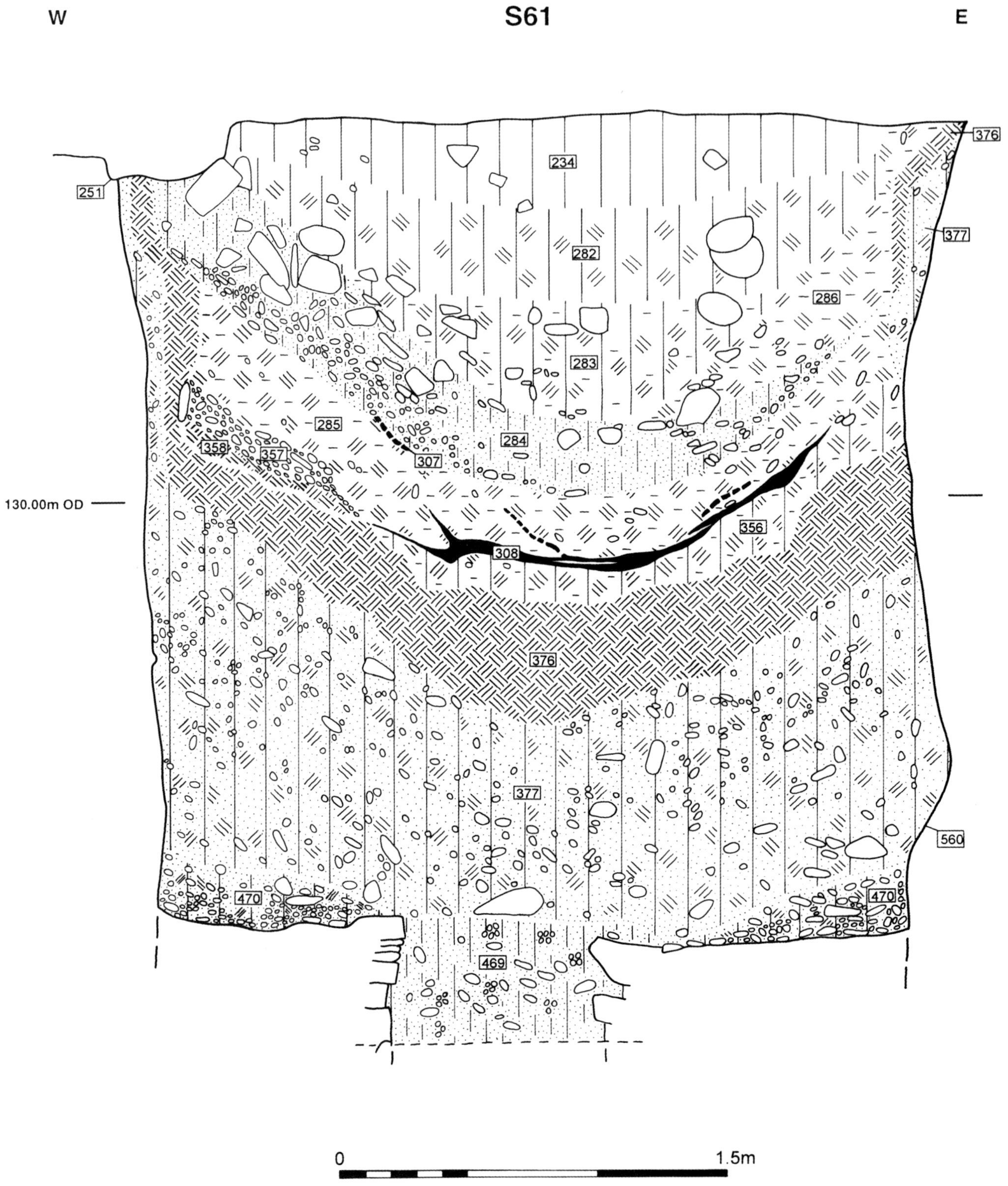

Fig. 2.21 Half-section across clay-lined cistern

to the top of a stone-lined well (see Figs 2.21 and 2.22). Fortunately, this new section also coincided with the centre of the well, thereby providing a detailed stratigraphic sequence through its successive fills, as far as the top of the stone collapse [558] where excavation ceased for safety reasons.

As will become apparent, the precise interpretation of these different features and their successive fills is not without its problems. Clearly the original operation on site had involved the excavation of a substantial, vertically-sided pit [560], measuring 3.50 by 2.90m (Fig. 2.23), which was traced to a depth of

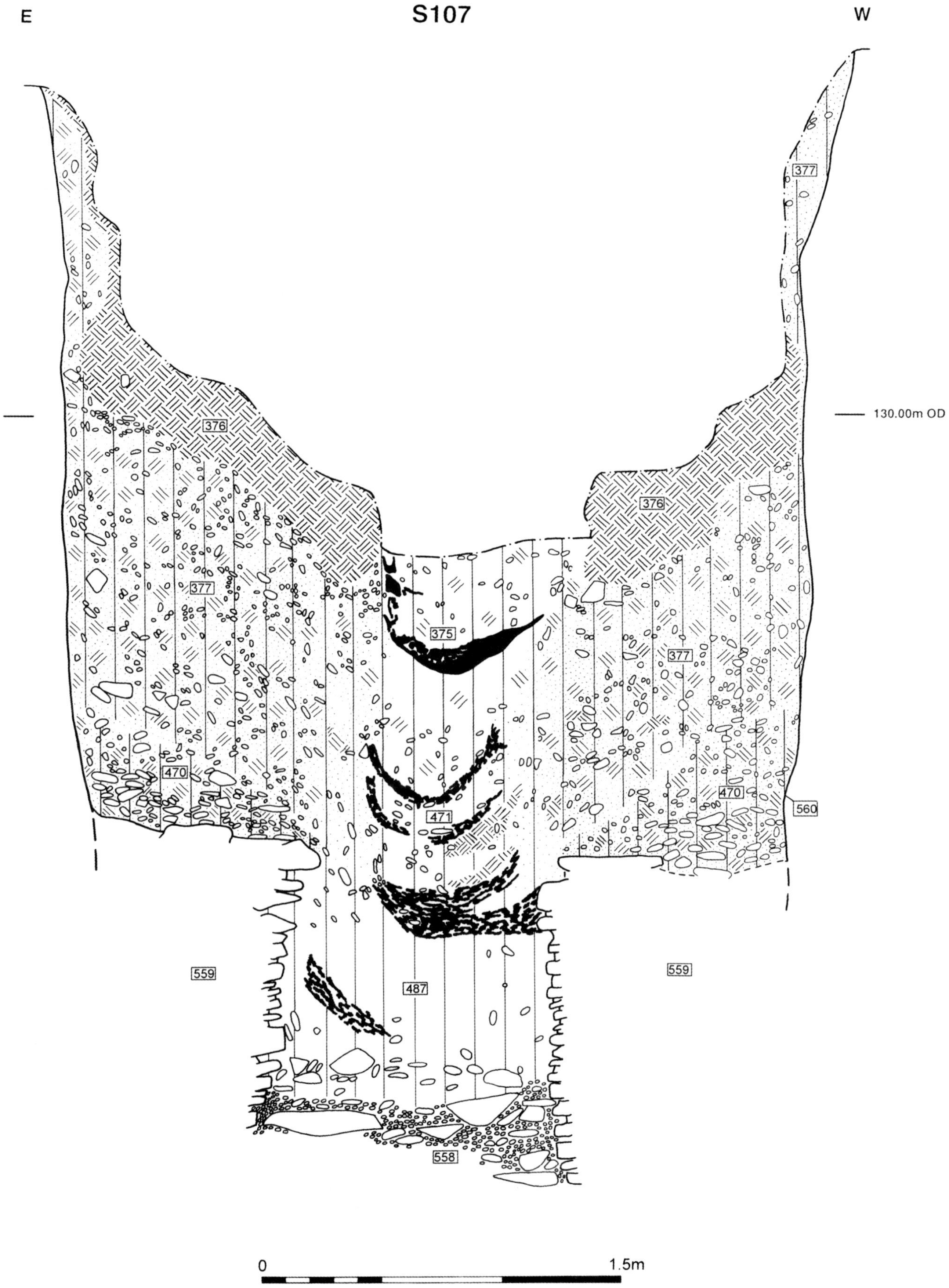

Fig. 2.22 Half-section across cistern and stone-lined well

Fig. 2.23 Main pit containing stone-lined well, looking north

3.40m from the surface, to the point where it met the top of the stone-lined well [559]. This pit must originally have been somewhat deeper, however, because there can be no doubt that it was dug to facilitate the insertion of the well, the lining of which was traced for a further 1.20m before excavation ceased. The well itself, which measured 1.00m in diameter, proved to have been constructed from roughly-laid blocks of shale (Fig. 2.24), with an irregular dump of large shale fragments and gravel behind to infill the original construction pit.

Somewhat surprisingly, this stone lining and its associated infill did not extend all the way up the pit as excavated, but terminated *c.* 3.40m from the top, at a point where its upper surface seemed to have been roughly levelled with large blocks of shale (Fig. 2.25). Above this, there was no definitive structural evidence to indicate whether the lining had ever continued up to the surface, or whether this was the original extent of the stone construction. Close examination of the stratigraphic sequence, however, sandwiched between the bottom of the clay lining and the top of the stone lining, makes it clear that some sort of lining must once have existed, for otherwise it would be difficult to explain the marked interfaces between the successive charcoal and clay lenses [471] which extend down the central part of the pit, and the mixed gravel and stone fills [377 and 470] which form an outer lining at this point. The simplest interpretation of such an interface would seem to be that, above a depth of 3.40m, the stone lining had been robbed out, leaving the original construction backfill in place, prior to the well shaft itself being infilled with successive dumps of material. This explanation would certainly account for the pronounced curvature of the charcoal, clay

Fig. 2.24 Stone-lined well, looking north

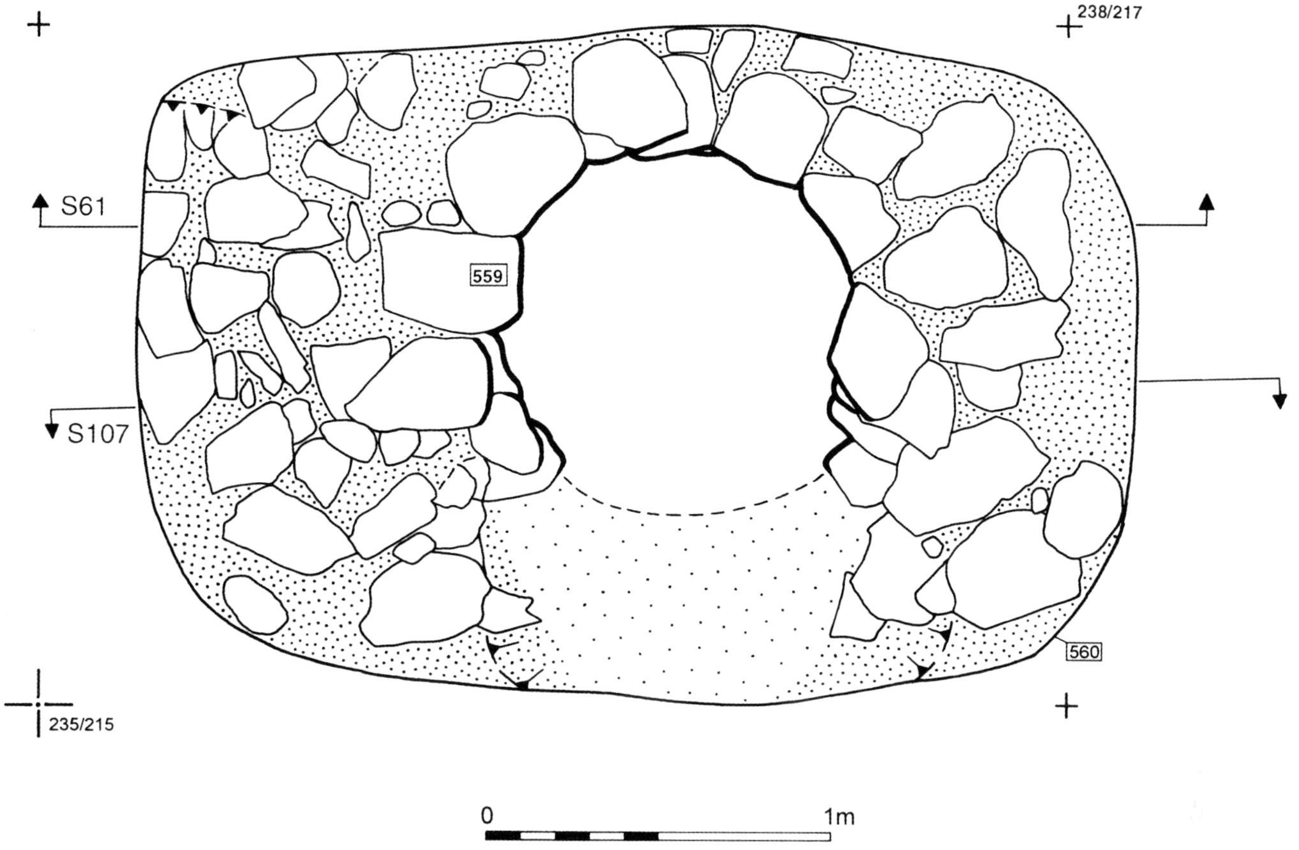

Fig. 2.25 Top of stone-lined well and surrounding fill

and gravel lenses [471], which is so reminiscent of sinkage into a shaft or well, and for the fact that the interface broadly corresponds with the upward projection of the underlying stone well. An almost identical sequence has been proposed for a 3rd-century well on the Bays Meadow site at Droitwich (Hurst, forthcoming).

An alternative, if less likely interpretation of the evidence, is the suggestion that the original lining did originally terminate part way up the pit, and that its roughly levelled upper surface formed a convenient platform for a timber construction which would have extended the well up to ground level. The demolition of such a structure and the infilling of the remaining well shaft, prior to the insertion of the clay-lined feature above, could equally account for the interfaces already noted in the stratigraphy. However attractive such an hypothesis might be, it is difficult to find any immediate parallels for this type of dual construction, except where timber elements were used in place of stone beneath the water table. Even less likely is the suggestion that the successive charcoal and clay lenses [471] represent the infilling of some sort of timber shaft extending downwards from the square sump at the bottom of the clay-lined feature. This would have necessitated the removal of any original lining and its associated backfill to the appropriate depth, prior to the insertion of the new structure and the infilling of the intervening space between its timbers and the sides of the re-excavated pit. In this model, the lower of the two surrounding fills [470] would be material which had fallen in from the sides during construction, prior to the insertion of the main fill [377]. Though plausible as an interpretation of the stratigraphic sequence, it is hard to envisage how such a clay-lined feature and shaft would have worked, let alone find any obvious parallels.

(ii) The clay-lined cistern (Figs 2.21 to 2.22, 2.26)
At some point in its life, the southern side of the stone lining had partly collapsed, thereby blocking the well with a mass of shale and gravel [558], only the very top of which could be safely examined in 1989. Contemporary with this collapse, a decision seems to have been taken to infill the well and, perhaps also, to convert the top part of the original pit into a clay-lined cistern.

Assuming that the stone lining had originally continued up to ground level, the process of infilling would seem to have involved two successive stages: (i) the removal of the stone lining to a depth of 3.40m, leaving the original mixed gravel fills [377 and 470] in place. Some of the latter may well have slumped at this stage into the lower part of the well, which could well account for the layer [469] of olive-brown loamy coarse sand (2.5 Y 4/4), containing up to 80% very small to medium-sized stones, which was encountered around the northern edge of the stone lining, sandwiched behind the lowermost fill [487]; (ii) the insertion of two successive dumps of material above the collapse. The lower of the two [487] proved to be a dull yellowish-brown loam (10 YR 4/3), containing only 10% small stones and charcoal flecks. By contrast, its upper counterpart [471] was far more mixed, with up to 20% coarse components and several distinct lenses of charcoal and clay throughout, the curvature of which clearly indicates considerable sinkage down the shaft. Over time this latter had also been compressed by the inward pressure of the original lining of the well [377].

Either contemporary with this infilling or soon after, the decision was taken to insert a clay-lined cistern into the upper part of the original pit. In either case this would have involved some recutting, hence the new pit proved to be *c.* 3.20 by 2.30m in size, slightly smaller than its predecessor, with near-vertical sides and a gently curving base, which extended to a maximum depth of 2.50m at its centre. It had been extensively lined with a layer [376] of bright yellowish-brown clay (2.5 Y 7/6), except at its base, where there were traces of a timber-lined sump or hole *c.* 1.00m square (Fig. 2.26). Around the sides of the pit, the thickness of the clay rarely exceeded 0.10m, but across the sloping base it proved to be more extensive, reaching a maximum thickness in places of 0.40m. By itself, such a clay lining would have been inherently unstable and likely to weather quickly once exposed to the elements, a point which was emphasised by a collapse along its northern side during the 1989 excavations. This makes it likely that there was originally some form of timber lining to hold the clay in place, which would help to explain the presence of several fragmentary carbonised timbers at various points around the circumference of the feature; not enough survived, however, to establish any pattern. The existence of a timber-lined sump or shaft at the base of the cistern was also somewhat surprising, because it bore no trace of any watertight clay lining. This makes it difficult to see how the feature was originally designed to function, let alone how its successive fills relate to the infill horizons immediately below.

At the end of its life, the clay-lined cistern was infilled with a number of successive layers (Fig. 2.21), some of them deliberate, others representing phases of weathering or natural silting. They include:

356 a dull yellowish-brown clay loam (10 YR 4/3) with up to 20% small to medium-sized stones, forming the lowest fill on the northern side of the cistern, immediately above the clay lining.

308/375 first identified as a layer [308] of dark reddish-brown sandy clay loam (5 YR 3/3) with a high ash and

Fig. 2.26 Clay-lined cistern, looking west

charcoal content, which dived over the edge of the sump to merge with a very mixed horizon [375] of charcoal, ash, daub, and brick/tile fragments forming the lowest fill inside the sump.

358 a thin band of yellowish-brown clay (2.5 Y 5/4) only recorded on the east side of the cistern immediately above the clay lining.

357 a thin layer of gravel and stones located at various points immediately above the primary fills on the sloping sides at the base of the cistern.

285 a deposit of brown silty clay (10 YR 4/6) with few coarse components, extending across the width of the cistern to a maximum depth of 0.40m.

307 a thin lens of charcoal sloping east-west on the north side of the cistern.

284 a fill comprising 95% small to large stones in a loose matrix of brown loamy fine sand (10 YR 4/4) apparently dumped into the cistern from its eastern side, where it proved to be up to 0.50m thick.

286 a layer of brown silty clay (10 YR 4/6) with some charcoal flecking against the western edge of the cistern.

283 a layer of dull yellowish-brown silty clay loam (10 YR 4/3) with up to 70% coarse components extending across the centre of the cistern to a maximum depth of 0.45m.

282 a spread of dull yellowish-brown clay loam (10 YR 4/3) containing less than 10% small to medium-sized stones, seen to extend across the whole width of the cistern to a maximum depth of 0.40m.

234 the uppermost fill in the cistern, comprising a dull-yellowish brown loam (10 YR 4/3) with 50–60% very small to medium-sized stones.

From the differences in character of these successive fills it is possible to reconstruct the different stages by which the pit came to be infilled. The lowest fill certainly seems to indicate a process of initial silting [356] derived from the clay lining of the cistern, followed by a phase of deliberate dumping involving significant quantities of demolition debris [308/375], no doubt derived from nearby buildings. Above this it is possible to identify a further sequence of layers derived either from episodes of natural silting and weathering [358; 285; 286], or from phases of deliberate dumping [357; 284], all of them incorporating exclusively Roman finds. After this, what remained of the hollow was apparently left to silt up naturally with three successive layers of loamy material [283; 282; 234], some or all of which could have accumulated after the end of the Roman period.

(iii) Dating evidence and chronological discussion

The chronology of these important features is, unfortunately, somewhat difficult to define, because all the relevant stratigraphic relationships with the associated rampart and towers to the north had been removed by the foundations of the later clom and stone-built structures which had cut right into the natural at the critical points. It is clear, nevertheless, that the well must belong early in the relative sequence of the site, sandwiched as it is between the butt-end of the primary clay rampart and the line of the east-west road, just inside one of the fort's gates. Quite when it was converted into a clay-lined cistern remains unclear, though it would be possible to argue that this occurred at the time of the reorganisation of the defences when the stone wall was added. It is also highly probable that the cistern went out of use before the end of the military occupation, because it can be demonstrated that the latest of the five phases of building on the 1989 site, represented by the east-west foundation trench [414], had cut into the clay apron along its southern side (see section 2.2.2(Gvi) below). This and the close proximity of the building and cistern make it most unlikely that both could have been in use at the same time.

The successive fills of the well and cistern all contained high concentrations of Roman finds, including pottery, glass, brick and tile, daub, slag and nails. The principal pieces include:

487 fill of stone-lined well: various samian sherds, including one of form 27g, two of 18, dated *c.* 80–100, and one 37, probably of *c.* 70–90 (Table 2.2).
joining handle sherds from a Dressel 20 amphora (not catalogued).
flange sherd from a mortarium, probably from Caerleon industry (CP 52); late 1st to early 2nd C.
base sherd from a Gillam 238 mortarium (not catalogued).
rim and joining body sherd from a bead rim jar (CP 53).
rim sherd from a greyware jar (CP 51).
rim sherd from a greyware carinated bowl (CP 54); Flavian to Trajanic.
sherd from a greyware lid (CP 55).
handle from a cylindrical or prismatic bottle (Gl 6).

377 fill behind sump: various samian sherds, including one form 37 (S 2), dated 80–100, a probable form 37 base, a probable Curle 11 bowl, and a possible form 29 wall sherd (Table 2.2).
flange sherd from a mortarium, probably from Caerleon industry (CP 38); late 1st to early 2nd C.
joining pieces from a long-necked glass jug (Gl 2); mid 1st to mid 2nd C.

471 fill of sump: various samian sherds, including forms 18 or 18R, 27, 29 and 37 (Table 2.2).
rim and body sherds from a greyware jar (CP 46); late 1st to early 2nd C.
rim sherds from a probable rusticated jar (CP 47); Flavian-Trajanic (*cf.* CP 42 from context [460]).
rim sherd from a rusticated jar (CP 48); Flavian to Trajanic.
joining body and rim sherds from a greyware jar, with high burnishing akin to 'terra nigra' (CP 49); probably Flavian.
joining rim sherds from a jar with rouletted decoration akin to samian form 29 (CP 50); possibly *c* . 70–90.
rim sherd from a greyware jar (CP 51).
rim and neck fragment from a cylindrical or prismatic bottle (Gl 6).
glass gaming counter (Gl 18).

376 clay lining: various samian sherds, including forms 29, 36 and Curle 11 (Table 2.2).

Lower fills of cistern

375 rim sherd of samian form 36 (Table 2.2).
rim sherd from a greyware jar (CP 36); late 1st to early 2nd C.
rim and body sherds from a greyware lid (CP 37).
joining pieces from a long-necked glass jug (Gl 2); mid 1st to mid 2nd C.

357 sherds of samian forms 18/31 and 35 (Table 2.2).
sherds from a jar with wavy line decoration, common in 1st to 2nd C (CP 32).
decorated sherd of egg-shell beaker or cup (CP 33); late 1st to early 2nd C.
body and base sherds from a Black-burnished jar (not catalogued).

356 various samian sherds, including forms 18/31 and 18 or 18/31 (Table 2.2).

308 rim, body and base sherds from a Malvern jar (CP 25); mid to late 1st C.
rim and body sherds from a greyware lid (CP 26).

285 denarius of Titus, AD 80 (Co 1).
basal sherd of samian form 29, rare after AD 85 (Table 2.2).
body sherds of Dressel 20 amphora (not catalogued).
rim sherds and handle from a Black-burnished lid (CP 17); mid 1st to late 2nd C.
rim fragment and body sherds from a Black-burnished jar (CP 18).
glass bead (Gl 10).

284 various samian sherds, including a probable form 29, dated *c.* 70–85, and a form 35 (Table 2.2).
body sherds of Dressel 20 amphorae (not catalogued).
rim sherd of a mortarium from Verulamium region (CP 13); AD 70–110.
spout sherds of a Gillam 238 mortarium (not catalogued).
rim and body sherds from a Malvern jar (CP 14); 1st C AD.
rim and body sherds from a Black-burnished jar (CP 15).
basal sherd from a flagon or jar (CP 16); AD 70 to mid 2nd C.
fragment from rim and neck of a small glass jar (Gl 1); 1st to early 2nd C.

Upper fills of cistern

286 rim, body and base sherds from a Black-burnished jar (CP 19); 1st or early to mid 2nd C.

283 rim sherd from a mortar-like bowl (CP 12); late 1st and early 2nd C.

234 body sherds of Dressel 20 amphorae (not catalogued).
glass beads (Gl 11 and 12).

Along with the material in the fills of the large pit [245] containing the burnt timber building (see section 2.2.2(Hiv) below), these finds are clearly fundamental to our understanding of the site's wider chronology. They are less useful, however, for any fine tuning of the dating of the well and cistern within such a narrow timescale, given the quality of the samian assemblage and the traditional date ranges assigned to most of the coarse wares (see section 2.2.4(C) below). On current evidence, the vast majority of the assemblage from the well/cistern can be assigned to the Flavian-Trajanic period, with only one or two problematical Black-burnished pieces potentially extending beyond this range (CP 15 and 18). One feature of particular interest, however, is the relatively late appearance of Black-burnished wares in the stratigraphic sequence, the known examples being found exclusively in several of the secondary fills [234, 284–6, and 357] of the cistern; similar trends have been recognised elsewhere on the site (see sections 2.2.2(Gvii) and (Hiv) below). This may have a bearing on the wider problem of their dating, as discussed by Webster (see section 2.2.4(Cii) below), and could provide a stratigraphic marker for correlating activity in different parts of the site.

(E) The east-west road and its associated drains (Figs 2.10, 2.27 to 2.28)

Immediately to the south of the rampart and the well the earliest Roman activity was represented by the fragmentary remains of the original *intervallum* road extending east-west across the site (Fig. 2.27),

Fig. 2.27 East-west *intervallum* road, looking east

composed almost entirely of small and medium-sized coarse gravel components with little or no matrix. This had been extensively disturbed by later activity, not least to the east of the post-medieval cellar, where only isolated patches of its gravel makeup [465] were recovered, little more than 0.10m thick; fortunately, west of the cellar, enough had survived to demonstrate that its gravel surface [434] was originally 5.00m wide, with a maximum thickness of 0.20m towards its western end. This gravel had obviously been laid as a single operation, sealing in the process the existing ground surface, a pair of earlier pits or tree holes [464 and 468], and a smear of clay derived from the construction of the rampart [381] (see Fig. 2.7).

Immediately south of the gravel makeup of the road lay a shallow rounded hollow [502], which was traced over a distance of *c.* 8.00m from west to east. Its width proved to be somewhat variable, between 0.30 and 0.50m, while its depth never exceeded 0.10m. On the east, it became so shallow that it eventually petered out altogether. This was originally interpreted as a shallow drainage hollow or drip gully, sandwiched between the road and the earliest buildings to the south, but it is possible that it actually represents part of an original wall line associated with the much deeper feature [493] to the east, with which it would have formed a convenient right angle; this possibility will be discussed further below in section 2.2.2(Gii). For most of its length, it was filled by a dull yellowish-brown silty clay loam (10 YR 5/3), with variable concentrations of small to medium stones [501].

More certainty attaches to the interpretation of the three gullies [416; 422; 427] which were located along the northern edge of the gravel spread [434] west of the cellar (Fig. 2.28A). The earliest was probably the shallow U-shaped gully [427], *c.* 0.40m wide, which was traced for a distance of 3.40m east of the main section. At its maximum it was 0.12m deep, cut through the edge of the gravel into the underlying clay. Its profile, however, became progressively more shallow towards its eastern end, where it eventually petered out in the vicinity of the later foundation trench [413]. Its fill [426] comprised a high proportion (*c.* 80%) of coarse gravel components and yellow-orange clay lumps, set within a matrix of dull yellowish-brown silty clay loam (10 YR 5/4), most of which was no doubt derived from the adjacent road and rampart surfaces.

Immediately to the north lay a second U-shaped gully [422], cut entirely into the clay rampart, which was only traced for a distance of 1.80m east of the section. It too was relatively shallow at its eastern end, but became progressively deeper towards the west, where it proved to be at least 0.25m wide and 0.10m deep. Its original depth may have been somewhat greater, however, because its profile had been partly cut away by the latest of the gullies [416] which ran on much the same alignment at a higher level. This latter gully had also removed any stratigraphic link between the earlier features and their respective fills. Throughout its length gully [422] was filled by a brown silty loam (10 YR 4/4) with a relatively high proportion (*c.* 60%) of coarse gravel components [421], the bulk of which represents silting from the adjacent road and rampart surfaces.

Both the lower gullies and their respective fills must have been originally sealed by two successive horizons [394; 417] which were variously recorded in plan and section (Fig. 2.28B). The first of these was a layer [394] of dull yellowish-brown silty clay loam (10 YR 5/4) with up to 10% coarse gravel components, which seems to represent erosion off the clay rampart. It was first recorded in plan, infilling a slight hollow in the surface of the gravel road [434], but was subsequently noted as well right along its northern edge beneath the secondary gravel spread [417]. This latter was composed almost entirely of small and medium-sized coarse gravel components, set within a yellowish-brown loam (2.5 Y 5/4), very reminiscent of the existing road surfaces. It had survived only as a narrow strip, 0.45m wide and 0.05m thick, but must have been originally more extensive to the north, where it had been cut away by a later gully [416]. It is best interpreted as a secondary gravel dump along the northern edge of the road, sealing the fills of the earlier gullies; alternatively, its relationship with the underlying silty clay loam [394] might suggest that both were originally more extensive and that [417] might be part of a wider resurfacing.

The latest gully [416] in this part of the site had been cut through the northern side of the secondary gravel [417] into the underlying horizons. It proved to be a shallow U-shaped trench, 0.45m wide and 0.10m deep at maximum, extending for a distance of 3.80m east of the main section. Like the earlier gullies it became progressively shallower towards its eastern end, appearing to terminate at the point where it had been truncated by a later foundation trench [413]. On the west, by contrast, the gully was not recognised in the main section, suggesting that it had begun to narrow and shallow out by that point. Its fill [415] comprised a high proportion (*c.* 90%) of coarse components, including some large rounded stones, set within a matrix of dull yellowish-brown silty loam (10 YR 4/3). This suggests a deliberate process of backfilling rather than natural silting.

All three gullies are best interpreted as drains, serving to drain excess rainwater away to the west, though they would have required regular maintenance to keep them operational if they were not in some way timber-lined. There is also the possibility that they represent a more localised feature, connected in some way with the thickening of the clay rampart at this point, though whether for drainage or some other structural reason remains uncertain.

The only datable finds associated with any of these

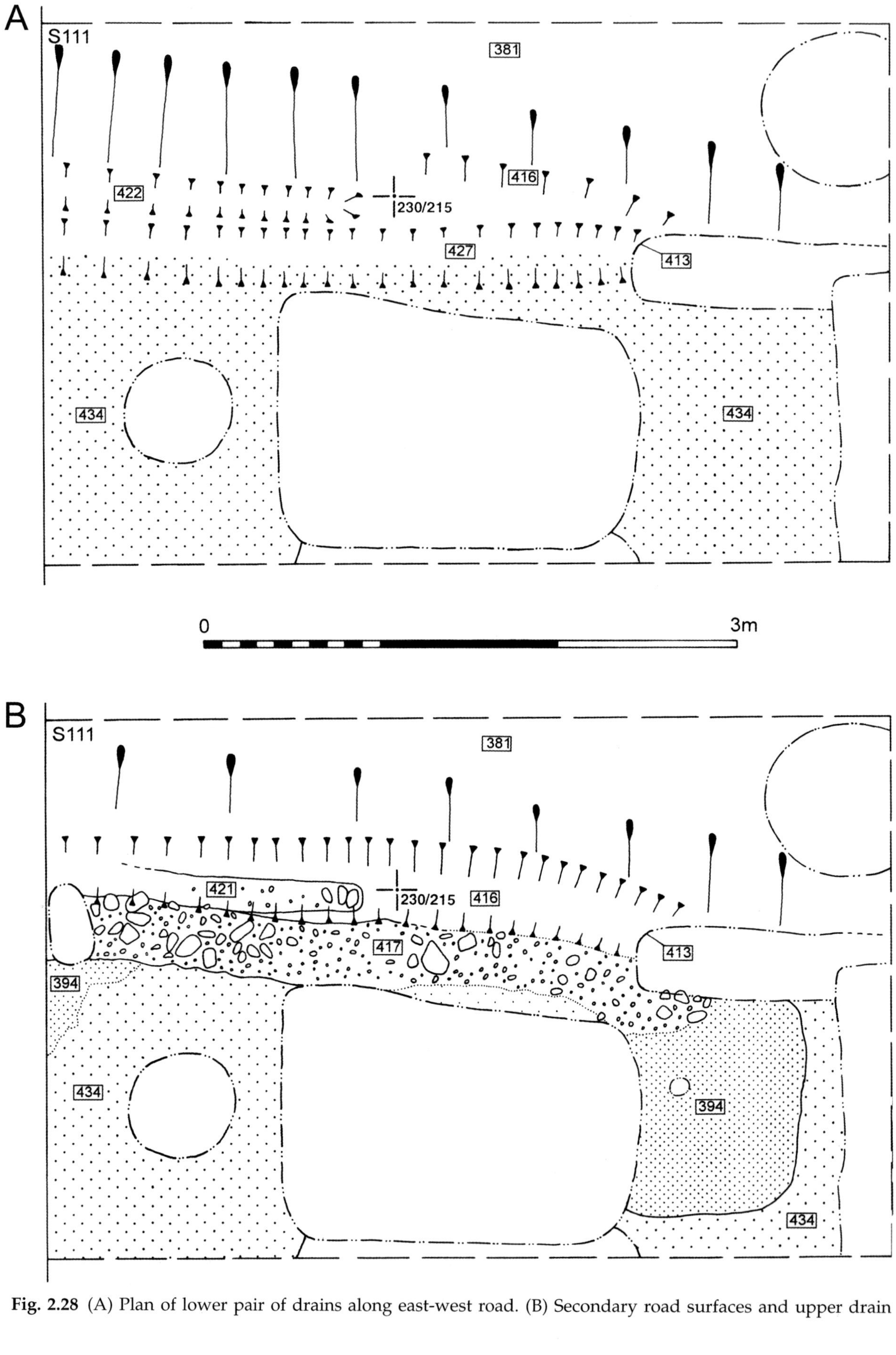

Fig. 2.28 (A) Plan of lower pair of drains along east-west road. (B) Secondary road surfaces and upper drain

layers were four fragments of a Gillam 238 mortarium (not catalogued), which were found in the makeup of the gravel road [434] west of the cellar; they can be comfortably assigned to the later first century.

(F) The north-south road (Fig. 2.29)
The east-west road represented by the gravel spreads [434 and 465] would originally have formed a junction with a north-south road, which was first identified by Jones and Little in 1973 immediately to the east of the 1989 site; unfortunately, any trace of this junction within the current area had been removed by later activity and by the post-medieval foundations. The only possible evidence relating to the original north-south road was located in the very south-eastern corner of the site, where there were two related patches of coarse gravel [425] set within a matrix of dull yellowish-brown silty clay loam (10 YR 5/3). They were traced over a distance of *c.* 7.50m along the eastern side of the trench, running parallel with the line of the known road and the adjacent buildings, though their original western edge had been cut away in part by a later gully [279] and in part by the large pit [245] containing the burnt timber building.

It was clear, however, that these gravel spreads could not have been part of the original road, because they were surprisingly shallow and overlay two successive layers [452; 454] which must be related in some way with the earlier building horizons discussed in the following sections. This would suggest that the original road lay slightly to the east, and that the gravel spreads described here represent a later widening or resurfacing. There were no associated finds to determine the date of this operation.

(G) The building sequence (Figs 2.10, 2.30 to 2.40)
(i) Problems of excavation and phasing
The 1989 excavations examined a complex sequence of foundation trenches across the central part of the site, representing at least four and possibly five phases of building activity. For convenience these can be divided into two groups: (a) those which lay entirely to the south of the east-west road (phases 1 to 3); (b) those which had been built at a later stage across the east-west road (phases 4 to 5). As will become clear, the definition of a detailed building sequence did not prove easy for a number of reasons, including (i) the area available for excavation, (ii) the absence or non-survival of any certainly associated floor levels, (iii) the tendency for successive buildings to re-use the same building lines, and (iv) later, post-Roman disturbances.

Of the earliest buildings (phases 1 to 3) only the northernmost 4 to 5.00m was available for excavation, sandwiched between the east-west road [434 and 465] and the later pit [245] containing the burnt timber building (Fig. 2.30). Such a restricted area clearly

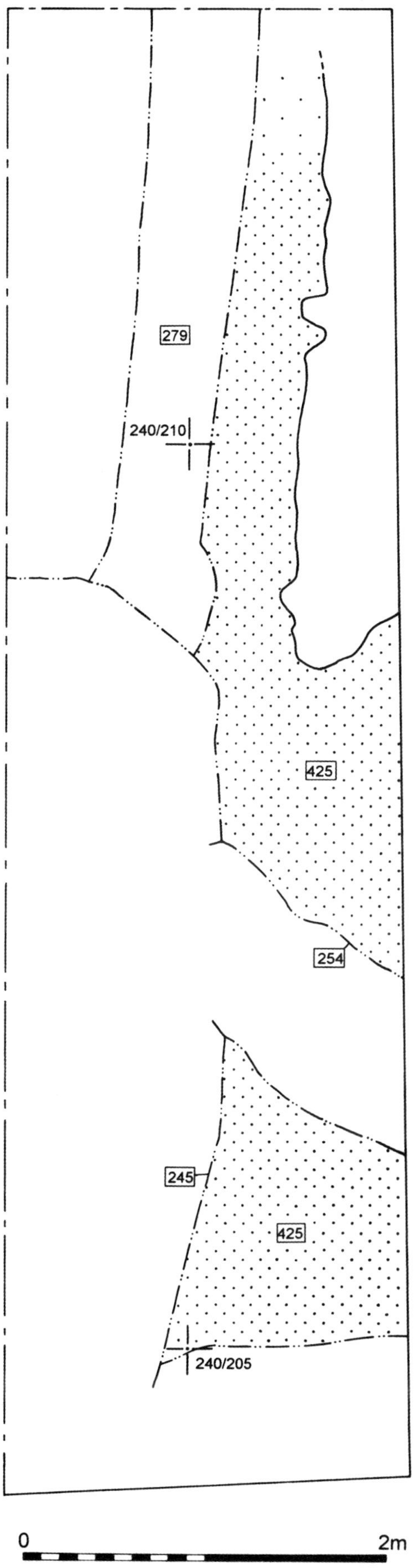

Fig. 2.29 Plan of north-south road

Fig. 2.30 Timber building Phases 1 to 3, looking east

made it difficult to establish any detailed plans for comparison, a situation which was further complicated by the obvious coincidence of some of the building frontages, not least during phases 2 and 3. This part of the site had also been subject to considerable disturbance, most likely in the post-Roman period, to such an extent that all the Roman levels were either churned or truncated. At the western end, for instance, a shallow rectangular hollow [244], *c.* 3.00 by 5.00m in extent, had removed all but the bottom of the successive foundation trenches - in some cases only as little as 0.10m had survived - leaving only a narrow band of intact archaeology between its northern edge and the east-west road [layers 472; 466; 459; 455] and along its eastern side [layers 456; 411; 342]. None of these could be identified, however, as floor levels, the two groups apparently deriving from activity on the site in the intervals between the different building phases.

Further east, even these few intact horizons were lacking. Instead the Roman levels had been churned up by later activity, resulting in the formation of an homogeneous silty clay horizon [359], which extended across the top of all the foundation trenches. In the face of such disturbances and the absence of any recognisable floor levels, the early building phases had to be defined on the basis of the coherence of their foundation alignments and a comparison of their relative depths.

Paradoxically, the later buildings (phases 4 to 5) were much easier to excavate and phase, despite the absence of related floor levels, the occasional reuse of the same building lines, and the disturbances already mentioned at their southern end. This was partly because a much larger area was available for excavation (up to 8.00 and 10.00m respectively along the north-south dimension), and partly because they had been dug through a distinctive gravel horizon at their northern end, where they had encroached onto the east-west road (see Fig. 2.27). For these reasons also, more of their plan was recovered for comparison.

(ii) Phase 1 (Figs 2.10 and 2.31)

This first phase presents a number of problems and is open to alternative interpretation on the evidence available. Two key features enter into the discussion. The first is the shallow rounded hollow [502], already mentioned in section 2.2.2(E), which was traced over a distance of *c.* 8.00m, running immediately to the south of the east-west road. This varied in width from 0.30 to 0.50m, with a maximum depth of no more than 0.10m; on the east it became progressively more shallow and eventually petered out altogether. There was nothing distinctive about its fill, which comprised a dull yellowish-brown silty clay loam (10 YR 5/3) with variable concentrations of small to medium stones [501].

The second feature is the vertically-sided, flat-bottomed trench [493] which was traced for a distance of *c.* 6.00m south of the east-west road, parallel with the postulated north-south road through the fort. Like all the other north-south foundation trenches it had been entirely cut away at its southern end by the later pit [245]. It had also been partly truncated along its western side by the later foundation trench [414]. Even so, excavation showed that it became progressively wider and deeper from north to south, the width increasing from *c.* 0.45 to 0.70m, while the depth increased from 0.45 to 0.65m (Fig. 2.32, S119 and S120). Its fill [492] principally comprised a bright brown silty clay loam (7.5 YR 5/8) containing 50–60% coarse components, very reminiscent of the natural subsoil in colour and character, though there were also patchy traces of a primary deposit of pale grey/brown silty loam towards the base of the feature.

Neither of these was originally identified as a structural feature forming part of an early building. Instead, the east-west hollow [502] was interpreted as a shallow drainage or drip gully much like those already described in section 2.2.2(E), while the much deeper north-south trench [493] was thought to be part of a main drain extending south across the fort. These were seen as being sandwiched between the

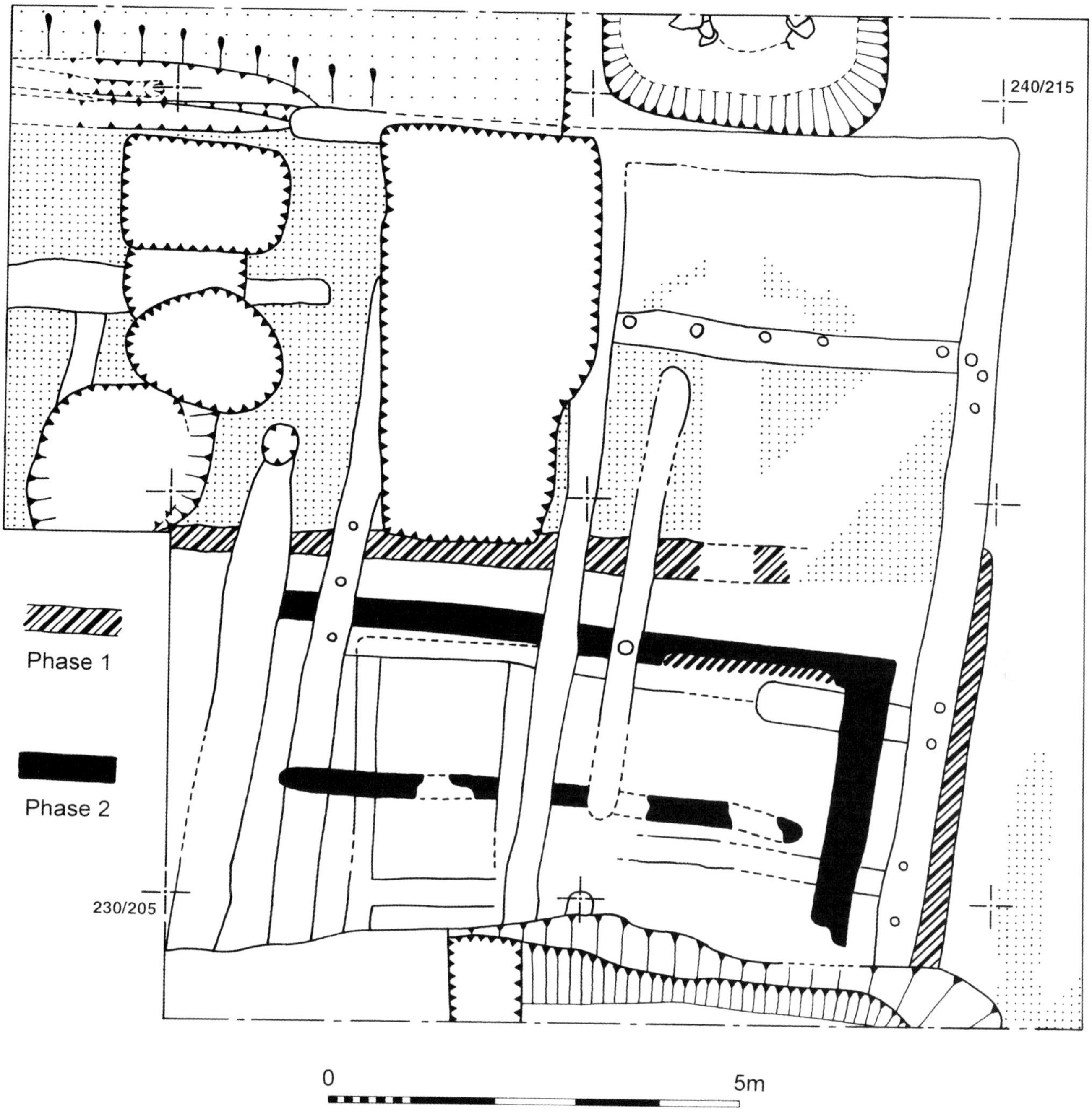

Fig. 2.31 Plan of Phase 1 and 2 buildings

earliest building frontage, represented by the right-angled foundation trench [480], and the east-west and north-south roads respectively. If the identification of the north-south trench as a drain is correct, then its surviving profile would suggest that it was originally timber-lined, the timbers being removed before the insertion of the relatively homogeneous infill.

The alternative hypothesis presented here is that the two features originally formed the north-eastern corner of a single building, extending at least 10.00m from east to west and 6.00m from north to south. This might help to explain the presence of the somewhat deeper, rounded terminal at the northern end of the north-south trench, which looks as if it represents the base of a substantial timber upright at the corner of a building. Such an interpretation is not without its problems, however, not least the marked differences in character between the two sides of the proposed structure. Hypothetically this could be because the shallow east-west hollow was misinterpreted in 1989, being in reality little more than slumpage into a much deeper feature, although this would seem unlikely since nothing was actually seen in the exposed sides of the numerous trenches which cut across its postulated line.

The existence of such a building running east-west

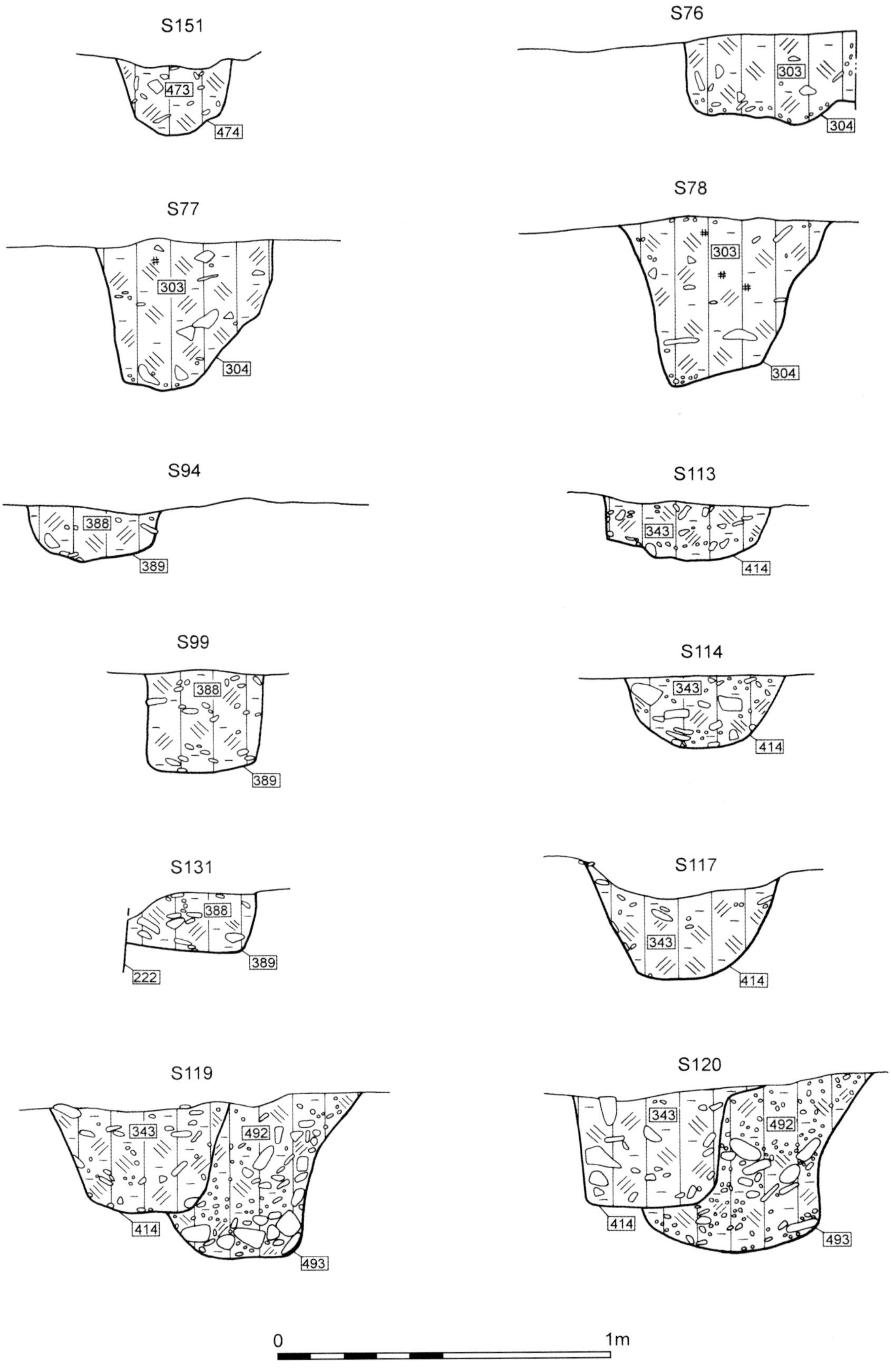

Fig. 2.32 Sections across foundation trenches

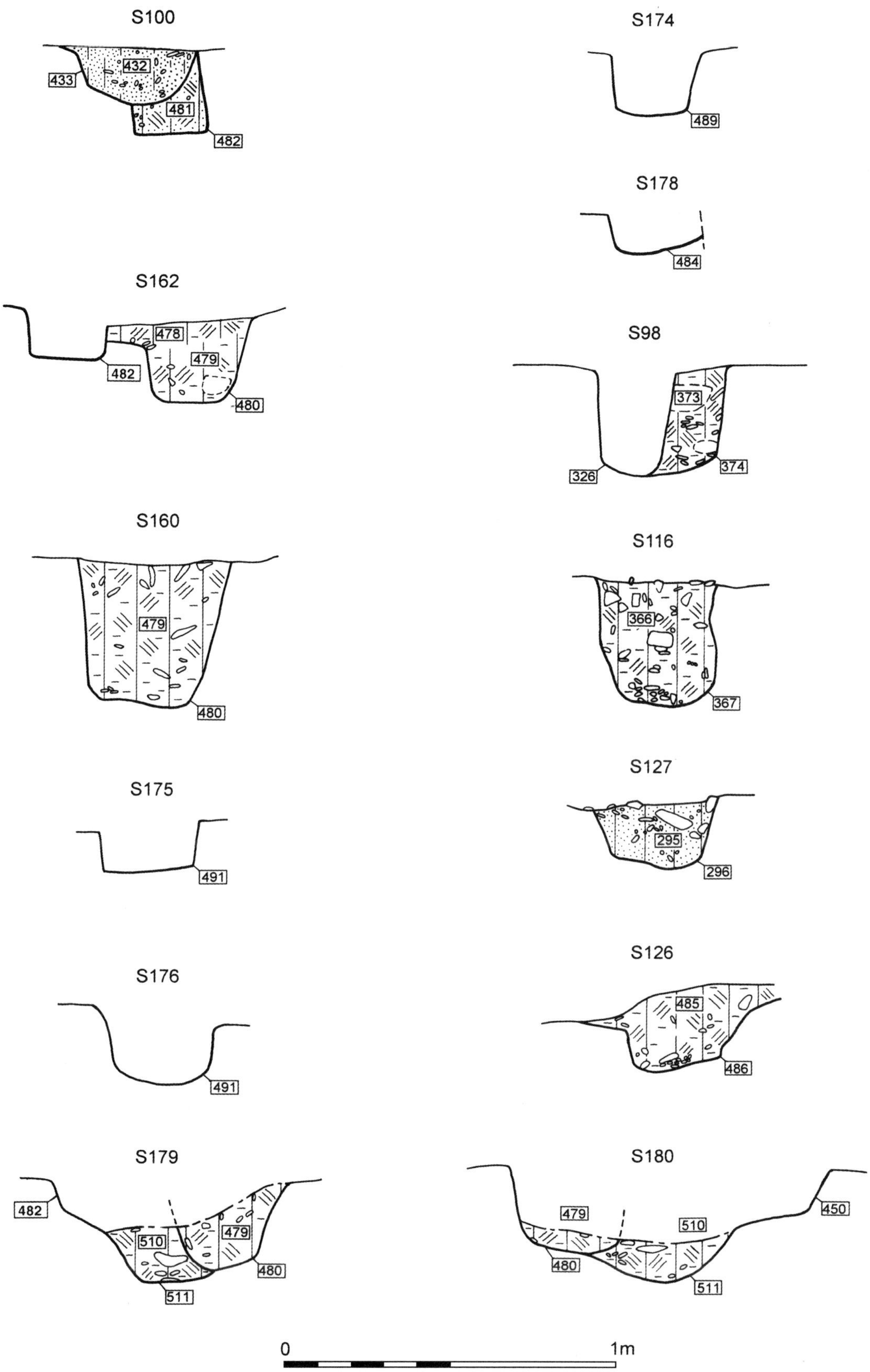

Fig. 2.33 Sections across foundation trenches

at this point, in contrast with the prevailing north-south alignment of all the later structural phases, has much to commend it, because it would directly parallel the sequence identified by Jones and Little (1973) during their excavations on the site of what is now the hotel car park (see section 2.2.1 above); here too, the earliest buildings were aligned east-west, prior to being re-aligned through 90°, which would mean that both sites exhibit a degree of internal regularity from the outset, reflecting their location in the same general sector of the fort. Unfortunately, there were no finds associated with this phase on the 1989 site.

(iii) Phase 2 (Figs 2.10 and 2.31)

Should the argument prevail that both features assigned to phase 1 were drains, then the earliest building frontage with which they could be associated would be the right-angled foundation trench [480]. This was traced over 3.50m along its eastern side, to the point where it had been truncated like everything else by the later pit [245] containing the burnt timber building. Along its northern side, it was examined over a distance of 7.50m, to a point on the west where it had been truncated by a later foundation trench [304]; since it could not be traced beyond this, it seems likely that the later trench had entirely removed the western side of the phase 2 structure which must, therefore, have run along the same north-south alignment. This would indicate a building *c.* 8.00m wide overall, running north-south like all the later phases.

Excavation of this right-angled feature was somewhat complicated by later disturbances, especially along the northern side, not least the tendency for later phases to re-use much the same alignment. Careful sectioning along this side, however, demonstrated that it was a steep-sided U-shaped feature, at least 0.30m wide and 0.30m deep (Fig. 2.33, S162, S179 and S180). A corresponding section on the east seemed to indicate that it was somewhat deeper, up to 0.50m, but it is possible that this reflects recutting from a higher level associated with a later foundation trench [482(?)], which was not otherwise recognised in 1989 (Fig. 2.33, S160); certainly the section seems to show slight variations in the character of the fill from top to bottom which might relate to this. Generally the infill [479] comprised a yellowish-brown silty clay loam (10 YR 5/6) containing a 50–60% admixture of small to medium-sized gravels.

Towards its eastern end, in an area of considerable complexity, this feature seemed to have cut into and disturbed the fills of two irregular sausage-shaped hollows [511], both running east-west (Fig. 2.34). Their fills [510] were identical, comprising a dull yellow-orange silty clay loam (10 YR 6/4) with a small admixture (10%) of stones and some charcoal flecks. Similar material was also found in limited quantities in the neighbouring fill of the foundation trench, suggesting that it had been disturbed and redeposited during construction work. One of the hollows was twice as long as the other (2.00 and 1.00m respectively), but otherwise they were both 0.40m wide and up to 0.25m deep in their truncated form. Neither of these features can be easily ex-

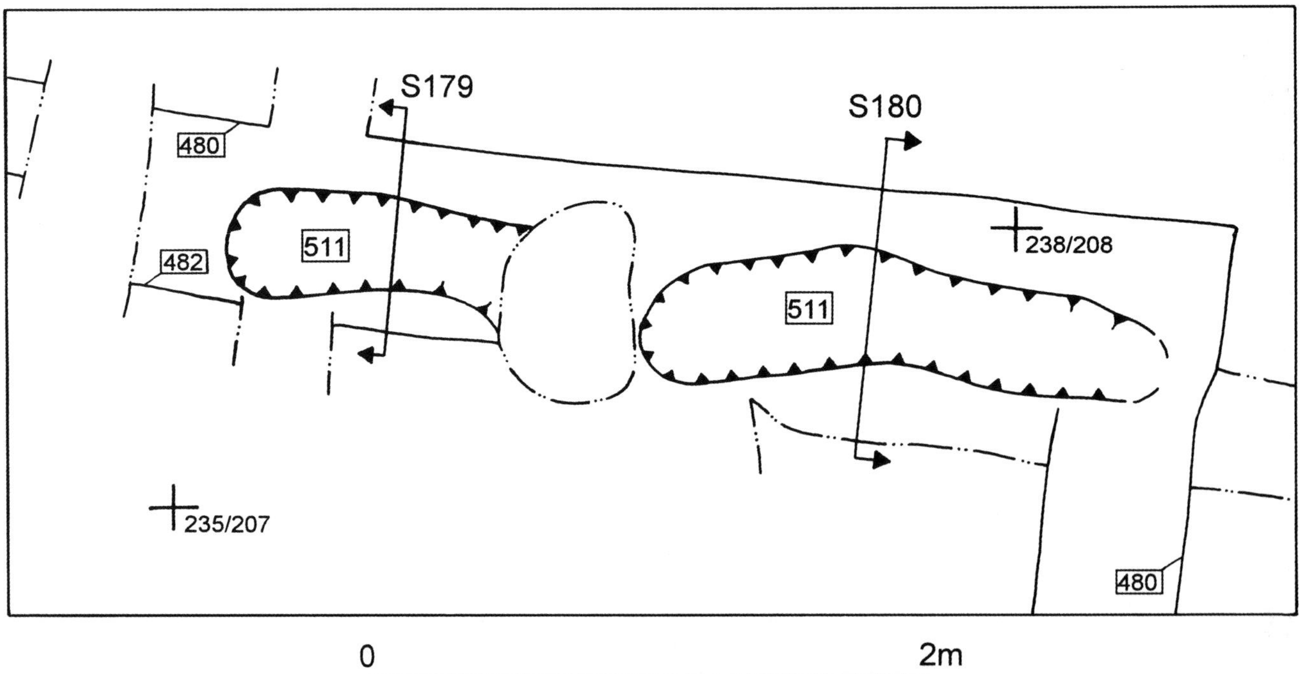

Fig. 2.34 Irregular sausage-shaped features at base of Phase 2 trenches

plained, since they are not obviously related to any of the main building phases as they are currently understood.

One further foundation trench [491], lying towards the southern end of the surviving buildings, ought to be connected with this phase. It proved to be a steep-sided flat-bottomed feature, 0.25 to 0.30m wide and up to 0.30m deep (Fig. 2.33, S175 and S176), which was traced over a distance of 6.30m from east to west. It was further characterised by the fact that it had rounded ends, which conveniently stopped 0.30 to 0.40m short of the known or postulated frontages of the phase 2 structure. It also ran parallel with its northern frontage, at a distance of *c.* 1.80m, all of which makes it likely that we are dealing with an internal partition rather than a separate structural phase. Its fill [490] comprised a dull yellowish-brown loamy coarse sand (10 YR 4/3) containing a 40% admixture of coarse components and some charcoal flecks. The possibility that this was only one of a series of internal partitions at roughly 1.80m intervals cannot now be checked because the next potential division would lie just inside the area truncated by the large pit [245]. As with phase 1, unfortunately, there were no finds associated with this phase.

Along its northern side, the fill of the right-angled foundation trench [480] was sealed by a thin layer [478] of distinctive dull yellowish-brown sandy clay (10 YR 5/4) containing 20% small and medium-sized stones. An almost identical layer [500] was found to overlie the fill of the shallow hollow [502] immediately to the north, already described in connection with phase 1. Both may well have been part of an originally more extensive layer which has only survived in a recognisable form at those points where it accumulated above the fills of earlier features. The only find from this phase was a fragment of a glass jar/vessel (Gl 3) from context [478].

(iv) Phase 3 (Figs 2.10 and 2.35)

Despite considerable problems during the excavation of the earliest buildings, it is clear that this phase comprised an east-west foundation trench [482] and two separate north-south trenches [484 and 489].

The first of these [482] was initially identified during the investigation of one of the foundation trenches [433] assigned to phase 4, at which point it was thought to be a deeper section running along its southern side (Fig. 2.33, S 100), even though this interpretation did not really explain why the latter appeared to terminate short of the western end of the main trench. This situation was subsequently clarified by excavation east of the known extent of the phase 4 feature, in the restricted area between two of the later north-south foundation trenches [296 and 389], which located the continuation of the deeper section as a separate vertically-sided, flat-bottomed trench, measuring 0.20m wide and 0.30m deep. East of this, its line could only be traced with considerable difficulty for a further metre or so, before it was entirely lost in the area of the intrusive post-like feature [477] and the later foundation trench [450]. Where best preserved, its backfill [481] comprised a brown sandy clay loam (10 YR 4/4) containing a mixture of very small to medium-sized stones (30%) and some charcoal flecks.

Both the north-south foundation trenches [484 and 489] were recognised at an early stage in the excavation of the building sequence, having been partly exposed in the bottom of the shallow rectangular hollow [244]. That on the west [489] proved to be a vertically-sided flat-bottomed trench, 0.28m wide and 0.20m deep (Fig. 2.33, S174), which extended southwards for a distance of 3.40m to the point where it had been cut away by the later pit [245]. Its counterpart [484] lay *c.* 1.50m to the east, immediately adjacent to a later foundation trench [389] which had truncated its eastern side. Even so, enough survived to demonstrate that it too had been a vertically-sided and flat-bottomed trench at least 0.30m wide and 0.18m deep (Fig. 2.33, S178); unfortunately, it had been so severely truncated by the hollow that it could only be traced southwards for a distance of 2.30m. Both trenches contained an identical fill [488 and 483 respectively] of brown sandy clay loam (10 YR 4/4), very reminiscent of the material found in the contemporary east-west trench, except for the inclusion of several lumps of red and yellow clay presumably derived from earlier buildings.

Despite the problems associated with the excavation of the east-west trench [482], it is still possible to reconstruct the original width of the building from the surviving evidence. There can be little doubt, for instance, that the rounded terminal at its western end represents its original extent; this coincides well with the line of the north-south trench [489], which must therefore have formed the western frontage. The situation is obviously less clear on the east, but there is nothing to contradict the suggestion that the east-west trench originally extended as far east as its phase 2 predecessor, and that its eastern frontage ran along exactly the same line as the north-south arm of trench [480]. This would certainly help to explain the slight variations in the fill of the north-south trench already noted in the preceding section, which would therefore be related to a partial recutting of the feature unrecognised in 1989. Such evidence would allow us to reconstruct a building *c.* 6.40m wide from east-west, somewhat narrower than its predecessor.

One further feature which is likely to belong with this phase needs to be considered here. This was a short length of east-west foundation trench [561] located towards the southern end of the two north-south trenches, immediately north of the line of the large pit [245]. It was seen as a faint soil mark in the base of the shallow hollow [244] at an early stage in

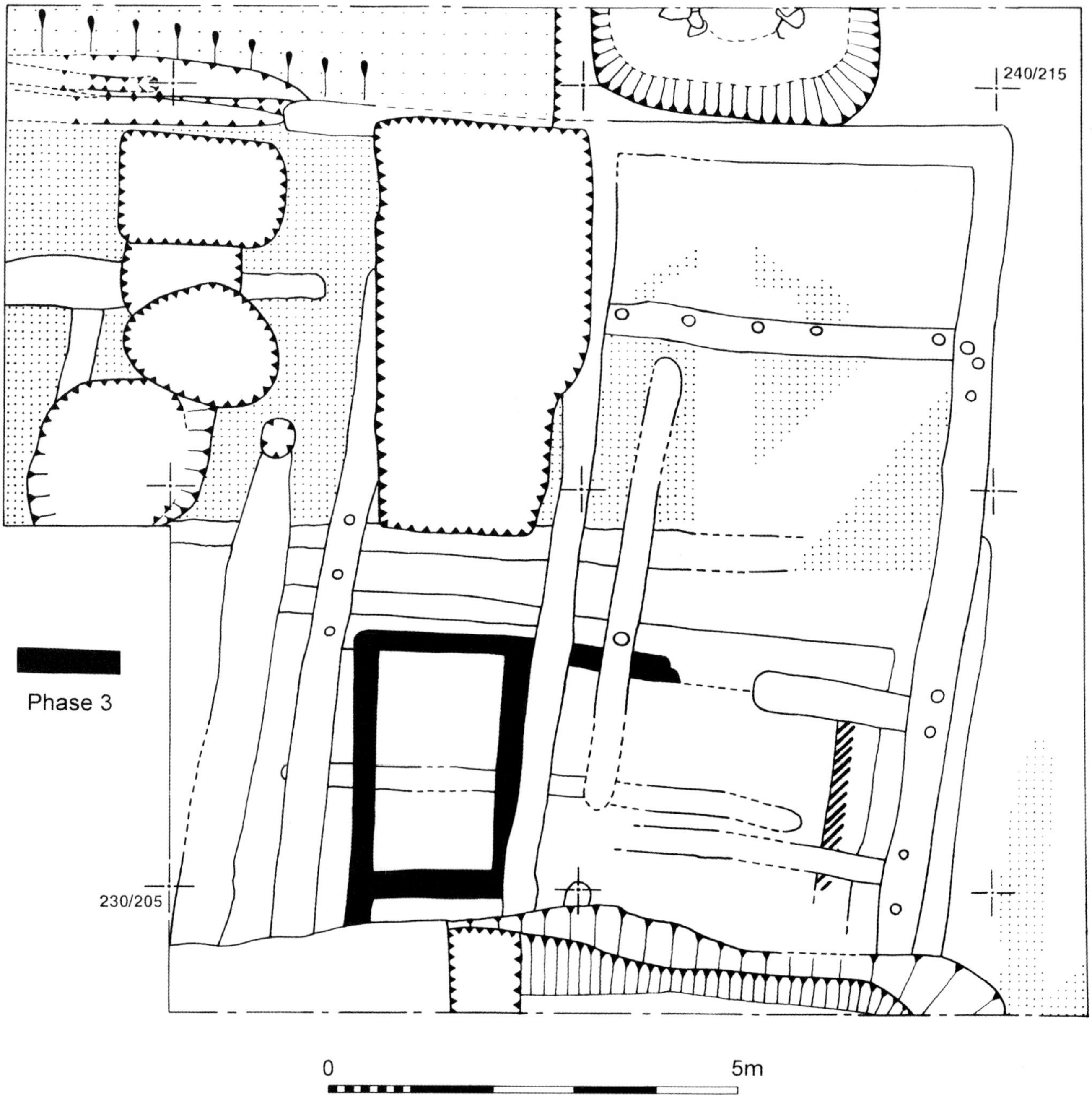

Fig. 2.35 Plan of Phase 3 building

the excavation, but proved to be so shallow that little survived to be excavated. From the available plans it would seem to have been an internal partition *c.* 1.50m long by 0.30m wide. Finally, it should be noted that this was the first phase to produce a closely datable find, in the form of a fragment from a dish in 'terra nigra' (CP 28) from context [481], probably dating to the Flavian era.

Various intact layers were found to overlie the phase 3 foundation trenches in the narrow area between the east-west road [434 and 465] and the shallow rectangular hollow [244]. These include (Fig. 2.36):

472 an extensive layer of clean yellowish silty clay (2.5 Y 5/4), 0.05m thick, with 10–20% medium to large rounded stones, which progressively changed in character towards the western end of the site, where it became almost pure clay.

466 a thin sausage-shaped spread of dull yellowish-brown silty clay loam (10 YR 5/3), up to 2.90m long and 0.60m wide, containing 20% coarse components and 15–20% charcoal fragments.

459/455 an extensive layer of very mixed and dirty silty clay loam, 0.05 to 0.12m thick, which varied in colour from brown [459] to dull yellowish-brown [455] (10 YR 4/4 to 10 YR 5/4) and contained a range of small to large-sized stones and charcoal flecks (30%); towards

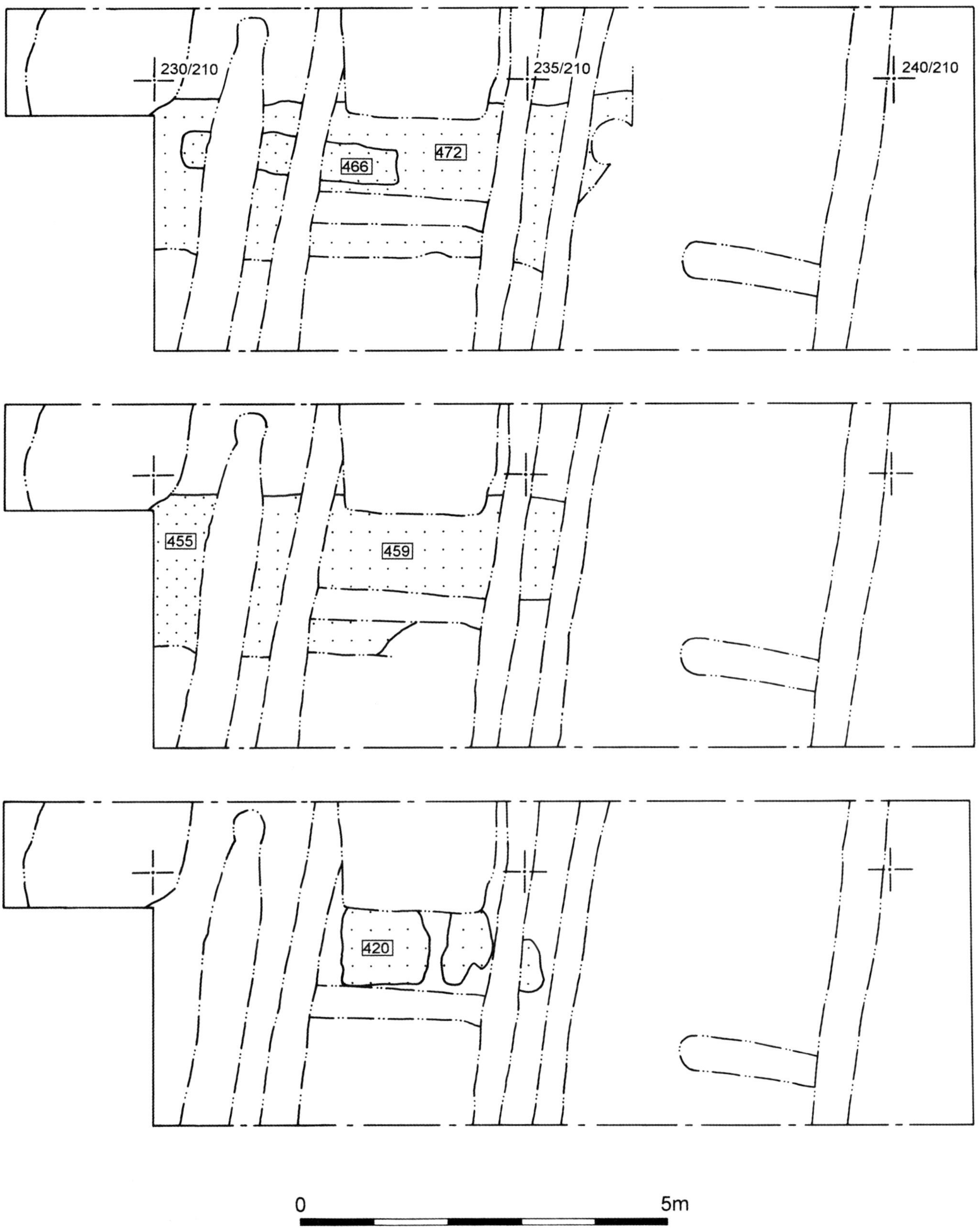

Fig. 2.36 Demolition deposits between Phases 3 and 4

the west this charcoal became more marked, as did the presence of several lumps of yellow clay.
420 a roughly square spread of bright reddish-brown silty clay (5 YR 5/8), 1.90m long, 1.85m wide, and 0.08m thick, containing less than 5% very small stones; in places this clay was banded and interleaved with charcoal.

The most extensive of these horizons were the silty clay [472] and the silty clay loam [455/459], both of which had obviously been truncated by later features. Because of this, their original extent to the east and south cannot now be determined, though the evidence would suggest that both were petering out close to their surviving eastern limit. None of these layers can be identified as *in-situ* floor levels; rather they seem to represent demolition debris, deliberately spread across this part of the site in order to raise it up to the level of the east-west road, prior to the construction of the phase 4 buildings.

Two similar horizons [452 and 454] were located at the south-eastern corner of the site, sealed beneath the gravel surface [425] associated with the widened north-south road (see section 2.2.2(F) above). The lower of the two was the more extensive, being traced for a distance of *c.* 5.50m along the eastern edge of the site. It comprised an almost pure, dull yellowish-brown silty clay (10 YR 5/3), containing 5% medium-sized rounded stones and charcoal fragments. This reached a maximum thickness of 0.10m in the section, but became progressively thinner and more mixed towards its extremities. Only a very small part of the upper horizon [452] proved to be available for excavation, since most of it obviously extended east of the 1989 site. It comprised a small patch of heat-reddened silty clay, dark reddish-brown in colour (2.5 YR 3/6), containing up to 5% small to medium-sized stones; in places, its surface was also covered by a thin discontinuous lens of charcoal. Although neither could be examined in detail, the simplest hypothesis about both these layers is that they represent demolition debris much like those already described.

There were few finds from these demolition horizons, with the exception of contexts [420] and [472], the first of which produced sherds from a Dressel 20 amphora, while the second yielded several samian sherds as well as fragments of a greyware jar (for a discussion of the dating evidence see section 2.2.2(Gvii) below).

(v) Phase 4 (Figs 2.10 and 2.37)

Of all the buildings excavated in 1989, those assigned to phase 4 proved to be the most informative in terms of their plans and constructional details. Two separate buildings were identified, both aligned north-south, and both encroaching up to 3.00m across the line of the east-west road, which must therefore have ceased to function.

Of the pair, that on the east was the better preserved, and clearly comprised seven distinct foundation trenches, two of them representing external walls [367; 374] and five being internal partitions [296; 433; 450; 474; 486]. Its western frontage was defined by a vertically-sided, flat-bottomed foundation trench [374], 0.40m wide and 0.33m deep (Fig. 2.33, S98), which was traced over a distance of 8.20m running north-south. It had been truncated on the south by the later pit [245] and likewise on the north by the post-medieval cellar [222], just at the point where it originally swung eastwards through 90° in line with the northern frontage [367]. Its backfill [373] comprised a dull yellowish-brown silty clay loam (10 YR 5/3), with up to 50% coarse gravel components in the vicinity of the east-west road, but as little as 20% further south. Towards the bottom of this fill, three separate post-impressions were identified as faint soil discolourations [390 to 392], all of them 0.05m deep and between 0.10 and 0.13m in diameter; their location makes it clear that, along this side, the original timber uprights were probably placed at regular intervals of 0.70m.

The corresponding northern frontage of this building was formed by a very similar vertically-sided, flat-bottomed foundation trench [367], up to 0.40m wide and 0.38m deep (Fig. 2.33, S116), which was traced for a distance of 4.40m east of the post-medieval cellar, up to the point where it had been truncated by a later north-south trench [414]. This contained a fill [366] of dull yellowish-brown silty clay loam (10 YR 5/3), with up to 50% coarse gravel components principally derived from the east-west road. Towards the base of this backfill there was evidence for a further five post-impressions [368 to 372], four of them characterised by a packing of small to medium-sized stones (Fig. 2.38); the fifth [372] actually opened up as a void, 0.20m deep, at an early stage in the excavation. Overall their diameters varied between 0.15 and 0.18m. To judge from the surviving evidence, the original timber uprights were clearly spaced at intervals of 0.70 to 0.80m.

Although no archaeological evidence was recovered relating to the building's eastern frontage, there can be little doubt that it originally ran along the same alignment as its phase 5 successor, represented by the somewhat deeper foundation trench [414]. This would allow us to reconstruct a building measuring 7.60 to 7.80m wide from east to west, by at least 8.40m from north to south.

Within this building there were several internal partitions, running both north-south and east-west. The most important, which apparently comprised two elements [296 and 486], effectively divided it longitudinally in half. The first of these [296] proved to be a U-shaped foundation trench, 5.70m long, 0.35m wide and 0.20m deep (Fig. 2.33, S127), which terminated in a characteristic rounded end some

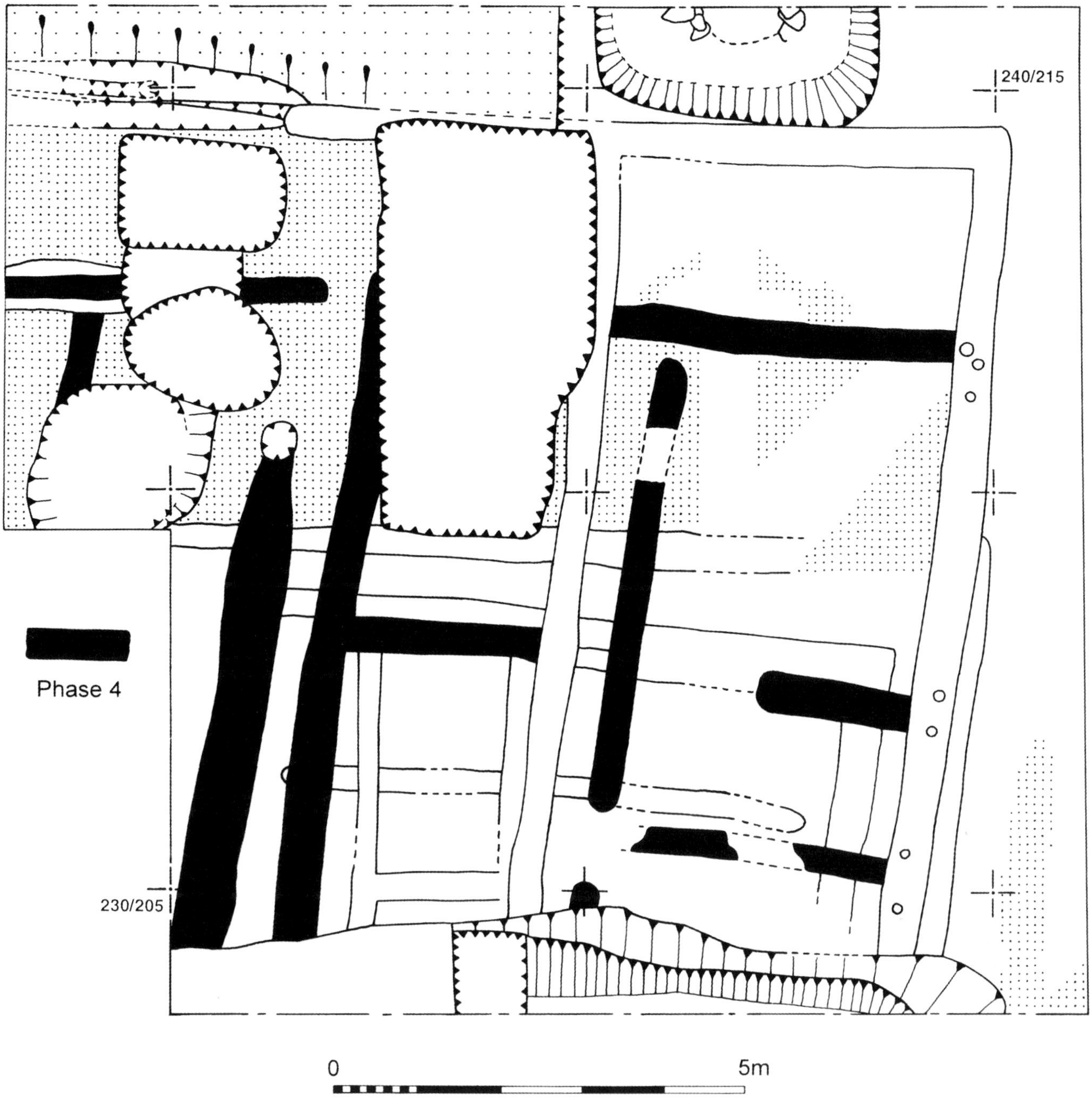

Fig. 2.37 Plan of Phase 4 buildings

0.20m short of the northern frontage. A similar rounded terminal was also identified at its southern end, in the base of a later rectangular intrusion [458]. Its fill [295] comprised a brown coarse sandy loam (7.5 YR 4/3), containing up to 40% small to medium-sized stones. Only one post-impression [424] was identified along its length, *c.* 3.50m from its northern end. This turned out to be 0.34m deep, with a diameter which tapered from 0.19m at the top to 0.11m at the base. It contained a fill [423] of dull yellowish-brown silty loam (10 YR 4/3), with much the same percentage of coarse components as the surrounding trench.

The second element in this north-south partition [486] only survived as a shallow U-shaped butt end, 0.32m wide (Fig. 2.33, S126), on the very edge of the later pit [245] *c.* 1.00m south of the terminal of its northern counterpart. Technically, therefore, it could be part of a small pit, post-hole or foundation trench, although the latter seems the most likely interpretation, continuing the alignment of the partition southwards along the spine of the building. Very little of its fill [485] was actually recovered, but it proved to be a silty clay loam, varying in colour from dull yellow to brown (2.5 Y 6/4 to 10 YR 4/4).

The remaining three partitions ran east-west, two

Fig. 2.38 Post emplacements in trench [367] of Phase 4 building, looking east

of them [450; 474] in the eastern half of the building and the third [433] to the west. The more southerly of the eastern pair [474] proved to be a U-shaped foundation trench, 0.30 to 0.34m wide and 0.20m deep (Fig. 2.32, S151), which was traced in two segments over a distance of 2.90m *c.* 1.00m north of the edge of the later pit [245]. It had been truncated on the east by a later foundation trench [414] and likewise on the west by a rectangular intrusion [458], though it must originally have terminated just inside this feature, because it could not be traced beyond it into the western half of the building. Most likely it ended in a rounded terminal close to the end of the north-south partition represented by trench [296]. This would have created a room measuring *c.* 6.00m north-south by 3.40m east-west at the northern end of the building. The trench itself contained a fill [473] of dull yellowish silty clay loam (10 YR 5/4), mixed with 15–20% very small to medium-sized stones.

The second partition on this side of the building [450] turned out to be very different in character from its southern counterpart, from which it was separated by a gap of 1.60m. There would seem to be little doubt about its attribution to this phase, however, because it had been cut away at its eastern end by the same north-south trench [414] as other features. It proved to be a vertically-sided, flat-bottomed trench, 0.50m wide and 0.30m deep, which only extended 1.90m into the building, terminating in a rounded end. Its fill [449] comprised a yellowish-brown loam (10 YR 5/6), containing quantities of charcoal and red and white flecked clay near the top. This effectively divided the main room into two separate sections, the larger one on the north measuring 4.00m north-south, while its southern counterpart was much narrower at 1.60m. The differences in character between the two partitions in this half of the building might suggest that they had a different load-bearing role, or that they were not contemporary.

The single partition [433] in the western half of the building lay *c.* 3.80m south of the northern frontage, almost directly opposite the shorter of the two partitions described above. Its excavation was initially complicated by the recognition of a deeper section along its southern side, but this was eventually shown to belong to an earlier foundation trench [482] assigned to phase 3. As fully defined, it could not be traced the whole distance between the western side of the building [374] and the north-south partition [296], which must surely indicate that it originally terminated short of the latter, in the area where it was subsequently cut away by a later foundation trench [389] belonging to phase 5. The surviving section proved to be 2.30m long, 0.28m wide and 0.26m deep, with vertical sides and a flat bottom (Fig. 2.33, S100). Its backfill comprised a dull yellowish-brown loamy coarse sand (10 YR 5/3), containing a relatively high percentage (60%) of very small to medium-sized stones. Effectively this created a room measuring some 3.80m north-south by 3.10m east-west.

By contrast with the building just described, only the north-eastern corner of its western counterpart lay within the excavated area, but even so this included two of its external walls [304; 557] and one internal partition [451]. It eastern frontage was formed by a regular U-shaped gully [304], up to 0.70m wide and 0.45m deep (Fig. 2.32, S76 to S78), which was traced for a distance of 6.50m north of the point where it had been entirely cut away by the later pit [245]. Throughout this length, it ran parallel with the west wall of its neighbour [374], at an average distance of 0.45m. At its northern end it must originally have ended in a rounded terminal, much of which was subsequently cut away by a later pit [281]. Its fill [303] comprised a dull yellowish-brown silt loam (10 YR 4/3), containing for the most part only 10–20% very small to medium-sized stones and charcoal flecks, though this figure rose to as much as 70% at its southern end.

When first excavated in isolation, the profile and character of this feature suggested its interpretation as a possible drainage gully, but this became less easy to maintain as work progressed on the other foundation trenches, not least the east-west example [557] forming the northern end of the proposed building. This was only recognised at a very late stage in the excavation, surviving as two separate segments to either side of one of the 'late' pits [328] in this part of the site. Both had been so severely cut into by later activity that only the bottom 0.10–0.15m was available for examination. The eastern segment proved to be a vertically-sided, flat-bottomed trench, 1.10m long by 0.30m wide, which terminated in a characteristic rounded end *c.* 0.45m short of the corresponding north wall [367] of its neighbour. It also coincided with the projected line of the U-shaped gully [304], already described, despite a gap of *c.* 1.60m, which strongly suggests that both formed part of the same building. Unfortunately, none of its fill was sampled.

The corresponding segment west of the 'late' pit was only really clear in section (see Fig. 2.7), where it measured 0.33m wide and 0.15m deep. It lay right at the bottom of an east-west gully [365] which had clearly been cut into it along its entire length. This latter proved to be a U-shaped feature, up to 0.60m wide and 0.30m deep, which very much resembled the profile and character of the eastern foundation trench [304]. It too was initially interpreted as a drainage gully, but the coincidence of alignments and the lack of logic in the resulting drainage patterns strongly suggests that the two foundation trenches had been redug at some point, perhaps during the demolition of the building at the end of its life. The two features combined would form a building at least 3.90m from east-west and 6.50m from north-south.

The U-shaped gully [365] had also truncated one end of a north-south foundation trench [451], which is best interpreted as an internal partition running parallel with the eastern wall of the building at a distance of *c.* 2.20m. This proved to be a steep-sided, flat-bottomed feature, 0.34m wide and 0.21m deep, which was traced over a distance of 1.20m; it had been entirely cut away at its southern end by a 'late' pit [351]. Its fill [447] comprised a brown silty clay loam (10 YR 4/4), containing a mixture of very small to medium-sized stones.

Very few finds were made in association with these buildings. They include a samian sherd of form 27 from context [295]; mica-dusted and greyware sherds from contexts [295 and 373]; a samian rim of form 37 from [373]; and fragments of bottle glass from [432] (for a discussion of the dating evidence see section 2.2.2(Gvii) below).

As with the earlier structural phases, there were no certainly associated floor levels or occupation horizons. In places, however, the fills of the foundation trenches had been cut into or overlain by later features, some of which could well pre-date the construction of the phase 5 structure (Fig. 2.39). These include:

458 a rectangular, vertically-sided pit, 1.30m long, 0.80m wide and 0.20m deep, which had cut into the southern end of the main north-south partition [296] in the eastern of the two structures. Its fill [457] proved to be very much the same as the overlying horizon [411], suggesting that they were both derived from the same operation.

411 a very mixed layer of clay or silty clay loam containing less than 5% coarse components and charcoal fragments, which was only located in a small area immediately to the east of the rectangular hollow [244]. It varied in colour from dull yellow to brown (2.5 Y 6/4 to 10 YR 4/4). The density of clay and charcoal progressively petered out to the east and south.

456 a very small and irregular patch of bright reddish-brown silty clay (5 YR 5/8) located immediately north-west of the rectangular pit [458] and partly overlying the fill of the north-south partition [296]. It contained few coarse components, but was interspersed with thin charcoal lenses.

342 an homogeneous layer of yellowish-brown silty clay loam (2.5 Y 5/4) containing some 40–50% coarse components, which was found to overlie the fill of the north-south partition [296] along part of its length, as well as the patch of silty clay [456] just described. It had obviously been much truncated by later features and so its original extent remains unknown.

477 a somewhat irregular post-like intrusion up to 0.70m long, 0.50m wide and 0.90m deep, which had been cut through the fills of the successive east-west foundation trenches [482 and 480] associated with phases 2 and 3. It had a roughly conical upper profile, dropping into a steeper-sided feature with a relatively flat base. Inside were two successive fills: both were dull yellowish-brown in colour (10 YR 5/4), the lower one [476] being a relatively homogeneous silt loam containing a single large stone, while the upper of the two [475] proved to be a loam with an admixture of clay and charcoal lumps.

Of these, only the silty clay [456] and the silty clay loam [342] seem to have had any definite stratigraphic relationship with the phase 5 structure, both of them having been cut away by the foundation trench [389] supporting its western wall. None of the other contexts can be so securely fixed in the relative sequence, which leaves the possibility that some or all of them could post-date the phase 5 building on this part of the site. There is also the distinct possibility that the silty clay loam [342] is less homogeneous than was originally thought, and that it actually represents the churned-up amalgam of two separate deposits, one pre-dating the phase 5 structure, the other thrown up during its construction; this would explain the inclusion of a fragment of a dish in 'terra nigra', which clearly joins

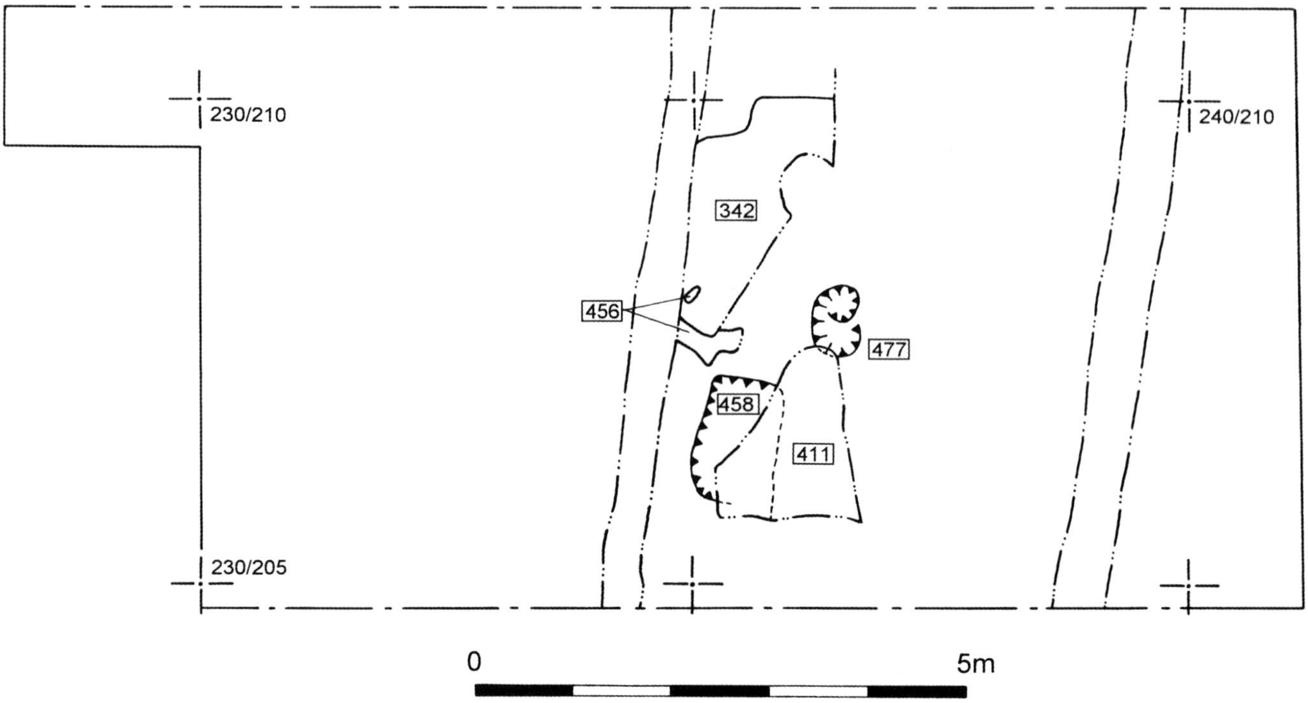

Fig. 2.39 Deposits/features post-dating Phase 4

a similar piece securely sealed in one of the fills [481] associated with the phase 3 structure. Other finds from these horizons included fragments of bottle glass and lead scrap from [411].

(vi) Phase 5 (Figs 2.10 and 2.40)
Following the demise of the phase 4 structures, the eastern one of the pair was replaced by a new building which was found to extend right across the line of the east-west road, almost as far as the southern lip of the clay-lined cistern [376]. It was defined by two separate external foundation trenches [389 and 414], without any obvious evidence for internal partitions, to which should be added a somewhat enigmatic projection [413] immediately west of the post-medieval cellar [222]. Its western frontage was represented by a vertically-sided, flat-bottomed foundation trench [389], up to 0.40m wide and 0.30m deep (Fig. 2.32, S94, S99 and S131), which was traced over a distance of 9.60m running south from its junction with the northern frontage [414] to the point where it had been truncated by the later pit [245] containing the burnt timber building. Its backfill [388] comprised a dull yellowish-brown silty clay loam (10 YR 5/3), containing a variable admixture of small gravel fragments throughout its length.

The corresponding northern and eastern frontages were formed by a right-angled foundation trench [414], with a fairly typical vertically-sided and flat-bottomed profile up to 0.50m wide and 0.40m deep (Fig. 2.32, S114, S117 and S119 to S120). It was traced over a distance of 10.30m along its eastern side, north of the point where it had been truncated by the later pit [245], and up to 5.00m along its northern side, as far as the edge of the post-medieval cellar [222] where it met the western frontage [389]. Its backfill [343] comprised a dull yellowish-brown silty clay loam (10 YR 5/3), with up to 30% very small to medium-sized stones, towards the base of which seven separate post-impressions were identified [435 to 441], with diameters varying between 0.13 and 0.18m. The majority were little more than faint soil discolourations up to 0.04m deep [435 to 439], but two of them opened up as voids, 0.35m deep, during excavation [440 to 441]. To judge from the limited evidence of their spacing along the eastern side, the original timber uprights must have been placed at irregular intervals of between 0.40 and 0.70m.

Such evidence would allow us to reconstruct a building measuring 5.00m wide from east to west, by at least 10.30m from north to south, which would make it somewhat narrower than its phase 4 predecessor. One enigmatic feature of the building, however, is the existence of a possible projection of the northern frontage beyond the line of the west wall; this is represented by a short length of U-shaped foundation trench [413], 0.40m wide and 0.36m deep, which was traced for a distance of 1.10m west of the post-medieval cellar [222]. Part of it must originally have survived along the northern side of the cellar, a point which was not immediately obvious prior to the decision to cut a trial trench through this part of the rampart (see section 2.2.2(Ci) above). It had been dug into the back of the clay rampart [381], truncating

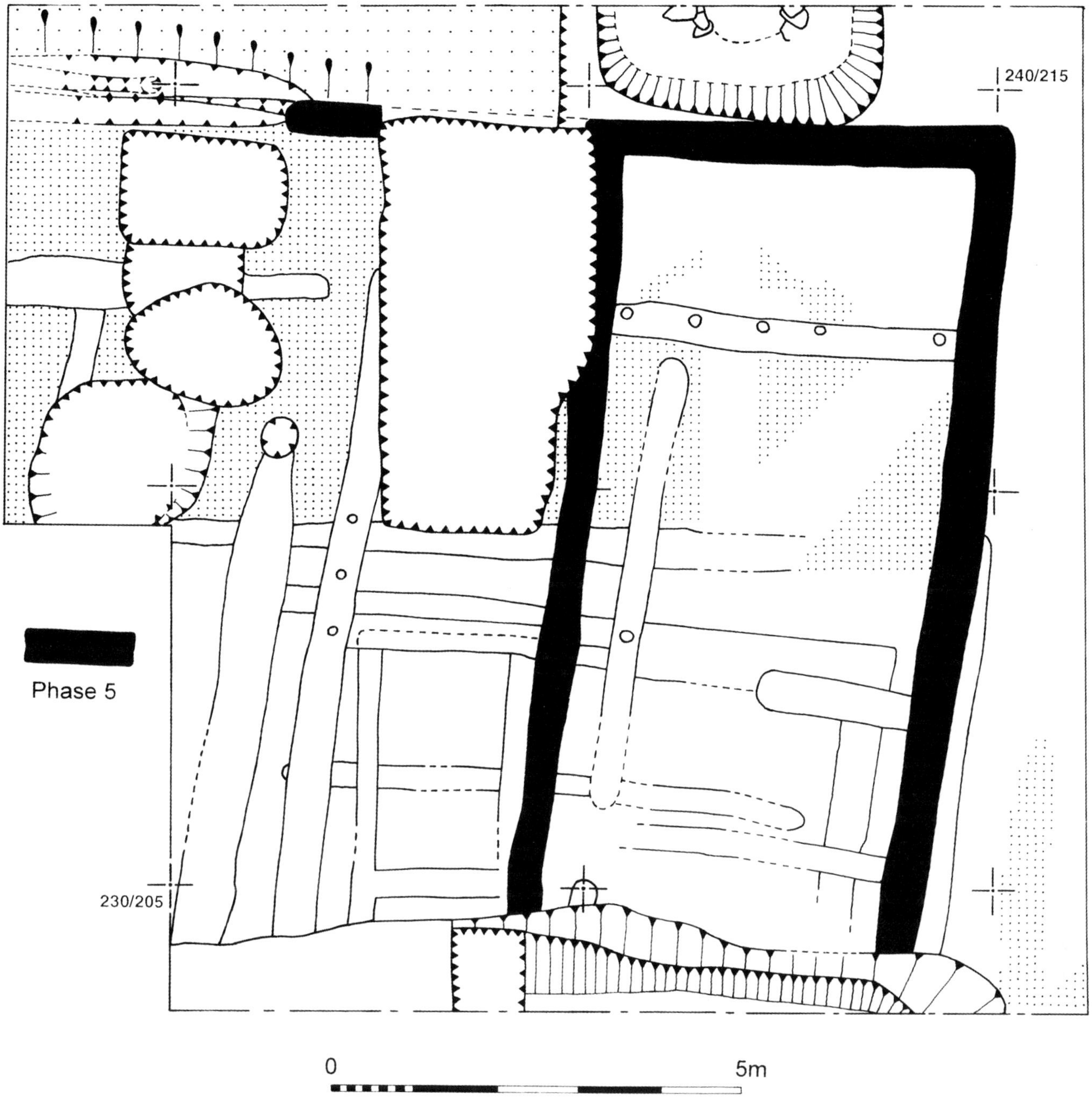

Fig. 2.40 Plan of Phase 5 building

in the process the infilled end of the latest east-west gully/drain [416]. Its fill [412] comprised a dull yellowish-brown silty clay loam (10 YR 4/3), containing some 40–50% very small to medium-sized stones as well as numerous lumps of yellow/orange clay.

This projection would have extended the original length of the northern frontage from 5.00 to 8.80m. Quite why this was considered necessary remains uncertain, because there are no obvious features with which it can be associated. The most plausible explanation might be that it was deliberately designed to close off access between the areas north and west of the building respectively, by extending the north wall to the point where it adjoined the widest part of the rampart. It would also be tempting to link this with the existence of the clay-lined cistern [376] north of the building, though the very proximity of the two features might argue against their contemporaneity. This point could not be ascertained for certain, beyond the fact that the northern foundation trench had cut into the cistern's clay apron [376], because all the crucial stratigraphic relationships had been removed by the levelling operations associated with the later clom and stone-built structures.

Only a small number of finds were made in association with this building. They include sherds of redware and greyware, a fragment of bottle glass and an indeterminate bronze object from context [343]; a samian sherd, possibly Curle 11, and a fragment of vessel glass from [388]; and a sherd from a redware flagon from [412] (this evidence is further discussed in the next section).

(vii) Dating evidence and chronological discussion

The absence of any obvious floor levels in association with these five structural phases inevitably means that the amount of dating evidence is somewhat limited, the key pieces being derived either from the fills of the foundation trenches or from the few intact demolition (?) horizons. Consequently, although the quantity of finds progressively increases after phase 3, much of the material is likely to be residual with only a few pieces having a direct bearing on the chronology of the overall sequence. This situation is further complicated by the narrow timescale for the site's occupation as proposed by Peter Webster (see section 2.2.4(C) below), which does not readily facilitate any fine tuning of the individual phases, given the quality of the samian assemblage and the traditional date ranges assigned to most of the coarse wares. The principal pieces include:

Phase 2/3

478 body fragment from a glass jar/vessel (Gl 3) (*cf.* context [388]).

Phase 3

481 fill of foundation trench: rim sherd from a dish in 'terra nigra' (CP 28); probably Flavian.

Demolition horizons between phases 3 and 4

420 handle and body sherds from a Dressel 20 amphora (not catalogued).

472 samian sherds, including one form 37 (Table 2.2). body sherds and a complete base from a greyware jar or bowl (not catalogued).

Phase 4

295 fill of internal partition: abraded rim sherd of samian form 27 (Table 2.2); probably Flavian.

303 fill of foundation trench: body sherd from near the base of a redware jar (not catalogued).

373 fill of foundation trench: samian rim sherd of form 37 (Table 2.2).
rim sherd from a mica-dusted jar or large beaker (CP 34); late 1st to early/ mid 2nd C.
rim sherd from a flanged, redware bowl (CP 35); probably Flavian-Trajanic.

432 fill of foundation trench: body fragments from a glass bottle (Table 2.9).

Horizons post-dating phase 4

342 joining rim sherds from a dish in 'terra nigra' (CP 28); probably Flavian.

411 body fragment from a glass bottle (Table 2.9).

Phase 5

343 fill of foundation trench: body sherds from a greyware lid and a redware jar (not catalogued).
body fragment from a glass bottle (Table 2.9).

388 fill of foundation trench: abraded samian sherd, possibly Curle 11 (Table 2.2).
body fragment from a glass jar/vessel (Gl 3) (*cf.* context [478]).

412 fill of foundation trench: sherd from a redware flagon (not catalogued); probably Flavian+.

Despite the paucity of material from the different phases, all of it can be satisfactorily dated within the period stretching from the Flavian era through to the early 2nd century. This is broadly in line with the chronology provided by the finds from the successive fills both of the well and cistern (see section 2.2.2(Diii) above) and of the pit containing the burnt timber building (see section 2.2.2(Hiv) below), but with the significant difference that none of the building phases incorporated any Black-burnished wares. Though not compelling, given the amount of material involved and the absence of floor levels, this may corroborate the relatively late date at which significant quantities of BB ware began to appear on site.

(H) The burnt timber structure (Figs 2.41 to 2.50)

(i) The large pit and its associated fills

Following the removal of the modern buildings and their associated features, a spread of brownish-black 'garden soil' [157] was encountered across the southern end of the site, apparently infilling an extensive area of disturbance. This was provisionally linked with gravel digging pre-dating the construction of the post-medieval clom building. In an effort to confirm this hypothesis, an initial trial trench was cut across the axis of the disturbance, from north to south, at the western end of the area which was subsequently cleared in plan. This soon revealed the edge of a vertically-sided pit [245], which had been cut down into the natural gravel to a depth of over 2.00m, through the foundation trenches associated with the known Roman buildings (Fig. 2.41). Subsequent work on the 'garden soil' further demonstrated that this pit extended at least 10.00m east-west by 5.00m north-south, making it a much more substantial feature than first envisaged for gravel digging. The trial trench also revealed an interesting sequence of fills within the pit, including what was obviously the remains of a burnt timber structure represented by carbonised timbers and quantities of destruction debris. As a result of this important discovery, an area within the pit measuring *c.* 5.00 by 4.00m was opened in plan, to allow a more detailed investigation.

Within the pit, three successive fills were found to overlie the destruction horizon [540] associated with the building (Fig. 2.42). The uppermost fill [157] proved to be a relatively homogeneous, brownish-

Fig. 2.41 Timber building Phases 1 to 3 and pit containing the burnt timber building, looking south

black 'garden soil', which had obviously accumulated gradually over a considerable period of time. Not surprisingly, it became progressively thicker away from the edges of the pit, such that at its maximum extent it reached a depth of 1.30m. It contained a range of 18th-century finds (pottery, glass, clay pipes and metal objects), as well as a few scraps of medieval pottery and a quantity of residual Roman material, including pottery and glass fragments. This horizon overlay a second, very different fill [224], up to 0.50m thick in places, which had apparently been dumped into the pit while it was still open, with the result that it sloped steeply downwards from the outer edge of the feature towards its presumed centre. No record was made at the time of its actual texture, but there can be little doubt that it was a relatively homogeneous silty clay with very few coarse components. This layer was originally thought to represent the levelled and degraded remains of the fort's turf-clay defences, but if so it contained a surprisingly high number of Roman finds when compared with their virtual absence from the surviving rampart features further north.

Beneath these two deposits lay a third, very mixed fill [460] containing very few coarse components but much charcoal. A close examination revealed that it was composed of numerous lenses of different material (mostly clay loam and silty clay loam), which varied considerably in colour from brownish grey (10 YR 6/1) through reddish brown (5 YR 4/8) to dull yellow (2.5 Y 6/4). Close to the edge of the pit, this layer sloped steeply downwards, before becoming almost horizontal across the bottom, where it reached a maximum thickness of 0.70m. The mixed character of this fill strongly suggests that it was a deliberate deposit, made up of material which is very reminiscent of demolished timber buildings and disturbed clay floors. The obvious source would have been the immediate vicinity of the pit, which could account for the lack of any deposits associated with the successive buildings described in this report. Like the overlying horizon [224], this layer produced one of the highest concentrations of finds in any Roman context, including a broad cross-section of pottery types, glass and metal objects. This demolition material sealed yet another distinctive fill [540], identified as the destruction horizon associated with the burnt timber building itself.

(ii) The destruction horizon

This proved to be a very mixed fill [540] of reddish-brown silty clay loam (5 YR 4/8), largely composed of ash, charcoal fragments and carbonised timber fragments, together with a considerable admixture of redeposited natural gravel which directly overlay the remains of the burnt timber building (Figs 2.43 and

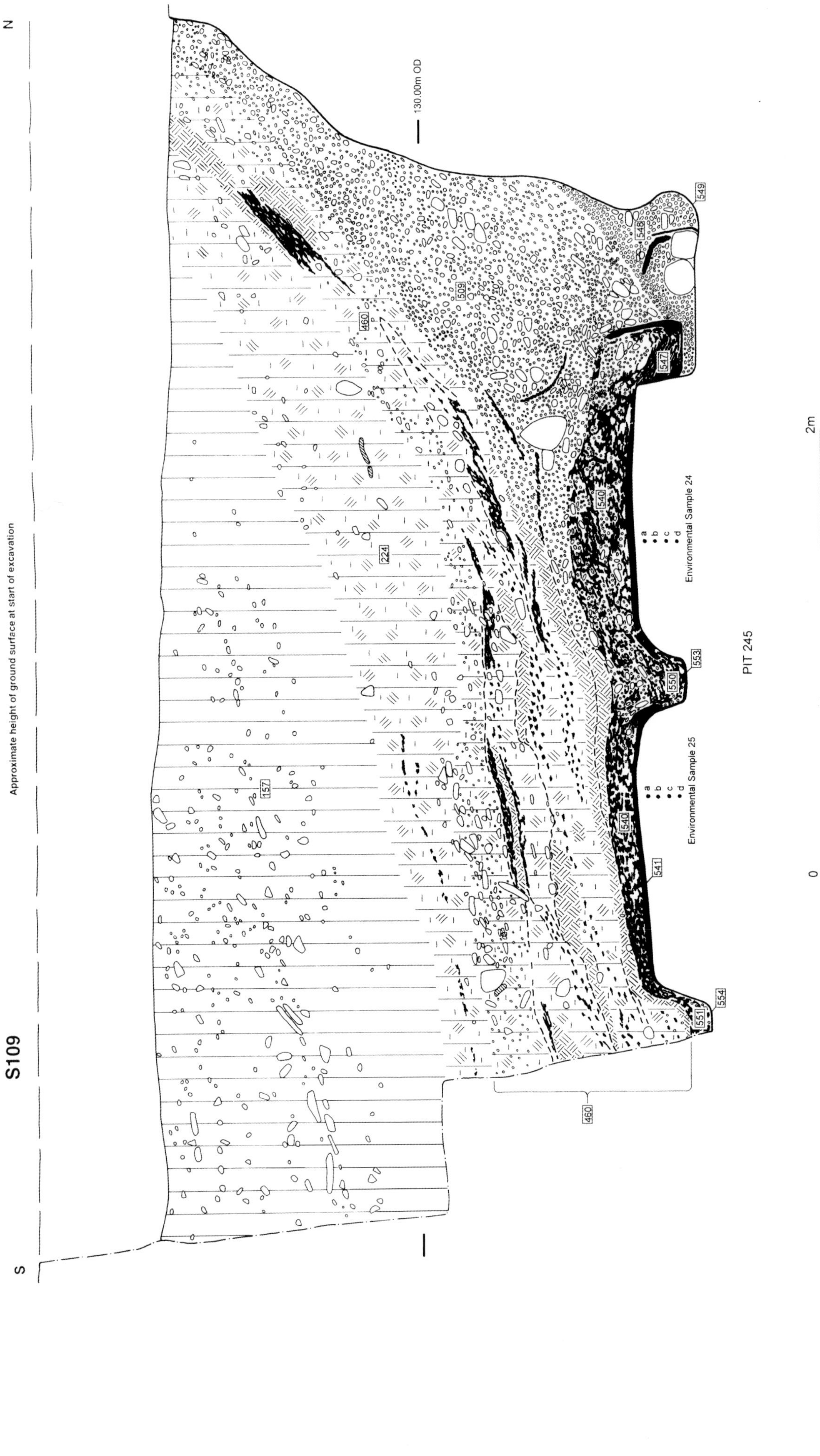

Fig. 2.42 West section through large pit

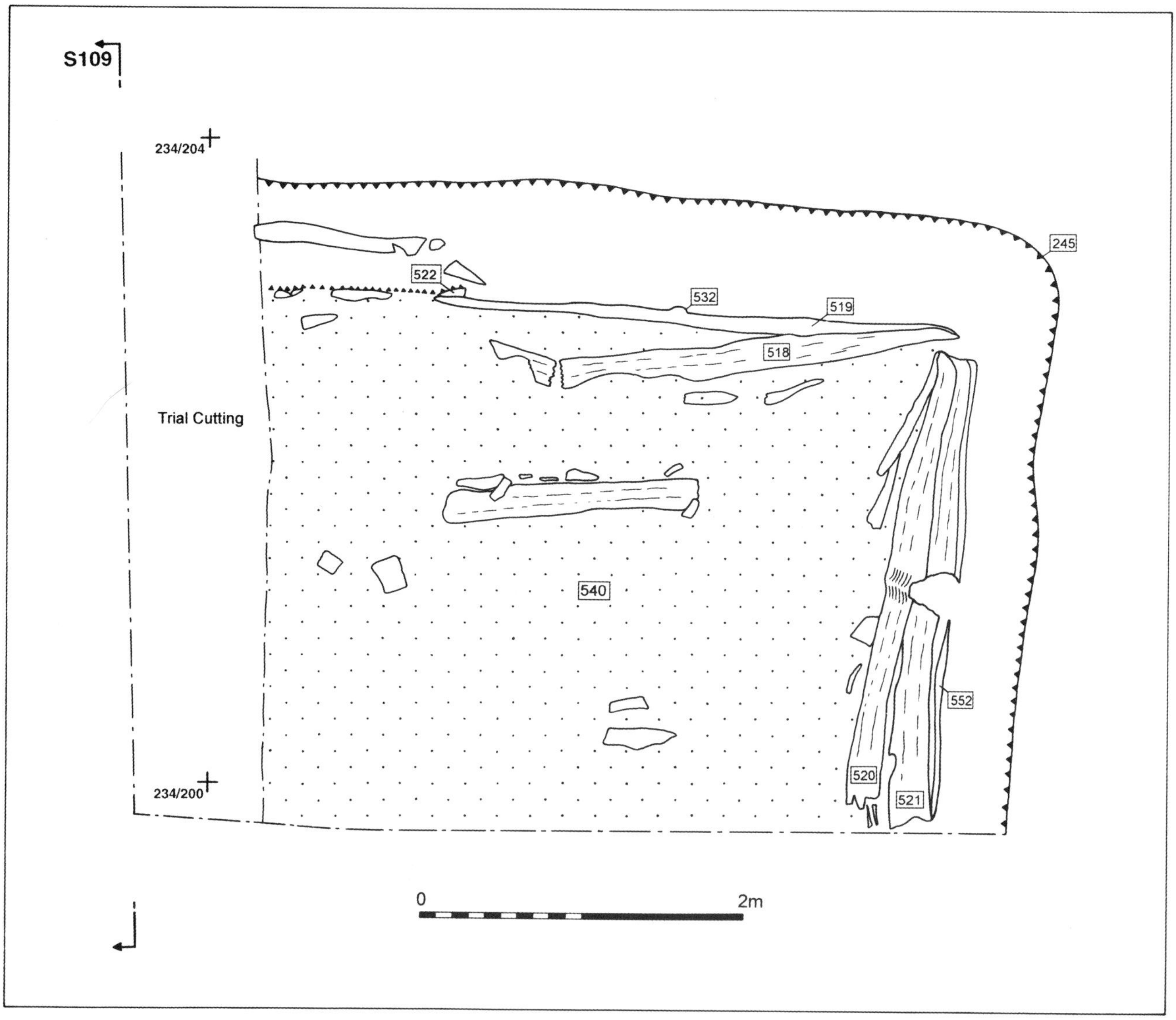

Fig. 2.43 Destruction horizons and burnt timbers as first exposed

2.44). Its thickness varied somewhat, from as much as 0.35m against the north and east walls of the structure to as little as 0.10m in the south-western corner of the excavated area in grid square 235/200. Within its makeup, excavation identified several large carbonised timbers (Figs 2.45 and 2.46), as well as many other fragments of planks and beams, most of them too small to plan or describe separately. The principal timbers include:

527 a section of beam, 0.55m long and 0.13 by 0.08m in cross-section, aligned north-south and lying horizontally. Associated with two short nails.
528 a section of plank, 0.55m long, 0.20m wide, and 0.03m thick, aligned north-south and lying horizontally.
529 a section of thick plank or beam, 1.00m long, 0.17m wide, and 0.05m thick, aligned north-east to south-west and lying horizontally.
530 a section of plank, 0.80m long, 0.17m wide, and 0.01m thick, aligned north-south and lying horizontally.
531 a section of post, 1.10m long and 0.19 by at least 0.07m in cross-section, aligned nearly north-south. For the most part it lay horizontally, but towards its northern end it seemed to be curving into an upright position.
542 a short section of post or beam, 0.22m long and 0.11 by 0.08m in cross-section (not illustrated).
546 a large section of two separate timbers, 1.40m long and 0.34 by 0.14m in cross section, aligned north-south and lying horizontally on the floor.
555 a section of beam, 0.36m long by 0.16m wide, aligned north-south.

The vast majority of these carbonised timbers must have fallen from the superstructure of the building during the final conflagration, thereby becoming incorporated into the build-up across the original floor. The one possible exception is the section of post [531], which may well represent part of a vertical upright associated with the north wall of the structure described in the next section. At a late stage in the process, what survived of the weakened north and east walls had collapsed inwards, which explains

Fig. 2.44 Burnt timber building after the removal of the later infill deposits, looking west

why some of the planking appears to overlie the destruction horizon. Few finds were made in this horizon, except for a number of bones, various iron objects, and a possible coin (Co 3).

(iii) The timber structure

The 1989 excavations allowed a detailed investigation of this remarkable structure over an area measuring 5.20 by 3.40m, including a significant stretch along its northern and eastern sides. A close examination of the tip lines in the standing sections, however, suggests that this might still represent only a quarter or less of the structure's original extent, for in both cases the overlying deposits slope steeply downwards from the outer edge of the pit, as if towards its centre, without any sign of their curving back upwards (for the western section, see Fig. 2.42).

The structure itself had been laid upon at least three horizontal beams [547; 550; 551], running east-west, which had been set into separate foundation trenches [549; 553; 554] dug into the natural gravel at the base of the pit (Fig. 2.47). Sadly, these beams rarely survived as recognisable carbonised timbers; instead, over much of their length, their place was taken by a thick concentration of charcoal and ash, which would seem to represent not so much their carbonised remains as subsequent slumpage into the voids left by the timbers as they rotted *in situ*. Their size was clear, nevertheless, from the dimensions of the surviving beam slots, which would indicate that they measured some 0.20m in cross-section. It was not possible, however, to determine the length of the original beams, as there was nothing to indicate whether a single long timber had been used or whether several shorter portions had been laid end to end; the latter would certainly not be impossible,

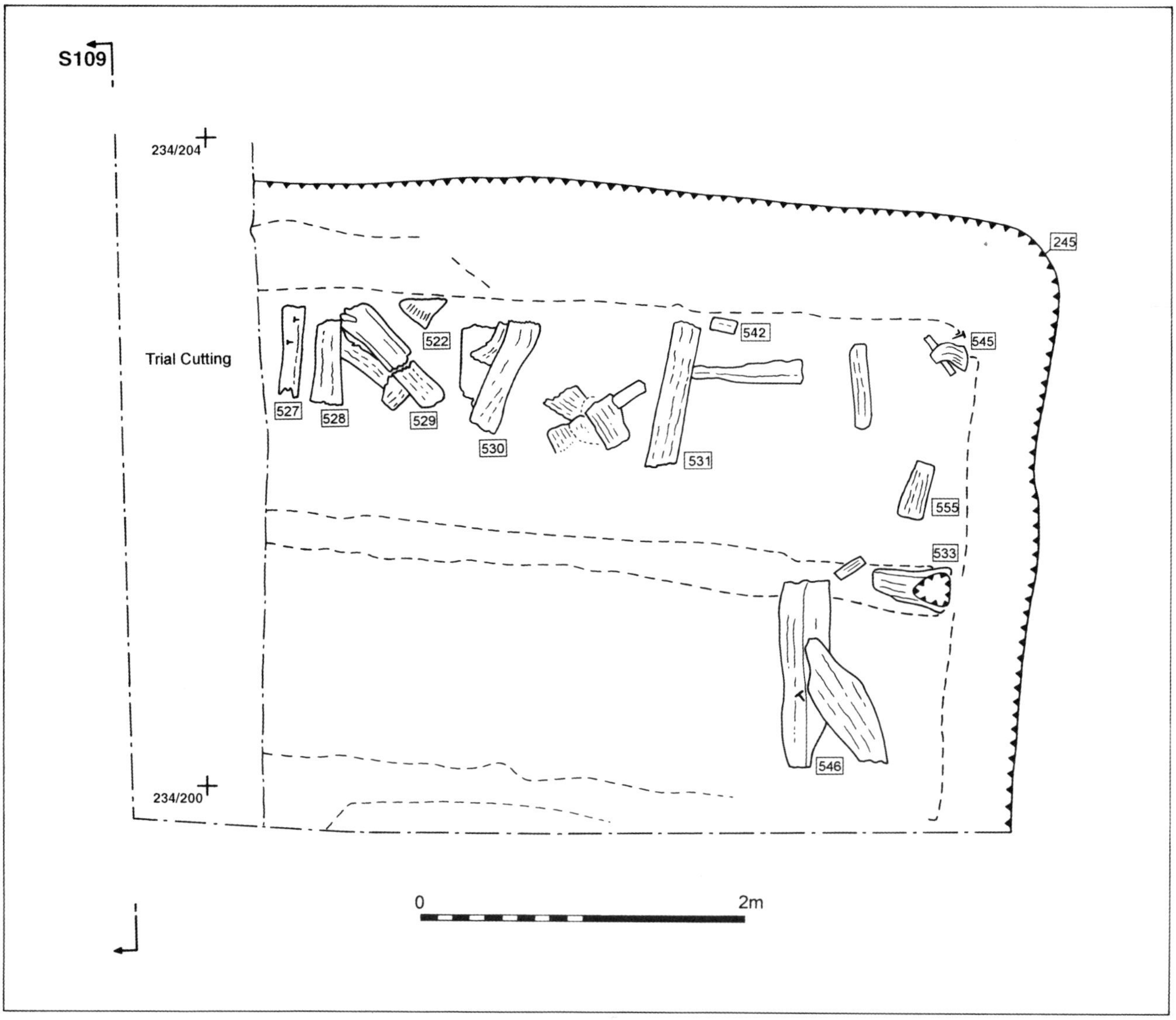

Fig. 2.45 Burnt timbers in makeup of destruction horizon

because in this position there would probably have been no loss of structural strength.

Of the three foundation slots, the central and southern examples [553; 554] proved to have been dug only as wide and as deep as was necessary to accommodate their respective timbers; thus they were both 0.20m deep, 0.20m wide at the base, and no more than 0.30m wide at the top. The only exception to this was the way in which the west end of the southern slot appeared to be widening out to more than 0.45m, right in the angle between the two standing sections; this may have been to accommodate some form of timber upright at the presumed centre of the structure. The northern foundation slot [549], by contrast, proved to be a much wider, flat-bottomed trench, up to 0.68m wide and 0.27m deep, which extended *c.* 0.40m beyond the eastern end of the building. The original timber beam [547] had obviously been laid along its southern side and then levelled, prior to the whole thing being refilled with a layer [548] of fine natural gravel. All this means that the spacing between the adjacent pairs of beams was a constant 1.50m.

These three beams supported a floor [541] constructed from closely-set planks, laid horizontally from north to south (Fig. 2.48). These varied in width between 0.18 and 0.28m and survived to a maximum thickness of 0.01m. They had been carbonised for the most part in the final conflagration and were, therefore, well preserved except where they had slumped into the voids left by the underlying beams as they rotted *in situ*. At their northern ends, the individual planks had been secured in place by driving large iron spikes [544] centrally through them into the horizontal beam; these spikes averaged 0.15m in length and had characteristic squared heads [nos 10,030–43 and

Fig. 2.46 Burnt timbers in makeup of destruction horizon, looking west

10,046–51]. No comparable spikes or nails were recovered from the central or southern slots, suggesting that either wooden pegs had been used or, more probably, that the timbers were laid across the beams without any form of attachment. In the excavated area, the individual planks extended right across the floor area, making them at least 3.20m long.

As excavation progressed, it became clear that the side walls of the structure had been constructed around a series of timber uprights [522; 531; 533; 545], placed at intervals along the northern and eastern sides of the building (Fig. 2.45). None of these was particularly well preserved, but enough had survived the final conflagration to establish the basic pattern. The most obvious lay at the north-eastern corner of the structure, where traces of a vertical timber [545] had survived to a height of 0.30m (Fig. 2.49). Its centre had completely rotted away, leaving just its outer carbonised skin, but this was enough to demonstrate that it had originally been *c.* 0.13m in cross-section. Possibly attached to it was a cross piece *c.* 0.25m long and 0.05m in cross-section, which lay at 45° to the main wall planks and may represent some form of internal brace.

Working south from this corner post, the next timber upright [533] was logically located at the end of the central beam slot. It survived as a carbonised timber *c.* 0.50m long, for the most part lying almost

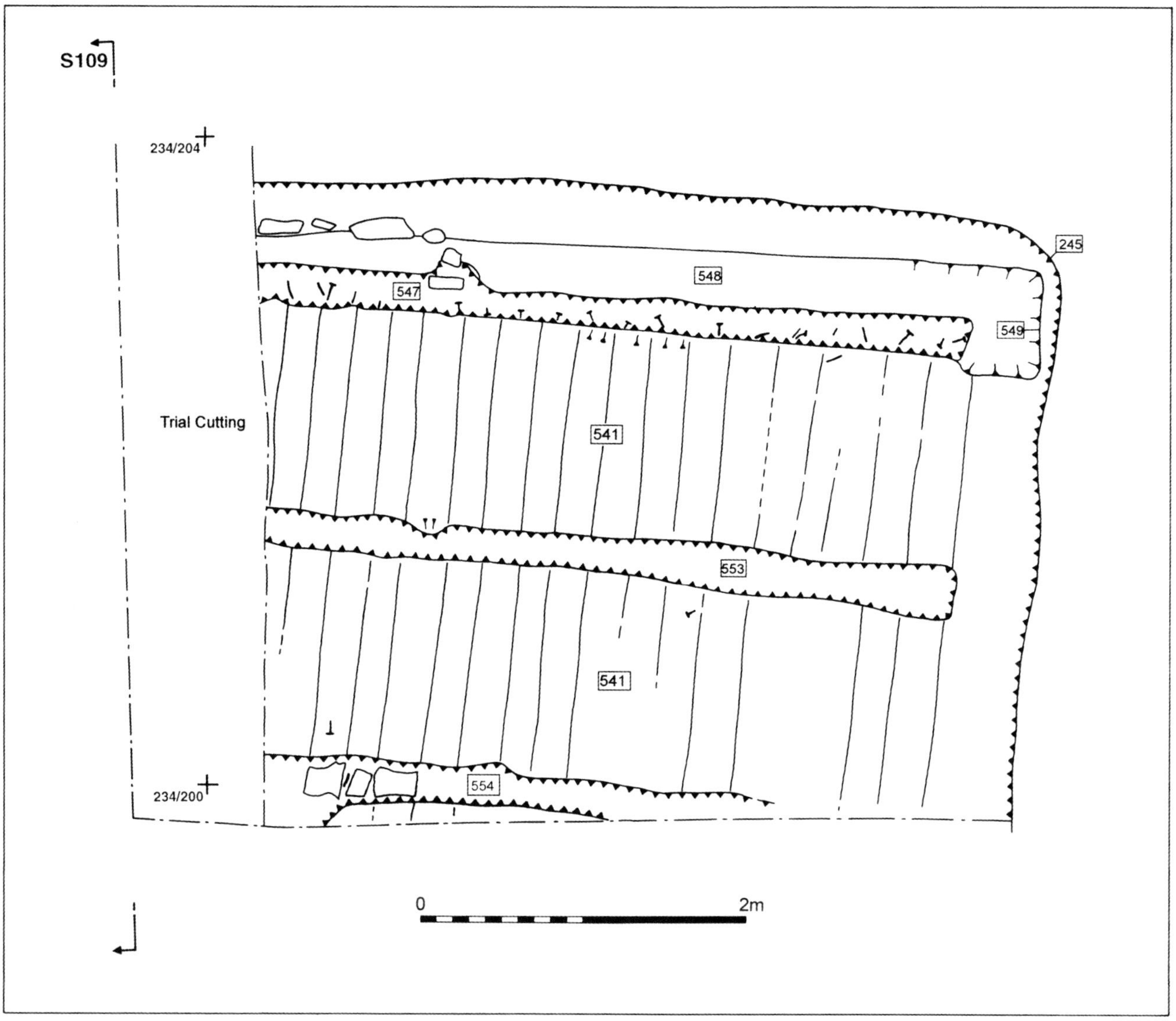

Fig. 2.47 Timber planking and beam slots

horizontally, but near vertical at its eastern end. Its heartwood had completely rotted away, but the outer carbonised skin was enough to demonstrate that it had originally been 0.23m square in cross-section. Where this timber had collapsed inwards, it was overlain by two of the planks [520 and 521] of the side walling. The same two planks also overlay traces of a further possible timber upright, which was tantalisingly located right on the edge of the excavation at the end of the southern beam slot.

Working west from the corner post, the first timber upright [531] along the north side of the structure was represented by a fragment of carbonised post *c.* 0.50m long, noted in the preceding section, most of which had apparently collapsed inwards to lie horizontally. At its northern end, however, it appeared to be curving into an upright position, just at the point where it coincided with the northern beam slot. Where it lay horizontally, it was overlain by two of the planks [518 and 519] of the side walling. Further west again lay the very fragmentary remains of another timber upright [522], which had only survived as a short carbonised stump little more than 0.20m wide.

From the surviving evidence, these uprights had clearly been regularly spaced at intervals of 1.50 to 1.60m along both sides of the building, the interval coinciding with the spacing of the main east-west horizontals. Unfortunately, the severity of the conflagration had destroyed all trace of the jointing methods used, but it would be plausible to suggest that the uprights were originally mortised into the main beams without the need for any spikes or nails. This, of course, only accounts for the lower ends; they must also have been firmly secured at their tops, but above the first metre or so, far too little of the superstructure had survived to establish any details.

From an early stage in the excavation, it was clear that the side walls of the structure had comprised a

Fig. 2.48 Timber planking and beam slots, looking south

Fig. 2.49 North-east corner post

Fig. 2.50 Section across three planks on east side

series of planks laid horizontally edge to edge, one on top of the other (Figs 2.43 and 2.50). At least three such courses, and fragmentary traces of a fourth, had survived in places along the northern and eastern sides [518/519/532 and 520/521/552 respectively], varying in width between 0.19 and 0.27m, with a maximum thickness of 0.01m; unfortunately their original lengths could not be established with any certainty, because most had apparently been broken across one end, though in several cases lengths of over 3.00m had survived. Their full details include:

518 section of the uppermost plank on the northern side, 0.27m wide, 0.01m thick, and at least 3.00m long, though it had clearly been broken at its western end. It was lying almost horizontally at this point, but became progressively more vertical as it approached the corner post [545].

519 section of the middle plank on the northern side, vertically-set and probably, therefore, in its original position. It measured 3.30m long, 0.19m thick, and 0.01m thick. The fact that it terminated next to the possible timber upright [522] might suggest that its full length has survived.

532 section of the lowest plank on the northern side, vertically-set and measuring 3.30m long, 0.23m wide, and 0.01m thick. Like [519] it terminated next to the possible timber upright [522].

520 section of the uppermost plank on the eastern side, at least 3.20m long, 0.20m wide, and 0.01m thick. Like [521] it was lying almost horizontally throughout its length, but retained a vertical profile at the point where it met the corner post [545]. As with all the other planks on this side, the break at its southern end coincided with one of the east-west beams [554].

521 section of the middle plank on the eastern side, apparently broken in two on either side of the central timber upright [533]. It measured 3.20m long, 0.24m wide, and 0.01m thick.

552 section of the lowest plank on the eastern side, set vertically into the natural gravel close to the ends of the horizontal beams. It measured at least 3.20m long and 0.01m thick.

These planks had obviously been laid against the outer edges of the corner post and the timber uprights, but no evidence had survived to demonstrate how they had been fastened together. In the absence of any metal spikes and nails at the appropriate points, either wooden pegs were used or else the planks were effectively held in place by the gravel infill [509], which had been dumped between them and the steep sides of the main pit [245]. From the position in which some of the surviving planks were found, it would appear that they had been forced inwards under the weight of this gravel infill, presumably when the uprights became weakened and collapsed during the final conflagration.

Within the excavated area there was no clear evidence for any form of access. There was, however, a rather puzzling irregularity towards the western end of the northern wall, which might have had some such function. At this point, an extra line of wall planking was located, set back from the main wall line by *c.* 0.40m towards the cut edge of the main pit [245]. It proved to be a single length of plank [not contexted] at least 1.00m long, which had apparently been accommodated within the wider foundation slot [549] cut into the natural gravel at the base of the pit. Unfortunately little else survived above this lowest course to allow an assessment of its original structural character.

Sandwiched between the destruction horizon [540] and the plank flooring [541] there was evidence for a discontinuous silty deposit [534], which proved to be little more than 0.01 to 0.02m thick. This is best interpreted as a layer of dust which had accumulated across the planks prior to the destruction of the structure. A similar deposit [543] was also located beneath the flooring, in what must originally have been the narrowest of spaces between the planks and the underlying gravels. More than likely this latter represents material which had seeped through the gaps between the planks, either during sweeping or washing. Both horizons were extensively sampled for environmental and laboratory analysis (see sections 2.2.4(Q) and (S) below).

(iv) Principal dating evidence and chronological discussion

Despite the importance of the burnt timber structure to the history of the site, it has not proved easy to establish any precise chronology for its construction, use and destruction. This is largely a consequence of the fact that none of the 13 samples submitted for tree-ring dating proved suitable, partly because of the condition of the wood and partly because of the shortness of their surviving ring sequences (see section 2.2.4(O) below). Dating has, therefore, been dependent upon the stratigraphic sequence of the main pit [245], the evidence provided by the finds recovered from the overlying contexts, and the calibrated radiocarbon dates.

The stratigraphic sequence at the south end of the site made it absolutely clear that the main pit [245] had been cut through all the identifiable Roman horizons, including the two north-south foundation trenches [389 and 414] associated with the latest of the buildings (phase 5). In its turn, once the structure had burnt down, the pit had been infilled with three successive horizons [460, 224 and 157], the lower pair containing exclusively Roman finds, while the uppermost, which had accumulated over a long period of time, included material down to the 19th century. Such evidence provided a *prima facie* case for the structure being Roman, an hypothesis which was further indicated by its alignment, which clearly perpetuated that of earlier Roman buildings, whilst at the same time respecting the line of the north-south road immediately to the east (see Fig. 2.6).

The successive fills of the pit all contained high concentrations of Roman finds, including pottery, glass, brick and tile, slag and nails. While these are obviously fundamental to our understanding of the site's wider chronology, along with the material in the successive fills of the well/cistern (see section 2.2.2(Diii) above), it was to be expected that their contents would also have an important bearing on the date of the burnt timber building. In the event, the story proved to be more complex. The principal pieces include:

509 gravel infill: body sherd of a rouletted redware jar (not catalogued).
melon bead (Gl 15).

460 various samian pieces, including a decorated form 29 (S 3), dated *c.* 65–80, and forms 18 or 18R, and 27 (Table 2.2).
body sherds from a Dressel 20 amphora (not catalogued).
rim and base sherds from a Malvern jar (CP 40).
rim sherds from a mortarium, from Gallia Belgica (CP 41); AD 50–85.
rim and body sherds from a rusticated greyware jar (CP 42).
rim sherd of ring-necked flagon (CP 43); mid to late 1st C.
rim and body sherds from an everted rim jar (CP 44); probably Flavian or Trajanic.
body sherds from a Black-burnished jar (CP 45).
melon bead (Gl 17).

224 various samian pieces, including forms 18 and 18R, a stamped 27 base (S4), a Curle 11 flange, and a Curle 11 or Ritt 12 flange, dated *c.* 70–90 (Table 2.2).
body sherds from Dressel 20 amphorae (not catalogued).
body and base sherds from a Gillam 238 mortarium (not catalogued).
rim and body sherds from an everted-rim, greyware jar (CP 5); mid 1st to early 2nd C.
rim and base sherds from a Black-burnished ware dish (CP 6); mid to late 2nd C.
rim and body sherds from an everted-rim, redware jar (CP 7); 1st to early 2nd C.
rim sherd from a Malvern tankard (CP 8); mid 1st to 2nd C.
rim sherd from a flanged, mortar-like bowl (CP 9).
rim and neck fragment from a cylindrical or prismatic bottle (GL 4).
melon bead (Gl 14).

157 (only the relevant Roman finds are included for discussion).
various samian pieces, including a form 37 (S 1), a late South Gallic 37, dated *c.* 90–100, and a 33 from Les Martres, dated *c.* 100–120 (Table 2.2).
rim sherd from a Dressel 20 amphora, Martin-

Kilcher type 27 (CP 2); early/mid to mid/late 2nd C.
flange sherd from a Mancetter-Hartshill mortarium (CP 4); 2nd C.
sherd from a small, Black-burnished jar (CP 3).
fragment from base of a glass bottle (Gl 8).

On current evidence, the vast majority of the assemblage from these horizons can be assigned to the Flavian-Trajanic period, with only three pieces (CP 2, 4 and 6) from the uppermost fills potentially extending beyond this date range (see below). Assuming that the lowest fill [460] in the pit represents site clearance at the time of the structure's destruction (see section 2.2.2(Hi) above), then some of the later material in its assemblage ought to be indicative of the building's period of use and its demise. If so, this would seem to indicate a construction date in the early 2nd century, which would not be out of line with the structure's place in the overall stratigraphic sequence, with a lifespan of no more than a decade or so to judge from the virtual absence of any later material. The only other alternative, to treat all the material in the overlying fills as a *terminus post quem* (henceforth TPQ) for the structure, would not seem an immediately obvious or attractive hypothesis, were it not for the radiocarbon dates discussed below.

In the hopes of achieving a higher level of precision in the dating of the structure, 13 samples were submitted for tree-ring analysis at the Dendrochronology Laboratory in Sheffield. When these proved unsuitable, four samples were identified for radiocarbon dating, all of them representing posts where there was a good chance that the results would be a close reflection of the date when the timber was felled (samples 5, 15, 21 and 22 from contexts [531], [542], [546] and [533] respectively). In the event, the resulting uncalibrated dates (CAR 1251 to 1254) reveal a surprising range (see section 2.2.4(P) below), to such an extent that they are clearly not statistically the same at the 95% confidence level. If one excludes the youngest sample (CAR-1252), however, on the grounds that it had been heavily penetrated with rootlets, suggesting possible contamination with younger material, the remaining three dates are statistically the same, with a pooled mean radiocarbon age of 1738±46 BP; this calibrates to AD 177 to AD 419 at the two sigma level. Although such evidence clearly lacks the sort of precision to which we are accustomed in the Roman period, it must surely indicate the presence of a timber structure on this part of the site at a date which ought to lie somewhere between 180 and 420 in round numbers.

Such a date is not without its problems, given the nature of the assemblage recovered from the overlying horizons and the complete absence of any material which can be assigned to the period after the 2nd century, whether from the 1989 site or from elsewhere in the fort. It would also mean that the site clearance which followed the demise of the building involved the dumping of large quantities of debris derived from much earlier structures, clearly indicating extensive operations which failed, however, to incorporate any contemporary material. This is not entirely beyond the bounds of possibility, if the activity associated with the building was impermanent or short-term, but the issue is now incapable of easy resolution.

Returning to the pottery assemblage, it is worth recording that Black-burnished wares were found in association with all the fills [460; 224; 157] above the destruction horizon [540], unlike the evidence from the well/cistern (see section 2.2.2(Diii) above). Moreover, in the two upper horizons [224 and 157], the BB wares (CP 1 and 6) were found alongside at least one Hartshill-Mancetter mortarium (CP 4) and a Dressel 20 amphora of Martin-Kilcher type 27 (CP 2), both broadly datable to the 2nd century. Such finds may well represent the latest pieces from the 1989 site, indicative perhaps of some limited later activity in the immediate vicinity.

(I) The 'late' pits (Figs 2.51 and 2.52)

Towards the western edge of the site, excavation revealed a group of four intersecting pits [302; 323; 328; 351], which extended roughly north-south across the line of the east-west road. The earliest of the group was probably the relatively shallow pit [328], which yielded the crushed remains of a Black-burnished 1 jar (CP 27). This pit had been largely cut away by two much deeper pits, one rectangular, the other oval, to the north and south respectively [302 and 323]. From the oval example came the 1st-century intaglio representing Nemesis (Gm 1). The oval pit had also cut into a somewhat larger and shallower rectangular pit [352], immediately to the south, which extended beyond the edge of the excavation. Their full details include:

(i) Pit 302

This was a steep-sided rectangular pit, measuring 2.10 by 1.45m and 1.65m deep, with evidence for four successive fills (Fig. 2.52, S73):

310 the lowest fill, comprising an olive-brown silty clay loam (2.5 Y 4/3) with 60–70% coarse components, including some charcoal flecking and small decomposed pieces of brick and tile.
301 an olive-brown clay loam (2.5 Y 4/4) with 60–70% coarse components, including a greater concentration of charcoal, tile and brick.
300 an olive-brown clay loam (2.5 Y 4/4) with only 15–20% coarse components, including a high percentage of charcoal flecking.
269 the uppermost fill, comprising a dull yellowish-brown silty clay loam (10 YR 5/3), with 50–60% coarse components.

The similarity of the lowest three fills suggests a

Fig. 2.51 'Late' pits, looking east. From left to right [302], [328] (truncated), [323] (with scale) and [351]

process of natural silting into an open pit, followed by a deliberate, shallow backfill at the end to create a level surface. With the exception of the lowest fill, all contained a relatively high concentration of finds, including a range of pottery types, brick and tile fragments, and a quantity of nails. The key pieces include:

301 rim and body sherds from a tazza (CP 24).
300 collar of a Hofheim type flagon (CP 22); mid 1st to mid/late 2nd C.
rim and body sherds from a Severn Valley jar (CP 23); mid 1st to mid/late 2nd C.
body sherd of Black-burnished lid (not catalogued).
269 one samian form 27 (Table 2.2). Flavian.
rim sherd from a Black-burnished jar (CP 11); typically early-mid 2nd C, but might be earlier in south Britain.

(ii) Pit 323

This was a steep-sided oval pit, measuring 1.90 by 1.50m and 2.40m deep, with evidence for nine successive fills (Fig. 2.52, S85):

341 the lowest fill, comprising a dull yellowish-brown silty loam (10 YR 5/4), with up to 20% coarse components and 30% charcoal fragments. In section, slight colour changes indicated that this fill contained several bands.
340 a yellowish-brown silty loam (2.5 Y 5/4), containing only 5% coarse components and a small amount of charcoal, apparently confined to the eastern side of the pit. Very similar to fill 319 on the west.
319 a yellowish-brown silty loam (2.5 Y 5/3), with up to 10% coarse components and charcoal fragments, apparently confined to the western side of the pit. Very similar to fill 340 on the east.
339 a dull yellowish-brown clay loam (10 YR 5/4), containing 15–20% coarse components and some charcoal. Thickest at the centre of the pit, it progressively faded out towards its edges.
318 a yellowish-brown clay loam (2.5 Y 5/3), containing 90% small to medium-sized stones, apparently dumped in the western side of the pit.
320 a dull yellowish-brown silty loam (10 YR 5/4), containing only 5% coarse components.
317 a light grey to dull reddish-brown clay (5 YR 8/1 to 5/3), with up to 5% coarse components, apparently confined to the western side of the pit.
316 a dull yellowish-brown clay loam (10 YR 5/4), containing 25% coarse components.
299 the uppermost fill, comprising a greyish-yellow-brown clay loam (10 YR 5/2), containing 10–15% coarse components.

These fills are likely to have been derived from a number of different processes, some natural and some deliberate. The lowest of the fills [319; 339; 340; 341], in particular, appear to represent natural silting

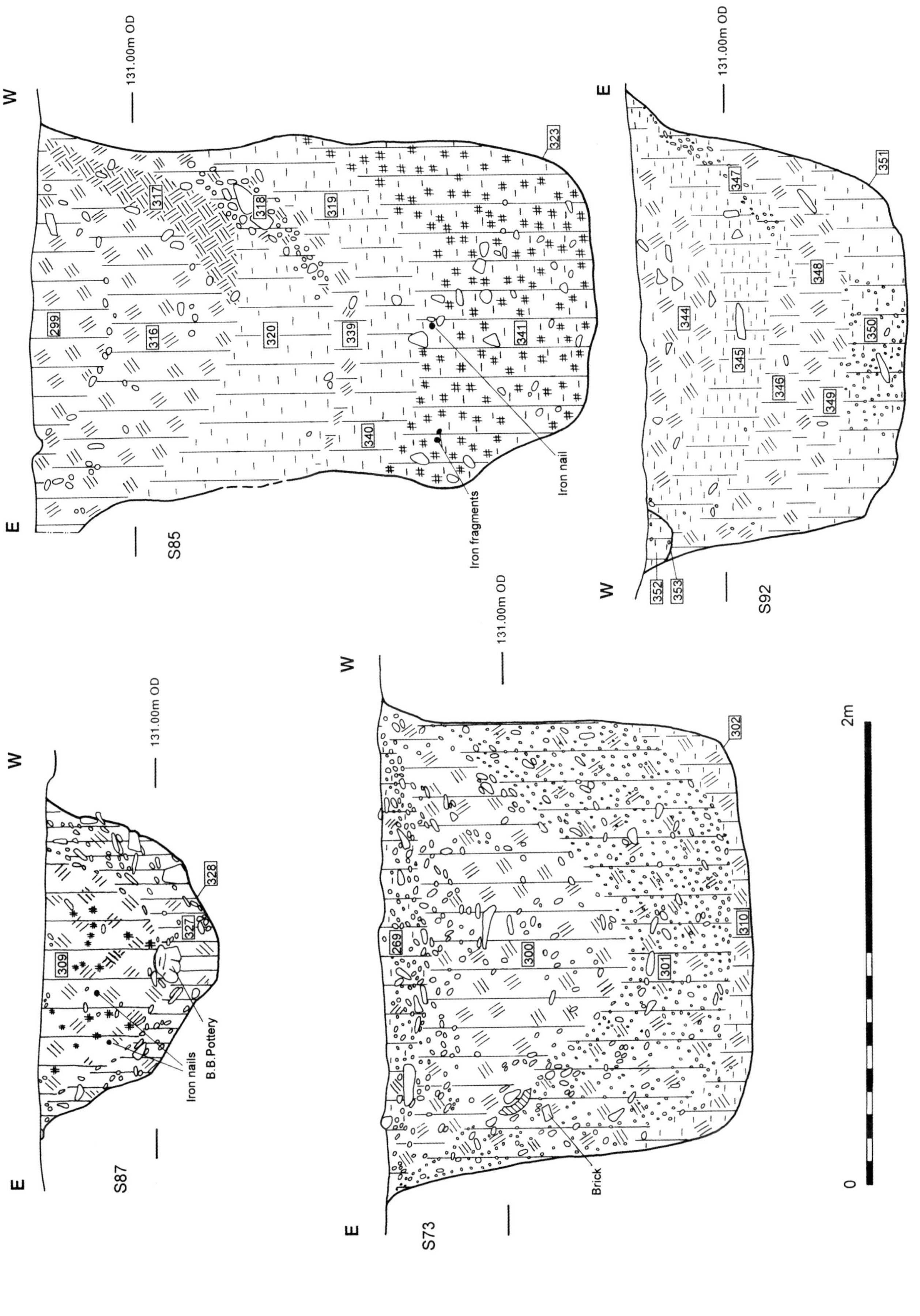

Fig. 2.52 Sections across 'late' pits

into an open pit, being derived in part from the sides of the feature and in part from the surrounding occupation horizons. This would explain the quantity of charcoal fragments and the presence of several interesting finds, including a possible lamp holder (Fe 4), an intaglio (Gm 1), and an Ae coin (Co 2) from [340]. Above these came three fills, none of which contained any finds; those [317 and 318] which were confined to the western side of the pit probably originated as deliberate dumps of material, while the third [320] represents a further period of natural silting. The two uppermost fills [299 and 316] both contained a small number of finds, but it was impossible to determine whether they were the result of natural silting or a deliberate episode of backfilling. The key pieces include:

340 sestertius of Domitian, AD 85 (Co 2).
339 neck and body sherds from Dressel 20 amphora (not catalogued).
316 samian rim of Curle 11 (Table 2.2). Flavian, probably AD 70–90.
body sherds from a Dressel 20 amphora and a Black-burnished jar (not catalogued).
299 fragment from the base of a square bottle (Gl 7).

(iii) Pit 328

This was a relatively shallow pit, with evidence for two successive fills, much truncated by later pits (Fig. 2.52, S87). Its surviving dimensions proved to be 1.38 by at least 0.60m and 0.80m deep:

327 an olive-brown silty clay loam (2.5 Y 4/3), containing some charcoal flecking and 60–70% coarse components. The Black-burnished jar was found in the upper part of this fill.
309 a dark olive clay loam (5 Y 4/3), with up to 20% coarse components and a higher concentration of charcoal flecks.

The nature of both these fills and the angle of rest of their coarse components strongly suggests that they were the result of natural silting into an open pit. Finds included various nails and the crushed remains of a Black-burnished jar (CP 27), apparently of the mid 1st to mid 2nd century, sandwiched at the interface between the two fills.

(iv) Pit 351

This was a steep-sided, roughly rectangular pit with evidence for seven successive fills, which extended beyond the south edge of the site (Fig. 2.52, S92). Its surviving dimensions proved to be 2.10 by at least 1.80m and 1.15m deep:

350 the lowest fill, comprising a dull yellowish-brown silt loam (10 YR 5/3), with charcoal flecks and a concentration of larger stones towards its base.
349 a yellowish-brown silty clay loam (10 YR 5/6) with charcoal flecks and small to medium-sized coarse components, located on the east and west sides of the pit, but petering out across its base.
348 a greyish-yellow clay loam (2.5 Y 6/2) with some very small to small sized coarse components, located across the base and the eastern side of the pit.
347 a greyish-yellow-brown silty loam (10 YR 5/2), with up to 80% coarse components, confined to the eastern side of the pit.
346 a dull yellowish-brown silty clay loam (10 YR 4/3), with some charcoal flecks and up to 60% coarse components, located across the base and the western side of the pit.
345 a dull yellowish-brown silty loam (10 YR 5/3), with some charcoal flecks and very small to medium-sized coarse components.
344 the uppermost fill, comprising a dull yellowish-brown silty clay (10 YR 5/4), with charcoal flecks and small to medium-sized coarse components.

The character of these successive fills strongly suggests that all of them represent natural silting into an open pit, partly derived from its sides and partly from the surrounding horizons. Only two [345 and 350] contained any finds, which included pottery and glass fragments, brick and tile, and various iron nails. The key pieces include:

350 neck and handle sherds from a Dressel 20 amphora (not catalogued).
rim sherd from a flagon (CP 31); 1st C.
345 two samian sherds, possibly from the same form 18/31, dated *c.* 90–110 (Table 2.2).
rim sherds from a Black-burnished lid (CP 29); mid 1st to late 2nd C.
wall sherds of a roughcast beaker (CP 30); Flavian to mid 2nd C.

The function of these four pits could not be established with any certainty. None appeared from the nature of their fills to have been a latrine pit (by comparison with that examined in the river section immediately south of the fort: Burnham and Burnham 1987a), nor, apart from a small amount in the oval pit [323], did they seem to contain the type of debris suggestive of clearance of a fort site, such as was found down the large pit [245] containing the burnt timber building. During excavation, the oval pit was thought to be a possible well, but this soon proved to be an unsatisfactory interpretation once it bottomed out at a depth of only 2.40m, well short of the recorded depth of the stone-lined well (see section 2.2.2(D) above). This leaves the possibility that they were dug through the gravel makeup of the road either in search of gravel or as rubbish pits, though the latter is not easy to reconcile with the character of their fills, which for the most part seem to represent natural siltings.

All four pits appear to have been dug through the makeup of the east-west road, following the demise of the western of the two structures assigned to phase 4, and so could well be contemporary with the final phase of building which lay immediately to the east. This would have located them conveniently in an open area in the angle between the building and the defences and would conveniently account for the exclusively Roman nature of their associated finds. A relatively late Roman date might also be indicated by the presence of Black-burnished wares at all levels, when compared with other parts of the site (for details see sections 2.2.2(Diii), (Gvii) and (Hiv) above). It is quite possible, however, that some of the pits or their respective fills could be of any date right up to the laying of the earliest of the modern farmyard surfaces [180/223], beneath which they are stratigraphically sealed. This would mean that most of the material associated with them could be residual, with the possible exception of the Black-burnished jar found in what is likely to have been the earliest [328]. Whichever is the case, these features could well be contemporary with a further large pit [410], the stratigraphic relationship of which to the back of the rampart remains uncertain (see section 2.2.2(B) above).

2.2.3 The Post-Roman levels

(A) The principal elements of the 1989 site

The main post-Roman elements on the Pumsaint site can be divided into three very broad groups: (a) early features which post-date the Roman activity but which underlie the main late 18th- to 19th-century clom (clay-walled) and stone buildings (Fig. 2.53); (b) later features either pre-dating or related to the clom building (Fig. 2.60); and (c) the buildings themselves, together with further features clearly associated with them (Figs 2.62 and 2.68).

The first group consists of a number of very disparate features, including a broad gully running diagonally across the site, six oval hollows of unknown purpose, metalworking hearths and related features, a north-south gully, some gravelled areas, and a number of miscellaneous features, few of which made much obvious sense. Although these features post-dated the Roman occupation and pre-dated the later building and its associated yards, it was in most cases virtually impossible to date them any more closely; with few exceptions, there were either no finds or merely residual Roman material. The second group of features underlay the makeup for the main floor of the clom building, and may well have been associated with a possible predecessor, or lay to its north at the same level as its north wall. All these features are discussed in detail below.

The buildings themselves were somewhat easier to deal with. The 19th-century stone building was still standing until shortly before the excavations began. Beneath its central part were found the remains of its immediate predecessor, the clom building, which had traces of other activity to its north and south. Most of this probably post-dates 1775, since a coin of this date (Co 9) was found beneath the floor. The discovery of the remains of a demolished post-medieval cellar in the yard to the west was something of a surprise.

Further information was obtained about the 19th-century stone building, which, combined with a report on its vernacular architecture by Darlington and Noake (see section 2.2.3(N) below), compiled before its demolition, allowed us to build up a picture of its development in three stages. The southern portion appears to have been built in *c.* 1850, as an extension to the earlier clom structure, which was itself replaced by the central section in the 1860s or 1870s, while the northern extension was added in the 1890s. Traces were also recovered of various internal arrangements belonging to both phases of building; the northern and central portions contained stalling for cattle, while the southern consisted of a cart-shed and dairy; in recent years there was a half-hayloft above the cart-shed, but the building can be seen to have had a full upper storey in *c.* 1900, and was probably built like that. The arrangement of the clom building may have been somewhat different, judging from the patterning of its pitched stone flooring.

Farmyard surfaces with related features, in particular, gatepost holes, belonging to each main building phase were identified. Alterations from more recent times, including extensive modifications to the milking parlour at the north end, could be dated from living memory. TPQ dates obtained from various of the contexts in the post-Roman sequence are set out in Table 2.28.

(B) Post-Roman features clearly pre-dating the clom building

(i) The broad gully (Figs 2.53 to 2.55)

The earliest of the post-Roman features underlying the floor of the clom building was a broad shallow gully [298], averaging 1.10m wide and 0.15m deep (Fig. 2.54). It ran at an oblique angle across the area, from north-east to south-west. The original depth and purpose of this gully was not clear, and there seemed to be some curious irregularities in its outline (Fig. 2.55, S74 and S75); it narrowed appreciably, to *c.* 0.70m, at its south-western end, where it was truncated by the later rectangular feature [244]. It contained a fill [297] of dull yellowish-brown coarse sandy loam (10 YR 4/3), with 10% very small and small stones. It also contained fragments of amphora, brick or tile, iron slag, iron nails and copper-alloy slag (not catalogued; object 10,021).

(ii) The oval hollows (Fig. 2.53)

Six oval hollows [294; 306; 315; 335; 361; 462], averaging 1.85 by 0.65m and 0.12m deep, and oriented north-east to south-west, were scattered across the portion of the site underlying the later building. A more enigmatic feature [258], discussed in section 2.2.3(Bvi) below, may also belong with this group. They were not initially very clear, since their

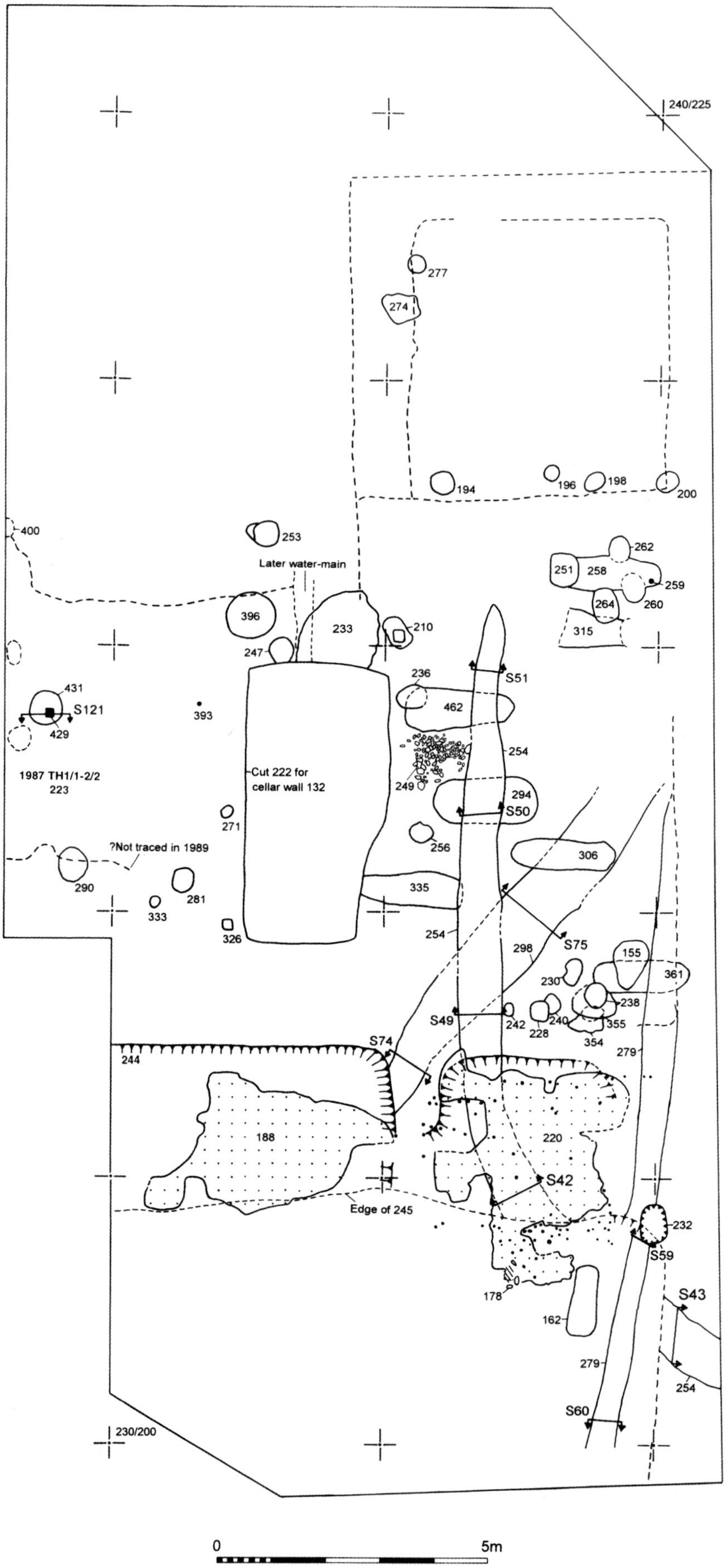

Fig. 2.53 Composite of earlier post-Roman features, predating clom building

Fig. 2.54 Broad gully [298], looking north-east

fill was in most cases broadly similar to the surrounding matrix, and to begin with they were not recognised as forming a coherent group. It was only with closer attention and further excavation that it was possible to define their extent and to distinguish them from other features. One of them [306] cut the fill of the broad gully [298] discussed above, another [315] that of the Roman cistern [284], while three of their fills [293; 334; 461] had been cut by the north-south gully [254] discussed in the following section. The remaining fill [360] was cut by the metalworking activity covered in section 2.2.3(Biv) below.

There was no distinctive material in their fill, although their north-east to south-west alignment, reflecting that of the later buildings, might suggest that they related in some way to earlier structures on the site. They do not, however, appear to form a single coherent plan as a group. The possibility that they may represent grave hollows was also considered, at a late stage of their excavation, on the strength of their size and orientation, despite their limited surviving depth. They may never have been any deeper, although it is tempting to suggest that they may have been truncated by a later episode of levelling across the site; there is however little evidence for anything so radical. Unfortunately their fills were not sampled for soil phosphorus.

294 Oval 1.85 × 0.75m, 0.12m deep. Filled by a dull yellowish-brown (10 YR 5/4) coarse sandy loam [293] with 60% very small and small stones. Contained two sherd fragments of redware/fineware.
306 Oval 1.88 × 0.58m, 0.12m deep. Filled by a dark greyish-yellow (2.5 Y 5/2) silty clay (loam) [305], with some small stones.
315 Truncated; >1.20 × 0.70m, 0.12m deep. Filled by a dull yellowish-brown (10 YR 4/3) silty clay loam [313], with small and medium stones and gravel; a series of upright stones [314] had apparently been set along the cut edge, more regularly on the northern than the southern side.
335 Truncated; >1.85 × 0.65m, 0.20m deep. Filled by a greyish-yellow-brown (10 YR 5/2) silt loam [334], with 20% small stones.
361 Oval 1.70 × 0.60m, 0.20m deep. Filled by a dull yellowish-brown (10 YR 5/4) silty clay loam [360], with 20% very small to medium small stones. Contained a copper-alloy slag fragment (object 10,022).
462 Oval 2.00 × 0.40m maximum width, 0.10m deep. Filled by a dull yellowish-brown (10 YR 5/3) silty loam [461] with 75% gravel, very small, small and medium stones. Contained Roman brick or tile fragments.

(iii) The north-south gully (Figs 2.53, 2.55 and 2.56)
A gully [254], basically V-shaped in cross section, ran for a total of 15.00m across the site (Fig. 2.56). From a point close to the Roman cistern, it continued

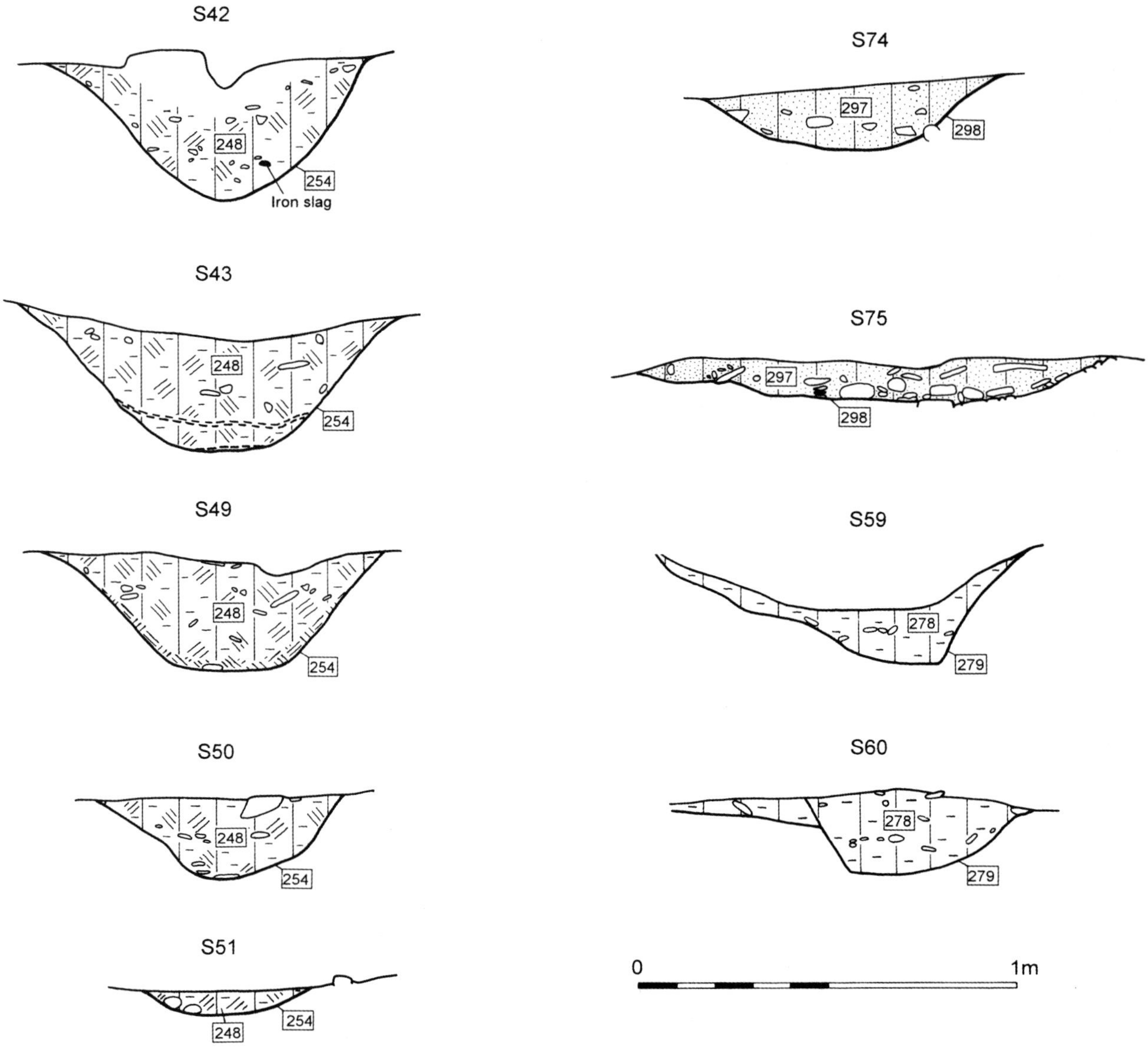

Fig. 2.55 Cross-sections of gullies [254], [279] and [298]

southwards as far as the edge of the pit [245] in which the burnt timber building lay; beyond this, it swung eastwards across the upper infills of [245] to cross the metalling of the Roman road in a virtually east-west direction. The gully was very shallow at its northern end and became deeper and wider towards the south, reaching a maximum width of 0.90m and depth of 0.50m where it met the eastern edge of the site (Fig. 2.55, S42 and S43, S49 to S51). Its destination is a mystery although the gradient strongly suggests that it was intended as a drain. It contained a fill [248] of greyish-yellow-brown silty clay loam (10 YR 5/2), with 20% small stones; there was slightly more gravel towards the base of the fill in places. The associated finds were all Roman with the exception of a single piece of medieval pottery (LP 5).

This gully cut the fills [293] and [461] of two of the oval hollows discussed above, and, like the metal-working features discussed below, was cut through a distinctive clay horizon [359]. A later clay deposit [226] appeared to seal its fill.

(iv) The metalworking hearths and associated contexts (Figs 2.53, 2.57 and 2.58)

Towards the south end of the area later occupied by the clom building was a considerable area of unevenly coloured clay (successively [359] and [226]), which was reddened in parts and appeared to have been subject to burning. This had been first seen in 1987 in a trial pit cut through the floor of the 19th-century building (see section 2.2.1 above). It contained a number of hollows [240; 355; 238; 228; 230] which appeared from their fills, including charcoal, ash, slag and possible lime flux, to be metalworking hearths or related features,

Fig. 2.56 General view looking north, showing north-south gully [254]

although their structure was not very clearly defined (Fig. 2.57). A post-hole [242] and a stone spread [338] at the lower level and a pit [155] at the higher also appeared to be part of this complex of features, and a number of possibly related stake-holes were identified at both levels.

The clay spread was apparently made up of two successive layers, [359] and [226], since features dug into [359] (including the long north-south gully [254] discussed above) had been covered by [226]. The lower of the two [359] was a dull yellowish-brown silty clay (10 YR 5/4), with a 20–30% admixture of gravel, very small, small and medium-sized rounded stones. It was fairly extensive across this part of the site, and many of the fills of the Roman construction trenches became visible only after its removal. It contained only Roman material, and may well represent Roman occupation and destruction debris, preserved here beneath the later buildings.

The earliest feature dug into this clay (apart, probably, from the north-south gully [254]) appeared to be [355]. This had a central hollow measuring 0.40 by 0.30m and 0.20m deep, surrounded on the south and west by a shallower, shelving area *c.* 0.08m deep, measuring 1.20 by 0.65m overall. The entire hollow was filled by a very mixed brownish-grey silty clay loam [354] with 20% very small to medium-sized stones, ash, clay, iron slag and charcoal. A concentration of iron slag and charcoal at the base of the deposit covered a copper-alloy object (Ae 1), while the north-west corner of the depression was lined with clay.

This hollow had been replaced, still apparently from the same ground level, by another similar hollow [238], centred only 0.30m to its north, and partially cutting it away (Fig. 2.58). The central hollow of [238] measured 0.55 by 0.40m, and was 0.25m deep; like its predecessor [355], it was surrounded by a shallower shelf on the south, giving an overall extent of 0.80 by 0.70m. The hollow was filled with a very mixed dark brown (7.5 YR 3/3) silty clay loam [237] with 10% very small stones, ash, clay, charcoal, including some quite large pieces, and some fragments of possible limestone or chalk flux. This fill also contained 17th- and 18th-century pottery of fabric types 3 and 7, part of a 17th-century clay pipe, and three fragments of 'mortar', possibly flux. Another feature [240], 0.80m from these, was a simple hollow measuring 0.40 by 0.30m and 0.12m deep; it was filled with a mixed dark brown (7.5 YR 3/3) silty clay loam [239] with 10% very small stones, containing the possible refined iron bloom (Fe 21).

Immediately to the north-west of the hollow [238] lay a roughly circular spread of medium-sized stones [338], rather scattered at the edges, measuring *c.* 0.70 by 0.60m, with a matrix of dull yellowish-brown silty clay loam (10 YR 4/3). There were signs of burning on the stones, and patches of charcoal among them; excavation revealed a denser but thin spread of charcoal underneath, with one fragment of iron slag. The 'post-hole' [242], 0.80m to the west of [240], also appears to have been cut from this level. At 0.28 by 0.19m and 0.20m deep, it seems a little small to be a hearth, and its straight-sided profile also tells against such an interpretation. Its fill [241], a loose reddish-brown loam (7.5 YR 4/3) with 50% very small and small stones, contained some charcoal flecks but no ash or 'mortar'.

The remaining features in this group appear to post-date the deposition of the upper clay spread [226], which was a very mixed layer, varying in colour between a strong red and a light buff. From this clay came post-medieval pottery, mainly 18th-century (Type 7), but with one possibly intrusive sherd dating to the late 18th century or beyond, and iron slag. This clay had apparently been cut along its south edge by the foundation trench for the clom building [156]. Through it were dug two further, more irregular, hollows [228] and [230], possibly related to metal-working. The first of these [228] was roughly square, with sides of 0.40m, and 0.20m deep; it contained a fill [227] of a mixed dark brown silty clay loam (7.5 YR 3/3), with 10% very small stones, generally ashy in appearance with pieces of charcoal and lumps of clay. This fill contained post-medieval pottery (Type 14) dating to the late 18th century or beyond. Some of

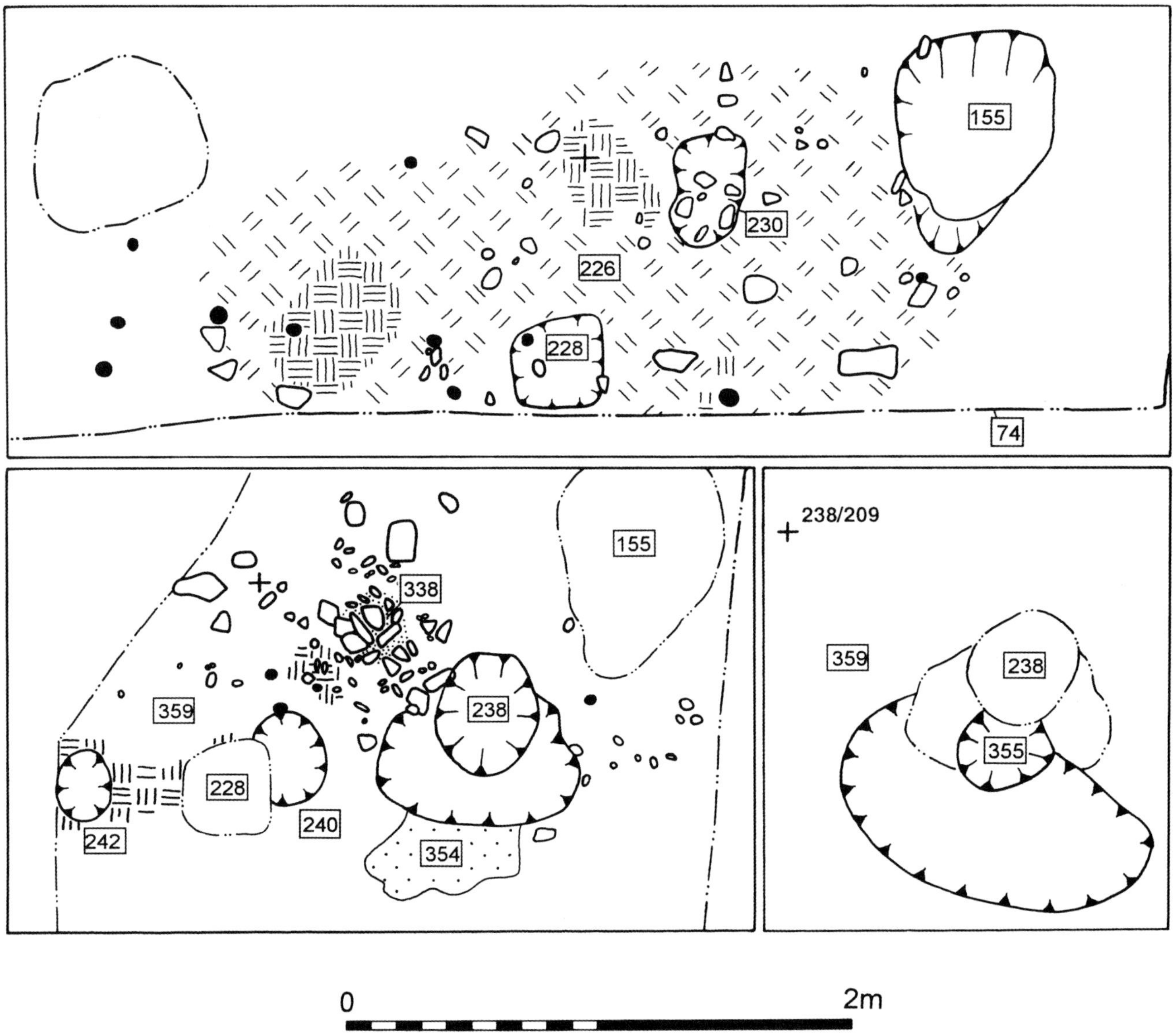

Fig. 2.57 Detail of metalworking hearths and associated features (earlier and later phases)

the deposit was more reddish in colour (7.5 YR 5/8 bright brown). The second hollow [230] was much more irregular in outline, roughly kidney-shaped, measuring 0.50 by 0.30m and 0.10m deep. Its fill [229] was a brown silty clay loam (7.5 YR 4/3) with 70% very small, small and medium-sized stones and some 'mortar' flecks. The material was generally ashy and there was a cluster of medium-sized stones in the top of the feature. This second hollow lay directly above the stone spread [338] discussed above, although it was difficult to define any relationship between them.

A short distance to the east of these two features lay a pit [155], measuring 1.00m north-south by 0.60m, with a shallow slope at its southern end. It was filled by a very mixed deposit [154] with 70–80% 'mortar' fragments and small and medium-sized stones, in a matrix of dark brown silty clay loam (10 YR 3/4). At the base of this fill was a very thin layer of coal dust.

(v) The gravelled areas (Figs 2.53 and 2.59)
Two irregular areas of gravel metalling [188] and [220] lay towards the southern end of the site (Fig. 2.59), roughly between the edge of the pit [245] containing the burnt timber building and a line parallel with the southern wall of the clom building [75/156]. These areas were very similar in consistency although they did not join and were hence numbered separately. That to the west [188] measured *c.* 4.20 by 2.40m, with the longer dimension running east-west; it lay within a slight rectangular hollow [244] and contained pottery dating to the 17th century or later (Type 11). The eastern spread [220], which contained pottery dating to the 18th century or later (Types 7 and 9; LP 28), measured *c.* 3.00m square; a very small part of it lay beneath the wall of the clom building, while it also overlay the fill [248] of the gully [254] discussed above. It also overlay, and the cut [244]

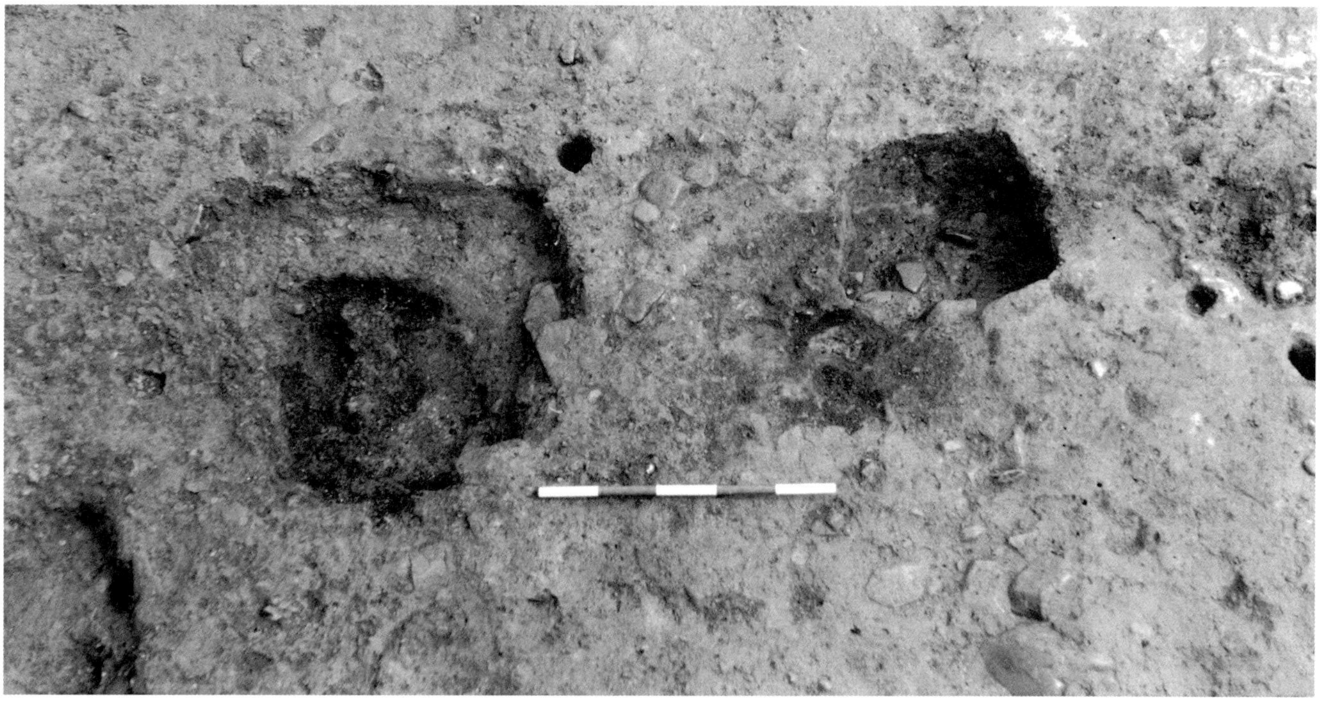

Fig. 2.58 Hearths [238] and [228], looking south

Fig. 2.59 Gravelled areas [188] and [220] with later drain [221], looking west

was cut into, a very mixed deposit of dark soil [157], which formed the uppermost fill in the pit [245] and extended for some distance beyond it. This soil [157] contained a considerable amount of medieval and post-medieval pottery (LP 1; 4; 6–8; 14–16; 20; 23–26; 29), numerous fragments of clay pipe (CTP 2–4; 6), post-medieval glass and iron nails, as well as a considerable quantity of residual Roman material. While the latest dated object from [157] was a clay pipe of *c.* 1880, this does not necessarily date the gravel surfaces, since the deposit was considerably more extensive and very mixed.

The interpretation of these surfaces is not entirely clear, but they seem to respect in large part the southern wall line of the later clom building. They may perhaps form part of the metalling of a lane or yard passing at this point into the area behind the main road frontages. A number of stake-holes in and around the eastern spread [220] may suggest animal pens or other makeshift structures, and there seems to be a line of such holes dividing the surface from the area of the infilled pit [245] to its south.

Curiously, the upper clay deposit [226] in the area of the metalworking features is confined to the northern side of the clom wall [75], despite being cut by its construction trench [156]. This suggests either that this area was already differentiated from that to the south at the time of the activity, or that it became so at some date before the construction of the wall, any traces of clay to the south being obliterated by activity in the lane or yard area. The clay may represent flooring in an earlier building not otherwise identified, or collapsed material from clay furnaces.

(vi) Other miscellaneous features (Fig. 2.53)

There were a number of other features, probably of a variety of dates, which appeared on stratigraphic grounds to post-date the Roman activity and to pre-date the clom and stone buildings and the successive farmyard surfaces probably associated with them; these latter surfaces [180/223] and [79] are discussed below (see sections 2.2.3(Giii) and (L) respectively). It is worth noting that the density of features in general appears greater towards the east of the site, suggesting that activity always tended to concentrate closer to the line of the route through the village.

In the area of the later yards, an irregular robber trench [331] at the north end of the site had followed the line of the Roman stone rampart in a rather haphazard fashion, thereby removing the surviving upper parts of its structure. It was filled by a dull yellowish-brown (10 YR 5/4) silty clay loam [330] with gravel, very small, small and medium-sized stones, and contained, besides residual Roman material, both post-medieval pottery of the late 17th century onwards (Type 6) and glass. Both the trench and its fill were stratified beneath only the most recent farmyard surface [32], but this part of the site had been truncated by machine before excavation began, and it seems reasonable to suggest that this robbing forms part of the same episode as that of the tower to the east, beneath the later building. Certainly the 17th- to early 18th-century dating of the fills of both robber trenches might suggest this.

The remaining features from the area of the later yards consisted of a variety of post-holes and hollows of different sizes which it was difficult to combine into definite groups. Most were sealed beneath the later farmyard surfaces [223] and [180], which have been treated as the same surface. Underlying [223], the western part of the spread, were two post-holes [431] and [290], which may perhaps have supported a pair of posts 3.00m apart, though this is marginally wider than the later gateways associated with the farmyard surfaces. The more northerly [431] was a circular pit 0.55m in diameter and 0.65m deep. It contained the impression of a post [429], 0.16m square and 0.48m deep, infilled by dull yellowish-brown (10 YR 4/3) silty clay loam [428] with some gravel, very small and medium-sized stones. The area around the post was filled by a yellowish-brown (10 YR 5/6) silty clay loam [430] with some gravel, very small and small stones, and a few large packing stones. Its counterpart to the south [290] was slightly oval, 0.60 by 0.56m, and 0.29m deep. It contained the irregular impression of a post [288], 0.175 by 0.15m and 0.13m deep, infilled by greyish-yellow-brown (10 YR 4/2) silty loam [287], with a mixture of small to medium-sized stones. From it came post-medieval pottery dating to the 17th century onwards (Types 7 and 11) and post-medieval glass. The area around the post was filled by a dull yellowish-brown (10 YR 4/3) loam [289] with a mixture of medium stones and some large packing stones.

Underlying [180], to the east, was a further batch of features (from north to south):

396 Circular pit, 1.05m in diameter and 0.95m deep. Filled by a very mixed lumpy dull yellow-orange (10 YR 6/4) silty clay [395] with virtually no coarse components, probably derived from rampart material.
393 Stake-hole, 0.10m in diameter and 0.20m deep. Filled with brown (10 YR 4/4) silty clay loam (not numbered separately).
271 Circular post-pit, 0.20m in diameter and 0.20m deep. Filled by a greyish-yellow-brown (10 YR 4/2) loamy coarse sand [270] with very small, small and medium-sized stones; three medium-sized stones pitched vertically.
281 Circular hollow, 0.44m in diameter and 0.10m deep. Filled by a greyish-yellow-brown (10 YR 4/2) silty loam [280] with very small, small and medium-sized stones, also a fair amount of charcoal. Clay pipe *c.* 1660 (Table 2.8).
333 Stake-hole, 0.14m in diameter and 0.25m deep. Filled by a greyish-yellow-brown (10 YR 4/2) silty loam [332] with 10–20% gravel, very small, small and medium-sized stones, and charcoal flecks.
326 Post-impression, 0.20m square and 0.32m deep. Filled by dull yellowish-brown (10 YR 4/3) silty clay loam [325] with 10–20% gravel, very small, small and medium-sized rounded stones; medium-sized flat stones vertically around edge.

Of these, [271], [333] and [326] could be part of a more tenuous feature later cut away by the post-medieval cellar [222], although [271] and [326] could plausibly relate to the latter in some way.

Two further features had been cut by the construction of the post-medieval cellar [222], and therefore presumably also belong early in the sequence. One was a layer [233], measuring 1.40m square and 0.30m thick, consisting of 70–80% large rounded stones in a matrix of dull yellowish-brown silty clay (10 YR 5/4). From it came Roman nails, slag and brick. This might represent fill over slumpage into the pit [410]. The other was a circular pit [247], 0.50m in diameter and 0.35m deep. It was filled by a dark brown (10 YR 3/4) silty clay loam [246] with 50% gravel, small and medium-sized rounded stones and fragments of coal. From it came two fragments of redware/fineware. This might form a pair with another similar circular pit [253], 0.48m in diameter and 0.38m deep, lying 2.25m away, although no clear relationship with [180] was recorded during excavation. This pit was filled by a greyish-yellow-brown (10 YR 4/2) silty clay loam [252] with 50% gravel, small and medium-sized rounded stones.

Within the area of the later buildings, a number of features were identified which clearly pre-dated the clom and stone structures. These were sealed by the surfaces [225] at the north end (see section 2.2.3(Ci) below) and [129] further south (see section 2.2.3(Cii) below). Although [225] and [129] were apparently largely confined to the areas of the later Rooms 1 and 2 respectively, this may simply be an artefact of the depth of the foundation trenches for the stone building, which had severed many relationships between the different parts of the site.

At the north end of the site the surface [225] sealed a robber trench [292], *c.* 1.00m wide, which followed the outline of the underlying Roman stone tower [494] and contained a fill [291] of dull yellowish-brown sandy clay loam (10 YR 4/3) with 50% small to medium-sized gravels. From it came post-medieval pottery of the 17th century onwards (Type 11), cinder, residual Roman material (CP 20), and some flint (Fl 3). The surface also sealed two further features:

274 Circular pit, 0.65m in diameter and 0.35m deep. Filled by a dark brown (10 YR 3/3) silty loam [273] with 20% gravel, very small, small and medium-sized rounded stones and some charcoal flecks. Three fragments of pottery dating to 16th century or later (LP 18).
277 Circular post-impression, 0.28m in diameter and 0.37m deep. Filled by a dark brown (10 YR 3/3) silty clay loam [276] with 20–30% gravel, small and medium-sized rounded stones. One iron nail.

Further south, four irregularly spaced features had been truncated by [201], a cut aligned with the south side of the clom building's north wall [63] (see section 2.2.3 (Cii) below), perhaps defining the northern end of a levelled area later largely cut away by further alterations. This cut was filled by [129], thereby sealing the four features below it. The late 18th-century or later pot from [198] may be intrusive. Their alignment and general size would suggest that they are broadly contemporary and perhaps relate to a structure or fence line at this point.

194 Circular post-pit, 0.38m in diameter and 0.30m deep. Filled by a dull yellowish-brown (10 YR 4/3) silty clay loam [193] with small, medium and large rounded stones, mortar fragments, high proportion of ash and cinder.
196 Circular post-pit, 0.36m in diameter and 0.28m deep. Filled by dull yellowish-brown (10 YR 4/3) silty clay loam [195] with small, medium and large rounded stones, mortar fragments, 50% ash and cinder.
198 Oval hollow, 0.40m in diameter and 0.10m deep. Filled by a dull yellowish-brown (10 YR 4/3) silty clay loam [197] with 20% small, medium and large stones. Post-medieval pot late 18th century + (Type 14).
200 Circular pit, 0.36m in diameter and 0.20m deep. Filled by a dull yellowish-brown (10 YR 4/3) silty clay loam [199] with 20% small, medium and (mostly) large rounded stones.

In the central area of the site, a number of features were sealed beneath [129] itself, although their relationships were rather difficult to unravel and they may date from more than one period. The first feature in the sequence was probably the stake-hole [259]. Its fill had been truncated by the elongated hollow [258], into the southern side of which a smaller oval hollow [260] had been cut. Excavation conditions did not allow the fills of these two features, and hence the order of their construction, to be distinguished, and their homogeneous fill was numbered [257]. A further hole [262], to the north of [258], had its own fill [261], but its edges were not sufficiently clear to define its relationship with [257]. Were it not for these doubts about the relationship of the hollow with these two post-holes, [258] might have been regarded as one of the 'oval hollows' discussed in section 2.2.3(Bii) above; the feature appears to cut natural, so that there is no stratigraphic help to be had there. Two further holes [264] and [251] on the west and south of [258] respectively, were undoubtedly secondary. Two other features, [249] and [256], were also sealed beneath [129]. With the exception of some redware/fineware in [251] there were no associated finds.

259 Stake-hole, 0.10m in diameter and 0.20m deep. Fill (not numbered separately) of mixed gravel and grey clay.
258 Elongated oval hollow running E-W, >1.47m × 0.58m, 0.08m deep. Filled by [257] (below).
260 Oval hollow running N-S, 0.70 × 0.44m, 0.30m deep. Filled by [257] (below).
257 Fill of [258] and [260]. Dull yellowish-brown loam (10 YR 5/4), with 80% small stones.
262 Oval hollow running N-S, 0.56 × 0.35m, 0.28m deep. Filled by a dull yellowish-brown (10 YR 5/4) loam [261] with 80% small stones; quite heavy iron-staining.
264 Oval hollow running N-S, 0.70 × 0.43m, 0.10m deep. Filled by a brown (10 YR 4/4) fine sandy loam [263] with 90% very small and small rounded stones.
251 Sub-rectangular pit running N-S, 0.62 × 0.50m, 0.20m deep. Filled by a dull yellowish-brown (10 YR 5/4) silty clay loam [250] with 50% gravel and small rounded stones. Two fragments of redware/fineware.
249 Roughly triangular cobbled area with no discernible matrix. 0.90 × 0.80m, 0.10m thick. Straightish edge on N, and on W where truncated by construction trench for stone building [74]; straggling edge on SE. Possibly patch on *via sagularis*, with which it would fit quite well.
256 Circular post-pit, 0.44m diameter, 0.30m deep. Filled by a brown (10 YR 4/3) silty clay loam [255] with gravel, small rounded stones, and four large rounded stones distributed across the top of the fill.

At the south end of the site, a variety of features were stratified above the topmost deposit [157] overlying the pit [245] containing the Roman burnt timber structure (see section 2.2.3(Bv) above). These lay to the south and east of the surfaces [188] and [220] already described in that section. Two of the features were sealed beneath the surface [128] (see section 2.2.3(Gi) below). One was an elongated pit [162] running north-south, measuring 1.30 by 0.40–0.50m and 0.50m deep. It was filled by a dark brown (10 YR 3/3) sandy clay loam [161] with 80% small to medium/large stones, the latter concentrated in the southern half. It contained pottery dating down to the late 18th century onwards (Types 3; 12; 14; LP 17). The other [178] was an area measuring 0.55 by 0.30m of brownish-black (10 YR 2/2) sandy clay loam with 20% very small stones; there were three larger stones around its edges, leading to its interpretation as a possible hearth. On the east of the area a gully [279] ran north-south, cutting the fill [360] of one of the oval hollows [361] (see section 2.2.3(Bii) above). It measured over 12.50m by 0.50–0.60m, was 0.30m deep, and was filled by a dull yellowish-brown (10 YR 4/3) silty loam [278] with 10% small stones (Fig. 2.55, S59 and S60). This fill was in turn cut by a subrectangular post-pit [232], running north-south, which measured 0.70 by 0.60m, and was 0.20m deep. It was filled by a brownish-black (10 YR 2/2) silty clay loam [231] with 70–80% gravel, small, medium and (mostly) large rounded stones, sealed by the surface [121] (see section 2.2.3(Gii) below).

(vii) Dating problems

While the features discussed in the foregoing sections were all clearly stratified between the Roman activity and the later buildings or their associated yards, it is difficult to say much about their individual dates. The presence of Roman finds was almost certainly residual, while many of the features contained no dating material at all. It is possible to put many of them into a broad sequence on stratigraphic grounds, as has been done above; the broad gully and the six oval hollows are undated but clearly early, while the north-south gully [254] could be medieval. The complex around [258], the metalling [249] and the post-pit [256] could belong at any date up to the 18th century, when the deposition of [129] sealed them. The remaining features would appear for the most part to be post-medieval. At the north end of the site, the backfills of the robber trenches over the Roman stone tower and rampart contained material which could date back to the 17th century, and may well mark the beginning of the building activity in the area. The pit [274] and post-impression [277] at the north end also appear relatively early in the sequence, while the overlying deposit [225] and some of the features cut into it may also be of 17th-century date (for details see the next section). The earlier set of metalworking features, dug into [359] at the southern end of the area later occupied by the clom building, could also be 17th-century. The bulk of the activity is probably 18th-century, apparently pre-dating the clom building, which can be dated to some time after 1775, and most probably to the early 19th century. Our hopes of recovering information about any medieval activity within the village were thus disappointed, although there are hints of a fair amount of post-medieval activity on the site before the first clearly identifiable building. A severe lack of evidence for anything much between the Roman period and the 18th century is, however, a not uncommon problem in rural Wales, where the population attested in the historical sources must have lived in buildings and used equipment which leaves no archaeological trace.

(C) Later features either pre-dating or related to the clom building

(i) Features to the north of the clom building (Figs 2.60 and 2.61)

A number of features lay in the area to the north of the clom building, later occupied by the late 19th-century Room 1. Many of these had been cut into the disturbed horizon [225], the same deposit into which the construction trench for the north wall of the clom building had been dug, and were therefore presumably either earlier than or contemporary with this structure (Fig. 2.61). They all appear to respect the line of the clom wall, although it is possible that its construction trench removed any which were immediately beneath it. These features may relate to a structure or successive structures, but no trace of any walls was found. Further features cutting the overlying deposit [92] are discussed in section 2.2.3(H) below.

[225] itself was a mixed layer of gravel, yellow clay and burnt clay, very variable in both colour and description, and without any finds. Some attempt has been made here to group the features cut into it, but there are few obvious patterns to be made from them. It can be tentatively suggested that the three pits [183; 175; 177] in the northern part of the later Room 1 are the earliest in the group:

183 Oval pit, oriented NE-SW, 0.55 × 0.50m, 0.30m deep. Filled by a dark brown (10 YR 3/3) coarse sandy loam [182] with 20% small to medium-sized stones; post-medieval pot 17th century + (Types 3 and 5); clay pipe *c.* 1680 (Table 2.8).

175 Pit, 0.60m in diameter, 0.45m deep. Filled by a dark brown (10 YR 3/3) coarse sandy loam [174] with 20% small to medium sized stones; medieval and post-medieval pottery down to 18th century (Types 1, 3, 4, 7; LP 2, 19); 17th-century clay pipe fragments (Table 2.8) and slag. One sherd joined with one in overlying deposit [92].

177 Irregular oval pit, oriented N-S, 0.90 × 0.40m, 0.30m deep; deeper at S end. Filled by a dark brown (10 YR 3/3) coarse sandy loam [176] with 20% small to medium-sized stones.

The fill of [175] was overlain by a scatter of large stones [133], 1.10 by 0.90m in extent and oriented north-south. Together with another scatter [136], measuring 0.69 by 0.45m and oriented east-west, this may represent flooring in the structure hypothesised for this area in section 2.2.3(H).

The fills of two other post-holes, [165] and its successor [167], are somewhat similar to those in the three northern pits just discussed, though their function remains equally unclear.

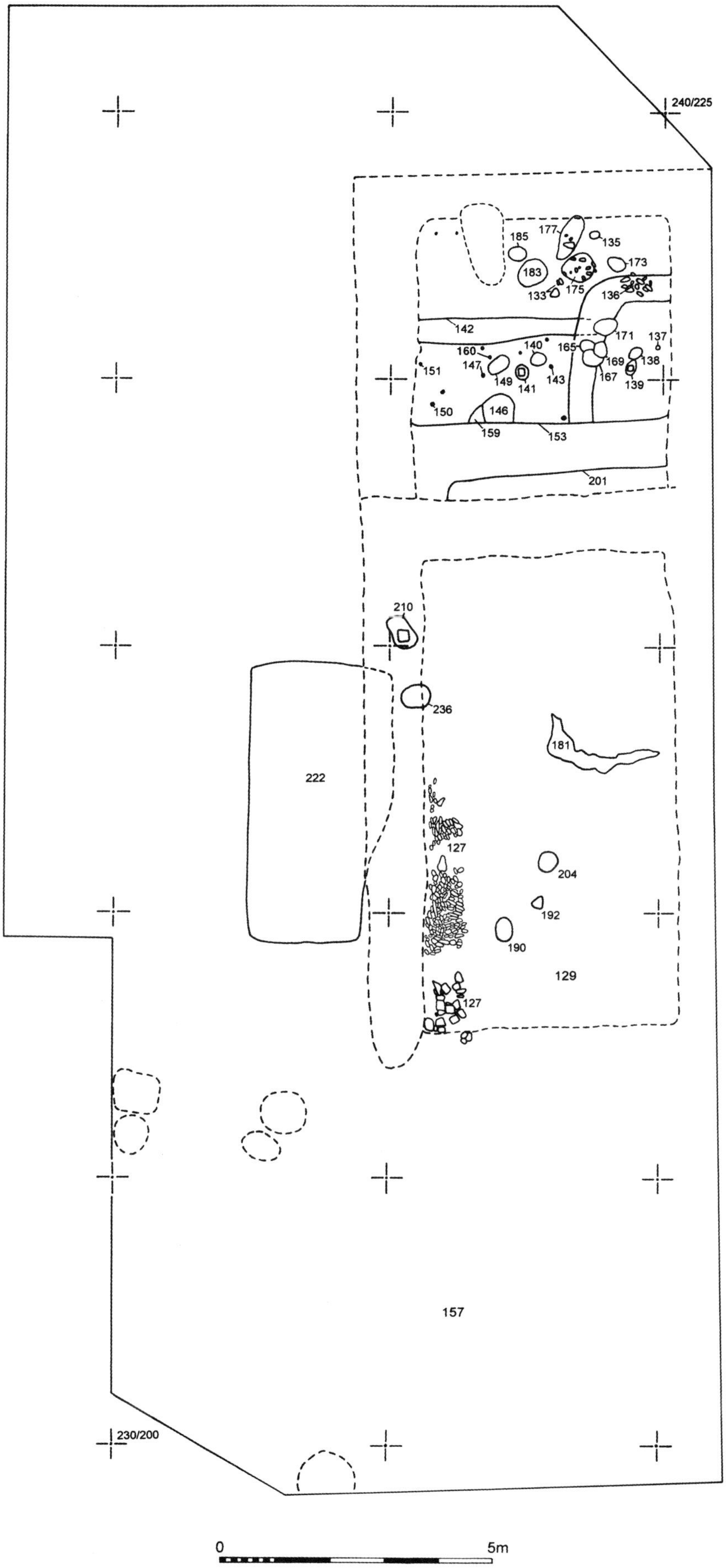

Fig. 2.60 Composite of later post-Roman features, predating clom building

Fig. 2.61 Area to north of building at level of [225], with cut [201] to right, looking east

165 Post-hole, 0.22m in diameter and 0.17m deep. Filled with a dark brown (10 YR 3/4) coarse sandy loam [164] with 5% small stones and some lumps of slag.
167 Oval post-hole, oriented N-S, 0.38 × 0.32m, 0.20m deep. Filled with a dark brown (10 YR 3/4) coarse sandy loam [166] with 5% small stones.

The fill of [167] was in turn cut by another post-hole [169], which has similarities in terms of shape and fill with those of [149], [171] and [173]. Post-holes [149] and [169] are 2.00m apart, as are the posts for the recent cattle stalls; they may represent stalling for cattle in the area to the north of the clom wall, although it is not clear how [171] and [173] might have been used in this layout.

169 Oval post-hole, oriented N-S, 0.43 × 0.33m, 0.40m deep. Filled with a greyish-yellow-brown (10 YR 4/2) sandy clay loam [168] with 20% small to medium stones.
149 Oval post-hole, oriented NE-SW, 0.45 × 0.28m, 0.35m deep. Filled with a greyish-yellow-brown (10 YR 4/2) silty clay loam [148] with 50% small and medium stones.
171 Oval post-hole, oriented E-W, 0.49 × 0.33m, 0.33m deep. Filled with a dark brown (10 YR 3/4) sandy clay loam [170] with 20% very small and small stones.
173 Oval post-hole, oriented NW-SE, 0.42 × 0.30m, 0.30m deep. Filled with a dull yellowish-brown (10 YR 4/3) sandy clay loam [172] with 40% small stones.

Perhaps the most inviting group comprises four hollows [138; 139; 140; 141]. These can be divided into two probably successive pairs, each about 2.00m apart, firstly [138] and [140], and later [139] and [141], with their surviving traces of decayed wooden posts. They were all filled with overburden [92]. These too may represent stalls in some structure, which has otherwise left no trace, against the outer face of the clom wall.

138 Hollow, 0.27m in diameter and 0.10m deep.
140 Hollow, 0.30m in diameter and 0.22m deep.
139 Oval hollow, oriented N-S, 0.25 × 0.22m, 0.08m deep.
141 Oval hollow, oriented NW-SE, 0.32 × 0.27m, 0.11m deep.

Two further features [185] and [135] form a less convincing pair, though they could possibly have fulfilled the same purpose against a hypothetical northern wall; the post-hole [173], discussed above, is also 2.00m from [185] and may have formed part of such a layout.

185 Oval post-hole, oriented NW-SE, 0.30 × 0.25m, 0.22m deep. Filled with dark brown (10 YR 3/3) coarse sandy loam [184] with 20% small to medium-sized stones.
135 Hollow, 0.24m in diameter and 0.08m deep. Filled with dark brown (10 YR 3/4) coarse sandy loam [134] with 30% very small and small stones.

If one or both of these groups were stalling, then the east-west trench [142] might have served as a drain between the two sets. This feature was 3.30m long and 0.55m wide, and between 0.01m and 0.08m deep, being deepest on the west and filled by overburden [92]. The length of about 1.50m to either side is not generous, but might have been adequate for early, unimproved beasts.

There are a number of stake-holes in the area [137; 143; 147; 150; 151; 160], many of them towards the south-west of the later Room 1. They do not form any particularly clear pattern, however, and may reflect *ad hoc* activity of no great duration. With the exception of [151] which was oval, 0.01 by 0.08m, the rest were all square, ranging in size from 0.05 to 0.12m. All were filled with overburden [92], except for [160] which had its own distinctive blackish fill. The latter was one of a group around post-hole [149], including [147], not all of which were contexted. It may also have distorted the profile of [149], in which case it may be later.

It is difficult to fit the two successive features [159] and [146] into any sequence at this level. They would seem to belong much more convincingly with the later group of features discussed in section 2.2.3(H) below, where they could be seen to mirror [109] and [85] towards the north of the area. It is not now clear whether this would have been stratigraphically possible.

159 Post-hole, 0.28m in diameter and 0.40m deep. Contained the remains of a square timber of 0.13m scantling [144], and a fill of dull yellowish-brown (10 YR 4/3) silty clay loam [158] with 20% small stones, containing a clay pipe of *c.* 1800–50 (Table 2.8).
146 Roughly semicircular pit against clom wall, 0.65m N-S × 0.60m, 0.45m deep. Filled with a dull yellowish-brown (10 YR 4/3) silty clay loam [145] with 20% small stones.

(ii) Features within the area of the clom building (Fig. 2.60)
All the features discussed here lie stratigraphically between [129] and the makeup [123/126] of the later pitched stone floor (see section 2.2.3(Eiii) below), within the area of the clom building. [129] was a dull yellowish-brown silty clay loam (10 YR 4/3) with an 80% admixture of small and medium stones, which may possibly have been truncated for the laying of the later pitched stone floor. The very mixed, mainly 18th-century, finds from it included pennies of 1730 and 1775 and another, unidentified, post-medieval coin (Co 9, 10, 11; see section 2.2.4(Aii) below), and a clay pipe fragment of *c.* 1880 (Table 2.8) which must be intrusive since the overlying 19th-century stone building was in place by that date. This layer, noticeably clayey in consistency, may well represent an area of flooring of much the same extent as that in the later clom building. It appears to rest in a cut [201], one edge of which ran close to the south side of the northern clom wall [63]. This cut may represent levelling of the floor area, and could relate to an earlier wall on the same alignment, or just possibly to the clom wall itself; unfortunately the foundation trench for the 19th-century stone building [74], a short distance to its south, made the relationships here less clear than they might be. All the features cut into [129] were rather scrappy and difficult to interpret:

190 Circular post-pit, 0.45m in diameter and 0.30m deep. Filled by a black (10 YR 1.7/1) silty clay loam [189] with small, medium and several large rounded stones, coal and mortar fragments.
192 Sub-circular post-pit, 0.25m in diameter and 0.15m deep. Filled by a black (10 YR 1.7/1) silty clay loam [191] with small, medium and a couple of large stones, coal and mortar fragments.
204 Circular pit, 0.35m in diameter and 0.10m deep. Filled by a black (10 YR 1.7/1) silty clay loam [203] with small and medium-sized rounded stones, and a small square patch of coal dust in its centre.
181 Spread of stones running E-W for 2.00m and N (at W end) for 0.60m. May form SW corner of something but very tenuous. Some stake-holes in area may be related.

Of more interest were three small areas of early pitched stone flooring [127], measuring (from the north) 0.90 by 0.60m, 1.60 by 0.70m and 1.10 by 0.65m, and 0.10m thick, laid irregularly in the south-west corner and along the southern part of western side of the area of the much later Room 2. This flooring incorporated a copper-alloy coin of 1773 (Co 8). It may perhaps reflect an area of wear in the surface [129], though it is possible, if unlikely, that an entire floor surface has been removed except for this portion.

Two further post-pits [210] and [236] were truncated by the construction trench [74] for the 19th-century stone building, leaving them without any clear relationship with [129]. They could relate to this phase or to the later clom building, or even pre-date [129] and the building activity altogether. It is however tempting to see [210] at least as a doorpost, or possibly a structural member, relating to the clom building, while [236] may be part of the partition [68] visible in the pitched stone floor [41] of that building (see section 2.2.3(Eiv) below).

210 Oval post-pit, running NW-SE, 0.70 × 0.40m, 0.35m deep. Contained post-impression [206], 0.22m square and 0.28m deep, infilled by a black (10 YR 2/1) silty clay loam [205] with small stones, charcoal fragments and traces of the original wood of the post. Area around post filled by a dull yellowish-brown (10 YR 4/3) silty clay loam [209] with 50–60% small, medium and (mostly) large rounded stones.
236 Oval post-pit, running E-W, 0.50 × 0.40m, 0.30m deep. Filled by a dark brown (10 YR 3/4) silty clay loam [235] with gravel, small, medium and large rounded stones, large fragments of shale and mortar. Contained post-medieval pottery of late 18th century +.

(D) The clom and stone buildings

(i) The principal phases of development

The identifiable remains of post-Roman buildings on the site fell into two main groups: those which were still standing until their demolition immediately before the excavation, and those which clearly predated them and were only identifiable from below-ground remains. The latter are referred to here as 'the clom building' and 'the post-medieval cellar' (Fig. 2.62) and the former as 'the stone building' (Fig. 2.68).

The sequence of construction was clearly somewhat more complex, as could be seen both from an analysis of the standing fabric of the stone building and from the archaeology. The clom building, apparently occupying roughly the area of Room 2 and the southernmost portion of Room 1 of the stone structure, was the earliest clearly definable post-Roman building on the site, although there are hints, discussed in section 2.2.3(Bv) above, that its southern wall perpetuated the line of a previously existing boundary. The post-medieval cellar, to its west, was assumed to be associated with it in some way, although any relationship between the two had been destroyed by the deep footings of the stone-built Room 2 and it is somewhat difficult to combine building and post-medieval cellar into a sensible plan.

The earliest part of the stone building, the southern section containing Rooms 3 and 4, was dated on architectural grounds to the mid 19th century, and appeared to have been built as an annexe to the clom building while the latter was still standing. A photograph dating to the early 20th century shows this end of the building with two storeys, but by the 1950s it had been reduced to a single storey (see Figs 2.71 and 2.72). At some point in the 1860s or 1870s, judging from the architectural evidence, the clom building was replaced by the stone-built Room 2, while at a later date again, probably in the 1890s, the northern extension was built. A number of features both inside and outside the building appear to relate to activity associated with its different periods; these are discussed in the appropriate places below.

(ii) The principal dating evidence

Relatively little direct dating evidence was obtained from the buildings themselves, although the associated features contained some helpful clues. Post-medieval pottery of various types, clay pipe fragments, glass and coins all contributed to the overall picture (see Table 2.28).

The clom building was given a TPQ by a penny of 1775 (Co 9), from a context beneath its floor; the foundation trench for its southern wall [156] contained pottery of 18th-century or later date (Type 12), although this was not convincingly sealed. The coin was in fact found in the makeup [129] underlying a fragment of earlier pitched stone flooring [127], from the base of which also came another similar coin of 1773 (Co 8); this fragment lay in the area of the south-west corner of the later, stone-built Room 2. Another deposit [123], which lay between the fragment and the main pitched stone floor [41] of the clom building, yielded a similar coin (Co 7) on which the date was illegible, though clearly of George II or III, as well as the remains of a copper-alloy button (not catalogued; object 10,003).

The dated coins are apparently associated with an earlier episode of flooring, which might just relate to the north wall [63] at least of the clom building, although the evidence for this is somewhat circumstantial (see section 2.2.3(Cii) above). The structure with the pitched stone floor [41], is not precisely dated, therefore, but the coin (Co 7) in the material [123] immediately underlying the main flooring, together with the need for rebuilding by mid-century, do however suggest that this phase of the building belongs to the early years of the 19th century at the latest. It is not shown on an estate map of 1803 (NLW Dolaucothi 2), though it is not clear whether all buildings extant at that time were included.

There was little specific evidence to date the post-medieval cellar. Its mixed fills, discussed in detail below, contained post-medieval pottery and fragments of clay pipe including one of *c.* 1900. The demands of the excavation did not require it to be bottomed, so that its flooring was not examined. Although it was tempting to try to relate it to the clom building, its eastern side does not marry up very happily with the most likely line of the building's western wall, which may well have overlapped it (although the alignment is the same); the remainder of the post-medieval cellar structure runs on a slightly different line. This might suggest that the post-medieval cellar was already out of use by the time the clom building was erected. The dating of the building elements of the later stone building is based on a report on the vernacular architecture of the standing buildings carried out before their demolition (see section 2.2.3(N) below).

(E) The clom building

(i) Construction details (Figs 2.62 and 2.63)

The clom building was visible in the excavation as two lengths of walling [63] and [75], 10.3 to 10.6m apart, consisting of large stones set in stiff buff clay, flanking the north and south edges of a spread of carefully-laid pitched stone flooring (Fig. 2.63). The walls and floor were only visible within the area of the later stone building, and no trace was seen of any east or west wall. While it is possible that the clom building was somewhat more extensive, the pattern on the pitched stone floor seemed to suggest a room of about the same width (5.20m inside the walls) as the later building, besides which it would be simplest to assume that the stone building had cut away the remaining walls. The structure would in this case

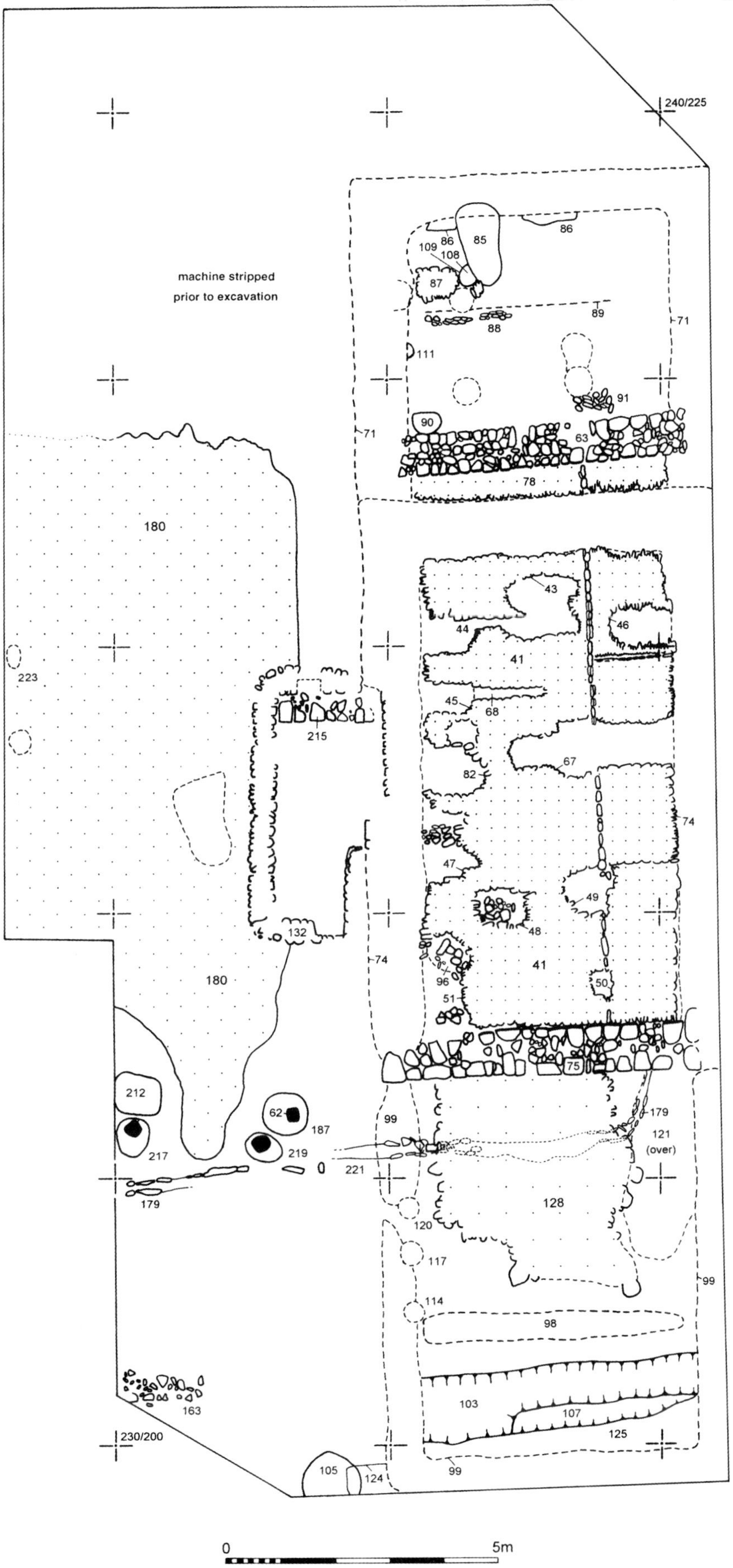

Fig. 2.62 Composite of clom building, post-medieval cellar and yard

Fig. 2.63 Clom walls and pitched stone floor, looking south

have been somewhat off square, as the clom walls lie at a slight angle to the later stone ones, while the design in the floor appears to conform more closely, though not completely, to the north-south alignment of the latter. The size of the southern extension is another factor suggesting that the clom building was similar in width to its successor.

(ii) The north and south walls (Fig. 2.62)

The north wall of the structure [63] was the first to be recognised, since it lay within the area of the later Room 1, its southern face being 0.50 to 0.80m north of the northern side of the later stone wall adjoining Room 2. It was 0.85m wide, and consisted of stones with slightly rounded edges set somewhat roughly in considerable amounts of stiff buff clay, with some traces of mortar. One sherd of 17th-century or later pottery (Type 11) came from its fabric. The clay material appeared to continue above the stones, being first recognised as a deposit of dull yellowish-brown (10 YR 5/4) sandy clay loam with 20% small stones [54], containing pottery of the late 18th century or later (Types 7, 11, 13, 14). This would therefore appear to be a clom, or clay-built, structure on stone footings. This wall was set in a foundation trench [153], which was only seen for 0.05 to 0.20m to the north of the wall, where it had been infilled with a brown (10 YR 4/4) sandy clay loam [152] with 70% small and medium sized stones, containing some fragments of post-medieval glass and iron.

The south wall [75] directly underlay the southern wall of the later 19th-century Room 2, and had been partially incorporated into it, with the majority of its extra width left on the north of the later wall except at its west end, where it projected slightly to the south (see Fig. 2.70); this compensated for the difference in angle between the two walls. It was of similar construction to the northern wall [63], 0.85m wide and bonded with considerable amounts of stiff buff clay and some mortar. It had been set into a shallow foundation trench [156] 1.00m wide, in which there was little obvious backfill apart from the wall itself. Finds from this trench were a fragment of 18th-century or later pottery (Type 11), and residual Roman material (CP1).

(iii) The pitched stone and mortar floors (Figs 2.62 to 2.64)

An extensive spread of carefully-laid pitched stone flooring, measuring 11.00m north-south by at least 4.50m east-west, ([41] in the later Room 2, [78] in Room 1) was clearly associated with the clom building, since it abutted both its northern, and the eastern half of its southern, wall (Fig. 2.63). The flooring had been cut away by the foundation trenches [74] of the later 19th-century east and west walls, and divided into two by the trench [74] for the northern stone wall of Room 2, hence the two numbers. The floor had been laid predominantly east-west, with the easternmost 1.50m divided off by a single band of stones set north-south. To the east of this band, the northernmost 3.50m and the southernmost 3.00m had been laid north-south, while the remainder was laid east-west. The flooring had been set into a gravelly makeup, numbered [126] to the north of the 19th-century foundation [74], and [123] to its south, consisting of 95% very small, small and medium stones in a matrix of brown loamy coarse sand (10 YR 4/4); this deposit yielded pottery dating to the later 18th century or later (Types 7, 14), with clay pipe fragments dated to 1750+ (Table 2.8) and post-medieval glass (Table 2.10) and iron, as well as the illegible coin (Co 7) and copper-alloy button mentioned above.

At a later stage, while the clom walls were still in use, the northern 5.00m of the pitched stone floor were covered in a spread of cement, forming a new floor ([39] in Room 2, [64] in Room 1) which ended on the south in a pronounced line (Fig. 2.64); this spread included a clay pipe fragment of *c.* 1850 (Table 2.8). To the south of this, an irregular band of softer mortar [40] extended for up to 1.5m.

(iv) The internal layout (Fig. 2.62)

The history of the clom building seems to fall broadly into two, or possibly three, phases, as reflected by the earlier pitched stone and later cement and mortar flooring. These floors themselves appear to give some clue to the internal layout of the building at their respective times, while additional information can be

Fig. 2.64 Mortar spread across pitched stone flooring of clom building, looking east

obtained from the disturbances in them. The interpretation of these disturbances is difficult because the surviving flooring of the succeeding later 19th-century Room 2 was very fragmentary, which makes it hard to be certain from what period many of them originated, since no later floor sealed them. Some of them appeared to belong with the later 19th-century stone building on the strength of their apparent relationship with the walls of that structure; their stratigraphic relationships were in most cases not very readily seen.

Three features [68; 67; 82] were undoubtedly sealed by either the cement floor [39] or the mortar floor [40]:

68 Gap running E-W, *c.* 1.35m long and 0.12m wide, in pitched stone floor [41] beneath cement [39] on W side, 4.20m from N wall [63]; traces of timber in void.

67 Linear disturbance running E-W, 3.00m long and 0.55–0.75m wide, in pitched stone floor [41] beneath and filled by mortar [40], running in from E side; N edge *c.* 5.00m from N wall [63]. Contained clay pipe *c.* 1800 (Table 2.8) and pottery of late 18th century + (Types 7, 14).

82 Linear disturbance running E-W, 1.30m long and 0.20–0.50m wide, in pitched stone floor [41] beneath and filled by mortar [40], running in from W side; N edge *c.* 5.00m from N wall [63]. Contained 18th-century pottery (Type 7), clay pipe *c.* 1850 (Table 2.8), and post-medieval iron and slag.

The gap [68], sealed by [39], probably represents a wooden partition, which is likely to be part of the original layout of the building relating to the pitched stone floor [41]. The post-hole [236], truncated and divorced from its context by the construction trench [74] of the later stone building (see section 2.2.3(Cii) above), may also be part of this feature. The pattern of the pitched stone floor suggests a layout with stalls at right angles to the western wall and an entrance on the east (see section 2.2.6(B) below).

[67] and [82], which were not of any appreciable depth, appear to be areas of wear in the pitched stone floor, which the mortar spread [40] may have been intended to repair. It is tempting to interpret [67] as wear around a central eastern doorway, while [82] may perhaps indicate the position of a further partition (regular spacing would demand one somewhere within it), although there was no obvious trace of one. Alternatively, it may represent wear around a central western doorway, although such an arrangement is not normal in buildings of the type suggested. The southern edge of the concrete floor [39] roughly follows the northern edge of these two disturbances, on the same alignment as the northern wall of the clom building. A post-hole (not numbered) at the north-eastern corner of [67] (0.25m to the north-west of the later [42h]) may be a doorpost or may relate to the southern edge of the concrete floor [39].

The east-west disturbance [45], 1.70m long and 0.10 to 0.50m wide, overlay the western end of [68]. It may perhaps represent the rotting and collapse of the timber, traces of which were found in the void further east, causing the concrete floor [39] to break up in this area; alternatively the floor may have respected an existing structure which remained in position when it was laid.

Two other areas of disturbance [44; 46] in the northern part of the floor appeared to relate primarily to the later 19th-century stone building (see section 2.2.3(Jiii) below), where posts had been set towards their southern sides. It seems likely, however, that [44] at least originated in this phase (features of different periods may have been conflated under the number), since the distinctive elongation at its eastern end would appear to define a partition similar to [68], and 1.35m from it. This is well within the size range for stalls for horses in this period, although rather narrow for a pair of cattle, even small ones. The apparent length of the intervening partitions would also tend to suggest stalling for horses. The disturbance [43], which cuts the concrete floor [39] is not readily explained, though it might be an area of wear. It ran east-west and measured 1.30 by 0.80m, adjoining [44] on its north. Its fill (not numbered separately) contained post-medieval pottery dated to the late 18th century onwards (Types 7, 14) and post-medieval iron.

There were six areas of disturbance in the southern part of the floor, [48] (discussed below), and five others:

47 Disturbed area *c.* 0.90m square. Lies immediately inside position of W door of later 19th-century stone building. Stones from the flooring have been relaid flat on the ground over the northern 0.30m of the area, suggesting repair of worn area.

96 Disturbed area, *c.* 0.60m in diameter, adjoining northern end of [51], although relationship with it not established. Fill (not numbered separately) contained post-medieval pottery dating to late 18th century onwards (Types 3, 7, 9). Some stones laid flat in fill.

49 Hollow running N-S, 0.70 × 0.65m, *c.* 2.60m from S wall. Fill (not separately numbered) contained post-medieval pottery of the late 18th century onwards (Types 10, 14) and post-medieval iron.

50 Hollow running N-S, 0.36 × 0.33m, *c.* 0.70m from S wall. Fill not separately numbered.

51 Irregular disturbed area, 1.70m N-S × 1.00m (measured from later walls) in SW corner of later Room 2. Some stones laid flat towards S edge.

Although there is little evidence to date these features, or to prove their relationship, it is possible to suggest that [47] and [96] may be part of a scheme of stalling at right angles to the western wall as already hypothesised (although the intervals between the partitions would vary slightly; these two may relate to a rearrangement of the features when the concrete floor was laid), with [49] and [50] representing wear at the entrances to the southern two stalls. [51] and [48] are however more difficult to fit into this scheme. [51] occupies the same portion of the south-west corner of the building as the southern part of the earlier surface [127], which itself had possibly been repaired in this area. It is tempting to see this as indicating the position of a doorway, an arrangement not unknown in early 19th-century farm buildings, which might suggest that the building did not contain a complete set of stalls along the west wall; it is hard to see why this part of a stall might become so worn. [50] becomes in this case less easy to explain; the area might have served as a stall at some stage in the building's life, or the damage might have been caused by foot traffic. Such a doorway might be convenient for access to the post-medieval cellar, always assuming that the two structures were in use together.

A regular set of stalls might suggest that another partition should lie in the portion of the floor at the north end cut away by the later 19th-century foundation trench [74], while a doorway on the south-west might have been balanced by another at the north-west. This would give up to seven stalls without the hypothetical western doorways, or five with them, although the southern part of the building might originally have incorporated one or more loose-boxes, and only later have been adapted as stalls, which might explain the irregular spacing of the disturbances. The northern part of the building might also have been laid out or later adapted as a loose-box, in which case [46] might mark the position of an eastern partition meeting the western one.

The concrete floor [39] does not conform to the suggested arrangement, since its edge would fall halfway across one of the hypothetical stalls. This is likely to indicate that the internal layout of the building had changed by, or was changed at, this stage, although it is difficult to identify what the new layout might have been, since even quite substantial timber stalling need not necessarily leave any archaeological trace. [48] might relate to a new layout, but it would equally fit into the later building (see below). It is possible that this rearrangement marks at least a partial change from horses (=arable) to cattle, a widespread trend in the mid to late 19th century. The stable may have related to the pub next door, and have been effectively replaced by the construction of the coach house and stables opposite in 1833, as commemorated on its date-stone.

(F) The post-medieval cellar (Figs 2.62 and 2.65)
The post-medieval cellar lay to the west of the later 19th-century Room 2, and the masonry walling [132] on its western, and parts of its northern and southern sides became visible as soon as the recent farmyard

Fig. 2.65 The post-medieval cellar, looking north

surface [32] was removed. The eastern side was revealed in the bottom of the later 19th-century foundation trench, by which it had been cut. The structure measured internally 4.10m north-south by a maximum of 2.00m east-west in its northern half; its southern half was reduced to only 1.30m wide as a result of an abrupt change of alignment of the eastern wall at this end; the extra width may have been intended to accommodate an access stair or ladder, probably from the south. The pit into which the walls were laid [222] also changed in alignment here. The walls were for the most part 0.50m thick, except on the northern half of the eastern side, where there appeared to be little more than a facing of stone against the side of the pit; this may have been aligned with the outer face of the clom building, though the other evidence might tend to suggest that this projected further to the west.

The feature was not bottomed, but the walls were followed down to a depth of *c.* 1.00m and later removed to this depth. There were traces of an alcove 0.40m wide and 0.30m deep on the north side, towards the western end, and a more substantial recess 0.60m wide and 0.40m deep occupying the eastern half of the southern side. This latter might have served as a light well, or alternatively have permitted access by ladder; the walling on its southern side was again little more than a facing. A further 0.5m-deep stretch of walling [215] had been built across the whole of the northern end, against the original wall; this was found only at a depth of *c.* 1.00m and was not pursued very much deeper. It was of similar construction to the walls of the clom building, being bonded with stiff buff clay, but it definitely met the outer walls in straight joints, so must have been an addition to the structure rather than the remains of an earlier phase. Assuming an overall depth to the structure of *c.* 2.00m, this would make a useful shelf, but no upper surface survived; such a surface could of course have been made of wood.

There was no clue as to how this post-medieval cellar had been roofed, or, indeed, as to whether there had ever been any superstructure above it. The irregular shape and the insubstantial nature of some parts of the walls would seem to militate against this, but a free-standing timber building might have left no trace.

As found, the post-medieval cellar was largely filled with a very mixed deposit [131], with a rather stonier deposit [130] at the highest level on the eastern side. Both these deposits might be interpreted as demolition debris; [131] might have come from the clom building.

131 90% small, medium and large stones in a matrix of dark greyish-yellow sandy clay loam (2.5 Y 4/2). Contained later post-medieval pottery (Types 3, 11, 14), clay pipe fragments including one of 1900 which may be intrusive (Table 2.8), bone, iron slag, pieces of brick and tile and a residual samian sherd.

130 95% very small, small and medium stones in a brownish-black matrix of coarse sandy loam (10 YR 3/2). Contained post-medieval pottery dating up to 18th century (Types 3, 7) and a 17th-century clay pipe fragment (Table 2.8).

(G) Features to the south and west of the clom building

(i) The stone spread and drain (Figs 2.62 and 2.66)
Adjoining the southern wall of the clom building on its south side was an irregular area of rough stone flooring with gravel infill [128] (Fig. 2.66). This measured *c.* 4.00m square, and occupied much of the central area of the later Room 3, although it lay beneath the deposit [121] and [122] into which the latter's foundations were dug. It overlay the earlier gravel spread [220] discussed in section 2.2.3(Bv) above, and it too may perhaps be seen as an area of hard standing; alternatively, it may be a floor within a precursor of the 19th-century stone structure. Such a precursor may be defined by the trenches [107] and [103], discussed in section 2.2.3(Gii) below, which lie a short distance to its south. From [128] came post-medieval pottery dating to the late 18th century or later (Types 9, 11, 13, 14), fragments of clay pipe (Table 2.8) and bottle glass (Table 2.10), and some residual Roman glass.

A 0.2m-wide drain [221] appeared to have been constructed in conjunction with this surface (see Fig. 2.59). Starting from a point near the south wall of the clom building, *c.* 1.00m to the west of the south-east corner of the later stone-built Room 2, it turned westwards to run *c.* 1.20m from the wall, gradually trending further southwards as it made its way to the western edge of the excavation. This drain had been cut into the underlying gravel spread [220] and the infill [243] (a dull yellowish-brown (10 YR 4/3) silty clay loam) across the gravel [188], and had been roughly lined with stones set on edge to either side [179]. It had been capped with rough, flat stones, presumably forming part of the surface. From it came post-medieval pottery (Types 3, 6, 7, 9, 11) and glass (Table 2.10), and clay pipe fragments dating down to 1780 or beyond (CTP 8).

(ii) Trenches and further deposits (Fig. 2.62)
Running east-west across the area of the later Room 4 was a trench [107], on a similar alignment to the clom walls [63] and [75], *c.* 5.50m to the south of the southern one. It probably ran the whole width (5.20m) of the later room (although it could only be seen for the easternmost 3.50m) but was not visible outside it. It was filled by a black (10 YR 2/1) silty clay loam deposit [106] with small and medium-sized stones, mortar and shale fragments, from which came post-medieval pottery dating down to the late 18th century or later (Types 3, 14; LP 11, 12). This trench had been truncated

Fig. 2.66 Stone spread [128], looking west

by [103], interpreted as a robber trench, which ran on the same alignment for much of its length, and was 0.70m wide and 0.10m deep, filled with [102] a brownish-black (10 YR 2/3) silty clay loam with small and medium-sized rounded stones, slate and mortar fragments. This fill yielded post-medieval pottery dating down to the late 18th century or later (Types 3, 14), and fragments of clay pipe (Table 2.8). This alignment may relate to the stub of walling [163], measuring 1.50 by 0.50m, which ran east-west at the south-west corner of the excavated area.

To the south of the trenches [107] and [103] lay a dense, compact pebble surface [125], which seemed to have been cut by them, although it was not apparently present to their north. This surface, which yielded pottery dating down to the 18th century, may have belonged to a pathway beyond the suggested building; its full extent was not seen as it ran into the southern and eastern edges of the excavated area.

Between the trenches and the southern wall of the clom building [75], within the area of the later building, lay two deposits which may relate to the hypothetical structure defined by the trench [107/103]. These were [121], confined to the eastern part of the area, with a western edge roughly corresponding to the eastern edge of [128], and [122], which covered the remainder of the area including the south-eastern corner.

121 Brownish-black loamy coarse sand (10 YR 3/1), with 30% small and medium-sized stones and a considerable proportion of pulverised coal. Contained post-medieval pottery dating down to late 18th century onwards (Types 1, 7, 14; LP 3), a residual fragment of clay pipe (Table 2.8), glass fragments (Table 2.10), a piece of iron slag and a possibly polished flat pebble.

122 Brownish-black silty clay loam (10 YR 3/2), with 60% small and medium-sized stones. Contained post-medieval pottery dating down to late 18th century onwards (Types 7, 9, 14; LP 27), fragments of clay pipes (Table 2.8), and post-medieval glass (Table 2.10).

(iii) The gravel yard surfaces and associated post-holes (Fig. 2.62)

The farmyard surfaces proved somewhat difficult to excavate, since they had clearly been repaired repeatedly. It was therefore not at all easy to define where each surface began and ended, both in horizontal and in vertical terms; they did not often lift neatly off one another. The sequence eventually established does, however, make sense. The first coherent surface consists of [180/223], the two numbers reflecting the excavation strategy, with [180] first identified towards the southern end, while [223] was a deposit in the 1987 trial trench on the west of the site which appeared on examination to merge into [180]. This surface, which consisted of small, medium and large rounded stones in an olive-brown

(10 YR 4/3) matrix, had been built up around the post-medieval cellar wall [132], but its relationship with the building on the east was not clear because it had been truncated by machine before excavation began. It contained pottery dating down to the 18th century or later (Types 3, 5, 7, 11, 12; LP 21), and residual clay pipe fragments (Table 2.8).

Four post-holes associated with this surface, [217], [219], [212] and [187], appeared to represent successive versions of the same gateway, towards the south of the yard area, roughly in line with, or a little to the south of, the south wall of the clom building [75]. [212] and [187] were probably the later of the two, since the post-pipe [62] in [187] survived to a high level and appeared to have remained in use during the following phase, when [212] was superseded in the same position by the post-hole [66] (see section 2.2.3(L) below). The drain [221], discussed above in section 2.2.3(Gi), crossed the area immediately to the south of this gateway.

217 Circular pit, 0.65m in diameter and 0.70m deep. Contained post-impression [208], 0.28m square and 0.35m deep, infilled by a brownish-black (10 YR 2/2) loam [207] with small stones. Area around post filled by a brown (10 YR 4/4) silty clay loam [216] with gravel, small and medium sized angular and rounded stones.
219 Sub-circular pit, 0.40m in diameter and 0.70m deep. Contained post-impression [214], 0.30m diameter, 0.50m deep, infilled by void [213], sealed beneath surface [79]. Area around post filled by a brownish-black (10 YR 3/1) silty clay loam [218] with gravel, small and medium sized stones.
212 Sub-circular pit, 0.86m in diameter and 0.70m deep. Filled by a mixed brownish-black (10 YR 3/2) silty clay loam [211] with 50% small, medium and two large stones, mortar fragments and coal specks. Contained post-medieval pottery of 17th century onwards (Type 11).
187 Circular pit, 0.80m in diameter and 0.80m deep. Contained post-pipe [62], 0.25m square and 0.85m deep, surviving as void to high level. Filled around post by a brownish-black (10 YR 2/2) silty clay loam [186] with 80–90% gravel, small and medium sized rounded and angular stones, mortar fragments and specks of coal. Contained post-medieval pot of 17th century onwards (Types 3, 11), of clay pipe *c.* 1660 (CTP 1), residual Roman material.

(iv) The stone platform (see Fig. 2.68)
A stone platform [80], 3.80m north-south by 1.60m and 0.20m thick, was found abutting the west side of the later 19th-century stone building, next to the southern wall of Room 2 and overlapping the area of Room 2 and Room 3 about equally. It was built of shale slabs and boulders in a matrix of brownish-black (10 YR 2/2) silty clay loam. This probably represents the footing for a flight of external steps. It appears to conform better to the line hypothesised for the clom building on this side than to that of the stone-built Room 2, which differs slightly, though the damage by the later water pipe trench [57], which cut it, makes it hard to be sure. If it does belong with the clom building, it points strongly to the existence of an earlier southern extension, as suggested by trench [107/103] discussed in section 2.2.3(Gii) above; it is not clear whether the steps served the clom building or the southern extension. It is likely that either the same stair, or at any rate one on the same base, later served the upper floor originally provided above the later stone-built Rooms 3 and 4, and possibly also Room 2. Stratigraphically, the base overlay the fill [186] of one of the gatepost holes [187] apparently associated with the clom building (see section 2.2.3(Giii) above) and was cut by the water pipe trench [57] (see section 2.2.3(M) below).

(H) Features to the north of the clom building (Figs 2.62 and 2.67)
In the area north of the clom building later occupied by the late 19th-century stone-built Room 1, a further group of features had been cut into the deposit [92], which had sealed the earlier group lying at the same stratigraphic level as the north wall of the clom building [63] (see section 2.2.3(Ci) above). These later features still clearly pre-date the overlying stone-built Room 1, the foundations of which had been cut into the makeup [52] which covered them. While both they and the overlying deposit [52] all appear to respect the clom wall, suggesting that they relate to the period when the clom building was still standing, it is theoretically possible that some of them may date to the period between the construction of the stone-built Room 2 in the 1860s or 70s and that of the northern extension in the 1890s.

The makeup [92] was a loose gravelly deposit consisting of 95% very small and small stones in a matrix of dull yellowish-brown loamy coarse sand (10 YR 4/3). It contained medieval and post-medieval pottery dating down to the late 18th century or later (Types 3, 4, 6, 12, 14; LP 19, 22), fragments of clay pipe (Table 2.8), a copper-alloy ring (Ae 2), and a residual fragment of amphora. The south side of this deposit appeared to adjoin the northern clom wall.

The majority of the features at this level appear to relate to a structure at this north end of the site, pre-dating the later Room 1. Whether this would have been the same structure to which the earlier features, discussed in section 2.2.3(Ci) above, belonged or not, is not clear, as no trace was found of any walls.

Many of the remains are isolated patches of flooring. There are two areas of pitched stone flooring [87] and [88], possibly associated with the tenuous traces of a timber partition [89], running east-west parallel with the clom wall and 2.30m from it. These may or may not be contemporary with the roughly-laid stone patch [91].

Fig. 2.67 Area to north of building at level of [92], showing clom and later stone walls, looking south

87 Patch of pitched stone flooring towards NW of later Room 1, oriented E-W, 1.10 × 0.70m. Laid N-S but finished along western side of northern edge by three stones laid E-W.
88 Line of pitched stone flooring running E-W to S of possible timber partition [89]. Laid E-W.
91 Small area of pitched stone flooring laid in irregular herringbone configuration, immediately to N of, but not obviously adjoining, clom wall [63].

Three areas of mortar flooring were also found, [90] adjoining the clom wall, and the comparable pair [86] along the northern side of the later Room 1, cut away by its foundation trench. There was some hint during excavation that [90] overlay the herringbone flooring [91]; this would suggest that the sequence on this side of the wall mirrored that on the south, where pitched stone flooring was succeeded by mortar.

90 Part of mortar spread adjoining N side of wall [63]. Originally extended further along the wall than planned, also in other patches further along.
86 Two patches of mortar spread, 0.60 × 0.20m (W) and 1.00 × 0.20m (E). Comparable with [90]. May lie just inside earlier wall on later line, but no trace of such remains.

The pitched stone flooring [87] had been cut by a post-hole [109]. This in its turn had been cut by a substantial pit [85], which had also cut the mortar spread [86], The purpose of this pit is not clear.

109 Oval post-hole, oriented N-S, 0.44 × 0.25m, 0.60m deep. Filled by a dull yellowish-brown (10 YR 4/3) sandy clay loam [108] with 60% very small and small stones. Central area of base 0.15 × 0.17m (truncated), 0.10m deeper than remainder, suggesting a post driven in.
85 Large oval pit, oriented N-S, 1.10 × 0.70m, 0.65m deep. Filled by remains of post [83], 0.20 × 0.13m, near southern end, and fill of dull yellowish-brown (10 YR 4/3) sandy clay loam [84] with 60% very small, small and medium-sized stones, containing one sherd of pottery dating to late 17th century onwards (Type 6; LP 22); this joined a sherd in underlying deposit [92], suggesting pit was backfilled with its own spoil. TPQ for [92] is late 18th century onwards.

One remaining feature was so severely cut away by the later foundation trench that little could be said about it.

111 Severely truncated by foundation trench [71] for 19th-century stone wall. Post-hole 0.25m in diameter, depth unclear. Filled by [110], consisting of 80% small stones in matrix of dull yellowish-brown (10 YR 4/3) loamy coarse sand.

All these features were overlain by a single layer [52] into which the construction trench for the late 19th-century stone-built Room 1 had been dug. It was very similar in description to [92], consisting of 95% very small and small stones and black slag in a

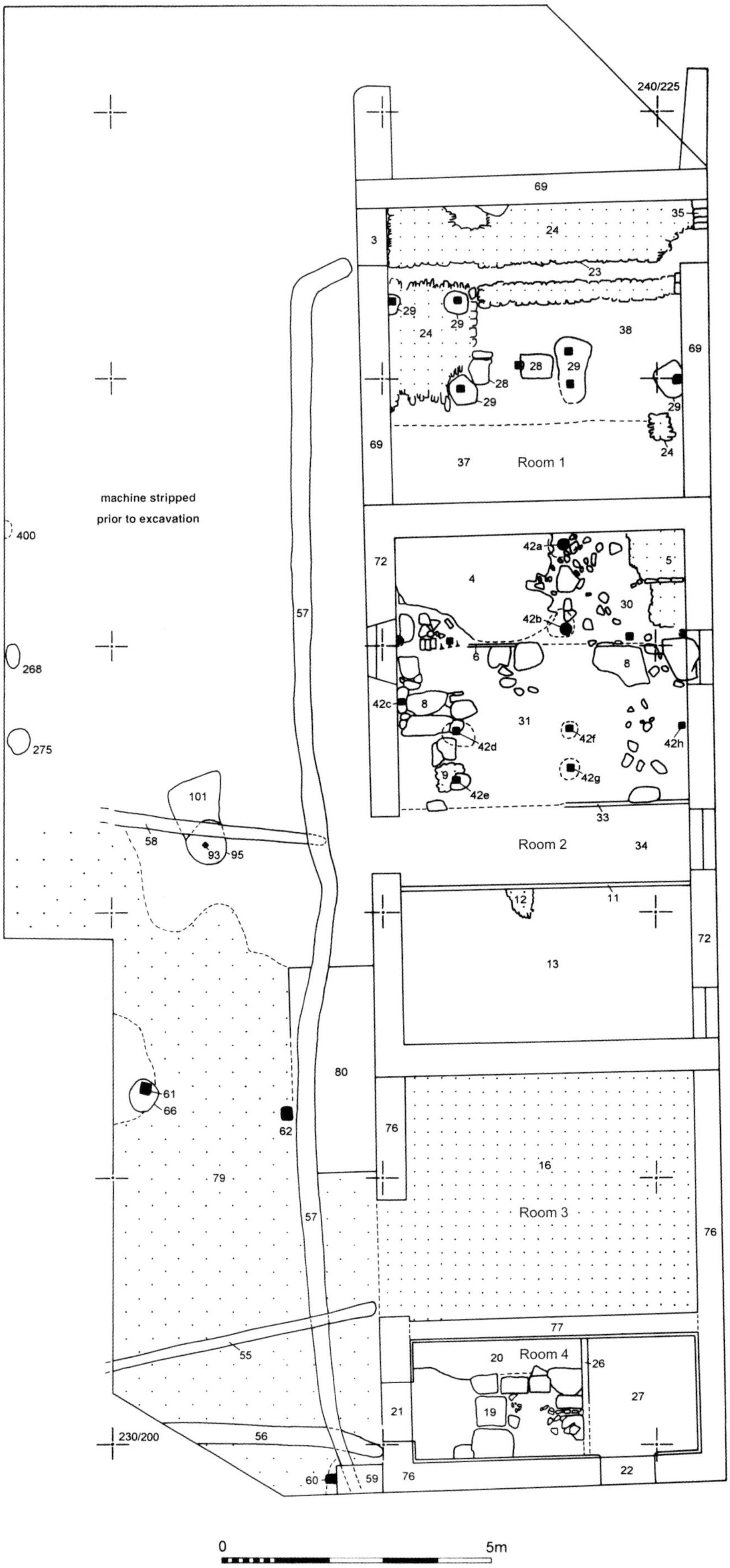

Fig. 2.68 Composite of later stone building and yard

matrix of dull yellowish-brown loamy coarse sand (10 YR 4/3). The admixture of slag increased towards the base of the deposit, with a concentrated layer *c.* 0.02m thick at the base itself. The deposit contained post-medieval pottery (Types 3, 14) and fragments of clay pipe, the latest of which dated to *c.* 1900 (CTP 10, 11). This deposit too appeared to respect the line of the clom wall; a slight overlap across it may reflect levelling when the wall was finally demolished. It is possible that this material was laid at this time in preparation for the construction of the new late 19th-century northern extension. The slag may reflect activity on the site, or may have been brought in from elsewhere.

(I) The southern extension to the clom building (Rooms 3 and 4)

(i) Construction details (Figs 2.68 to 2.72)

The archaeological evidence suggests that the southern part of the recently standing stone building, measuring 7.20m by 5.30m internally, was built as an extension to the clom structure, while the latter was still standing. The work may have involved the replacement of the southern clom wall with a stone one. Darlington and Noake in their vernacular report (see section 2.2.3(N) below) make this assumption, which appears to be borne out by the photographic evidence (see Figs 2.4 and 2.69), though the archaeology seen after demolition was inconclusive (Fig. 2.70). At the western end of the dividing wall, the quoins belonging to the footings of the clom building [75] appeared to remain *in situ*, but at the eastern end the position was less clear-cut. The photographs show that the portion of the western wall of the stone-built Room 2 which was later refaced in blockwork clearly started to the north of the cross-wall, suggesting that the failed portion of walling had been added to an existing corner. On the east, meanwhile, the northern corner of the southern extension is perforce neatly finished since for much of its surviving height it forms the southern jamb of the southernmost door of the later Room 2. The masonry in this area does, however, appear to have been repointed and possibly altered at some stage.

Fig. 2.70 Clom and later stone south walls, looking east

Fig. 2.69 Stone-built cowshed/combination range before demolition in 1989, looking west

The vernacular report suggests that this part of the stone building was the oldest, probably dating to the mid 19th century. There are some signs, discussed in section 2.2.3(Gii) above, that an earlier, slightly shorter, structure previously stood in this position, possibly also related to the clom building. A photograph of *c.* 1900 (Fig. 2.71) shows this part of the building as having two storeys, and it is likely that it was originally so constructed; latterly it was basically single-storeyed with storage in the roof-space (Fig. 2.72).

The building was built of quarried local shale laid in rough courses in mortar. The outer wall [76] stood in a foundation trench [99] which ran around most of the west, south and east of the structure, with the exception of a gap on the west corresponding with the northern half of the later entrance (see Fig. 2.62). From this trench came pottery dating to the late 18th century or later (Type 14). The wall itself was built 0.55m thick on the south and west, and 0.45m thick on the east, which may reflect the lack of windows, apart from the 'arrowloop' ventilator, at ground floor level on this side. The photograph of *c.* 1900 appears to show two windows on the east side at first floor level, as well as the gable door and window on the south believed by Darlington and Noake to be a later addition. The dividing wall on the north, whether erected in this phase or later, was up to 0.58m thick and overlay the stones [75] at the base of the southern wall of the clom building.

On the west, the post-holes [114; 120; 117], which may respectively have supported doors and held a doorstop, appeared to be cut by the foundation trench [99] and to belong relatively early in the sequence; the fact that the pitched stone floor [16] laid in Room 3 appeared to respect the northern and central timbers suggests that they may have been part of the original plan, although another doorstop had later been provided in a better central position.

114 Post-hole, 0.40m in diameter and 0.25m deep. Filled by the remains of a rotted timber post 0.10m square [112], and a brownish-black (10 YR 2/2) silty clay loam [113] with 70–80% small, medium and large rounded stones and mortar fragments. Most of this fill consisted of large packing stones for the post.
117 Post-hole, 0.40m in diameter and 0.25m deep. Filled by the remains of a rotted timber post 0.10m square [115], and a brownish-black (10 YR 2/2) silty clay loam [116] with small, medium and large rounded stones and mortar fragments. Most of this fill consisted of large packing stones for the post.
120 Post-hole, 0.40m in diameter and 0.25m deep. Filled by the remains of a rotted timber post 0.10m square [118], and a brownish-black (10 YR 2/2) silty clay loam [119] with small, medium and large rounded stones and mortar fragments. Most of this fill consisted of large packing stones for the post.

The threshold [21] for the small western door into Room 4, like that for the gable door [22], was set on the walling. The relationship of the western threshold to the internal flooring could not be established, as the surface had been damaged at this end, while the threshold of the gable door appeared to relate to the late concrete floor [17] discussed in the next section, suggesting that it may indeed be a later addition. This doorway does, however, open onto a part of the room which appears to have been partitioned off at an early stage.

(ii) Internal features and layout (Figs 2.68, 2.73 and 2.74)
This part of the building had been sub-divided into a northern room (3) measuring 5.30 by 4.50m and a southern room (4) measuring 5.25 by 2.30m (internal dimensions). The 0.35m-wide dividing wall [77] was constructed using roughly-coursed shale slabs with white mortar, set in a shallow foundation trench [98]. This wall met the outer wall of the extension at a straight joint on both east and west, suggesting a secondary feature, although the flooring adjoining it was very different in each of the resulting rooms, which may imply that it was part of the original plan of the extension. There was no sign of any doorway through the standing wall as seen in 1989.

In the northern of the two rooms, Room 3, a pitched stone floor [16], running east-west, and extending across the threshold on the west, had been neatly laid on a makeup [97] of dull yellowish-brown loamy coarse sand (10 YR 4/3) containing very small grits and stones (Fig. 2.73). This makeup was confined to the room, and overlay the earlier deposits [121] and [122] into which the foundations of the building had been dug. It contained post-medieval pottery dating down to at least the late 18th century (Types 3, 6, 9, 14) and fragments of clay pipe (Table 2.8), while the flooring itself yielded sherds of late 19th- or 20th-century glass (Table 2.10). These are likely to have become embedded during the life of the building, since this flooring remained in use until its demolition.

In the southern room (4), the picture was rather more complicated. The floors had been laid on a makeup [36] consisting of 80% mortar, bone fragments, and small and medium-sized stone in a matrix of very mixed dark greyish-yellow material (2.5 Y 4/2). This makeup also contained considerable quantities of broken bottles, some of them pulverised, including examples dating to the late 19th century or later (Table 2.10). This may relate to an episode of improvement in the early 20th century (see section 2.2.6(Ci) below), although no obvious trace remained of any earlier arrangement; such layers of broken bottles were apparently intended to deter rodents (see section 2.2.4(Eii) below). Other finds included post-medieval pottery (Types 3, 6, 7, 11, 14; LP 32), fragments of clay pipe (Table 2.8), and three larger post-medieval bottle fragments (Gl 20, 21, 22). This layer overlay the fill of the robber trench [102] of the possible earlier extension.

Fig. 2.71 The building in the early 20th century, looking north

Fig. 2.72 The same scene in 1989

Fig. 2.73 Pitched stone floor 16 in Room 3, looking east

A

B

Fig. 2.74 Flooring in Room 4: (A) earlier phase, looking west. (B) later phase, looking east

There were traces around the edges of the floors of a cement layer, apparently against the walls. This may represent the remains of rendering on the walls, or possibly of a surround to ease the cleaning of wet floors.

Before any of the flooring was laid, but subsequent to the cement layer just mentioned, a 0.10m-thick wooden partition [26] had been provided. This divided the area into two smaller rooms which were differently floored, that on the west measuring 3.15 by 2.30m, and that on the east, 2.30 by 2.00m. The partition itself ran for 1.20m from the northern wall, leaving an entrance between the rooms at its south end; there was a corresponding timber upright against the south wall. The eastern room was served by the gable end door, while the gable window lit the western.

There were two main phases of flooring in each room (Fig. 2.74). In the western room, the rather fragmentary earlier floor consisted of shale flagstones [19], which seemed to have been laid across only the southern 1.70m of the room; against the north wall was a band of concrete [20], 0.60m wide, which was covered, except for its easternmost 0.60m, in a white layer which appeared to be plaster. In the eastern room, the earlier floor [27] was a very sandy concrete with a thin skim of smoother concrete on the top. The flooring in the western room appeared to have been overlain at a later stage with a layer (possibly two) of concrete [18] which was recovered in a very fragmentary condition, while that in the eastern room was covered by a further layer of similar if somewhat harder material [17].

(J) The later 19th-century stone building (Room 2)

(i) Construction details (Fig. 2.68)

The building which superseded the clom structure, dated on architectural grounds to the 1860s or 70s, measured 9.60 by 5.30m internally. Its walls [72] were of roughly coursed shale slabs in white mortar. They had been set in a substantial foundation trench [74] on the north, west and east, while the wall on the south, which may indeed have been built as part of the southern structure already discussed, stood on the remains of the masonry at the foot of the clom wall [75]. The northern, and the northern 8.00m of the western, wall were 0.53m thick, the remainder of the western wall and the eastern wall 0.48m, while the southern wall reached a thickness of 0.58m. The foundation trench contained a mixed fill [73], visible only inside the room, consisting of 80% small and medium shale fragments in a matrix of olive-brown sandy clay loam (2.5 Y 4/4). From the wall and the deposit immediately around it came sherds of post-medieval pottery (Types 7, 10, 14) and glass (Gl 23), together with fragments of clay pipe including one of *c.* 1850–80 (CTP 12).

The building originally had three doorways on the east, all later reduced to windows, and a doorway and two narrow vertical ventilation slots on the west. The southernmost eastern doorway was at the extreme south corner of the room, and the remaining two lay at intervals of 3.30m between centres to its north. The doorway on the west was opposite the middle one of the three eastern doorways, while one of the ventilators lay opposite each of the other doors. The northern ventilator was later adapted as a window, while that on the south became hidden in relatively recent times (probably the 1960s) when the outer face of the western wall was refaced with concrete blocks (the failure of the wall might have been caused by the proximity of the post-medieval cellar on this side). The yellow brick arches above the doors provided the evidence for the dating. A pitching door, presumably for access to a hayloft in the southern section of the building, had been provided high in the southern end wall of the room.

There were traces, seen in 1989, of joist holes towards the tops of the standing walls of the building on the east and the west, which, together with an offset at much the same level on the southern wall, may suggest that it originally had a loft or upper floor. These holes lay at a slightly lower level than the trusses present in 1989. The photograph of the building in *c.* 1900 (Fig. 2.71) shows the walls extending externally above the brick door arches by about double the height of the bricks; even allowing for the effect of the addition of guttering in more recent years, some of this masonry would appear to have been removed. The pitching door may have been altered at a later date, since its threshold as recorded in 1989 lay below the offset, on which a joist would presumably have rested.

(ii) Relationship to clom walls (Figs 2.68, 2.70 and 2.75)

The southern wall of this building stood, as has already been mentioned, on the remains of the southern of the two clom walls (see Fig. 2.70). The earlier wall lay at a slight angle to the later work, with the excess thickness lying on the inner side of the later wall over much of the eastern half of its length. Towards the western end, however, the earlier wall tended to emerge to the south of the later, although this was not so pronounced; the lowest quoin stone of the south-west corner appeared to be part of the earlier clom wall.

The outer face of the north wall of the stone-built Room 2 lay between 0.50m (on the west) and 0.80m (on the east) to the south of the inner face of the earlier northern wall (Fig. 2.75). The substantial foundation trench [74] of the stone building had cut away a band between 1.10 and 1.20m wide of the pitched stone [41/78] and concrete [39/64] floors of the clom building.

(iii) Internal features and layout (Figs 2.68 and 2.75)
It is somewhat difficult to be precise about the layout of the stone building, prior to its most recent phases, since it had little coherent flooring to act as a guide. There was little obvious sign, however, that it ever contained the full complement of cattle stalls suggested by the design of the exterior.

Towards the northern end, evidence for a partition came from disturbances in the earlier floors (see Fig. 2.62); [44] and [46] appeared to form a pair, possibly indicating the line of one or more partitions across the building between 1.65m and 1.95m from its north end. Both these features contained holes for timber uprights, although these may have been secondary. It has already been suggested (see section 2.2.3(Eiv) above) that the narrow eastward continuation of [44] may in fact belong with the clom building, and it is possible that both features originated at that stage. Interestingly, however, the possible partition line suggested by the narrow eastward extension of [44], although it did not align with the uprights, did run clear of the edge of the northern doorway on the east of the later stone building, while the most recent division [6], to which the uprights may have related (although it appeared to run fractionally to the south of them), had partially encroached across the doorway, presumably after it had been blocked.

44 Disturbed area, 3.05m long and 0.15–0.75m wide, cut through concrete floor [39], on W side. Contained the remains of two wooden posts, beside the wall [72] and 1.00m from it; both posts 1.95m S of northern wall [72].
46 Disturbed area, 1.45m long and 0.70m wide, cut through concrete floor [39], on E side. Contained the remains of two wooden posts, beside the wall [72] and 1.00m from it; both posts 1.95m S of northern wall [72].
6 Timber partition, 4.10m long and 0.10m wide, 2.10m from north wall of room, with gap at W end. Divides concrete flooring [4] and pitched stone flooring [5] from flagged area [8].

Immediately before its demolition, Room 2 was open from end to end, but the flooring at the northern end suggested that the portion to the north of the partition [6], mentioned above, had been divided off and treated differently. This layout may perpetuate an earlier arrangement involving another partition between [44] and [46]. The line of this partition corresponds reasonably well with the position of one of the roof trusses in the building as it stood in 1989, though, as already mentioned, that does not appear to be the original roof of the building.

The flooring which was recovered at the northern end of the building was fairly fragmentary, but it appeared to indicate that the partitioned area had been divided unequally into two pens or boxes, *c.*

Fig. 2.75 Pitched stone floor of clom building with clom and later stone walls and post-medieval cellar, looking south

3.10m by 2.10m on the west and *c.* 2.20m by 2.10m on the east. The western pen or box had been floored with concrete [4], while the eastern had a pitched stone floor [5] running north-south, with traces of a central drain running east-west across its eastern half. Two post-holes [42a] and [42b], containing the remnants of round timbers set in concrete, may well have supported a partition between the pens or boxes, while the pairs of timbers within [44] and [46] respectively may represent doorways. These may be secondary to the construction of the partition [6] which seems originally to have been broken only on the west. It seems unlikely, considering their different methods of construction, that the two partitions (east-west and north-south) were contemporary; the east-west partition [6] is likely to have been the earlier, in some form. Certainly the 0.30m-thick makeup [30] beneath both floors, consisting of 80% small, medium and large stones in a matrix of dark brown (10 YR 3/3) loamy coarse sand, was indistinguishable. It contained post-medieval pottery dating to the late 18th century or later (Types 9, 10, 11, 14), and fragments of post-medieval glass (Table 2.10); a coin of 1957 (Co 4) may reflect a loss through a gap in the flooring, or might suggest that, as in the area to the south, the cement was laid in the 1960s. At some point, a drain had been laid through the western wall of the building from the broken south-west corner of the cement floor [4].

To the south of the partition [6], a fragmentary shale flagstone floor [8] had been laid on a makeup [31] 0.30m thick, consisting of 80% small, medium and large stones and some tile fragments in a matrix of very mixed dark brown sandy clay loam (10 YR 3/3). This makeup contained sherds of post-medieval pottery (Types 11, 14), a fragment of late 19th-century glass (Table 2.10) and a piece of clay pipe (Table 2.8). The flagstones [8] abutted, and therefore must have post-dated, the blocking wall of the northernmost doorway on the east. They survived best, and may only ever have been laid, to the north and west of the area defined by the post-holes [42c-h], which indicate the position of a set of stalling for cattle.

These stalls would appear to have faced north, though at first sight they do not appear to be far enough north of the cross-passage to have accommodated the full length of the beasts. Immediately south of the western window, however, an additional concrete-set post [42c] against the western wall of the building appears to relate to these stalls, all but the easternmost post of which [42h] were also set in concrete ([42d–g]). Post [42c] may well mark the front line of the stalls, the front face of which could otherwise have been effectively free-standing, or secured only to the walls. This would leave the remaining post-holes to represent only the lines of the partitions to the rear. Such an arrangement would have had sufficient room for five beasts, though the paving [8] suggests that the position of the fifth standing on the western side was occupied by a passage, at any rate after the northern door on the east was blocked.

A set of stalls for cows facing north, probably, to judge from the evidence of the concrete floor [7] (see below), the attenuated remains of the same set, was recorded before demolition in 1989, when they were tentatively dated to the late 19th or early 20th century (see Fig. 2.80; see section 2.2.3(N) below). These stalls consisted of a partition of edge-to-edge timber planks running east-west, *c.* 1.00m high, roughly in line with the southern edge of the northern door. This partition extended from the eastern wall across the width of two paired stalls, leaving a gap on the west, and was anchored at the top into the eastern wall. Halfway along and at its western end, subsidiary partitions, also of edge-to-edge planks, at right angles to the main line, ran southwards for about 0.40 to 0.50m, each finishing in a stouter post at the south end (probably the posts [42d] and [42f]). The floor area between the small partitions, to the rear of the main line, was occupied by a low manger, which had been rebuilt in concrete but would probably originally have been of wood or of stone, like that in the northern extension (see section 2.2.3(Kii) below). The rear of this manger met the eastern wall at another upright post, probably [42h].

A small area, 0.50m square, around the south-western post of the stalls [42e], of which the concrete bedding was particularly prominent, had been finished in pitched stone flooring running east-west [9]; it is not clear whether this was a repair or an isolated remnant of an earlier surface. The lower part of [42d] was inadvertently renumbered as [81] at a lower level. At a later date, overlying [9] and some of the flagstones [8], a surface of concrete [7] had been laid between the west of the stalls and the west wall, and across the area of the stalls themselves. As seen after the building was demolished, it occupied much of the area of the western pair in more or less broken form, but it is clear from the photograph in the vernacular report (see Fig. 2.80) that it did previously extend right across both sets of stalls; clearly it had been truncated at the time of demolition. Darlington and Noake state that the stalls were reset into a concrete floor in the 1960s, though the work seems less ambitious than alterations carried out in the northern extension at that date, and may date back to the 1950s or even earlier. The concrete certainly seemed to have been laid around the posts [42d–h] while they were still standing, which suggests that the rear projections of the stall partitions ([42e] and [42g]) were still in position at the time this work was done. The apparent removal of the western part of the shale band [33] may date to the laying of this concrete surface. No evidence was found of any earlier stalls, although it is possible that a previous set has left no archaeological trace (see section 2.2.6(Cii) below).

The area of the cross-passage, between the central doors, was defined by two kerbs 1.45m apart, made of shale slabs set on edge; on the north was [33], of which only the eastern 2.30m was recovered, while on the south, [11] ran the full width of the building. The stones of both were 0.30m deep and 0.10 thick. The space between them was filled with [34], an infill with 80% small and medium stones in a matrix of olive-brown coarse sandy loam (2.5 Y 4/4), containing post-medieval pottery dating down to the late 18th century or beyond (Types 7, 11, 14). Above the infill, a hard concrete floor [10], 1.45m wide and 0.10m thick, had been laid; this floor extended from the blockage of the eastern doorway out across the threshold of the western door, and may well also have been laid in the 1960s.

To the south of the cross-passage, with the exception of a fragment [12] of pitched stone flooring *c.* 0.50m square, laid north-south adjacent to the shale partition [11], no coherent traces of flooring survived above the makeup [13] by the time the excavation started. The vernacular report, however, (see section 2.2.3(Ni) below) refers to part of a flagstone floor in the south-west corner of the building, while the excavators recollect there being some constraint on the size and position of TH7 at the time of the trial trenching in 1987 (see section 2.2.1 above). This suggests that the surface was truncated during the demolition of the building. The makeup [13] was 0.30m thick, as elsewhere in the building, and consisted of 60% large stones, brick and tile fragments with some smaller stones, in a mixed matrix of dull yellowish-brown sandy clay loam (10 YR 4/3). There were a number of mortar lenses within this deposit, which yielded post-medieval pottery (Types 7, 12, 14) and two fragments of glass dating to the late 19th or 20th century (Table 2.10), which may have become trampled in or may reflect a possible early 20th-century phase of improvement (see section 2.2.3(M) below). It is possible that [48], a disturbance *c.* 2.40m from the south wall, running east-west and measuring 0.70m by 0.60m, relates in some way to this building. It was only recognised in the pitched stone flooring [41] beneath [13], and its fill was not separately numbered. The traces of a post towards its north-east corner would be in about the right place, two cows' width from the western wall, to form part of a southern set of stalls for which there is otherwise no evidence; perhaps suggestively, this post-hole lies beneath the patch of flooring [12].

(K) The northern extension (Room 1)

(i) Construction details (Fig. 2.68)

The clom wall [63] may have remained visible in some form after the construction of its later 19th-century stone successor [72], in the 1860s or 70s, since the makeup between the two walls [53] appeared distinct from that [52] lying at the same level to the north. If this is so, it is theoretically possible that some of the features dug into the makeup [92], described in section 2.2.3(H) above, may in fact be associated with activity at this time. [53] was an olive-brown coarse sandy loam (2.5 Y 4/4) with 80% small to medium-sized stones, and mortar fragments, yielding pottery dating down to the late 18th century or later (Types 7, 10).

The construction trench [71] for the extension of the 1890s, had cut into both [52] and [53] and the remains of the wall [63] between them. Within it had been built a wall [69], 0.52m wide, of roughly coursed irregular shale slabs set in copious amounts of white mortar, butting onto Room 2 with straight joints at both its southern ends. Doorways were provided at the northern end of the western wall and at the northern and southern ends of the eastern wall, while windows lay towards the southern end of the western wall and between the two doors on the east; the latter had themselves been reduced to windows at a later stage. The northern door on the eastern side had a rough threshold [35] of stone blocks set on edge, which appeared to have been inserted secondarily. The construction trench had been infilled with [70], an olive-brown deposit of loamy coarse sand (2.5 Y 4/4) with 90% very small and small stones, mortar flecks and some black slag, containing pottery down to the 18th century (Types 3, 7). There is no evidence for this building ever having been more than a single storey high.

(ii) The pitched stone floor and stalling arrangements (Figs 2.68 and 2.76)

Overlying the infill of the construction trench [70], confined to the new room, were further infills [37] and [38]. The infill [37] comprised a dull yellowish-brown sandy clay loam (10 YR 5/4) with 30% small and medium stones, containing several items of post-medieval pottery (Type 14) and ironwork, and glass dating to the late 19th century or later (Table 2.10). It broadly covered the remains of the clom wall [63] and the fill [53] to its south, although it appeared to end on a straight east-west line about 1.50m into the room, conformable with the new plan. With one minor exception, a small area (0.50m square) of pitched stone flooring [24] set east-west at its north-east corner, no trace of any flooring was found overlying this deposit, though the area was somewhat damaged at the time of demolition. If, as seems virtually certain, the plan seen before demolition perpetuated the earlier one in broad terms, with cattle facing south, this area would have lain in front of the stalls and could have been used for storing fodder.

To the north of [37], at the same level, lay another infill [38]. This was a dark brown loamy coarse sand

(10 YR 3/4) with 95% very small and small stones, like coarse sugar, and reminiscent of processing debris from the nearby gold-mines; it too contained a number of items of post-medieval material (pottery Types 10, 11, 14; glass, Table 2.10), the latest of which was a clay pipe fragment of *c.* 1850 (Table 2.8). Onto this infill a pitched stone floor [24] had been laid, which survived surprisingly complete to the north and west of what was presumably the stalls (Fig. 2.76). It ran north-south to the west of them, as far north as the drain [23] which ran along the northern cross-passage, and east-west elsewhere; only the line of the drain, which may relate to the 1960s alterations (see section 2.2.3(M) below), and a small area against the north wall were missing. The western passage was finished in a neat edge on its east side, though damaged on its south, while the southern extent of the flooring of the cross-passage was not so well defined, though marked by a roughly straight line. This flooring, however, looks as though it was originally laid in conjunction with a rather narrower set of stalls than that represented by the post-holes [29], some of which cut the flooring of the western passage. No trace remained of any structure associated with these hypothetical earlier stalls.

The timber uprights for the stalls [29] had been set into concrete, and were 0.15m square. They defined stalling for two pairs of cows (or just possibly, for five beasts) facing south. The eastern stall adjoined the east wall of the building, and was divided from the western by a short partition running 0.65m northward from the front (south) wall of the stalls; the western side of the western pair was marked by a longer partition 1.70m deep. This left a fifth space 1.15m wide on the west of the stalls, which must have been used as a passage, at least once the eastern doors had been blocked; another post-hole lay against the west wall opposite the northern end of the western partition of the stalls. These stalls seem to have been of a slightly different design from those in Room 2, with posts defining primarily the front face and the rear of the partitions; the rear of the central partition falls at the rear of the manger. The flagstones [28] appear to represent the remains of a crude manger; one stone, at the north-western corner of the western pair, was even still standing on its side to form a lip to the trough, which was sub-divided at the halfway point to define separate areas for the two animals. Such partitions were often provided to discourage cattle taking each others' food, and indeed this, together with the short partition between the two pairs, helps to confirm that the fittings are for cattle, not horses, although the generous provision of windows might initially suggest a stable. At a later stage these stalls appear to have been concreted around, and in about the early 1960s the shed was more radically refitted (see section 2.2.3(M) below).

Fig. 2.76 Pitched stone flooring [21] and associated features in Room 1, looking south

(L) Features to the west of the stone building (Fig. 2.68)

The later yard [79], in reality made up of a number of different surfaces probably reflecting repeated repairs, appeared to relate to the later 19th-century stone building, butting up against the wall [76] and floor [16] of Room 3. It was rather mixed, but generally dull yellow-orange (10 YR 6/4) in colour, and 0.10m thick at its thicker southern end. It yielded numerous fragments of medieval and post-medieval pottery (Types 3, 4, 7, 9, 12, 13, 14; LP 9, 30), some of post-medieval glass (Table 2.10), and clay pipes of which the latest dated to *c.* 1850 (Table 2.8).

There were a number of features associated with it, including a pair of gateposts which perpetuated a pair associated with the underlying surface [180]; indeed, the actual post [62] (see section 2.2.3(Giii) above) appears to have continued in use. Its partner was reconstructed in a fresh sub-circular post-pit [66], which measured 0.55–0.65m in diameter and was 0.80m deep. It contained a void post-pipe [61], sealed by the late layer [32]. The area around the post was filled by a greyish-yellow-brown (10 YR 4/2) sandy clay loam [65] with 60% small to medium-sized stones. From it came pottery fragments dating to the 18th century onwards (Type 12).

At the south end, in line with the south wall of Room 4, a stub wall [59], 0.90m long and 0.50m wide, had been constructed in a shallow trench [124] 0.20m deep. At the western end of this wall another post [60], represented by a void post-pipe, had been set in a circular post-pit [105], 1.00m diameter and 1.00m deep. The post had been filled around by a brownish-black (10 YR 2/2) silty clay loam [104] with 20–30% small and medium-sized rounded stones, which contained post-medieval pottery down to the late 18th century onwards (Types 3, 5, 6, 7, 11, 14; LP 10, 31), and clay pipes (CTP 7). The post had been sealed by a very late deposit of concrete (unnumbered). The post-pit underlay the stub-wall, and had been dug from a point before the surface [79] was fully complete.

Other features within the farmyard area included a curious double pit supporting a post, [101/95], and two further post-holes which may form a pair [268] and [275].

101 Subrectangular pit, 1.00m × 0.80m, 0.45m deep. Filled by a brownish-black (10 YR 2/2) silty clay loam [100] with small, medium and large stones. One fragment of post-medieval pot dating to 16th century onwards (Type 3).

95 Subcircular post-pit, 0.80m in diameter and 0.60m deep, cutting [100]. Contained *in-situ* timber post [93], 0.10 × 0.08m, 0.60m deep. Filled around post by an olive-brown (2.5 Y 4/4) silty clay loam [94] with small and medium-sized stones. Fragment of post-medieval pot (Type 3), and glass dating to mid-late 18th century (Gl 19).

268 Oval post-pit, truncated by section, 0.55 × >0.30m, 0.27m deep. Contained post-impression [266], 0.27m in diameter, 0.25m deep, filled by a greyish-yellow-brown (10 YR 4/2) loam [265] with 20% very small, small and medium stones. Area around post filled by a greyish-yellow-brown (10 YR 4/2) loamy coarse sand [267] with 40% very small, small and medium stones.

275 Irregular pit, 0.40m diameter, 0.17m deep. Filled by a greyish-yellow-brown (10 YR 4/2) silty clay loam [272] with 50–60% gravel, small and medium stones.

(M) 20th-century developments

At some fairly late stage in the history of the building, a hole [15], 1.10m by 0.70m, had been made in the pitched stone floor of Room 3 [16], and an area of concrete [14] inserted. Local information suggested that this may have been hard standing for a generator.

In Room 1, there appear to have been two main episodes of concreting. The first appears to have been laid around the concrete-set timber stalling while it still stood, as was done in Room 2, though the flooring itself was more solid and had been given a striated finish; the concreting of the western passage, finished in a herringbone pattern pointing towards the cross-passage, seemed to respect the timber which would have stood at the north end of the partition on its east. The area within the stalls [1] was built up somewhat, leaving a step to the rear, the edge of which was incised with a criss-cross pattern. It is not clear what arrangements were made for a manger at this stage, since the concrete found here belongs with the later remodelling described below. The surface of the western passage adjoined the fluted concrete surface [2] of the cross-passage, which had a drain running along it to a grating near the eastern side, through which material would have dropped into the channel [23] below. As seen before demolition, this drain ran eastwards to an inspection cover on the pavement outside the building; the final arrangement could only have been completed after the road was moved in the early 1970s, though the cover could have been in this position earlier. The insertion of this drain probably explains the secondary appearance of the masonry in the threshold area of the blocked doorway. A new concrete threshold [3] was provided for the western door.

At a late stage, the stalls themselves were completely remodelled, with the passage now running on the east, rather than the west; unfortunately this area had been damaged by machine before excavation began. The area of the passage on the west was built up to match the step running across the remainder of the width of the building, the wooden stalls were removed, the areas of the old manger and the post-holes concreted over, and the new stalls built with metal partitions, while semicircular water bowls, fed by piped water and activated by the animal's tongue, were provided for the cattle. This work took place in the early 1960s, according to the farmer who used it at the time; unfortunately he could not recall how the room had been before the work was done. It is not clear how quickly the final arrangement followed after the first concreting; this in turn was more ambitious than that carried out in Room 2. It is therefore likely that the concreting in the western pen, the stall area and the cross-passage (see above) of that room actually dates back to the 1950s or even earlier. Certainly Room 2 seems to have been out of regular use by the time the more radical alterations were carried out in Room 1.

The later 19th-century farmyard surface [79] (see above) had been cut by several drainage and service trenches. One of these [56], a trench 0.35m wide and 0.40m deep, ran east-west and contained a ceramic and cast iron drainage pipe. This served an external sump 0.50m deep against the west wall of Room 4, fed by a pipe through the wall of the building. There was also a water supply from the south to this point. Another trench, [58], 0.30m wide and 0.20m deep, contained a ceramic drainpipe leading west from the entrance to Room 2.

Both these trenches had been cut by [57], a trench 0.30m wide and 0.30m deep, which ran north-south parallel with the side of the building to reach Room 1 at the south side of its doorway. This trench contained a plastic water pipe, and may well have been laid in conjunction with the alterations here in the early 1960s mentioned above. It had in turn been cut by [55], a trench 0.30m wide and 0.12m deep, carrying a plastic drainage pipe away westwards from the entrance to Room 3.

In addition, two post-holes cut by the western section of the site had apparently been dug from a high level (see Fig. 2.7), immediately below the late surface [32]. These may relate to another building at the north-west of the yard, believed to have been a Dutch barn, marked on the OS 1:2500 survey of the early 1960s but long since gone.

400 Pit, truncated by section, >0.34m diameter, 0.40m deep. Filled by a greyish-yellow-brown (10 YR 4/2) silty clay loam [399] with gravel, very small, small and medium-sized stones.
398 Pit, 0.70m diameter, 0.75m deep. Filled by a greyish-yellow-brown (10 YR 4/2) silty clay loam [397] with 20–30% gravel, very small, small, medium and large stones.

All these features had been overlain by the most recent farmyard surface [32], a mixed layer with about 50% gravel, small, medium and large stones, boulders, fragments of brick and tile and specks of mortar, in a matrix of black loamy very fine sand (10 YR 2/1). There were more large stones and boulders towards the southern end, some of which may have come from the partial collapse, apparently in the 1960s, of the west wall of Room 2 where it overlay the post-medieval cellar. The yard, for which coins of 1917 and 1921 (Co 5 and Co 6 respectively) provided a date, had been disused for some time before the excavation began. It had become largely covered with turf [25], some of which had been removed by machine during the demolition of the standing building.

(N) Vernacular report on the standing building and pigsties (by Graham Darlington and Penny Noake)
[This report is included here in full as a matter of public record, even though both buildings were demolished in the course of the 1989 operations. The original wording has been retained unchanged.]

(i) The standing building (Figs 2.77 to 2.81)
Description: A long single-storey cowshed/combination range built in three phases during the second half of the 19th century (Fig. 2.77). The earliest part is the unusual lofted two-bay unit and dairy at the southern end. At a slightly later date, the yellow brick door arches suggesting some time in the 1860s or 70s, the central cowshed part was constructed. This seems to have had a transverse stalling arrangement with access for cattle through the doors by the road on the eastern side. Finally, a cow-house or stable extension was erected at the northern end. This has purple engineering brick arches, indicative of a construction date in the last decade of the century (Fig. 2.78). The structure is in a reasonable condition, though part of the western elevation was unfortunately refaced with blockwork in the 1960s (see Fig. 2.4). The building is no longer used for anything other than general storage purposes.

Walls: The southern part has mixed, angular quarried local shale stone laid to occasional rough courses in lime mortar. Joints are relatively fine and well-pointed, though the southern gable is rather smeared. The central part has slightly more irregular masonry while at the northern end the stones are much rounder and water-worn river stones are included. Joints tend here to be much wider. Quoin stones are stressed in the southern part but much less so towards the northern end.

Roof: The loft at the southern end was not inspected. Other parts have tie and collar beam trusses with sawn and adzed softwood The collar has a slight camber and a halved joint secured with iron cramps. Each apex is halved and crossed with a bolt fixing and a plank ridge and king block arrangement. All are typical mid to late 19th-century features.

Details: The southern unit has a two-bay area of unknown function with a granary(?) loft above. A splaybacked ventilator slot with a recessed cross space creating a cruciform shape, is set into the southern end of the eastern elevation. The side door on the western elevation is probably an insertion. A small integral dairy is set to the south. This has concrete surfaces replacing slate slab benches on brick piers by the northernmost wall. Originally, entrance was from the west, the door and window on the gable being later insertions (Fig. 2.79). The cowshed in the centre of the range is quite dilapidated. However, traces of the original layout and a single stall partition still survive (Fig. 2.80). The latter dates

Fig. 2.77 Cowshed/combination range before demolition in 1989, looking south-west

Fig. 2.78 Northernmost arch of central section (Room 2) and southernmost of northern section (Room 1) showing contrasting treatment, looking west

from the late 19th or early 20th century, having been reset into a concrete floor in the 1960s. The stalls and the door positions reveal that three rows of six cows, tied in pairs, were faced towards the northern end. Part of the original flagstone floor has been retained in the south west corner. The pitching door high in the southern wall in this part (Fig. 2.81) gave access to the loft above the southern unit. This cowshed was formerly lit by two small, stone-dressed, splay-backed ventilator slots set into the western elevation (one is now converted into a window). Part of this wall has been refaced with blockwork in the 1960s. The smaller two-bay part at the northern end may have been either a stable or a cow-house; the use of small windows with purple brick segmental heads hinting at the former but the proportions of the unit tending towards the latter use. It seems probable that the original stalls were arranged in a transverse manner similar to the concrete ones found in place during the survey.

Fig. 2.79 The range looking north-east, showing the door and window in the gable end

Fig. 2.80 The stall partition visible in 1989 within the cowshed in the centre of the range (Room 2), looking north-east

Fig. 2.81 The pitching door in the south wall of the cowshed in the centre of the range (Room 2), looking south

First impression of historical value: This range is not like any other combination ranges on the Lloyd Johnes Estate, where the housing for cattle tended to be kept separate from other functions. It also seems unusual in that it is quite a large building to be associated with a public house if its intended purpose was purely for housing stock. The present condition of the unit is reasonable, though alterations have occurred both internally and externally. It is a late, slightly atypical range and therefore of moderate historical value.

Most important features: There is nothing of direct historical importance. The eastern elevation has visual interest and the surviving internal transverse partition is of some note. The dairy contained within a farm building is a late feature and represents the last transfer from the vernacular house of agricultural functions. However, alterations inside this part mean it now has reduced importance.

(ii) The pigsties (Fig. 2.82)
[These lay to the south-west of the standing building (see Fig. 2.5). In their original form, they are likely to have operated in close conjunction with the dairying activities presumed to have taken place in the main building (discussed further in section 2.2.6(Ci) below).]

Description: Originally conceived as a detached pair of pigsties with, unusually, yards in the entrances to the sties placed on the side of the structure rather than on the longer elevation. The use of yellow brick segmental arched heads to openings suggests a construction date in *c.* 1860 to 1870. The units were converted to public conveniences in the mid 20th century.

Walls: Irregular, quarried local shalestone and round water-worn rubble laid uncoursed in mortar. Quoin stones are angular and well stressed.

Roof: The structure is unrecorded. Covered in medium sized regularly coursed thin North Wales slates. Verges are boarded, eaves oversail slightly. The building has the usual one and a quarter pitch design.

Details: The original small openings are blocked with stone. Yellow brick segmental arches are still visible along with rough stone dressings. Later full height doors with red brick jambs are mid 20th century insertions. There is no sign of a hen coup ever having been set into the loft space. The interior has been reorganised and converted into two public toilets. The external yard walls remain, the southern one retaining a rough stone coping, the northern one being replaced with a round cement top.

Fig. 2.82 The pigsties before demolition in 1989, looking south-west

First impression of historical value: An unusual variation of the typical 19th-century pigsty. Moderate importance.

Most important features: Plan type is of similar proportions to most small mid to late 19th-century pigsties but this positioning of yards has not been noted elsewhere. The materials and door details are normal and typical of the period.

2.2.4 The Finds and Specialist Reports

The material from the 1989 site does not represent an especially large collection, particularly for the Roman period, when compared with both the 1970s excavations and with the assemblages from other military sites across Wales. Given that so little material has been published from the site to date, however, the chance is taken here to include as representative a sample as possible, primarily for the pottery. The bulk of the post-excavation assessment and preparation of the related reports was undertaken by Dee Brennan, assisted by Helen Burnham, before key categories were despatched for detailed treatment and report by the various specialists. Most texts were finalised by the end of 1998, with limited checking and updating for key categories during the final editing in 2003. For ease of reference a concordance of contexts, finds and principal plan/section numbers is provided as an appendix at the end of the specialist reports (Table 2.27).

(A) Coins

(i) Roman (by Edward Besly)

Three Roman coins were recovered from the excavation, all of them in poor condition and only two readily identifiable to emperor. The single silver coin (Co 1) came from one of the fills [285] in the clay-lined cistern [376], while the two copper-alloy coins came from a fill [340] in one of the 'late' pits [323] (Co 2) and from the destruction horizon [540] above the burnt timber building in pit [245] (Co 3) respectively.

Co 1 Titus (79–81), denarius, AD 80. *Rev.* [TR P IX IMP XV COS VIII PP], elephant to left; *RIC* 22a. Little detail is visible, but the reverse design is distinctive. It is not possible to assess the state of wear of this coin. Context **[285]**. Find no. 10,013.

Co 2 Domitian (81–95), sestertius, COS XI = AD 85. *Rev.* emperor sacrificing in front of shrine of Minerva; *RIC* 256. The coin is corroded, but appears to show some signs of wear. Context **[340]**. Find no. 10,020.

Co 3 Corroded irregular copper-alloy disc, maximum diameter 20mm. Little, if any, of the original surface survives. The size would be appropriate for a smaller module Claudian counterfeit of the later 50s (see Boon 1982, 11–14), but in this state it would be unwise to treat this as a positive identification. Context **[540]**. Find no. 10,028.

Table 2.1 Post-Medieval coins

Cat no	Context no	Small find no		Date
Co 4	30	–	Copper alloy Elizabeth II halfpenny	1957
Co 5	32	–	Copper alloy George V penny	1917
Co 6	32	–	Copper alloy George V penny	1921
Co 7	123	10,002	Copper alloy George II or III penny	?
Co 8	127	10,004	Copper alloy George III penny	1773
Co 9	129	10,005	Copper alloy George III penny	1775
Co 10	129	10,007	Copper alloy George II penny	1730
Co 11	129	10,008	Not identifiable	
Co 12	220	10,009	Not identifiable, reverse shows Britannia	

(ii) Post-Medieval (by Dee Brennan)
A total of nine post-medieval coins were found, including several which provided useful dating evidence for the post-medieval buildings (Table 2.1).

The most informative sequence came from the horizons pre-dating the clom building. Of these, the undated George II or III penny (Co 7) came from the infill [123] beneath the building's pitched stone floor [41]; the two pennies, Co 9 (1775) and 10 (1730), and the unidentified Co 11 came from the underlying layer [129]; and the 1773 George III penny (Co 8) came from a small patch of pitched stone flooring [127] at the south-west corner of the later building, which lay stratigraphically between [123] and [129]. Together, these provide a very coherent TPQ for the date of construction (see section 2.2.3(Dii) above and Table 2.28). Elsewhere, the 1957 Elizabeth II halfpenny (Co 4) came from the fill [30] underlying the most recent surfaces [4] and [5] at the north end of Room 2, though unfortunately it is not clear whether it was sealed by any of the fragmentary surviving flooring. The 1917 and 1921 George V pennies (Co 5 and 6) came from the recent farmyard surfaces [32] to the west of the various buildings. The remaining, unidentified, post-medieval coin (Co 12) was found in [220], a gravel spread lying to the south of the clom building and apparently pre-dating it.

(B) Gemstone (by Martin Henig)
The only gemstone from the site (find no. 10,018) came from one of the fills [340] of the 'late' pit [323]. Martin Henig provides the following catalogue entry and commentary (Fig. 2.83):

Gm 1 Pale amber-coloured cornelian ringstone. Oval, with a flat upper face (shape F1), measures 13 by 10 by 2.5mm. The subject is the goddess *Nemesis* in profile to the right (impression described). She makes a characteristic gesture with her right hand, plucking the chiton at her breast in order to spit into it; the hem of the garment billows out at the back. In her left hand she holds the branch of a tree. She has the large eagle-like wings of Victory, which are frequently shown on representations of *Nemesis*, and her hair is tied in a small chignon on the crown, again in the manner of that deity. At her feet is a serpent. There is a short ground line.

The bold and neat cutting is typical of the Classicising Style of the mid 1st century AD. Somewhat similar but rather less refined is a plasma intaglio from Newstead where, however, the goddess has a wheel at her feet rather than a snake (Henig 1978, no. App. 144; Elliot and Henig 1982, 297, pl. 24a no. 10). The serpent more properly belongs to *Hygeia*, goddess of health, shown in art holding a serpent and often feeding it from a cup. Intaglios from Aquileia (Sena Chiesa 1966, nos 640, 641) show *Nemesis-Hygeia* holding a serpent. Stylistically even closer to the Pumsaint gem is an intaglio in The Hague (Maaskant-Kleibrink 1978, no. 840), where the goddess's chiton billows out around her ankles.

The prototype of the image was the statue of *Nemesis* by Agorakritos, a pupil of Pheidias, set up at Rhamnous in Attika, depicting the deity holding a bowl and an apple branch (Richter 1970, 184–5; citing Pliny, *N.H.* 36, 17 and Pausanias I.33.2). *Nemesis* was the deity of righteous retribution and personified the moral reverence for law. The action of spitting, especially onto one's breast, repelled

Fig. 2.83 Gemstone

the Evil Eye, a belief still current (Opie and Tatem 1989, 373) and was a natural addition to the iconography of *Nemesis*. Moreover, in Hellenistic and early Roman times she acquired a range of attributes from other deities including two shown here. It was natural to conflate her with Victory; the connection with health is less obvious but presumably as the restorer of the mean, she shares the task of bringing health to mankind (for a useful bibliography see Pannuti 1983, 61–2, also Casal Garcia 1981).

As a goddess of fate (and her most familiar attribute is the wheel of *Fortuna*) she came to be associated with amphitheatres; at Chester a centurion, Sextus Marcianus, set up an altar to her as the result of a vision (Thompson 1976, 167–9), while the dedicator of a *defixio* from the Caerleon Ludus prays that *Domina Nemesis* will punish an enemy or rival (RIB 323).

(C) Pottery

(i) The samian from the 1989 excavations (by Peter Webster)

Soil action has not been kind to the samian from the 1989 excavations. It is in a noticeably poorer state of preservation than the better pieces from 1972 and in some cases has come close to being totally dissolved by its environment. This makes identification of sources more difficult, though it seems reasonably certain that, with little exception, all the samian listed is from South Gaul. Table 2.2 attempts to indicate the number of vessels present, *not* the number of sherds. All are from South Gaul (SG) unless otherwise stated. ? indicates uncertain attribution. A total of 69 or 70 vessels appears to be represented.

The samian evidence would accord well with a site occupied within the main Flavian advance into west Wales. Form 18 is much more common than form 15/17 and, although form 29 is present, predominantly or wholly pre-Flavian forms such as 24/25 and Ritterling 8 and 9 are not. Chronologically, the presence of only a single identified piece from Les Martres must surely mean that activity of any significance after *c.* AD 100 is unlikely in the area excavated in 1989, as it seems improbable that activity much after this date could fail to have produced any Central Gaulish pottery. There seems little doubt, therefore, that most, if not all Roman phases on the 1989 site (with the exception of the burnt timber building) need to be fitted within the period *c.* AD 75–100. This contrasts with the 1972 excavations where activity up to *c.* AD 125 may be suggested. Even so, there is no doubt that the bulk of the activity in both areas was in the 1st century.

The contribution of the samian to the dating of specific parts of the site proved rather limited, since so much of it was poorly preserved. By far the largest concentration of vessels (27 in Table 2.2) came from the infill [487] of the stone-lined well [560], its projected upward extension [471], and the successive fills [234; 284; 285; 356; 357; and 375] of the overlying clay-lined cistern [376], as well as from the infill [377] beneath the clay lining and from the lining itself. Much of this material must have been derived from clearance operations connected with the demise of the well and cistern respectively, but unfortunately its dating proved insufficiently sensitive to shed much light on the chronology of the different phases.

A second concentration (12 in Table 2.2) came from the successive deposits [460 and 224] above the burnt timber building, but its value for dating is limited by the fact that most of the pieces are likely to be in secondary contexts. Similar problems surround the few pieces which came from the fills of the various 'late' pits, including [269; 300; and 301] in pit [302], [316] in pit [323], and [345 and 350] in pit [351]. Equally, little can be gleaned from the single fragment found in the secondary infill [380] between the turf-clay rampart [381] and its stone successor [382/496].

The remaining Roman contexts with samian are all related to the successive building phases, although fragments found in the construction trenches could date either to the construction of the buildings or their demise. The contexts concerned are a silty clay spread [472] between the phase 3 and 4 buildings, and the fills of various trenches associated with phases 4 [295 and 373] and 5 [388].

The remainder of the samian came from post-Roman contexts where it was clearly residual (131; 156; 157; 180; 248; 291; and 330]. Ironically, the single piece from Les Martres came from the 'garden soil' [157] at the south end of the site, a context which also produced a fragment of Hartshill-Mancetter mortarium (CP 4) and a Dressel 20 amphora of Martin-Kilcher type 27 (CP 2).

CATALOGUE

S 1 (Not illustrated). Form 37, South Gaulish. Panel decoration. The upper panel shows the feet of a figure facing right, the lower the head of a running hare facing left. For a similar design see Bregenz Cellar (Jacobs 1912, fig. 32). *c.* AD 80–100. Context [**157**].

S 2 (Not illustrated). Form 37, South Gaulish. An abraded sherd from a bowl which was once repaired with a lead rivet. The ovolo has a tongue with multiple end, possibly the four tongues of the ovolo used by M Crestio and Crucuro (Knorr 1919, textbild 5). Below is zonal decoration divided by wavy lines. The upper zone consists of a large rosette of *c.* 18 petals and the lower a 'tree' with branches terminating in groups of buds (possibly as used by Germanus, Knorr 1952, taf. 27, B. Flavian, *c.* AD 70–100. Context [**377**].

S 3 (Not illustrated). Form 29, South Gaulish. The base and a small fragment of the decoration of a vessel. The design includes a small six-armed rosette and a festoon with a fragmentary and abraded figure, possibly a running dog. *c.* AD 65–80. Context [**460**].

Table 2.2 Samian from the 1989 excavations

Context no	18 18R	18/31	27	29	35	36	37	Curle 11	other	comments
U/S	-	1	-	-	-	-	1	-	-	18/31 *c.* 90–110
131	-	-	-	-	-	?	-	-	-	Dish fragment
156	-	-	1	-	-	-	-	-	-	27g base
157	-	2	1	-	-	-	3	-	1	33 probably Les Martres (*c.*AD 100–120) 37 probably late SG (*c.* 90–100) 37 **Cat no S 1**
180	2	-	-	-	-	-	1	?	-	One example 18R?
224	3	-	2	-	1	-	-	2	-	18 base 18 *c.* 80–100 One example 18R 27 with eroded stamp on base **Cat no S 4** Curle 11 or Ritt 12 flange (*c.* 70–90) Curle 11 flange, very flat and early for the type (*c.* 70–90)
234	?		-	-	-	-	-	-	-	Chips ?18 or 18/31
248	-	-	-	-	-	-	1	-	-	37 SG with panel decoration but very abraded, probably *c.* 80–100
269	-	-	1	-	-	-	-	-	-	Flavian
284	-	-	-	?	1	-	-	-	-	SG wall sherd probably of form 29, with a medallion and what may be the arm of a figure. Probably *c.* AD 70–85
285	-	-	-	1	-	-	-	-	-	Basal sherd, probably 29; form rare after *c.* AD 85
291	-	-	-	-	-	-	-	-	1	Bowl sherd
295	-	-	1	-	-	-	-	-	-	Probably Flavian
300	2	-	-	-	-	-	-	-	-	18 and ?18R
301	1	-	-	-	-	-	-	-	-	
316	-	-	-	-	-	-	-	1	-	Flavian, probably AD 70–90 but typologically later than the Curle 11 from [224]
330	-	-	-	-	-	-	-	?	-	Bowl sherd, ?Curle 11
345	-	2	-	-	-	-	-	-	-	18/31 base 18/31, possibly same vessel. *c.* AD 90–110
350	1	-	-	-	-	-	-	-	-	
356	?	?2	-	-	-	-	-	-	-	One example 18 or 18/31
357	-	1	-	-	1	-	-	-	-	
373	-	-	-	-	-	-	1	-	-	Rim
375	-	-	-	-	-	1	-	-	-	
376	-	-	-	1	-	1	-	?	-	29 rim Bowl base, probably Curle 11. Another frag from behind clay probably part of same vessel.
377	-	-	-	?	-	?	?2	?	-	Base, probably 37 Wall sherd with abraded scroll decoration, possibly 29 Bowl, probably Curle 11 37 **Cat no S 2**
380	?		-	-	-	-	-	-	-	18 or 18/31 base
388	-	-	-	-	-	-	-	?	-	Bowl, possibly Curle 11
460	1	-	2	1	-	-	-	-	-	18 or 18R 29 **Cat no S 3**
471	1	-	1	1	-	-	1	-	-	
472	-	-	-	-	-	-	1	-	-	
487	3	-	1	-	-	1	1	-	-	27g 18, 2 vessels *c.* AD 80–100 37, the narrow upper rim area suggests an early date *c.* AD 70–90
	?3									
Totals	14	7	10	6	3	5	12	8	2	

Potter's stamp (by Brenda Dickinson)

S 4 (Not illustrated). Base of a Form 27 cup, South Gaul. The stamp is too eroded to be legible, but appears to read I\, presumably IV[, though there are too many potters in IV.. to make a positive identification. Flavian. Context **[224]**.

(ii) The coarse pottery from the 1989 excavations (by Dee Brennan and Peter Webster)

All the pottery from the excavation has been examined and listed for archive purposes with detailed fabric descriptions and outline references (by DB). The archive list will be deposited with the pottery. The list formed the basis for the selection of pottery for drawing and cataloguing (PVW) and for all the general comments made below.

Throughout, it should be borne in mind that the soil conditions at Pumsaint cause severe deterioration of fabrics. A measure of hardness is, therefore, rarely practicable and surfaces, particularly slips, do not always survive.

Fabrics and sources: The overall amount of pottery recovered was not great as will be gauged by the fact that it proved possible to illustrate about 90% of all vessels represented by rims. Quantification has not, therefore, been attempted and indeed is hardly necessary as the catalogue represents a fairly full picture of the material recovered. It may be useful, however, to offer a summary of published vessels in the form of a Table (2.3) and to make some general comments on the fabrics, listed in an order governed by the approximate distances of the fabric source from Pumsaint.

Local reduced fabric: Comparatively well represented (*c*. 16% of all vessels) is a reduced fabric in light grey with traces of a darker grey surface. A range of jars and bowls appear in this fabric (along with some other forms such as lids). The forms suggest a Flavian to Trajanic date and it seems reasonable to assume that the source is comparatively local. The ware is not, however, so well represented for it to be certain that manufacture was being undertaken within the environs of the fort. It is possible that the fabric is either the product of some slightly more distant pottery or that production was not continuous throughout the period of occupation. The latter seems more likely and there seems some slight evidence for the ware being more prevalent in lower levels on the site.

Severn Valley ware found its way to west Wales throughout the Roman period. Forms represented in the present assemblage are mainly or wholly jars and tankards, with tall, narrow-necked storage jars being the most common (more are represented in the archive than are catalogued below). In view of the small amount of Severn Valley ware observed in south-east Wales outside the conquest period, it seems likely that we are seeing here evidence for trade along a land route through central Wales rather than the more obvious Severn/Bristol Channel route.

Malvern jars: In common with other fort sites in west Wales, Pumsaint yields numbers of hand-made jars in a crude and apparently pre-Roman tradition. These are generally termed 'Malvern jars' following petrological work by David Peacock (1967). More recent work has tended to increase the number of fabrics

Table 2.3 Summary of published vessels from 1989 excavations

Fabric	Flagon	Jar/ Bkr	Tank-ard	Bowl	Dish	Tazza	Lid	Mort	Total	% of Total
Local grey		6					2		8	16
Other reduced		5							5	10
Oxidised	4	6		3		1			14	28
Cln Mortar								1	1	2
Severn Valley		1	1						2	4
Malvern		4							4	8
BB1		8			1		2		11	22
Verulam Mortar								1	1	2
Terra Nigra					1				1	2
N. Gaul Rough-cast		1							1	2
G238 Mortar								1	1	2
Eggshell		1							1	2
Totals	4	32	1	3	2	1	4	3	50	100
% of total	8	64	2	6	4	2	8	6	100	

identified (for example, Rees in Woodiwiss 1992, 48) and it is less surprising, therefore, to find that the 'Malvern' ware from Pumsaint appears to be from more than one source. In the catalogue below, vessels described as containing crushed rock have plentiful angular rock elements often of the metamorphic type familiar from Peacock's work. Those described as containing stone have a higher percentage of rounded gravel perhaps from a different source. Malvern jars seem more common in central and west Wales than in the south (see, for instance, Webster in Manning 1993, 232). If we assume that these vessels were sold for their contents (salt?) rather than their usefulness as pots, as seems reasonable on the grounds of their crudity when compared with other contemporary ceramics, then the distribution pattern seems likely to be revealing a trade pattern of interest.

Black-burnished ware is represented by the usual range of BB1 cooking wares (jars, bowls and dishes) with lids also being present. Bowls and dishes are, however, somewhat scarcer than one would expect (and no bowls were sufficiently complete to allow illustration). The dating of this material forms a major problem. In the past it has been assumed that the ware appeared only from the Hadrianic period (*cf.* Webster in Burnham and Burnham 1987a). However, it has recently become apparent that BB1 is present throughout south Wales from the conquest period onwards (Webster 1992) and this has led to a re-assessment of the 1985 Pumsaint dating (*ibid.*, 120). Stated briefly, the samian from the 1972–3, 1985 and 1989 excavations at Pumsaint are all in agreement in showing a preponderance of Flavian and Flavian-Trajanic material with only occasional pieces of samian from Les Martres-de-Veyre or Central Gaul which might push the date of abandonment beyond the turn of the first and second centuries. It seems most reasonable, therefore, to accept that the BB1 from all the excavations is contemporary with the samian and arrived, with other Black-burnished ware, in south Wales in the Flavian or Trajanic period. Certainly, none of the BB1 forms represented would be out of place in southern England within this date range (although a few show some characteristics which would normally be taken as indicative of a later date). In general, it has seemed unwise to place great weight on the Black-burnished ware evidence for dating purposes as it seems more likely that Pumsaint can contribute to the dating of early BB1 forms than *vice versa*. In this respect, it may be noted that most BB1 from the 1989 site appears to be derived from contexts which are late within the general stratigraphic sequence (for details see sections 2.2.2(Diii), (Hvii) and (I) above).

Mortaria: Mortaria from a variety of sources were present. Most common was the class generally known as *Gillam 238*, thought to originate in the Pas de Calais region of North Gaul, followed by those from the Verulamium region. Both these fabrics are likely to have reached the site in the Flavian and Trajanic periods. Few other sources are present, although the single mortarium from Mancetter-Hartshill (CP 4) is presumably of 2nd-century date and has, therefore, a chronological significance.

Amphorae: All but two of the amphora fragments recovered were from the common globular Dressel 20, South Spanish olive oil amphorae. The only exceptions were a few scraps of Gauloise 4, South Gaulish wine amphora (from context [321]) and the base of a possible amphora of unknown shape and source (CP 21).

Finewares are poorly represented, if we except samian. A single North Gaul/Argonne beaker has been noted (CP 30) along with a few 'Terra Nigra' pieces (CP 28 and 49). Even allowing for the fact that samian is present in some quantity, this seems a low total of fineware sources and may be indicative of a relatively low status for the occupants of the area excavated.

Other sources have not been individually identified and probably could not be so without a much wider analysis (including petrological work) than the quality of the assemblage warrants. What is, however, observable is that an unusually large range of unidentified sources of what would generally be termed 'self-coloured ware' are represented (seen as 'oxidised' and 'other reduced' in Table 2.3 and representing *c.* 38% of coarseware). The title covers a range of non-fineware which is, in the main, kitchen-ware such as would normally be supplied by just one or two very local sources. That this has not occurred may, in itself, testify to the lack of a specific Pumsaint pottery during all or part of the occupation. This has already been tentatively suggested (under 'local reduced fabrics' above). The site would seem to be drawing on a wider range of sources than usual, even for the Flavian and Trajanic period. Whatever the situation concerning a local pottery, this might also, of course, indicate a site with a greater than usual throughput of personnel, a feature which would not be unexpected on a fort site with a function rather different from that of the average auxiliary fort.

Function: Table 2.3 seems to reveal a rather greater number of jars in comparison with other forms than one might expect (jars and beakers form well over half of the vessels catalogued). This may simply be a product of a rather small sample, but if not, then it must presumably denote some oddity in the function of the site excavated or perhaps in the diet of the occupants. Would a largely vegetarian diet, for instance, necessitate more cooking pots and fewer bowls and dishes, the latter being presumably used primarily for casseroles?

Chronology: The date range of the Pumsaint assemblage must rely heavily on the samian evidence. This suggests that major occupation within the area excavated in 1989 is unlikely after the first decade of the second century. The coarse pottery is less specific (particularly if we accept the argument concerning the Black-burnished ware outlined above) but there is little doubt that, with the possible exception of the BB1, the assemblage is predominantly Flavian to Trajanic, although a very few later pieces are present (nos 2, 4 and 6). It may be noted that the Black-burnished ware from 1989 all comes from levels late within the stratified sequence or from even later levels. There can be little doubt, therefore, that it was arriving in Pumsaint later in the history of the fort. If we are to put that date of arrival on the 1989 site later than the late first century, then we either have to account for the fact that BB1 vessels which *could* be second century are present without other certainly 2nd century forms (and without Lezoux or East Gaulish samian) or to accept the fact that the BB1 vessels must be contemporary with all other pottery and, therefore, at the earlier end of their possible date range in southern Britain.

Viewed overall, the 1989 pottery assemblage would support a foundation date contemporary with the Flavian conquest of Wales in the mid 70s AD. The date of abandonment is more problematical and needs to be considered alongside other evidence from the site. It is argued elsewhere (Webster 1992) that the possible demolition pit excavated in 1985 should be dated *c.* AD 100. The marked lack of certainly 2nd-century material from the 1989 site might suggest that the demolition involved was that of the fortlet but a consideration of the 1989 and 1972 material together (see section 2.2.4(Ciii) below) makes this improbable. It is, therefore, suggested that the fort ceased to be occupied *c.* AD 100. The fortlet appears to have been occupied at least into the 120s, but the small amount of Central Gaulish samian from either site would not suggest that this occupation lasted long after the advent of Lezoux samian (*c.* AD 120/125). An abandonment of the fortlet by the mid-120s seems probable.

It must be noted, however, that this does not necessarily mark the end of all occupation at Dolaucothi, as the later material from Melin-y-Milwyr has long indicated (Lewis 1977). The 1997 material from the bath-house site would similarly seem to contain several pieces significantly later than that from the fort/fortlet site (see section 5.3.1 below). The combined evidence from all these sites will be considered further in the overview at the end of this volume (see section 6.1 below).

CATALOGUE

In the catalogue which follows, the pottery is broadly arranged by context (for details, see the concordance in Table 2.27). All those vessels with rims which are sufficiently well preserved to allow meaningful illustration have been included, together with various other intrinsically interesting or significant (if unillustrable) pieces, paying particular attention to the needs of dating and overall site chronology (Figs 2.84 and 2.85).

CP 1 Jar in Black-burnished ware with wavy line decoration on the neck. Context **[156]**.

CP 2 Dressel 20 amphora, Martin-Kilcher type 27. Early/mid - mid/late 2nd century. Peacock and Williams 1986, 138. Context **[157]**.

CP 3 Small jar in Black-burnished ware. Context **[157]**.

CP 4 (Not illustrated). Flange fragment of Mancetter-Hartshill mortarium. 2nd century. Context **[157]**.

CP 5 Jar in light grey fabric with darker surface. For the general type, see Usk, Manning 1993, fig. 110, 25.3. Mid 1st to early 2nd century. Context **[224]**.

CP 6 Dish in Black-burnished ware with inverted chevron decoration, *cf.* Usk, Manning 1993, fig. 130, 33.3. Mid to late 2nd century. Context **[224]**.

CP 7 Everted rim jar in light red fabric with a burnished buff outer surface and lid seating. The rim form and high shoulder suggest a 1st- to early 2nd-century date. Context **[224]**.

CP 8 Tankard in Severn Valley ware, light orange with a grey core and burnished outer surface, burnt black; *cf.* Usk, Manning 1993, fig. 134, 16.2. Mid 1st to mid 2nd century. Context **[224]**.

CP 9 Mortar-like bowl in a fine red fabric with a tendency to laminate. The surface has been mica-dusted. Mortar-like bowls were popular in the late 1st and early 2nd century. The very deep flange may indicate a 1st-century date. Contexts **[224]** and **[460]**.

CP 10 Flanged bowl in light orange fabric, probably one of the Flavian-Trajanic flanged and carinated bowls. Contexts **[246]** and **[378]**.

CP 11 (Not illustrated). Black-burnished jar with wavy line decoration on the neck, burnt pink in places; *cf.* Gillam 1976, no. 1. The type is dated early-mid 2nd century by Gillam, but might be earlier in southern Britain. Context **[269]**.

CP 12 Mortar-like bowl in light orange fabric with traces of white paint on the rim. Mortar-like bowls were popular in the late 1st and early 2nd centuries. Context **[283]**.

CP 13 Mortarium in granular buff fabric with white, grey and black trituration grits from the Verulamium region; *cf.* Usk, Manning 1993, fig. 187, no. 4. AD 70–110. Context **[284]**.

CP 14 Malvern jar, brown fabric with ground stone inclusions and black surface. Non-joining sherds show vertical burnished lines on the external wall and burnished lattice externally on the base; *cf.* Peacock 1967, fig. 1, no. 3. 1st century. Context **[284]**.

CP 15 Jar in Black-burnished ware. The rim resembles Gillam 1976, no. 4, but the sherd lacks the lattice decoration which would confirm such a late date. Context **[284]**.

CP 16 Body sherd form near base of a flagon or jar in

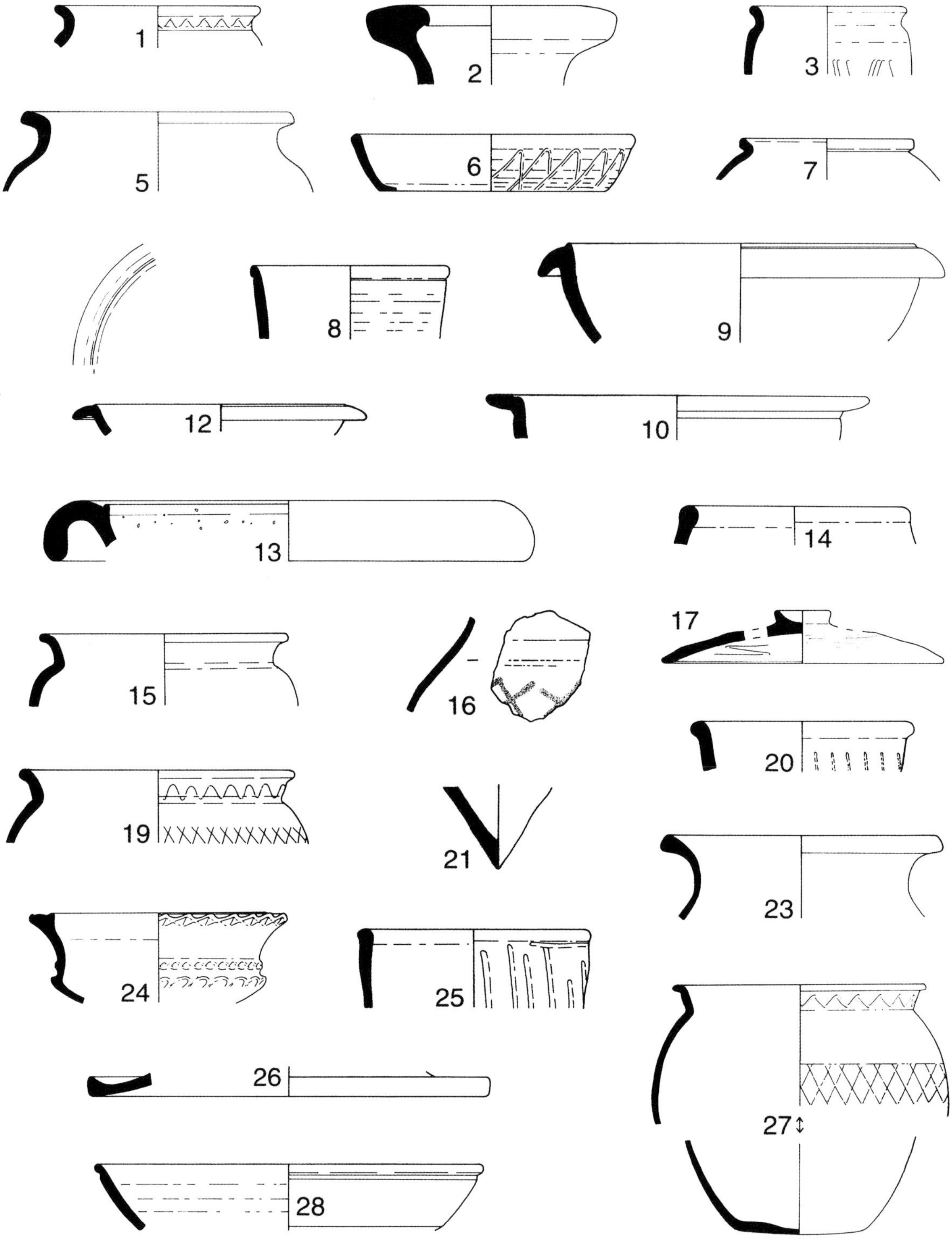

Fig. 2.84 Roman coarse pottery. Scale 1:4

granular cream fabric with traces of a red-painted lattice decoration. Apparently from the Verulamium region although the painted decoration is unusual. AD 70 to mid 2nd century. Context **[284]**.

CP 17 Lid in Black-burnished ware. The underside has a burnished zigzag decoration, *cf.* Wallace and Webster 1989, no. 10. Mid 1st to late 2nd century. Context **[285]**.

CP 18 (Not illustrated). Black burnished jar with wavy line decoration on the neck, similar to CP 11. Context **[285]**.

CP 19 Jar in Black-burnished ware with wavy line on the neck. For the rim see Gillam 1976, nos 1 and 3. The tight acute-angled lattice suggests a 1st- or early/mid 2nd-century date. Context **[286]**.

CP 20 Malvern jar in grey to brown fabric with crushed

stone inclusions and black surface with vertical burnished lines externally; *cf.* CP 40 from context [460]. Context **[291]**.

CP 21 Vessel base in granular orange fabric with grey core. ?Amphora base of unknown origin. Context **[297]**.

CP 22 (Not illustrated). Collar of a 'Hofheim' type flagon in fine soft pale buff fabric. The rim type probably did not last long into the Flavian period but does appear at Caerleon; *cf.* Usk, Manning 1993, fig. 148, no. 1. Mid 1st to mid/late 2nd century. Context **[300]**.

CP 23 Jar in Severn Valley ware, light orange with a grey core; *cf.* Usk, Manning 1993, fig. 132, no. 3.3. Mid 1st to mid/late 2nd century. Context **[300]**.

CP 24 Tazza in creamy-buff fabric with abundant inclusions. Faint signs of burning internally. Tazze are relatively common at the legionary site of Caerleon, *e.g.* Nash Williams 1932, fig. 61, nos 408–423 and fig. 62, nos 424–427, but rare on other Welsh sites. Context **[301]**.

CP 25 Malvern jar in brown fabric with crushed stone filler and black surface with burnished lines; *cf.* Peacock 1967, fig. 1, no. 7. Mid to late 1st century. Context **[308]**.

CP 26 Lid in light grey fabric with darker surface. Context **[308]**; a fragment from context [375] may be part of the same lid (*cf.* CP 37).

CP 27 Jar in Black-burnished ware, *cf.* Gillam 1976 no. 1, but with more pronounced bead to the rim. Probably mid 1st to mid 2nd century in this area. Contexts **[309]** and **[327]**.

CP 28 Dish in 'terra nigra', a light grey highly micaceous fabric with dark burnished surface; *cf.* Greene 1979, fig. 49, nos 15–16. Probably Flavian. Contexts **[342]** and **[481]**.

CP 29 Lid in Black-burnished ware, *cf.* CP 26 from context [285]. Mid 1st to late 2nd century. Context **[345]**.

CP 30 (Not illustrated). Wall sherd(s) of a roughcast beaker in light buff fabric with a grey-brown surface. A product of north Gaul or the Argonne, *cf.* Anderson 1981, fig. 19.3 and Symonds 1990, figs 2–3. Flavian to mid 2nd century. Context **[345]**.

CP 31 Flagon in fine peach-coloured fabric with considerable accretions acquired after burial. The type is represented at Usk, see Manning 1993, fig. 3, no. 1 and pp 11–18. In the Usk report Greene shows the continental affinities and demonstrates a 1st-century date. Context **[350]**.

CP 32 (Not illustrated). Jar in light grey fabric with a brown core. The shoulder has been decorated with a comb to produce a wavy line decoration, a common feature in south Wales, particularly in the first two centuries AD. Context **[357]**.

CP 33 (Not illustrated). Wall sherd of a ?beaker in thin creamy-white hard 'egg-shell' ware, decorated with lines of rouletting produced by a toothed wheel, *c.* 2mm wide, giving indentations 4mm apart. The ware is found at both Caerleon and Holt; for a summary of the evidence and references, see Usk, Manning 1993, 315–16. The findspots suggest a date range from the mid 1st to the earlier 2nd century. Context **[357]**.

CP 34 Jar or large beaker in granular pink fabric with mica-dusted surface surviving on the rim and interior with traces on the exterior. The sandy fabric may ally this with vessels produced in Cheshire in the late 1st and early/mid 2nd centuries, and ultimately derived from north Gaul/Argonne beakers; *cf.* Hartley and Webster 1973, nos 23–37, although this is unlikely to be a Wilderspool product. Context **[373]**.

CP 35 (Not illustrated). Plain flange from a redware carinated bowl. Flavian or Flavian-Trajanic. Context **[373]**.

CP 36 Jar in hard grey fabric. Context **[375]**.

CP 37 (Not illustrated, but see CP 26). Lid in light grey fabric with darker surface. Context **[375]**; a fragment from context [308] may be part of the same lid.

CP 38 (Not illustrated, but see CP 52). Mortarium flange in pink-buff fabric with a white slip. Probably a product of the late 1st to early 2nd century Caerleon industry, *cf.* Zienkiewicz 1992, nos 19–21. Context **[377]**; a fragment from context [487] may be part of the same vessel.

CP 39 Jar in a light red fabric with a grey core. The compete vessel may have resembled Usk, Manning 1993, fig. 110, 25–1. Context **[403]**.

CP 40 Malvern Jar, *cf.* Peacock 1967, fig. 1, no. 5. Context **[460]**.

CP 41 Mortarium in cream fabric, heavily abraded. Probably from a vessel as Usk, Manning 1993, fig. 185, no. 7, from Gallia Belgica. AD 50–85. Context **[460]**.

CP 42 Jar in light grey fabric with darker external surface and rusticated decoration of raised blobs. Gillam 1970, nos 97–8 show the general type; see also Usk, Manning 1993, fig. 109, types 15 and 16. The decoration seems most popular in the Flavian and Trajanic periods. Context **[460]**.

CP 43 Ring-neck flagon in light red-buff fabric, with traces of a creamy-white slip. Fairly even rings like this are generally taken as a 1st-century feature; *cf.* Usk, Manning 1993, fig. 3, no. 4 and Gillam 1970, nos 1–2. Mid-late 1st century. Context **[460]**.

CP 44 Everted rim jar in orange fabric with a blackened exterior. There seems to have been some attempt at creating a neck around part of the vessel, which must have resulted in a rather lopsided appearance. A Flavian or Trajanic date seems most probable. Context **[460]**.

CP 45 (Not illustrated). Wall of a Black-burnished jar with acute-angled lattice decoration. Context **[460]**.

CP 46 Jar in light grey fabric with darker surface. The high grooved shoulder suggests a pear-shaped vessel typical of late 1st and early 2nd century. Context **[471]**.

CP 47 Jar in light grey fabric with darker surface. There is a faint indication of rustication. Very similar to CP 42 from context [460] and probably from the same hand, though not the same vessel. Context **[471]**.

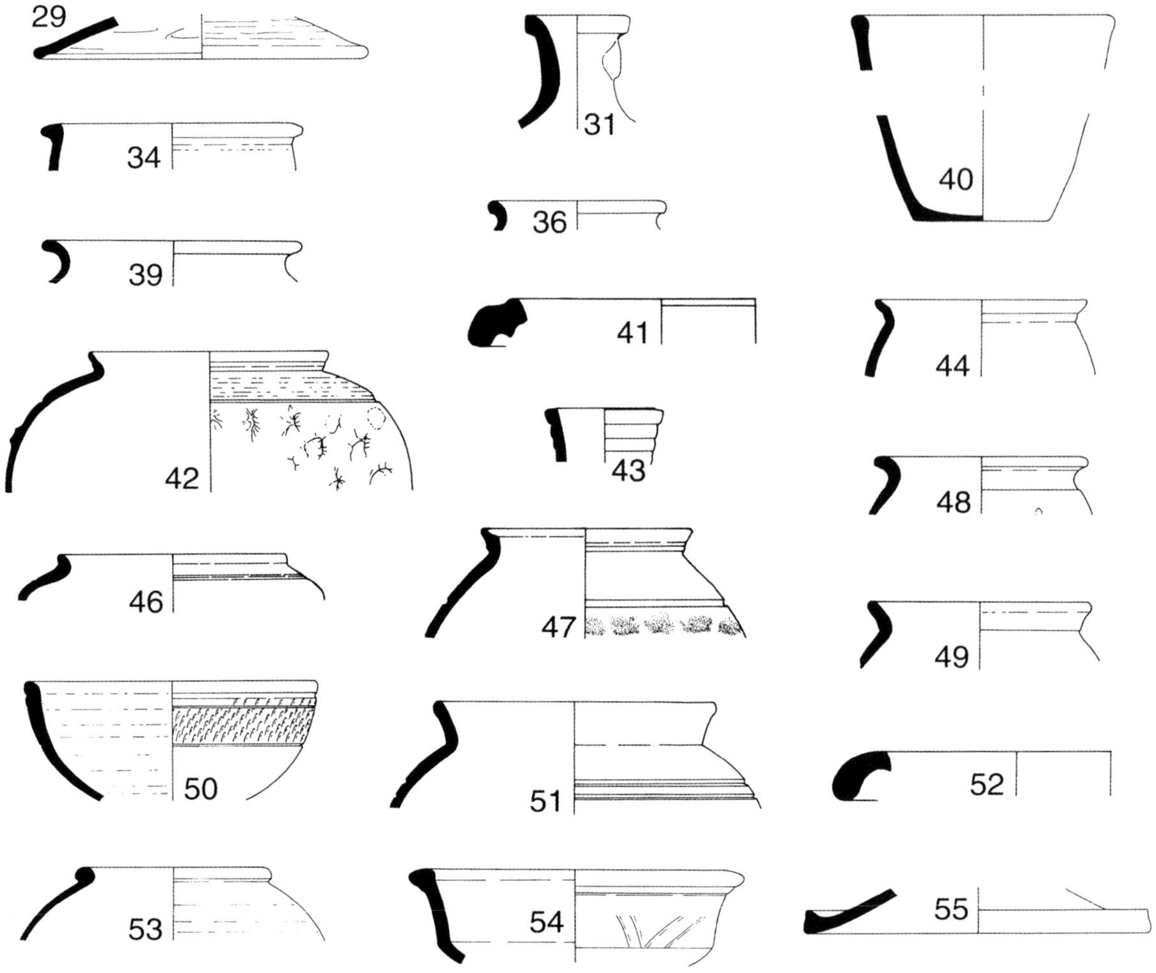

Fig. 2.85 Roman coarse pottery. Scale 1:4

CP 48 Jar in hard light grey fabric with blackened patch externally. The single raised blob is presumably from rusticated decoration, suggesting a Flavian or Trajanic date. Context [**471**].

CP 49 Jar in light grey fabric with darker surface. The exterior and rim have been burnished almost to a polish, allying the vessel to 'terra nigra'. This suggests a Flavian date. Context [**471**].

CP 50 Bowl in granular pink-buff fabric with a cream core and rouletted decoration in a zone externally. It is probably meant to be reminiscent of the samian form 37 but the position of the rouletting would better suit the earlier form 29. If significant, this might place the bowl in the period *c.* AD 70–90, when both samian forms were in use. Context [**471**].

CP 51 Jar in smooth light grey fabric with darker surface, burnished externally. Context [**471** and **487**].

CP 52 Mortarium flange in pink-buff fabric with a white slip. Context [**487**]; a fragment from context [377] may be part of the same vessel (*cf.* CP 38)

CP 53 Bead rim jar in hard grey fabric with a smooth burnished surface externally. Context [**487**].

CP 54 Flanged bowl in granular grey fabric. The sherd is abraded and has acquired a lime deposit, overlain by black encrustation after breakage. Presumably a carinated bowl of Flavian to Trajanic date. Context [**487**].

CP 55 Lid in light grey fabric with a darker surface. A popular local form, *cf.* CP 17, 26 and 29 from contexts [285], [308] and [345]. Context [**487**].

(iii) Roman pottery from the 1972–3 excavations

The publication of the 1987–99 excavations provides a convenient opportunity to revise conclusions on the material from the 1972–4 campaigns. Much of this was seen (by PVW) at the time of excavation but it has subsequently become clear that the dating offered then was too late and based upon an assumption that Black-burnished ware did not appear in south Wales before *c.* AD 120. That this is not the case is confirmed by a re-examination both of the site assemblage as a whole and of the extant samian ware.

Pottery from the 1972 excavations was kindly made available by the Carmarthen Museum (the coarse pottery) and by the late Professor Barri Jones (the samian). In due course all will be deposited in the Museum. Unfortunately, the much smaller quantity of pottery which resulted from the 1973 excavations has not been located in either Carmarthen or Manchester and is not currently, therefore, available for study. Comparison of extant sherds with notes made at the time of excavation suggest that a few pieces of decorated samian (including some of form 29) are also missing.

The samian (by Robert Hopkins)[1]: This report is arranged in four parts: (1) a tabulation of the samian by trench (T1, T2, etc; for location see Fig. 2.1); some sherds were, however, unbagged and without markings, or their contexts were illegible; (2) a tabulation of the total number of each vessel form; (3) a catalogue of the decorated ware (not illustrated); and (4) a report on the name-stamps and a graffito. An attempt has been made to match profiles of individual forms, to give an approximate number of vessels rather than overestimating each form by assuming individual sherds are from separate vessels. Joins and sherds from the same vessels have been noted, but otherwise each sherd may represent a separate vessel.

The soil conditions have badly eroded many of the sherds, with all but a few losing their slip, and all are powdery. Because of this, some decorated pieces were impossible to interpret. The majority of the fabrics represent late South Gaulish ware, *c.* AD 80–110, and only seven Central Gaulish examples, with a maximum date range of *c.* AD 120–190. Only two Les Martres-de-Veyre products (*c.* AD 100–120) were identified.

(1) Samian from the 1972 excavations by trench (Table 2.4)

Table 2.4 Samian from the 1972 excavations by trench

Trench T1

Context	Form	Source	Date	Comments
T1 2	15/17 or 18	SG	70–90	
T1 15A	18/31	SG	80–110	
T1 16	27?	SG	70–110	
"	1 Frag	SG	-110	Flange
"	2 Frags	SG	-110	
T1 25	33	CG	120–190	
T1 31	33	CG	120–190	Burnt
T1 32	1 Frag	SG	-110	Large bowl footring
"	Frag	SG	-110	
"	Frag	SG	-100	Flange?
T1 33	18	SG	70–90	

Trench T1 B

Context	Form	Source	Date	Comments
T1 B+	37	CG	120–190	**Cat no 1**
"	37	CG	120–190	Abraded ovolo
T1 B2	37	SG	75–110	Ovolo

Trench T2

Context	Form	Source	Date	Comments
T2+	15/17 or 18	SG	70–90	
"	15/17 or 18	SG	70–90	
"	15/17 or 18	SG	70–90	Rim
"	18/31	SG	80–110	
"	18/31	SG	80–110	
"	18/31	SG	80–110	Rim
"	18/31?	SG	80–110	Base
"	37	SG	75–110	**Cat no 2**
"	37	SG	75–110	
"	4 Frags	SG	-110	
T2 10	15/17 or 18	SG	70–90	
T2 18	36	SG	70–110	Thick body
T2 23	27	SG	70–110	Rim
T2 29	15/17?	SG	70–90	?Rim
"	15/17 or 18	SG	70–90	

[1] The author is indebted to Peter Webster for access to the pottery and unpublished material; Gavin Evans, Carmarthen Museum; Brenda Dickinson, for the name-stamp report, and the Carmarthenshire Antiquarian Society, for funding that report.

"	18	SG	70–90	Rim and junction
"	18	SG	70–90	
"	18	SG	70–90	
"	27	SG	Neronian/early Flavian	Early, flat rim
"	27	SG	70–110	Rim
"	27	SG	70–110	Rim
"	27	SG	70–110	Rim
"	27	SG	70–110	Rim
"	27	SG	Flavian	Name-stamp, graffito **Cat no 13**
"	27	SG	70–110	
"	27?	SG	70–110	
"	27g	SG	70–85	Footring
"	27g	SG	70–85	
"	27g	SG	70–85	
"	37	SG	75–110	**Cat no 3**
"	37	SG	75–110	**Cat no 4**
"	4 Frags	SG	-110	?Bowls
T2 56	27	SG	70–110	Rim
"	35	SG	70–110	Rim

Trench T2a = T9

Context	Form	Source	Date	Comments
T2a 1	18/31	CG	120–150	

Trench 3B+

Context	Form	Source	Date	Comments
3B+	18	SG	70–90	Rim and junction
"	18	SG	70–90	
"	67	SG	70–85	**Cat no 5**
"	37	SG	75–100	**Cat no 6**

Trench T4

Context	Form	Source	Date	Comments
T4 1	15/17 or 18	SG	70–90	
T4 2	15/17 or 18	SG	70–90	
"	37	SG	75–110	Abraded ovolo
"	2 Frags	SG	-110	
T4 9	29 or 37	SG	60–110	
T4 11	27	SG	70–110	Rim
"	Bowl	SG	60–110	Large footring
T4 18	15/17 or 18	SG	70–90	
"	18	SG	70–90	
"	18	SG	70–90	
"	27	SG	70–110	Semi-burnt
"	27	SG	70–110	
"	27	SG	70–110	
"	27	SG	70–110	
"	27	SG	70–110	
"	33	SG	70–110	
"	37	SG	75–90	**Cat no 7**
T4 21	15/17 or 18	SG	70–90	
"	27	SG	70–110	
T4 25	15/17 or 18	SG	70–90	
T4 28	Curle 11	SG	70–90?	
"	15/17 or 18	SG	70–90	
"	18	SG	Early Flavian	Portion of stamp **Cat no 14**
"	18	SG	70–90	
"	18	SG	70–90	
"	35	SG	70–110	Not T4 49
"	37	SG	75–110	Abraded decoration
T4 32	18	SG	70–90	Rim
"	42	CG	120–150	Rim
"	Curle 15	SG	80–110	

"	Frag	SG	-110	Bowl
T4 34	27	SG	70–110	
T4 37	15/17 or 18	SG	70–90	Footring
"	29	SG	70–85	**Cat no 8**
"	37	SG	75–110	Rim
T4 40	15/17 or 18	SG	70–90	
T4 44	15/17 or 18	SG	70–90	
"	27	SG	70–110	
T4 48	37	SG	75–90	**Cat no 9**
T4 49	35	SG	70–110	Not T4 28
"	37	SG	75–90	**Cat no 9**
"	37	SG	75–90	**Cat no 10**
"	46	SG	80–110	O&P Pl. LV No.4
"	Frag	SG	-110	
T4 51	27/33	SG	70–110	Footring of either form
"	37	SG	75–110	Part ovolo

Trench T5

Context	Form	Source	Date	Comments
T5+	Frag	SG?	-110	

Trench T6

Context	Form	Source	Date	Comments
T6 12	18/31	Les Martres	100–120	

Trench T7

Context	Form	Source	Date	Comments
T7 6	Frag	SG	-110	Bowl

Trench T8

Context	Form	Source	Date	Comments
T8 1	18	SG	70–90	
"	18	SG	70–90	
"	18	SG	70–90	
"	18	SG	70–90	
"	18	SG	70–90	
"	27	SG	70–110	
"	37	SG	75–110	Ovolo
"	Curle 11	SG	70–95?	Flange

Trench T9

Context	Form	Source	Date	Comments
T9 1	27	SG	70–110	
"	36	SG	70–110	Not T9 10
"	37	SG	70–85	**Cat no 11**
"	37	?Les Martres	100–125	**Cat no 12**
"	37	SG	75–110	
"	6 Frags	SG	-110	
T9 10	36	SG	70–110	Not T9 1

Trench T10

Context	Form	Source	Date	Comments
T10 8	15/17	SG	70–90	Rim

Trench T13

Context	Form	Source	Date	Comments
T13	15/17 or 18	SG	70–90	

Unstratified material

Context	Form	Source	Date	Comments
Unstratified	27 frag	SG	70–110	
"	27 frag	SG	70–110	

(2) Totals of each vessel form (Table 2.5)

Table 2.5 Totals of each vessel form

Form	SG	SG?	CG	CG?	Total
15/17	1	1	-	-	2
15/17 or 18	16	-	-	-	16
18	17	-	-	-	17
18/31	4	1	2	-	7
27	22	2	-	-	24
27 or 33	1	-	-	-	1
27g	3	-	-	-	3
29	1	-	-	-	1
29 or 37	1	-	-	-	1
33	1	-	2	-	3
35	3	-	-	-	3
36	3	-	-	-	3
37	17	-	2	1	20
42	-	-	1	-	1
46	1	-	-	-	1
67	1	-	-	-	1
Curle 11	2	-	-	-	2
Curle 15	1	-	-	-	1
Totals	95	4	7	1	107

(3) Catalogue of the 1972 Decorated Ware (none illustrated)

No 1 Form 37 Lezoux, Central Gaul. Vine scroll motif with vine leaves and spiral tendrils. *c.* AD 120–190. **[1B+]**.

No 2 Form 37 South Gaul. 4 abraded sherds. The ovolo is possibly trifid tongued, and to the right. Below is a horizontal ?bead row border. To the left, the panel is divided into an upper and lower. The upper is an animal (?hound) running left, with a ?bead row border below. The bottom panel has dog running right. The panels are bounded on the right by a ?bead row with rosetted junctions. The centre panel is composed of a central, composite column, to the left a ?nude woman facing the column with raised right arm. To the column's right, and facing it is a 'semi-seated' ?nude male, with right hand under his chin. The male's pose is identical to O.541 and O.542, but is a different size, while the female figure has not been identified; both appear, however, on an unpublished vessel from the Jenkins Field III barracks at Caerleon (Webster, *pers. comm.*). Beneath the panels is a wavy line border, and a basal wreath of S-shaped gadroons. The other sherds are of gadroons (2), and an upper panel with hound running right. *c.* AD 75–100. **[T2+]**.

No 3 Form 37 South Gaul. Rim sherd with ovolo, trifid tongue to the right and bead row border below. The decoration appears to be a winding scroll. *c.* AD 75–110. **[T2 29]**.

No 4 Form 37 South Gaul. Body sherd with the lower lobe of a winding scroll motif, within which is a grass tuft composed of a trifid-leafed motif stamped thrice in a semicircle. The winding scroll seems to have become less common in the 90s. *c.* AD 75–100. **[T2 29]**.

No 5 Form 67 South Gaul. Decoration consisting of a nile goose facing left, a scroll and a saltire composed of opposed trifid leaves tied by an astragalus. The leaf is commonly used as a basal design *e.g.* Frere 1972, fig. 87, no. 63. *c.* AD 70–85. **[3B+]**.

No 6 Form 37 South Gaul. Ovolo with rosette-tipped tongue to the left, used by Mercator (Mees 1995 taf. 134) and Frontinus I (Mees 1995, taf. 59). The decoration appears to be composed of two horizontal bands, divided by bead row borders. The upper zone is a hunting scene, a dog chasing a stag left, with grass tufts below. This is a common design attributable to Mercator (Mees 1995, taf. 133, 136 etc.). The lower zone appears to be a vine scroll motif, with a bottle bud terminal. Attributable to Mercator or his associates. *c.* AD 75–100. **[3B+]**.

No 7 Form 37 South Gaul. Body sherd with a horizontal panel division. The decoration in the upper panel has not been identified, but in the lower there is a nile goose facing left, and the upper left portion of a large 'open' leaf. Probably *c.* AD 75–90. **[T4 18]**.

No 8 Form 29 South Gaul. Basal wreath composed of an inverted leaf motif, unusual in that the motif is normally sideways, not inverted as in this example. *c.* AD 70–85. **[T4 37]**.

No 9 Form 37 South Gaul, 2 joining sherds. The main decoration consists of a winding scroll, the upper lobes of which contain tendrils ending in large lanceolate leaves. The lower lobes consist of a composite column which had tendrils springing from the base, both left and right, and ending in elongated leaves. The general scheme is reminiscent of form 29 (*e.g.* Knorr 1919 taf. 37, K and 64, H and 87). There is possibly a bead row horizontal between the scroll and the chevron basal wreath. *c.* AD 75–90. **[T4 48** and **49]**.

No 10 Form 37 South Gaul. Ovolo with oval tongue to

the right, and turning slightly to the left. Below, a horizontal bead row. *c.* AD 75–90. [**T4 49**].

No 11 Form 37 South Gaul. Body sherd with two horizontal bands of decoration. The upper has a vine scroll motif with rosettes and leaf terminals. The bottom appears to be leaf tufts alternating with animal figures. The identifiable figure is of a bear walking left, and occurs on form 29s by Masculus and Pass(i)enus (Mees 1995, taf. 108; 158); form 30s (Mees 1995, taf. 108) but not on form 37s by Masculus. *c.* AD 70–85. [**T9 1**].

No 12 Form 37 ?Les Martres-de-Veyre. Worn sherd with panelled decoration. The panels are demarcated by very fine, squat bead row verticals, a typical Martres feature, while the horizontal row is indistinct. At the junction is a circular rosette. The left hand panel appears to be full length, with indistinct decoration.. The right hand panel is divided. The figure in the upper part remains unidentified, but the figure in the lower might be a hare facing left. *c.* AD 100–125. [**T9 1**].

(4) Name stamps and Graffito (by Brenda Dickinson)
Both vessels are from the La Graufesenque kilns.

No 13 Form 27, South Gaulish. The potter's stamp is illegible. The footring has an external groove, as on pre- and early-Flavian examples of the form. A graffito, X, has been cut under the base, after firing. Flavian. [**T2 29**].

No 14 Form 15/17 or 18, South Gaulish, stamped OF..[. The F is followed by a vertical stroke, but there are too many potential candidates. Flavian, probably early in the period. [**T4 28**].

The Coarse Pottery (by Peter Webster): The coarse pottery from the 1972 excavations was examined as part of a course in Cardiff for undergraduates and Continuing Education Students in 1996–7. This is not the place to attempt a full publication of the material as this would require a full structural report. It does seem reasonable, however, to present a summary of the 1972 material in such a way as to make it comparable with the 1989 assemblage and this has been done through a Table (2.6) similar to Table 2.3 above.

It is interesting to note that the high percentage of jars noted among the 1989 assemblage is not present in the 1972 one, whereas bowls, poorly represented in 1989, are much more common among the 1972 material. It may be that we are seeing here some indication of differing functions for buildings within the two areas or that different types of cooking were undertaken in different areas.

Most of the fabric figures are broadly comparable between the two sites, with the exception of the Black-burnished ware which is noticeably more numerous among the 1972 material. It is argued below that BB1 tends to increase in quantity as one gets later in the fort/fortlet sequence and this is reinforced by observations made in 1989 that BB1 is present only, or mainly, in the later levels on that site. It is hard to escape the view that the greater incidence of BB1 on the 1972 site is because more later material was found on that site. Quite why this should be when both sites appear to lie within the fortlet, the latest military

Table 2.6 Summary of coarseware vessels from 1972 excavations

Fabric	Flagon	Jar/ Bkr	Bowl	Dish	Tazza	Lid	Mort	Total	%	1989 %age
Local grey	-	5	4	-	-	2	-	11	13	16
Other reduced	-	1	1	-	-	1	-	3	3	10
Oxidised	2	6	3	-	1	3	-	15	17	28
Cln Mortar	-	-	-	-	-	-	2	2	2	2
Cln Ware		-	1	-	-	-	-	1	1	-
Severn Valley	-	6	-	-	-	-	-	6	7	4
Malvern	-	8	-	-	-	-	-	8	9	8
BB1	-	12	14	1	-	4	-	31	35	22
Verulam Mortar	-	-	-	-	-	-	4	4	5	2
Gallo-B Mortar	-	-	-	-	-	-	1	1	1	-
Local Mortar	-	-	-	-	-	-	1	1	1	-
Terra Nigra	-	1	-	1	-	-	-	2	2	2
N.Gaul Rough-cast		3	-	-	-	-	-	3	3	2
Totals	2	42	23	2	1	10	8	88	99	
%	2	48	26	2	1	11	9	99		
1989 %age	8	64	6	4	2	8	4			

phase of the site, is unknown but there does seem to be a discrepancy in the evidence for post *c*. AD 100 activity on the two sites.

Chronologically, two things can be observed from the 1972 coarseware. Firstly, the presence of a single vessel of Caerleon ware is of interest. The ware is generally dated to the early to mid 2nd century and it is argued elsewhere that a date range of *c*. AD 110–160/70 would be appropriate (Webster, in Manning 1993, 256–7). The particular vessel represented is unlikely to be an early Caerleon ware product, but, as a vessel from unstratified levels (T5+), it may do no more than suggest some later Roman activity on a fortlet site already abandoned. It may, however, be compared with a similar vessel from the bath-house site (see section 5.3.1(B) below).

Of greater interest is the presence of BB1 flanged (or 'flat-rim') bowls in the 1972 assemblage where they account for the considerable increase in the number of BB1 bowls recorded. These are missing from the 1989 material. The class does not appear among the late iron-age forms from which Romano-British Black-burnished ware developed. In its earliest form it clearly mimics the common mid 1st- to early 2nd-century flanged and carinated bowls but the developed BB1 form may well be 2nd century in date. Of the examples noted among the 1972 assemblage all appear to be either from the last phase of the fort (contexts 4.11 and 12, 5 examples in all) or from the fortlet (contexts 1.15a, 1B.2, 1B.8 and possibly 2.19 and 4.49, 9 examples in all). Also of possible chronological significance is a bowl of Gillam 1976, form 52 which appears in south Wales from the early 2nd century (*cf.* Zienkiewicz 1992, no. 91).

The 1972 assemblage appears, therefore, to have a few elements in it which would make it slightly later in date than the 1989 one. For greater precision we must turn to the samian evidence summarised by Robert Hopkins above. This suggests activity on the fortlet site into the 120s. The coarse pottery evidence would support this but does not suggest that such activity lasted long into the reign of Hadrian.

(iv) The Post-Roman pottery from the 1989 site (by Dee Brennan)[1]

A total of 645 sherds of post-Roman pottery were found on the site, consisting largely of post-medieval material associated with the clom and later stone buildings. There are four sherds of medieval pottery of uncertain date: two pieces in an unidentified fabric (Type Fabric 2, LP 5 and 6) and two abraded (?)jug sherds of 'local' gravel-tempered fabric (Type Fabric 4). These few sherds and the small quantity of later 15th- to 16th- (Type Fabric 1, LP 1–4) and 17th-century wares (Type Fabric 11, LP 31 and 32) are an indication of earlier periods of occupation on the site. Only one sherd, however, (LP 5) is related to a feature of possible medieval date, namely the fill [248] of the linear gully [254].

The collection comprises an extensive range of products which are drawn from a number of sources. They include continental imports from France and Germany, and non-local English imports, most of them from North Devon, Staffordshire and Bristol. This material came from a broad range of contexts, the details of which are tabulated by type fabric in the accompanying Table 2.7. Together with the other post-Roman finds, the pottery provides one of the main means of dating the post-medieval phases on the site, especially those of the clom and later stone buildings, although many pieces were clearly residual. For ease of reference, the TPQ evidence is collated in Table 2.28; of the contexts listed there, only [127], [158] and [280] did not produce pottery.

TYPE FABRICS

1 Beauvais wares:
- A. Green-glazed, late 15th–16th century. LP 1 and 2
- B. Sgraffito, 16th century. LP 3
- C. Stoneware, 16th century. LP 4

2 Unidentified fabric, medieval, date uncertain. LP 5 and 6

3 North Devon wares:
- A. Calcareous, gravel-free, late 15th-early 17th century. LP 7
- B. Gravel-tempered, 16th–18th century. LP 8–18
- C Gravel-free, 17th–18th century. Not catalogued.
- D. Gravel-free slipware, 17th-mid 18th century. Not catalogued.

4 Dyfed gravel-tempered ware, (?) 12th–16th/early 17th century. LP 19

5 Tin-glazed wares:
- A. Unidentified fabric, import ?, date uncertain. LP 20
- B. English, 17th–18th century. LP 21

6 Bristol/Staffordshire type slipware, late 17th–18th century. LP 22–24

7 Bristol/Staffordshire iron-glazed wares, 18th century. LP 25

8 Westerwald stoneware (German), late 17th-early 18th century. LP 26

9 English stoneware, 18th and 19th century. LP 27–30

10 Banded creamware with mocha decoration, late 18th-early 20th century. Not catalogued.

11 Black-glazed red earthenware, 17th–19th century. LP 31 and 32

12 Miscellaneous brown- and yellow-glazed red earthenware, 18th–19th century. LP 33

13 Miscellaneous unglazed red earthenware, 19th–20th century. Not catalogued.

14 Whiteware (porcelain, china and earthenware), late 18th–20th century. Not catalogued.

---

[1] Thanks to Cathy O'Mahoney (now Cathy Freeman) for help in identifying some of the earlier pieces in the assemblage, most notably the imported 15th- to 16th-century Beauvais wares.

Table 2.7 Summary of Post-Roman Pottery by context and type fabric

	Type Fabric nos															
Context no	1	2	3	4	5	6	7	8	9	10	11	12	13	14	**Total**	Cat nos
13	-	-	-	-	-	-	1	-	-	-	-	1	-	14	**16**	-
16	-	-	-	-	-	-	-	-	-	-	-	-	2	3	**5**	-
25	-	-	1	-	-	-	3	-	1	-	-	-	-	-	**5**	-
30	-	-	-	-	-	-	-	-	1	2	1	-	-	7	**11**	-
31	-	-	-	-	-	-	-	-	-	-	2	-	-	1	**3**	-
32	-	-	11	-	-	1	-	1	2	4	-	10	3	20	**52**	LP 8
34	-	-	-	-	-	-	1	-	-	-	2	-	-	4	**7**	-
36	-	-	7	-	-	1	1	-	-	-	1	-	-	8	**18**	LP 32
37	-	-	-	-	-	-	-	-	-	-	-	-	-	1	**1**	-
38	-	-	-	-	-	-	-	-	-	1	2	-	-	9	**12**	-
39	-	-	-	-	-	-	-	-	-	1	-	-	-	-	**1**	-
41	-	-	-	-	-	-	-	-	-	1	-	-	-	3	**4**	-
43	-	-	-	-	-	-	3	-	-	-	-	-	-	2	**5**	-
49	-	-	-	-	-	-	-	-	-	1	-	-	-	1	**2**	-
52	-	-	5	-	-	-	-	-	-	-	-	-	-	1	**6**	-
53	-	-	-	-	-	-	1	-	-	1	-	-	-	-	**2**	-
54	-	-	-	-	-	-	-	-	-	-	1	1	-	11	**13**	-
57	-	-	-	-	-	-	1	-	-	-	1	-	1	1	**4**	-
63	-	-	-	-	-	-	-	-	-	-	1	-	-	-	**1**	-
65	-	-	-	-	-	-	-	-	-	-	-	2	-	-	**2**	-
67	-	-	-	-	-	-	2	-	-	-	-	-	-	5	**7**	-
70	-	-	1	-	-	-	1	-	-	-	-	-	-	-	**2**	-
71	-	-	1	-	-	-	-	-	-	-	1	-	-	-	**2**	-
72	-	-	-	-	-	-	1	-	-	-	-	-	-	-	**1**	-
73	-	-	-	-	-	-	-	-	-	1	-	-	-	4	**7**	-
76	-	-	5	-	-	-	-	-	-	-	1	-	-	-	**6**	-
79	-	-	34	1	-	-	3	-	4	-	-	2	4	10	**58**	LP 9, 30
82	-	-	-	-	-	-	1	-	-	-	-	-	-	-	**1**	-
84	-	-	-	-	-	1	-	-	-	-	-	-	-	-	**1**	LP 22
92	-	-	26	8	-	1	-	-	-	-	-	1	-	7	**43**	LP 19, 22
94	-	-	1	-	-	-	-	-	-	-	-	-	-	-	**1**	-
96	-	-	2	-	-	-	2	-	1	-	-	-	-	-	**5**	-
97	-	-	3	-	-	1	-	-	2	-	-	-	-	3	**9**	-
99	-	-	-	-	-	-	-	-	-	-	-	-	-	1	**1**	-
100	-	-	1	-	-	-	-	-	-	-	-	-	-	-	**1**	-
102	-	-	18	-	-	-	-	-	-	-	-	-	-	1	**19**	-
104	-	-	4	-	1	1	1	-	-	-	1	-	-	1	**9**	LP 10, 31

Table 2.7 Summary of Post-Roman Pottery by context and type fabric

						Type Fabric nos									
Context no	1	2	3	4	5	6	7	8	9	10	11	12	13	14	**Total**	Cat nos
106	-	-	18	-	-	-	-	-	-	-	-	-	-	2	**20**	LP 11, 12
121	1	-	-	-	-	-	3	-	-	-	-	-	-	2	**6**	LP 3
122	-	-	-	-	-	-	3	-	1	-	-	-	-	2	**6**	LP 27
123	-	-	-	-	-	-	1	-	-	-	-	-	-	4	**5**	-
125	-	-	9	-	-	-	1	-	-	-	-	-	-	-	**10**	LP 13
128	-	-	-	-	-	-	-	-	1	-	1	-	1	3	**6**	-
129	-	-	5	-	-	-	1	-	-	-	-	-	-	-	**6**	-
130	-	-	12	-	-	-	3	-	-	-	-	-	-	-	**15**	-
131	-	-	7	-	-	-	-	-	-	-	2	-	-	12	**21**	-
156	-	-	-	-	-	-	-	-	-	-	-	1	-	-	**1**	-
157	2	1	46	2	1	3	41	3	1	-	9	2	-	3	**114**	LP 1, 4, 6-8, 14-16, 20, 23-26, 29
161	-	-	6	-	-	-	-	-	-	-	-	1	-	2	**9**	LP 17
174	1	-	28	1	-	-	1	-	-	-	-	-	-	-	**31**	LP 2, 19
179	-	-	1	-	-	1	3	-	1	-	1	-	-	-	**7**	-
180	-	-	1	-	-	-	1	-	-	-	1	1	-	-	**4**	-
182	-	-	1	-	2	-	-	-	-	-	-	-	-	-	**3**	-
186	-	-	1	-	-	-	-	-	-	-	1	-	-	-	**2**	-
188	-	-	-	-	-	-	-	-	-	-	3	-	-	-	**3**	-
197	-	-	-	-	-	-	-	-	-	-	-	-	-	1	**1**	-
211	-	-	-	-	-	-	-	-	-	-	2	-	-	-	**2**	-
220	-	-	-	-	-	-	1	-	1	-	-	-	-	-	**2**	LP 28
223	-	-	-	-	6	-	-	-	-	-	-	-	-	-	**6**	LP 21
226	-	-	-	-	-	-	4	-	-	-	-	-	-	1	**5**	-
227	-	-	1	-	-	-	-	-	-	-	-	-	-	2	**3**	-
235	-	-	-	-	-	-	-	-	-	-	-	-	-	2	**2**	-
237	-	-	1	-	-	-	2	-	-	-	-	-	-	-	**3**	-
248	-	1	-	-	-	-	-	-	-	-	-	-	-	-	**1**	LP 5
273	-	-	3	-	-	-	-	-	-	-	-	-	-	-	**3**	LP18
287	-	-	-	-	-	-	1	-	-	-	1	-	-	-	**2**	-
291	-	-	-	-	-	-	-	-	-	-	1	-	-	-	**1**	-
311	-	-	1	-	-	-	-	-	-	-	1	-	-	-	**1**	-
330	-	-	-	-	-	1	-	-	-	-	-	-	-	-	**1**	-
Total	**4**	**2**	**261**	**12**	**10**	**11**	**90**	**4**	**16**	**12**	**36**	**22**	**11**	**154**	**645**	

CATALOGUE (Figs 2.86 and 2.87)

Type 1 A–C: The earliest identifiable post-Roman pottery found on the site are the imported wares from the Beauvais region of northern France. Four sherds represent four different vessels. Two small sherds are from green-glazed jugs (Type Fabric 1A, LP 1 and 2), one is from a dish or bowl with (?)sgraffito decoration (Type Fabric 1B, LP 3), and one is the base of a vessel, rather crudely made, in a plain grey stoneware (Type Fabric 1C, LP 4). All four vessels may be dated to the first half of the 16th century. 16th-century Beauvais wares are not uncommon in Welsh ceramic assemblages and a recent survey of medieval and later pottery in Wales lists their wide distribution (Papazian and Campbell 1992, 21).

LP 1 (Not illustrated) Small fragment from rim of jug in fine white fabric with mid green glaze externally, glaze extending partly over the rim internally. Context **[157]**.

LP 2 (Not illustrated) Fragment from body of jug in buff fabric with green glaze externally streaked brown. The interior is glazed partly red-brown, partly dark brown. Context **[174]**.

LP 3 Rim of dish or bowl in fine buff fabric. Red-brown slip with white over, under a clear lead glaze, exterior unglazed. Context **[121]**.

LP 4 Base of (?)jug in greyish-buff stoneware with light grey surfaces. Similar in profile to a vessel found in Exeter (Allan 1984, 172, no. 1931, fig. 84) with a group of pottery dated *c.* 1550–80. Context **[157]**.

Type 2: Two small cone-shaped (?)feet for vessels or (?)bungs were recovered. They are in an unidentified unglazed fabric which is hard-fired, orange to reddish-brown in colour and tempered with quartz sands and occasional red inclusions. Both examples are fire-blackened and sooted along three-quarters of their length from the narrow end. A medieval date seems likely, their precise function is uncertain.

LP 5 Vessel foot or bung? Circular sectioned cone with rounded tip, fire-blackened and sooted. Length 39mm, maximum diameter 20mm. Context **[248]**.

LP 6 (Not illustrated) As LP 5 but incomplete. Length as extant 32mm, maximum diameter 19mm. Context **[157]**.

Type 3 A–D: Well over half the coarseware vessels found on the site are from North Devon. Four different fabrics are present (for fabric description and forms see Allan 1984, 148–9).

Type 3A: The earliest North Devon ware found on the site is the rim of a jar in a fine calcareous gravel-free fabric. At Exeter the fabric is first present in a group of late 15th-century pottery. Of all 16th-century North Devon wares found there, 90% were in this fabric and all identifiable vessels were jars (Allan 1984, 129–131).

LP 7 Rim of lid-seated jar, partial olive-green glaze externally, interior unglazed. Context **[157]**. A comparable vessel (Grant 1983, type 14, 136 and fig. 40) from Exeter comes from a late 16th-century group of pottery (Allan 1984, 173, no. 1957, fig. 84). The type is usually glazed internally.

Type 3B: 258 sherds of North Devon gravel-tempered ware were found on the site. A wide range of kitchen wares in this fabric were exported to south and west Wales from the beginning of the 17th century right through to the start of the 20th (Evans 1979, 19), but most types from excavated groups fall within the 17th- to 18th-century date range, a date most suited to the Pumsaint vessels. Any attempt at closer dating is hampered by the fact that types were made over a long period of time (Allan 1984, 132).

The identifiable Pumsaint vessels include jars, bowls, basins and at least one colander or strainer. All are glazed internally in shades of olive green through to brown. One possible jug from context [157] (not catalogued), evidenced by a single body sherd, is glazed on both surfaces.

LP 8 Two joining rim fragments of a handled jar, intermittent thumb-pressing on rim exterior. Olive-green glaze internally. Contexts **[32]** and **[157]**.

LP 9 Rim of bowl with brown glaze internally. Context **[79]**.

LP 10 Rim of bowl with yellowish-green glaze internally. Context **[104]**.

LP 11 Rim of bowl with olive-green glaze internally, glaze extending over the upper part of the rim externally. Context **[106]**.

LP 12 Rim of jar with olive-green glaze internally. For the rim form *cf.* Grant 1983, type 11 (136–7, fig. 40). Context **[106]**.

LP 13 Two joining rim fragments of a bowl with olive-green glaze internally. Context **[125]**.

LP 14 Rim of lid-seated jar with olive-green glaze internally. Context **[157]**.

LP 15 Rim of bowl, no glaze surviving internally. For the rim form *cf.* Grant 1983, Type 3D (136–7, fig. 40). Context **[157]**.

LP 16 Rim of bowl with green-brown glaze internally. Context **[157]**.

LP 17 Handled basin with olive-green glaze internally. Context **[161]**.

LP 18 Fragment from base of colander or drainer with green-brown glaze internally, pierced holes *c.* 5mm diameter on floor of base. Context **[273]**.

Type 3C (Not catalogued): A single body sherd of North-Devon gravel-free fabric came from the makeup [52] immediately pre-dating the construction of Room 1. The sherd has a light brown glaze internally and is unglazed externally. The date and range of forms in this fabric are the same as those in the gravel-tempered fabric.

Type 3D (Not catalogued): North Devon gravel-free slipware is represented by a single body sherd from [125], a pebble surface at the south end of the site. The form is not identified. A white slip on both surfaces has a clear overglaze, appearing yellow. North Devon plain slipwares were produced into the mid 18th century (Allan 1984, 132).

Type 4: 12 sherds of 'local' Dyfed gravel-tempered ware represent three vessels. The fabric defined by O'Mahoney (1985a, 20) is characteristically gritty containing frequent shale and sandstone gravels and only a little quartz. Slight variations in fabric and form seen from material excavated within Dyfed indicate more than one centre of production. In a recent study of pottery from Carmarthen Greyfriars the first attempts have been made at distinguishing the different local wares (O'Mahoney 1995).

Both the starting and terminal dates for the production of Dyfed gravel-tempered wares are unclear. The earliest occurrence of the fabric is at Laugharne Castle where it is found in 12th-century contexts. The latest occurrence is from a 16th-century context from the Tudor Merchant's House, Tenby (O'Mahoney 1985b, 28–9). It has been suggested that production continued into the 17th century after which the local market was flooded with North Devon wares (see Papazian and Campbell 1992, 56).

The one identifiable vessel from Pumsaint is a lid-seated jar, internally glazed with a thumb-pressed cordon around the rim externally (LP 19). An unglazed body sherd from context [79], the 19th-century farmyard surfaces, is of indeterminate form. Another two abraded body sherds from the garden soil [157] have traces of glaze externally and may be from a jug of medieval date.

LP 19 Joining rim and body fragments of a lid-seated jar with thumb-pressed cordon around the neck of the rim, dark olive green glaze internally. Contexts **[92]** and **[174]**.

Type 5A: A single sherd of tin-glazed earthenware of uncertain form, origin, and date came from [157], the garden soil filling the top of the timber building pit [245].

LP 20 Lug handle in a cream-buff fabric, tempered with fine sands and sparse red inclusions. A discoloured (?)blue-tinted glaze is much decayed and heat damaged. The smallest trace of decoration consists of a pale blue spiral pattern. A probable import. Context **[157]**.

Type 5B: Nine sherds of English tin-glazed earthenware were found on the site. Two vessel types are present but only one is identified. This (LP 21) is a small 'drug pot' (Albarellos) with speckled purple glaze externally. Not enough of the vessel survives to reconstruct a profile for illustration. Other pieces not catalogued are a body sherd of a 'drug pot' with the same speckled glaze and possibly from the same vessel, from [104], the fill of the gatepost pit [105], and two plain body sherds of indeterminate form from [182], the fill of [183] an oval hollow in the area of Room 1. A 17th- to early 18th-century date seems most likely for these vessels.

LP 21 (Not illustrated) Rim and five body fragments of a small 'drug pot' with speckled purple (manganese) glaze externally. Context **[223]**.

Type 6: 11 sherds of Bristol/Staffordshire type slipwares were found on the site. These are in off-white or pinkish-buff fabrics. Decorated surfaces are coated in a dark red slip with a white slip over, appearing brown and yellow beneath a clear lead glaze. Types of decoration include combing and slip-trailing. One piece (LP 23) has additional 'jewelled' decoration where dots of white slip are applied over brown trailed slip.

The identified forms include sherds from large press-moulded dishes with piecrust edges (LP 22), plates with short overhanging rims (LP 23) and small hollow forms (LP 24). The date for these is late 17th to mid 18th century.

LP 22 Two joining rim fragments of press-moulded dish with piecrust edge. Combed decoration on interior, exterior undecorated and unglazed; *cf.* a dish from Exeter (Allan 1984, 214, no. 2710, fig. 122), from a pit group dated shortly after 1760. Contexts **[84]** and **[92]**.

LP 23 Rim of plate with short overhanging rim. Trailed-slip and 'jewelled' decoration on interior, red slip on unglazed exterior. Context **[157]**.

LP 24 Rim of bowl or cup with combed decoration on exterior, white slip on interior with clear over glaze, appearing yellow. Context **[157]**.

Type 7: Bristol/Staffordshire type iron-glazed wares recovered include 61 sherds from ale-tankards and 29 sherds from miscellaneous vessels; most are indeterminate body sherds (not catalogued).

All are in buff-coloured fabrics with a thick brown glaze inside and out. Dark brown streaks ('mottle') in the glaze are the result of iron-staining. The underside and lower portion of tankards are usually unglazed externally.

Ale tankards (LP 25) in this type fabric are decorated with bands of horizontal grooves above the base and at the point where the handle meets the body, both at the upper and lower attachment. See a typical half-pint tankard from Exeter (Allan 1984, 209, no. 2550, fig. 115), dated to the first quarter of the 18th century. At Pumsaint there are several body sherds with bands of horizontal grooves but the few recovered base sherds have only a single external groove. The date for these wares is 18th century though some pieces may be early 19th century.

LP 25 Two joining base fragments of a tankard with a single external groove. Mid brown streaky glaze inside and out, underside and lower portion of base unglazed. Context **[157]**.

Type 8: Four sherds of German stoneware are from

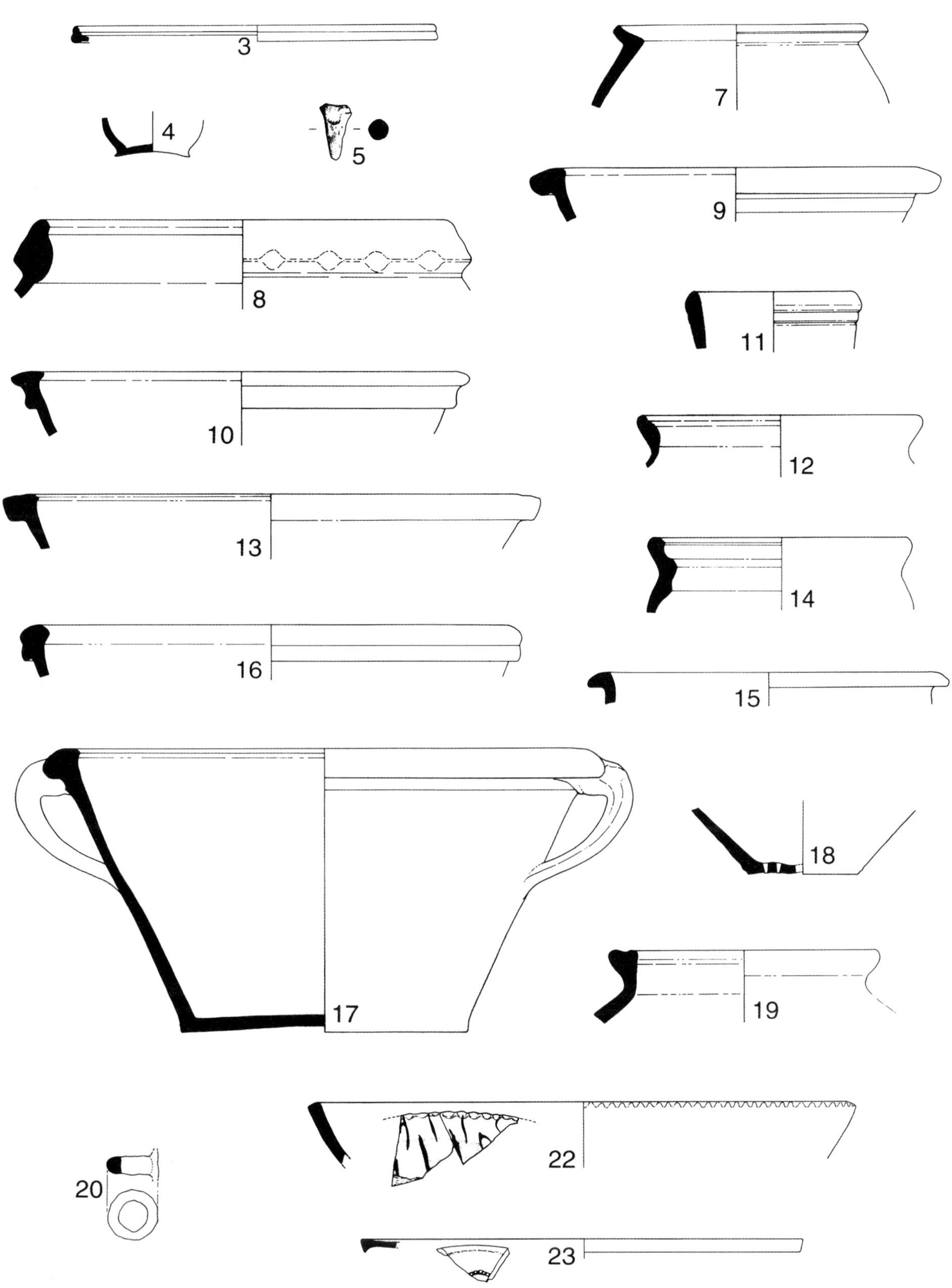

Fig. 2.86 Post-Roman pottery. Scale 1:4

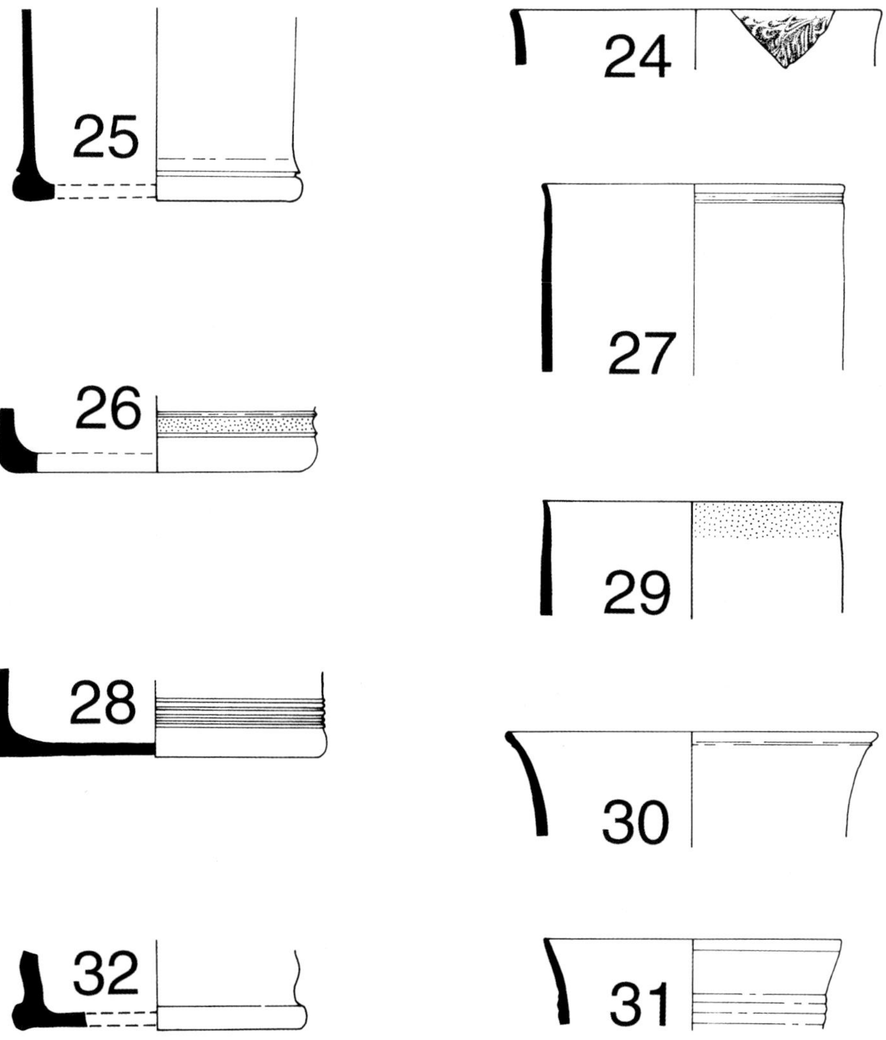

Fig. 2.87 Post-Roman pottery. Scale 1:2

Westerwald tankards of late 17th- or early 18th-century date.

LP 26 Fragment from base of tankard in light grey stoneware with grey-blue salt-glazed surfaces. Decorated externally with cobalt blue bands. Westerwald. Context **[157]**.

Type 9: 16 sherds of English stoneware represent a range of 18th- and 19th-century types. The 18th-century pieces are all from tankards. LP 27 and 28 are unsourced, LP 29 and 30 are probably Staffordshire products. The 19th-century vessels of uncertain source (not catalogued) include the handle of a large flask (?Bristol), two bottle rims, the base of a black-glazed teapot and part of an inkwell.

LP 27 Rim of tankard in buff stoneware with dark brown salt-glaze inside and out. English. Context **[122]**.

LP 28 Base of tankard in buff stoneware with light brown salt-glaze inside and out. English. Context **[220]**.

LP 29 Rim of tankard in light grey stoneware. Dipped white salt-glazed surfaces with a brown band of iron-oxide slip at the rim. ?Staffordshire, early 18th century; *cf.* a tankard from the Williamsburg excavations (Noël Hume in Atterbury 1980, 18, fig.4, left), discarded *c.* 1770. Context **[157]**.

LP 30 Three joining rim and body sherds of tankard in white salt-glazed stoneware. ?Staffordshire, 18th century. A comparable tankard from Williamsburg (Noël Hume in Atterbury 1980, 18, fig.4, right), was discarded *c.* 1759. Context **[79]**.

Type 10 (Not catalogued): 12 sherds of banded creamware were found on the site. These are in a white-bodied earthenware decorated externally in bands of colour beneath a clear glaze. All are similarly coloured with mid blue and dark brown bands above and below a broad grey zone. Five sherds showing the grey band have mocha decoration, a distinctive tree-like pattern achieved by applying colour in liquid form onto a wet surface which is then allowed to spread (see Atterbury 1980, 240–4).

Banded creamwares with or without mocha

decoration were made from the late 18th century to the early part of the 20th. The two most common forms are tankards and jugs which were made in standard measure sizes for use in public houses.

The Pumsaint sherds are from jugs and tankards of probable 19th-century date.

Type 11: 36 sherds of various black-glazed red earthenwares are from vessels of 17th-, 18th- and 19th-century date. The earliest types are fragments from handled cups in a hard-fired red or purple fabric with an all-over black glaze (LP 31 and 32). These wares are of unknown source but most would suit a 17th-century date. The later 18th- and 19th-century vessels (not catalogued) include sherds from bread crocks and other large storage vessels. The small hollow forms are not identified as only small body sherds were recovered.

LP 31 Small fragment from rim of cup in a hard-fired purple fabric with pitted black glaze inside and out. 17th-century. Context **[104]**.

LP 32 Fragment from base of cup in a hard-fired purple fabric with thick black glaze internally, underside and lower portion of the base unglazed externally. 17th-century. Context **[36]**.

Types 12 and 13 (Not catalogued): 22 sherds of miscellaneous yellow and brown-glazed red earthenwares (Type 12) and 11 sherds of unglazed red earthenwares (Type 13) consist mostly of indeterminate body sherds of 18th-, 19th- and 20th-century date. Identifiable vessels include bowls, jugs and plant-pots. None is sourced but all are typical of late red earthenwares produced at many small country potteries.

Type 14: 154 sherds of mass-produced 'china' were found on the site comprising everyday tablewares and utility vessels for use in the kitchen. Very few pieces were kept and none is catalogued.

18th-century cream and pearlwares were not recognised but a handful of sherds of late 18th-century porcelain were found. The majority of sherds were white-bodied earthenwares of 19th-century and more recent date; a little modern china was also recorded. This late material included transfer-printed, sponge-decorated, painted and plain wares. None of the recovered pieces had a maker's mark but it is likely that the main source of supply was the Staffordshire potteries.

(D) Clay tobacco pipes (by Gill Evans)

The clay tobacco pipes discussed in this report date from *c.* 1660 to *c.* 1900, with the majority made between 1660 and 1720. Stem fragments have been discounted.

The earliest, made *c.* 1660–80, are the 'Hereford' types with the 'Wheel' and 'Rose and Crown' stamped heads (Peacey 1985). Slightly later are the 'Broseley 5' types with their distinctive long 'tailed' heels (Atkinson 1975). There are strong parallels with the assemblage from the Carmarthen Greyfriars excavations (Evans 1996), with the domination of Hereford 'Rose and Crown' and 'Wheel' marks, and Edward Taylor of Much Wenlock as the maker of the 'Broseley 5' types. The 18th century is represented by one fragment of a long thick spurred pipe, and the 19th century by two popular designs and a beautifully made pipe with a mould-imparted decoration of a ram's head on the front.

Although this is a small assemblage the connection with the pipes from Carmarthen Greyfriars is very interesting, and merits further research. As Pumsaint was a drovers' herding point for cattle and other animals being driven from west Wales to the English Borders in the 17th to early 19th centuries, it is likely that the pipes were discarded by drovers. Clay tobacco pipes made by the Hughes and Taylor families are certainly found on other sites in south Wales with droving connections, such as Llanmaes (Newman and Wilkinson 1997), Brecon (Locock 1996) and Carmarthen.

The pipe fragments came from a variety of contexts, the details of which are tabulated in the accompanying Table 2.8. Together with the other post-Roman finds, the clay pipes provide one of the main means of dating the post-medieval phases on the site, especially those of the clom and later buildings; for ease of reference, the TPQ evidence is collated in Table 2.28.

CATALOGUE (Fig. 2.88)

CTP 1 Well-made bowl of white, weathered clay, back lefthand side and most of top missing. Milling medium distance from rim of bowl. Medium sized round heel with relief 'wheel' mark. Bowl type similar to Hereford 'J' (Peacey 1985) and many pipes from Hereford have a 'wheel' mark on heel. These pipes, however, are usually of a greyish colour and are highly burnished. Could be a Carmarthen product, see almost identical Carmarthen Greyfriars no. 25 (Evans 1996). Date *c.* 1660–1690. Context **[186]**.

CTP 2 Base of bowl and round heel only, made of creamy clay with a smooth finish. Round heel has a relief 'Rose and Crown' with the initials I to the left and V to the right. Although this type of mark is widely found in Hereford, and therefore presumed to have been used by makers working in Hereford, the initials I.V. are not known from Hereford (Alan Thomas *pers. comm.*). Once again, this could be a Carmarthen maker – see Carmarthen Greyfriars no. 35 (Evans 1996). Date *c.* 1660–1690. Context **[157]**.

CTP 3 Bowl fragment with relief 'Rose and Crown' stamp and initial O to right (left missing). Similar to Carmarthen Greyfriars no. 34 (Evans 1996). This mark is known from Hereford (Peacey 1985) – possibly Richard Overton of Birtley, near Hereford, working 1679–1700. Context **[157]**.

Table 2.8 Summary of clay tobacco pipes

Context no	No of frags	Description and comment	approx date	Cat no
25	1		*c*.1660	-
31	1		*c*.1700	-
32	18		1660–1800	-
"	1	Top of bowl only remaining	*c*.1660	-
"	3	1 fragment with long thick spur	*c*.1750	-
36	1		*c*.1700	-
38	1		*c*.1850	-
39	1	'Tail' of heel – similar types made by Pardoe of Nantgarw, Ring of Bristol and Kennicott of Swansea	*c*.1850	-
52	2	2 identical part bowls, bases and spurs missing.	*c*.1830+	CTP 11
"	1	'Button' mouthpiece. Widely made.	*c*.1900	CTP 10
"	5		1700–1850	-
67	1		*c*.1800	-
70	1	part only	*c*.1660	-
73	1	Bowl with Ram's head on front.	*c*.1850–1880	CTP 12
"	2		1700–1800	-
79	19		*c*.1700–*c*.1850	-
"	1	Bowl frag only	*c*.1700	-
82	1		*c*.1850	-
92	12	Stems	1600–1700	-
97	5	Stems	1600–1700	-
102	3	Stems	1660–1730	-
104	5	Stems	1660–1730	-
"	1	'Chubby' Broseley 5 bowl with part large ?tailed heel , stamped 'B'.	*c*.1680–1720	CTP 7
121	1		*c*.1660	-
122	2		1660–1730	-
"	1	Milling round stem at break.	*c*.1660–1680	CTP 9
123	2		1750+	-
125	1		1660–1730	-
128	3		1660–1730	-
129	5		1660–1880	-
130	1	Well-smoked	*c*.1660	-
131	3		1600–1900	-
157	59	8 pre-1680; 46 *c*.1680; 5 *c*.1880	1660–1880	-
"	1	Faded circular relief stamp, Rose and Crown with I.V, ?local.	*c*.1660–1690	CTP 2
"	1	Bowl fragment, circular relief Rose and Crown with O.	1679–1700	CTP 3
"	1	-OHN/HVG/-ES relief square frame.	*c*.1680–1720	CTP 4
"	1	E*D/TAY/-O- poor mark, relief square frame.	*c*.1680–1720	CTP 6
158	1	2 joining pieces	*c*.1800–1850	-
174	5	1 bowl, frag only	*c*.1660–1680	-
179	5		*c*.1680	-
"	1	Part bowl only with long thick spur.	*c*.1780	CTP 8
180	3		*c*.1680	-
182	1		*c*.1680	-
186	1	Bowl with wheel relief on base of heel, ?local.	*c*.1660–1690	CTP 1
188	1	ED-/-A-/-O- relief on base of heel.	1680–1720	CTP 5
"	3		*c*.1660–1730	-
220	1	Bowl fragment	*c*.1660	-
"	2		*c*.1680	-
237	1	Similar in fabric and finish to Hereford	*c*.1660	-
280	2		*c*.1660	-

CTP 4 Well-made complete bowl with long-tailed heel, similar to Broseley type 5B (Atkinson 1975). Medium milling with smooth unburnished finish. Kite-shaped heel has 3–line relief stamp -OHN/HVG/?ES in square frame. John Hughes of Broseley *c.* 1680–1720. See also Carmarthen Greyfriars no. 65 (Evans 1996). Context **[157]**.

CTP 5 Well-made bowl with righthand side missing, long-tailed heel. Burnished finish, medium milling at rim, similar to Broseley type 5B (Atkinson 1975). Long-tailed heel has part of 3–line relief stamp ED-/-A-/-O-. Edward Taylor of Much Wenlock *c.* 1680–1720. See Carmarthen Greyfriars nos 61 and 66 and Priory Street nos 236–240 (Evans 1996). Context **[188]**.

CTP 6 Similar to CTP 5. Most of bowl missing. Long-tailed heel with another mark of Edward Taylor of Much Wenlock – 3–line relief stamp E*D/TAY/-

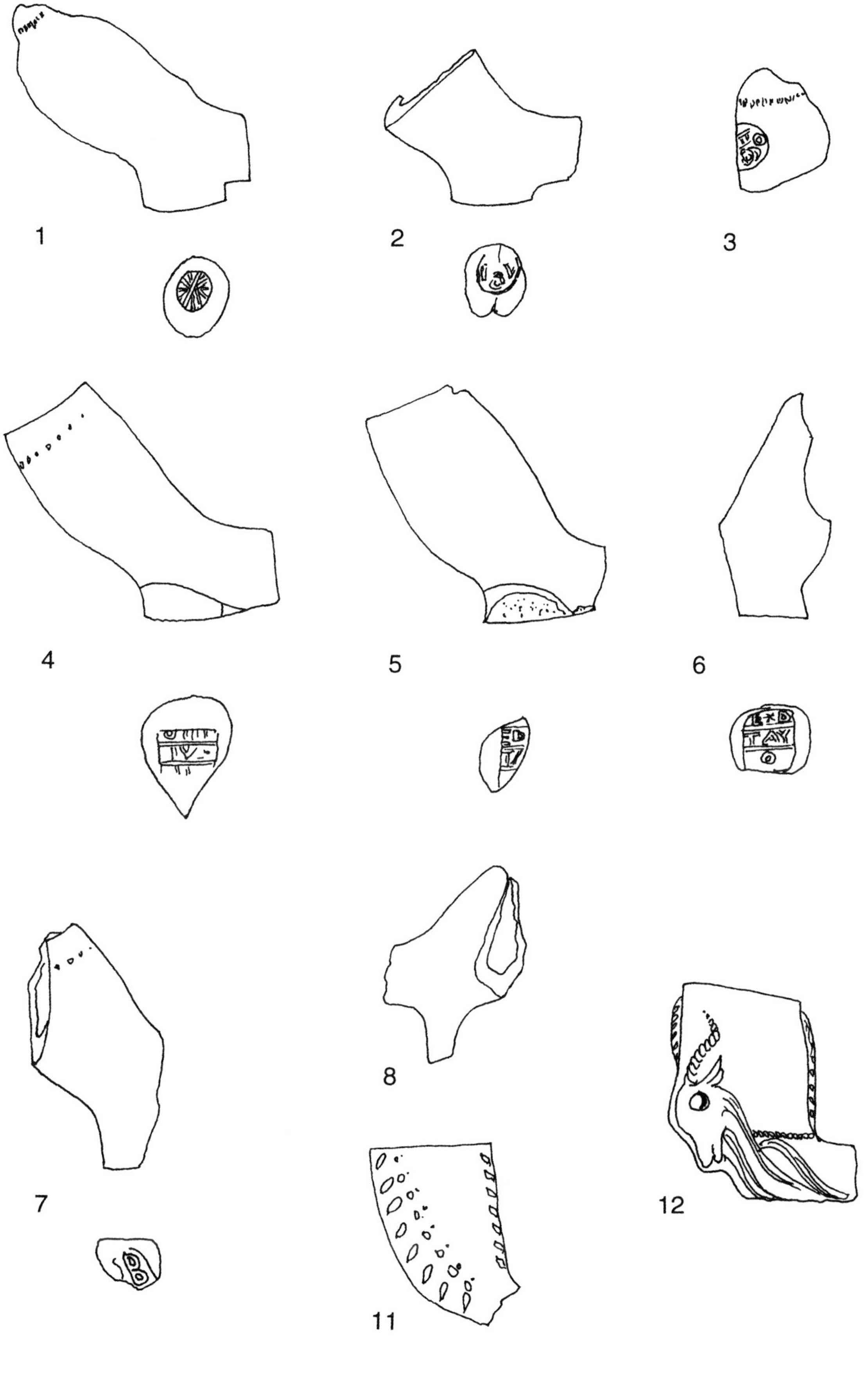

Fig. 2.88 Clay tobacco pipes. Scale 1:1

O-. See Carmarthen Greyfriars nos 61 and 66 and Priory Street nos 236–240 and Spilman Street no. 258 (Evans 1996). Date *c.* 1680–1720. Context **[157]**.

CTP 7 'Chubby' Broseley 5 type bowl (see Atkinson 1975), weathered and well smoked or burned. Low milling. Part long-tailed heel only with large relief ?B (lefthand side missing). Bowl style is similar to other 'local' Carmarthen 'Broseley 5' pipes (Evans 1996). Provenance uncertain, *c.* 1680–1720. Context **[104]**.

CTP 8 Base of bowl with long, thick spur only. Unmarked. *c.* 1780. Context **[179]**.

CTP 9 (Not illustrated). Stem fragment with milling at break. This type of stem fragment is often found in Hereford (*pers. obs.*). Also similar decoration found on Dutch and southern English pipe stems *c.* 1660–80. Dutch stems usually finer and burnished. Context **[122]**.

CTP10 (Not illustrated). 'Button' mouthpiece *c.* 1900. Context **[52]**.

CTP11 Two almost identical near-complete bowls, spurs missing or not present. Mould-imparted decoration, leaves and acorns on front mould line, leaves on back mould line of one; the other has back missing. Widely made in 19th century, mass produced 1830 onwards. Context **[52]**.

CTP 12 Bowl with beautifully executed Ram's Head around front mould line and sides of pipe. Raised and decorated back mould line. Date *c.* 1850–80. Context **[73]**.

(E) Glass (by Dee Brennan)

(i) Roman

The excavation produced a total of 65 fragments of Roman vessel glass, one fragment of window glass, eight beads and one small counter or gaming-piece, the details of which are summarised in the accompanying Table 2.9.

The vessel glass covers a period from the mid 1st century through to the mid/late 2nd century. All but two of the vessel fragments are of natural coloured bluish-green glass. 48 are from bottles, household containers used for the transport and storage of liquids. The total number of bottles could not be calculated; they do however outnumber all other glass vessels found on the site. The remaining bluish-green vessel fragments represent a limited number and range of everyday tablewares. They are a small jar (Gl 1) and a conical-bodied jug (Gl 2). A further nine bluish-green body fragments (not catalogued) are featureless and could not be identified. The only fragments of deliberately coloured glass come from the lower body and base of a pale 'lime' green jug, jar or bowl (Gl 3). There are none of the very brightly-coloured vessels of the early to mid 1st century and there is a complete absence of fine colourless tablewares, which become increasingly popular from *c.* AD 65–70. Strongly-coloured vessels are rarely found after this period (Harden and Price 1971, 320–3).

Table 2.9 Summary table of Roman glass

Context no	Small find no	No of frags	Comments	Cat no
bottle/vessel glass				
32	-	1	bottle	-
128	-	1	bottle	-
157	-	5	bottle	Gl 8
224	-	3	bottle	Gl 4
"	-	1	vessel	-
248	-	3	bottle	-
283	-	1	vessel	-
284	-	1	vessel	Gl 1
"	-	3	bottle	-
299	-	2	bottle	Gl 7
301	-	2	bottle	-
308	-	1	vessel/bottle?	-
330	-	2	bottle	-
343	-	1	bottle	-
345	-	1	bottle	-
357	-	2	bottle	-
359	-	1	vessel/bottle?	-
375	-	4	vessel: joins [377]	Gl 2
"	-	2	vessel/bottle?	-
"	-	2	bottle	-
377	-	1	vessel: joins [375]	Gl 2
"	-	4	vessel	-
388	-	1	vessel: joins [478]	Gl 3
411	-	1	bottle	-
432	-	2	bottle	-
460	-	1	bottle	-
471	-	3	bottle	Gl 5
478	-	1	vessel: joins [388]	Gl 3
487	-	3	vessel	-
"	-	9	bottle	Gl 6
window glass				
224	-	1		Gl 9
beads				
32	10,001	1		Gl 13
70	10,011	1		Gl 16
224	10,026	1		Gl 14
234	10,012	1		Gl 11
"	10,014	1		Gl 12
285	10,016	1		Gl 10
460	10,029	1		Gl 17
509	10,027	1		Gl 15
gaming piece				
471	10,025	1		Gl 18

None of the Pumsaint glass is provenanced to a known production area although the most likely places of manufacture are the Mediterranean area and the north-west Provinces. The one fragment of window glass (Gl 9) comes from a matt/glossy pane. This type of window glass was in use from the 1st to the early 3rd centuries.

Among the objects there are three annular beads (Gl 10–12), all common types, none of which can be closely dated. There are also five melon beads of turquoise glass paste (Gl 13–17) and one small monochrome gaming counter (Gl 18). Paste melon beads and monochrome counters occur on Roman sites at all periods but are most frequently found in 1st- and 2nd-century contexts.

Most of this glass was recovered from the fills of the various Roman features across the site. By far the largest concentration (53.3% of the assemblage) came from the infill [487] of the stone-lined well [559], its projected upward extension [471], and the successive fills [234; 283; 284; 285; 308; 357; and 375] of the overlying clay-lined cistern [376]; glass was also recovered from the infill [377] beneath the clay lining, one fragment even joining with a piece from the base of the sump [375]. Surprisingly, far fewer pieces (only 12%) came from the deposits [224 and 460] above the burnt timber building and from the gravel [509] redeposited behind its planking, while even fewer (only 6.7%) were recovered from just three fills [299; 301; and 345] of the 'late' pits. Such figures seem somewhat low, given their likely interpretation as demolition or rubbish deposits.

Individual finds (totalling 8%) had also become incorporated in some of the fills of the Roman foundation trenches, most notably those associated with phase 5 [343 and 388], and in one of the demolition layers [411] post-dating phase 4; that in [388] proved to join with a fragment from a silt layer [478] overlying the fill [479] of a phase 2 trench [480]. Two fragments, assigned in the table to a fill [432] associated with phase 4, could equally have come from one [481] associated with phase 3, since the two trenches coincided across this part of the site. Unfortunately, few of these pieces are sufficiently diagnostic to help date the successive phases.

The remaining fragments came from disturbed or post-Roman contexts, where they were clearly residual [32; 70; 128; 157; 248; 330; and 359].

CATALOGUE (Fig. 2.89)

Vessel glass; tablewares

Gl 1 Fragment from rim and neck of small jar; bluish-green. Tubular collar rim. Rim edge rolled in then bent out and down. Blown. Rim diameter 60–65mm; wall thickness 1mm. Context **[284]**.

This piece is the rim of a jar (Isings 1957, form 67b or 67c) with a globular or ovoid body. The type had either a ribbed (Charlesworth 1972, fig. 76, nos 25 and 26 from Verulamium) or plain body (Charlesworth 1979, fig. 70, no. 23 from Exeter) and a cut-out open base-ring like no. 3 below. Collar rim jars of this form were made in the Seine-Rhine region (Price 1978, 74) during the mid 1st century and are fairly common on 1st- and early 2nd-century sites in Britain and the north-west Provinces.

Gl 2 Four joining fragments from neck and upper body of conical jug; bluish-green. Long cylindrical neck with tooled constriction at its base. Body expanding outwards. Optic-blown vertical ribs on neck, ribs spiralling diagonally on upper body. Blown. Diameter at top of neck 15mm; maximum neck diameter 19mm; maximum diameter of body 70mm; wall thickness 1.5–3mm. Contexts **[375]** and **[377]**.

These fragments are from a long-necked jug with conical or rounded conical body (Isings 1957, form 55a or b). Long-necked jugs with a globular, discoid (see Price 1977, 155–8) or conical body (see Harden 1967, 238–40) were in use from the later 1st to the early 2nd century on Romano-British sites (Cool and Price 1995, 120). Like the collar-rim jars (Gl 1) they were produced in the Seine-Rhine region (Price 1978, 74) and are found on sites throughout the north-west Provinces. All of these jugs had a long cylindrical neck, angular-ribbed handle and either a simple concave or cut-out open base-ring. Small variations on the type are recently discussed with reference to finds from Colchester (see Cool and Price 1995, 120–130).

The neck and upper body of the Pumsaint jug is similar to a late 1st-century jug in *Glass of the Caesars* (Harden 1987, 141, pl. 69), probably from Gaul. It is bluish-green and is described as having a bell-shaped body with a carination above an open pushed-in base ring. Another jug with rounded conical body, but without the carination, was found at the Bartlow Barrows, Essex (Gage 1834, 5, pl. 2.1). Unfortunately not enough of the body of no. 2 survives to determine the exact body profile.

Gl 3 Two joining fragments from lower body and base of jar, jug or bowl; translucent lime-green. Part of pushed-in open base-ring. Blown. Base diameter *c.* 80mm; thickness 1–2.5mm. Contexts **[388]** and **[478]**.

These fragments are part of the lower body and base of a jar or jug, like nos 1 and 2, or alternatively it might be from a bowl. The type of base was common on a number of vessels during the 1st and early 2nd centuries.

Vessel glass; bottles

Gl 4 Fragment from rim and neck of cylindrical or prismatic bottle; blue. Rim folded upwards and inwards, top flattened downwards. Rim diameter (outer) 40mm; (inner) 20mm. Context **[224]**.

Gl 5 Fragment from rim and neck of cylindrical or prismatic bottle, bluish-green. Rim folded out and flattened downwards. Part of upper attachment of handle extant. Horizontal scratches on neck, below rim. Rim diameter (outer) 38mm; (inner) 21mm. Context **[471]**.

Gl 6 Near-complete handle of cylindrical or prismatic bottle; bluish-green. Applied, multi-ribbed angular handle. Ribs form fine pointed claws which grip the body. Part of folded upper attachment, neck and shoulder extant. Height 45mm; maximum width 37mm. Context **[487]**.

Gl 7 Fragment from base of square bottle; bluish-green. Parts of two concentric circles on underside of base. Context **[299]**.

Gl 8 (Not illustrated) Fragment from base of square bottle, blue. Part of one raised concentric circle on underside of base. Context **[157]**.

Gl 4–8 are bottle fragments. Bluish-green bottle glass very often comprises the major part of Roman glass assemblages. These very common containers were first introduced during the second half of the 1st century and are extremely common during the 2nd century. There is some variety in shape and size, but they all have similar rims, neck, shoulder and handles. Square (Isings 1957, form 50) and cylindrical (Isings 1957, form 51) bottles occur most frequently.

Cylindrical bottles were free-blown, square and other prismatic bottles were either free-blown and subsequently

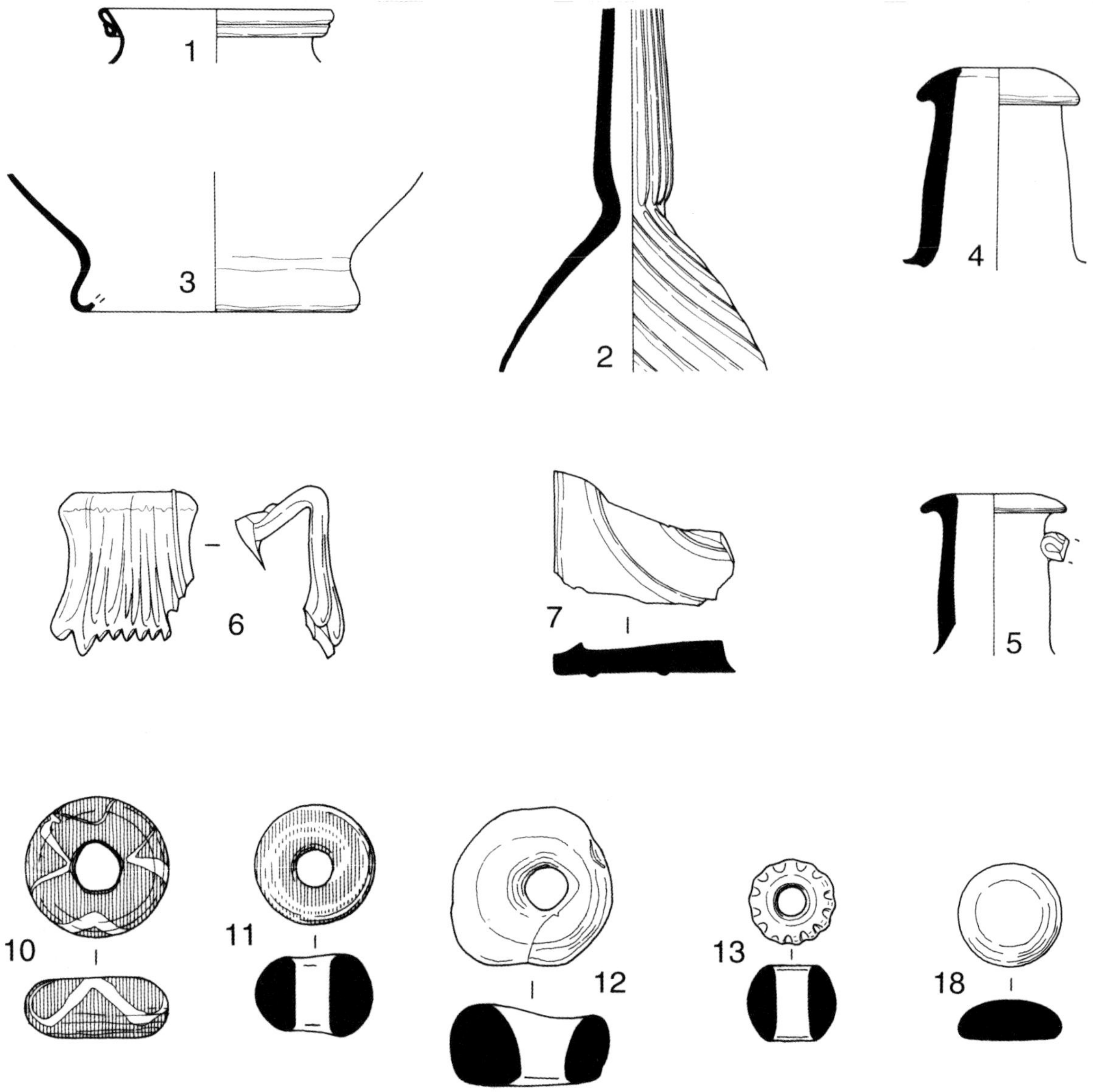

Fig. 2.89 Roman glass. Scale 1:2 (nos 1 to 7); 1:1 (nos 10 to 13 and 18)

shaped or blown into a body mould (Cool and Price 1995, 179). The latter almost always have some form of moulded base design. There are very many designs, concentric circles (Gl 7 and 8) being the most common. In a recent study, it has been suggested that bottles reaching sites during the Flavian period are generally smaller than later examples, the earlier types perhaps designed to take liquids. The later larger types with a wider neck and mouth were very possibly produced to contain solid goods (Cool and Price 1995, 190).

Window glass

Gl 9 (Not illustrated) Window glass; natural coloured green. Matt/glossy type, cast moulded. Maximum thickness 3.5mm. Context [**224**].

This is the only fragment of window glass, its presence alone is significant and indicates the existence of at least one window. Any other windows were probably painstakingly removed on abandonment of the site.

Beads

Gl 10 Complete annular bead; translucent deep blue opaque white wave, marvered flush with the surface. Height 8mm; diameter 19mm; perforation diameter 8mm. Context [**285**]. Find no. 10,016.

Gl 11 Complete annular bead; opaque deep blue with white swirls, streaky in appearance. Height 11mm; diameter 16mm; perforation diameter 5mm. Context [**234**]. Find no. 10, 012.

Gl 12 Complete annular bead; bluish-green. Irregular profile with one flattish surface and a-central perforation. Height 8–11mm; maximum diameter 23mm; perforation diameter 5mm. Context [**234**]. Find no. 10, 014.

Beads such as these occur as occasional finds on most Roman sites. The deep blue bead with marvered wave (Gl 10) turns up in contexts from the late iron age right through to the Saxon period (*cf.* Guido 1978, Group 5a, 128–133). Gl 11 and 12 are both common types but not closely datable.

The bluish-green bead with an irregular profile was probably made from recycled glass.

Gl 13 Complete small melon bead; turquoise glass paste with original bright glaze surviving. Height 11mm; diameter 12mm; perforation diameter 5mm. Context [**32**]. Find no. 10,001.

Gl 14 (Not illustrated). Complete small melon bead, turquoise glass paste. Heavily abraded. Height 12–14mm; diameter 15mm; perforation diameter 5mm. Context [**224**]. Find no. 10,026.

Also three melon bead fragments, all of turquoise glass paste. (Not illustrated).

Gl 15 About one half extant. Height 16mm. Context [**509**]. Find no. 10,027.

Gl 16 About one third extant. Height 19mm. Context [**70**]. Find no. 10,011.

Gl 17 About one quarter extant. Height 13mm. Context [**460**]. Find no. 10,029.

Melon beads such as Gl 13–17 are common throughout Britain and the Roman Empire, but are especially common on 1st- and 2nd-century sites.

Gaming counter

Gl 18 Complete monochrome gaming-piece or counter, opaque white. Plano-convex with a flat, slightly pitted base and a polished domed top. Maximum height 5mm; diameter 13mm. Context [**471**]. Find no. 10, 025.

This is a small round disc of opaque white glass. Glass gaming-pieces or counters are very common on sites throughout the Roman period. Monochrome counters are mostly opaque white or 'black', though other colours and occasionally decorated examples are also found.

(ii) Post-Medieval

A total of 508 fragments of post-medieval glass were found on the site, the details of which are summarised in the accompanying Table 2.10. As none of the pieces was of any great importance, only a representative sample was kept (43 fragments).

The assemblage was dominated by wine bottle fragments in olive-green glass. The earliest of these were three fragments from early 18th-century bottles of 'onion' form. Another 382 fragments were from late 18th- and 19th-century 'cylindrical' bottles with both free-blown (Gl 19) and machine-made types (Gl 20, 21, 22, 23) represented, mainly of Bristol manufacture. All other types of drink-associated bottles were of late 19th- and 20th-century date. They included 29 fragments from mineral water bottles in natural coloured blue and bluish-green glass, one fragment from a (?) beer bottle in brown glass and three fragments from a soft drinks bottle in bright green glass.

Among the 42 pieces of colourless vessel and bottle glass, some of which was indeterminate, were several fragments from stemmed drinking glasses and tumblers.

A limited range of small household bottles included the rim of a chemist's phial in olive-green glass, six fragments from medicine bottles in bluish-green glass, an ink bottle and a bottle for smelling salts, both in colourless glass. The only domestic item recovered was part of a lamp chimney for an oil or paraffin lamp. Two gaming marbles were both modern.

37 fragments of mixed 18th- and 19th-century window glass were found scattered across the site, but none was sufficiently large to determine the shape and size of the pane.

The greatest concentration of glass (219 fragments) came from the infill [36] beneath the western floors in Room 4; 83 fragments recorded from [19] above it were probably part of the same deposit. Broken glass was often used beneath floors as a rodent deterrent. Appreciable numbers of fragments also came from the recent farmyard surface [32] and the garden soil [157] extending across the infilled pit [245] at the southern end of the site. The great majority of fragments were thus associated with the 19th-century stone building and the makeup for its floors, but a few (including those from [157]) came from earlier contexts, such as the infills [123] and [129] beneath the pitched stone floor of the clom building [41], the gravel spreads [188] and [220] at the south end of the site, and the robber trench [330] above the stone rampart at the north end. These latter are clearly indicative of intensified activity in the area, predating the known structures.

CATALOGUE (Fig. 2.90)

Gl 19 Rim and neck of 'cylindrical' wine bottle in olive-green glass. Down-tooled string-rim close to the mouth. Free-blown. The type of rim would indicate a mid to late 18th-century date; *cf.* Noël Hume type 20 (1961, 104–5, fig. 5), dated 1750–1770. Context [**94**].

Gl 20 Rim of 'cylindrical' wine bottle in olive-green glass. Square string-rim with lip marvered downwards. Free-blown; *cf.* Noël Hume type 22 (1961, 105, fig. 5), dated 1790–1820. Context [**36**].

Gl 21 Rim and neck of 'cylindrical' wine bottle in olive-green glass. Deep cone-shaped lip with projecting lower collar. Manufactured in a three-piece mould with vertical mould seam visible on the neck; *cf.* Noël Hume type 23 (1961, 105, fig. 5), dated 1814–1853. Context [**36**].

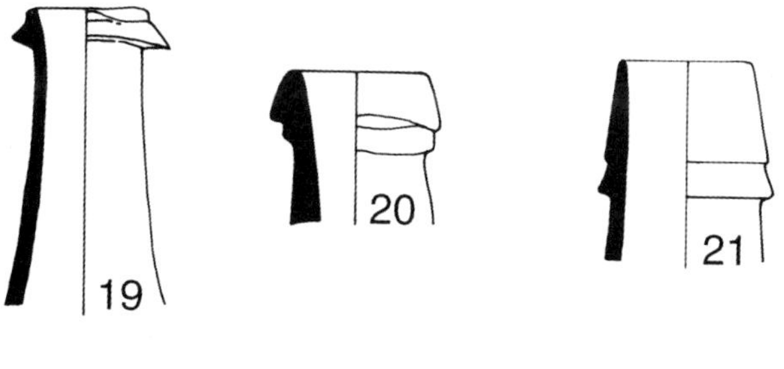

Fig. 2.90 Post-Medieval glass. Scale 1:2

Table 2.10 Summary table of post-medieval glass

Context no	onion wine bottle	cylind wine bottle	Blue-green bottle	Coloured bottle	Clear vess/ bottle	Lamp funnel	Marble	Window	Total found	Kept	Cat no
13	-	-	1	1	-	-	-	-	**2**	2	-
16	-	6	2	-	1	-	-	-	**9**	-	-
19	-	83	-	-	-	-	-	-	**83**	-	-
25	-	1	1	-	-	-	-	-	**2**	1	-
30	-	1	5	-	-	-	-	-	**6**	1	-
31	-	-	1	-	-	-	-	-	**1**	1	-
32	-	55	1	-	14	-	1	2	**73**	1	-
36	-	185	15	-	17	2	-	-	**219**	3	Gl 20, 21, 22
37	-	-	2	-	-	-	-	-	**2**	-	-
38	-	-	7	-	-	-	-	1	**8**	-	-
53	-	1	-	-	-	-	-	-	**1**	-	-
54	-	-	-	1	1	-	-	2	**4**	-	-
63	-	1	-	-	-	-	-	-	**1**	-	-
72	-	-	-	-	-	-	1	-	**1**	-	-
73	-	2	-	-	-	-	-	-	**2**	-	Gl 23
79	-	5	-	-	-	-	-	3	**8**	3	-
84	-	3	-	-	-	-	-	-	**3**	-	-
92	-	3	-	-	-	-	-	-	**3**	-	-
94	-	1	-	-	-	-	-	-	**1**	1	Gl 19
97	-	1	-	-	-	-	-	-	**1**	-	-
102	-	3	-	-	-	-	-	-	**3**	-	-
121	-	-	-	-	7	-	-	-	**7**	-	-
122	-	2	-	-	-	-	-	1	**3**	-	-
123	-	2	-	-	-	-	-	4	**6**	4	-
128	-	1	-	-	-	-	-	-	**1**	-	-
129	-	1	-	-	-	-	-	2	**3**	2	-
131	-	-	-	-	2	-	-	-	**2**	-	-
152	-	-	-	2	-	-	-	-	**2**	-	-
157	1	17	-	-	-	-	-	14	**32**	15	-
179	1	3	-	-	-	-	-	1	**5**	2	-
188	-	-	-	-	-	-	-	2	**2**	2	-
220	-	-	-	1	-	-	-	-	**1**	-	-
287	-	-	-	-	-	-	-	2	**2**	2	-
289	-	-	-	-	-	-	-	2	**2**	2	-
311	1	2	-	-	-	-	-	-	**3**	1	-
330	-	3	-	-	-	-	-	1	**4**	-	-
Total	**3**	**382**	**35**	**5**	**42**	**2**	**2**	**37**	**508**	**43**	

Gl 22 (Not illustrated). Fragment from base of a machine-made 'cylindrical' wine bottle in olive-green glass. A low basal kick is embossed 'H. RICKETTS'. A product of Henry Ricketts of Bristol. Of a similar date to Gl 21. Context [**36**].

Gl 23 (Not illustrated). Two joining fragments from base of machine-made 'cylindrical' wine bottle in olive-green glass. Embossed '[P]R & F. BRISTOL'. A product of Powell, Ricketts and Filer. Of a similar date to Gl 21 and 22. Context [**72/73**].

(F) Silver (by Dee Brennan)

Apart from the Roman coin listed in section 2.2.4(Ai) above, only one other silver object was recovered, a finger ring of uncertain date from the extensive spread of 'garden soil' [157] at the south end of the site.

Ag 1 (Fig. 2.91). Finger ring of D-shaped section, made from a band of metal with terminals forming a possible soldered(?) side joint. Transverse moulded decoration on the exterior. External diameter 23mm, internal diameter 20mm. Context [**157**]. Find no. 10,006.

Hilary Cool comments as follows:

Simple trinket rings with transverse moulded decoration similar to this piece are not uncommon in 4th-century contexts, as for example those from the Lankhills cemetery (Clarke 1979, fig. 87, no. 401) and the Godmanchester hoard (Green 1957, 87). There is limited evidence to suggest that some were also in use in the 3rd century. This piece, however, is on a slightly larger scale than undoubted Roman examples, besides which the type is not known in silver from Romano-British contexts. Taken together with the absence of any other late finds from the site (somewhat surprisingly given the proposed dating of the burnt timber building), and the nature of its context within a 'garden

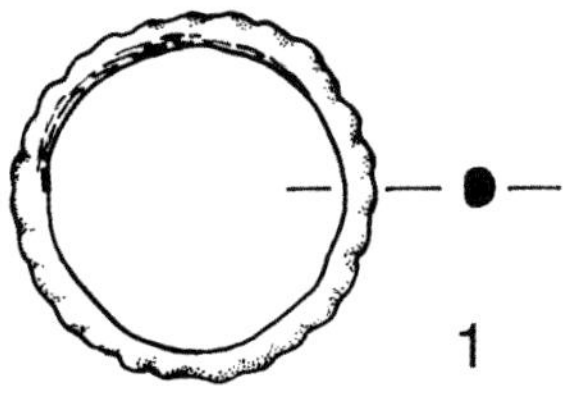

Fig. 2.91 Silver ring. Scale 1:1

soil' with material extending down to the 19th century, a Roman date seems unlikely for this particular example; more probably it is of relatively recent origin.

(G) Copper alloy (by Dee Brennan)
(i) Objects
Apart from the coins discussed in section 2.2.4 above, only three other copper-alloy objects were recovered from the site. One (Ae 1), which was not readily datable, came from the fill [354] of a possible metalworking hearth [355]. The other two were post-medieval: a ring object (Ae 2) from the infill [92] north of the clom building, and a button fragment (find no. 10,003), not worthy of detailed comment, from the makeup [123] beneath its flooring [41].

Ae 1 (Fig. 2.92). Fitting? Round-headed ?rivet with flattened tapering shank, clenched at tip. Length 33mm, maximum width 27mm. Context **[354]**. Find no. 10,019.

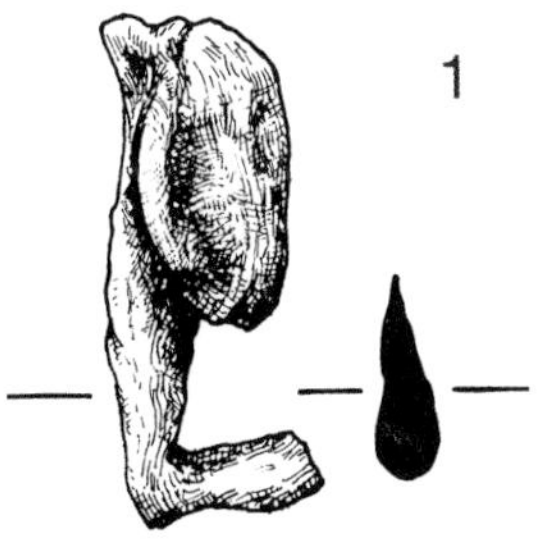

Fig. 2.92 Copper-alloy fitting. Scale 1:1

Ae 2 (Not illustrated). Ring, function uncertain. Complete, of solid circular section. External diameter 26mm. Context **[92]**. Find no. 10,010.

(ii) Slag and waste material
Only four pieces of copper-alloy slag and one fragment of sheet metal waste were recovered from the site. Of the slag, only that from the infill of the secondary rampart [448] is firmly datable to the Roman period (find no. 10,024), as was the sheet metal offcut (Ae 3) from the fill [343] of a phase 5 building trench [414]. The rest of the slag must either be residual or later, two pieces at least coming from the fills [297 and 360] of a post-Roman gully [298] and hollow [361] respectively (find nos 10,021 and 10,022); the remaining piece was unstratified (find no. 10,015).

Ae 3 (Not illustrated) Sheet metal fragment with fold marks and tear lines. Offcut? Dimensions 20 × 17 × 1mm. Context no. **[343]**. Find no. 10,023.

(H) Iron Objects and Nails (by Dee Brennan)
(i) Roman
Iron objects: A total of 12 Roman iron objects were identified from X-radiographs (Fe 1–12), the X-rays revealing detail obscured by corrosion. One other unidentifiable X-rayed object (not catalogued) may also be Roman. This small assemblage comprises items of weaponry (Fe 1 and 2), armour (Fe 3), domestic equipment (Fe 4 and 5), structural ironwork (Fe 6, 7 and 8) and miscellaneous pieces of uncertain function (Fe 9, 10, 11 and 12).

Only two pieces, Fe 6 and 10, can be regarded as in any way *'in situ'*, coming from the destruction deposit [540] associated with the burnt timber building; these may have formed part of its structure. Most of the remaining fragments were found in contexts which suggested site clearance at varying stages: thus, Fe 1 and 2 were recovered from the extensive demolition deposit [460] overlying the burnt timber building; Fe 7, 8, 9 and 12 all came from the fill [375] in the sump of the clay-lined cistern, while Fe 11 came from a higher fill [284] in the same feature; and three other pieces were recovered from fills in the 'late' pits, Fe 4 and 5 from [340] and [316] respectively in pit [323], and Fe 3 from [309] in pit [328]. The only other piece, which was not catalogued, came from the fill [303] of a phase 4 construction trench [304].

CATALOGUE (Fig. 2.93)
Nos 1, 2, 3, 4, 5, 6, 8, 11 and 12 are all conserved.

Fe 1 Catapult bolt head with damaged pyramidal head (the short pyramidal head is visible on the X-radiograph), long neck and broken conical socket. Length 60mm. Context **[460]**.

This is a Manning Type 1 artillery bolt head (1985, 170–171), a common find on Roman military sites. Comparable examples are to be found among a large illustrated group from Hod Hill, Dorset (*ibid.*, V141–V249, pls 82–85), all from mid 1st-century AD contexts.

Fe 2 Flat-bladed bolt head or arrowhead, incomplete. Triangular blade with missing socket or tang. Length as extant 54mm. Context **[460]**.

This is possibly an artillery bolt head of Manning Type IIB (1985, 176–77, V259–V270, pl. 85). Alternatively, it may be an arrowhead of Manning Type I (*ibid.*, 177, V280 and V281, pl. 85). The latter are apparently uncommon on Roman military sites and when found they usually turn up in late contexts (*ibid.*, 177).

Fe 3 Lorica squamata, incomplete. Fragment consisting of parts of two overlapping square scales. Two small piercings for wire ties are visible at the sides where the plates overlap. All other irregular-shaped holes are caused by corrosion. 29 × 25mm. Context **[309]**.

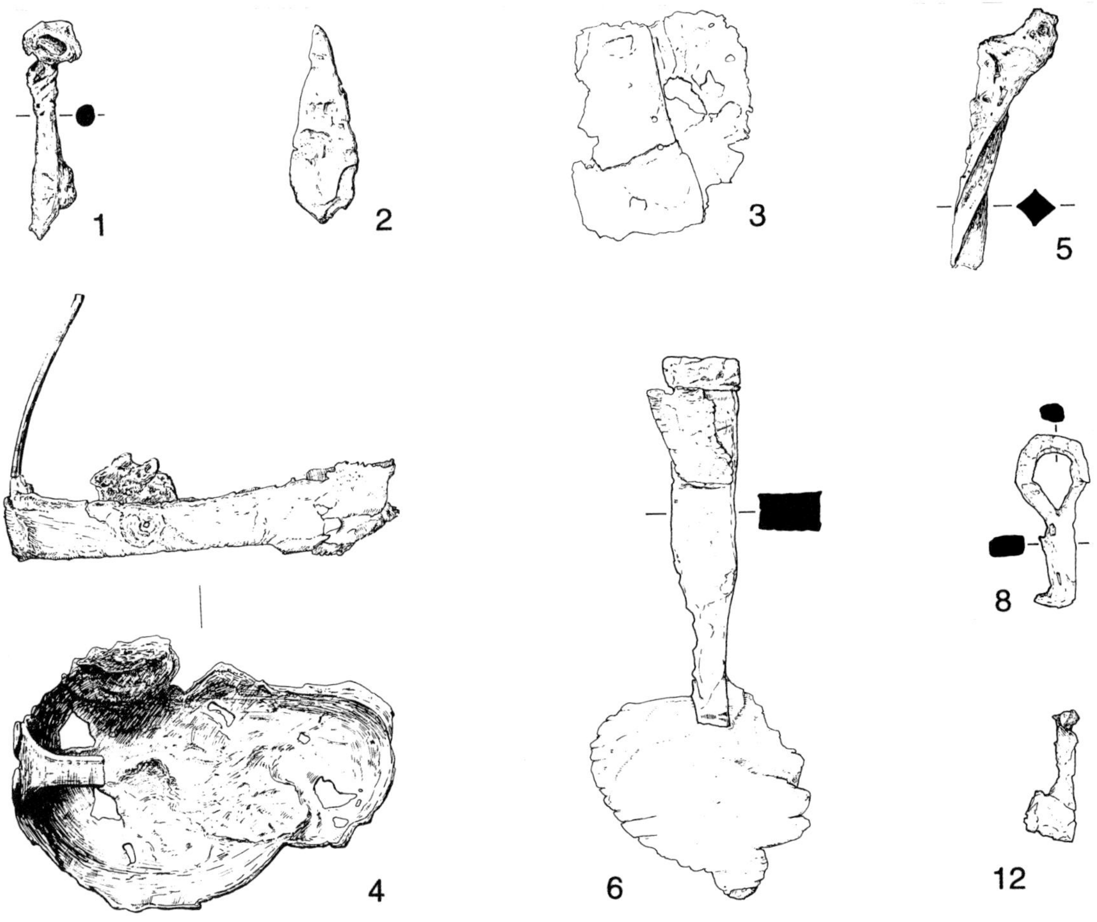

Fig. 2.93 Iron objects. Scale 1:2 (except no. 3 at 1:1)

Small overlapping scales of this type in bronze or iron were attached individually to a backing using wire ties. Larger surviving pieces from Chester (Manning 1985, 146, T1, T2 and T3, pl. 71) demonstrate the method of mounting and show clearly the form of overlap both vertically and laterally. The Chester finds were all from early 2nd-century contexts.

Fe 4 Open hanging lamp, almost complete. Shallow pear-shaped bowl with a curved strap handle (?)riveted to the back of the bowl externally. Handle ends in a simple perforated terminal through which a fragment of a looped swivel remains. The swivel is missing its (?)domed head enabling the surviving fragment to slip through the hole. Length of bowl 97mm, height from top of handle to base of bowl 88mm. Context [**340**].

Open lamps of this type are not uncommon. They are frequently found in burials. Comparable examples come from Bayford, Kent (Manning 1985, 99, P4 and P5, pl. 44), where a mid 2nd-century date is indicated from associated samian ware. One other similar lamp complete with its two-part hanger (hooked rod and swivel) was found in a Flavian burial at Baldock, Hertfordshire (Westell 1931, 247, pl. 1B).

Fe 5 Lamp hanger ?, incomplete. Spirally-twisted rod of square section with part of a rectangular-sectioned loop at the bottom. Length 74mm. Context [**316**]. Find no. 10,017.

Compare a complete lamp hanger from Water Newton, Cambridgeshire (Manning 1985, 100, P6, pl. 45). Complete examples have a spiked hook just below the top for attachment to a beam. The looped eye at the bottom took a hooked or looped swivel which then passed through a hole in the lamp handle (*ibid.*, 99 and fig. 26). This is almost certainly the hanger belonging to the open lamp already described (Fe 4). Both are from the fill of the same 'late' pit.

Fe 6 Joiner's dog or staple, incomplete. Rectangular-sectioned bar with mineralised wood surviving at broken end. The X-radiograph shows it to be corroded and bent slightly at the point where it penetrates the adhering mineralised wood. Length as extant 153mm. Context [**540**]. Find no. 10,052.

A common item of structural ironwork used for joining pieces of timber.

Fe 7 (Not illustrated). Staple, incomplete. Part of right-angled bar of rectangular section. Length as extant 80mm. Context [**375**].

Fe 8 Hook, complete. Loop-headed hook with rectangular-sectioned shank and very short out-turned tip. Length 47mm. Context [**375**].

A comparable, slightly larger, hook comes from Hod Hill, Dorset (Manning 1985, 129, R 22, pl. 59), from a mid 1st-century context.

Fe 9 (Not illustrated). Bar, incomplete. Fragment of

flat-rectangular section. Length as extant 55mm, width 18mm. Context [**375**].

Fe 10 (Not illustrated). Bar, incomplete. Fragment of square-sectioned hollow bar (possibly once solid), one end starting to curve. Possibly part of a U-shaped staple or similar. Length as extant 48mm, width 10mm. Context [**540**].

Fe 11 (Not illustrated). Unidentified object, complete. Short rod of solid circular section. Length 20mm, diameter 5mm. Context [**284**].

Fe 12 Unidentified object, incomplete. Square-sectioned fragment with expanded end, hammered flat. Length as extant 36mm. Context [**375**].

Nails: The total number of nails and nail fragments recovered from Roman contexts was 632. They were submitted for X-ray (University of Wales College at Cardiff) prior to being classified using Manning's typology (1985, 134–5). As most nails were poorly preserved and only a few were conserved, classification has been largely dependent on an examination of the X-rays.

Five different nail types were identified according to Manning's typology (Table 2.11). The most common of these are general purpose nails (Type 1B; Fe 13, 14, 15), with complete examples ranging in size from 40 to 85mm long. A small group of nails with shanks exceeding 100mm long are tentatively classed as 1A type nails (Fe 16, 17). Whilst none is complete, the largest recorded example is over 177mm long. The few nails of this type were recovered from the burnt timber building in pit [245]. These large-sized nails have mineralised wood and charcoal preserved on their heads and shanks.

Other common types of nails found in smaller numbers include tacks (Type 7; Fe 18) and hobnails (Type 10; Fe 19), the latter frequently found in corroded clusters and some with traces of mineralised leather beneath the head. One other unclassified type is an incomplete nail which has an unusually large head (Fe 20). Indeterminate nails are mostly shank fragments and fragments so poorly preserved as to make their identification impossible. Many of the nails are characteristically bent, indicative of use. Those nails which appear to be straight are mostly incomplete, their shanks being broken.

Perhaps the most interesting collection of nails came from the pit [245] containing the burnt timber building (Table 2.12). One particular group was clearly derived from the fabric of the structure itself; of these, 22 [contexted as 544] were found *in situ* along the north side of the building, each one roughly coinciding with the end of an individual plank, two more were associated with a fallen carbonised timber [527] from its superstructure, while another six came from the destruction horizon [540]. By contrast, the 57 nails recovered from the overlying fills [460 and 224] are more likely to have come from elsewhere on the site, most probably reflecting extensive clearance operations.

Such activities probably also account for the large number of nails (42.9% of the assemblage) from the 'late' pits: *viz.* 150 in the fills of pit [302]; 58 in the fills of pit [323]; 32 in the fills of pit [328]; and 31 in the fills of pit [351]. Such totals contrast with the presence of only eight nails in the fills of the large pit [410] to the rear of the rampart, which must surely indicate a difference either in date or function.

A second large concentration of 213 nails (33.7%) came from the infill [487] of the stone-lined well [559], its projected upward extension [471], and the successive fills of the overlying clay-lined cistern [376] and the sump at its base; here again they probably reflect clearance activities, as do the 26 nails from the demolition layers [342; 411; 420] associated with the phase 4 buildings. Not surprisingly, far fewer nails came from the fills of the Roman foundation trenches, the 10 examples being found exclusively in those associated with phases 4 and 5. Taken together, however, such evidence indicates that an appreciable number of nails had been used in the construction of the different buildings.

The few remaining nails came from post-Roman contexts, where they were clearly residual.

CATALOGUE (Fig. 2.94)

Fe 13 (Not illustrated). Type 1B, complete. Round flat head and square-sectioned tapering shank. Length 58mm. Context [**375**].

Fe 14 (Not illustrated). Type 1B, complete. Length 65mm. Context [**375**].

Fe 15 Type 1B, incomplete. Length as extant 35mm. Context [**540**].

Fe 16 Type 1A, almost complete. Sub-square head and

Table 2.11 Classification of Roman nail types

Nail type	No of nails or fragments	% of total
1A	12	1.9
1B	148	23.4
7	14	2.2
10	90	14.2
other (not classified)	1	0.2
indeterminate (mainly shanks)	367	58.1
Total	632	100%

Table 2.12 Distribution of Roman nail types

Context no	Type 1A	Type 1B	Type 7	Type 10	Unclass	Indeterminate	Total	Cat no
STONE-LINED WELL 559								
487	-	3	-	-	-	3	**6**	-
471	-	1	-	1	-	7	**9**	-
CLAY-LINED CISTERN 376								
234	1	12	1	-	-	4	**18**	-
283	-	1	-	-	-	3	**4**	-
284	-	5	-	2	-	17	**24**	-
285	-	12	7	1	-	33	**53**	Fe 18 (T 7)
308	-	2	1	-	-	7	**10**	-
357	-	2	-	4	-	17	**23**	-
375	-	23	-	3	-	40	**66**	Fe 13 Fe 14 (T 1B)
PHASE 4 FOUNDATION TRENCHES								
303	-	-	-	-	-	1	**1**	-
364	-	-	-	-	-	1	**1**	-
366	-	1	-	-	-	-	**1**	-
373	-	1	-	1	-	1	**3**	-
432	-	-	-	-	-	2	**2**	-
PHASE 4 DEMOLITION LAYERS								
342	-	-	-	-	-	2	**2**	-
411	1?	2	-	2	-	5	**10**	-
420	-	4	-	1	-	9	**14**	-
PHASE 5 FOUNDATION TRENCHES								
343	-	-	1	-	-	1	**2**	-
PIT 302								
269	-	14	-	2	-	61	**77**	-
300	-	6	1	13	-	35	**55**	-
301	-	7	-	1	-	10	**18**	-
PIT 323								
299	-	6	1	3	-	8	**18**	-
316	-	2	-	10	-	-	**12**	-
339	-	-	-	6	-	1	**7**	-
341	-	4	-	13	-	4	**21**	-
PIT 328								
309	-	5	-	2	-	22	**29**	Fe 19 (T 10)
327	-	1	-	-	-	2	**3**	-
PIT 351								
345	-	-	-	6	-	7	**13**	-
350	-	2	-	12	-	4	**18**	-
PIT 410								
233	-	-	1	4	-	-	**5**	-
378	-	1	-	2	-	-	**3**	-
PIT 245, CONTAINING BURNT TIMBER BUILDING								
224	-	10	-	-	-	20	**30**	-
460	-	12	-	1	-	14	**27**	-
527	-	1	-	-	-	1	**2**	-
540	-	2	-	-	1	3	**6**	Fe 15 (T 1B) Fe 20 (Uncl)
544	10	-	-	-	-	12	**22**	Fe 16 Fe 17 (T 1A)
POST-ROMAN CONTEXTS								
157	-	2	-	-	-	2	**4**	-
186	-	-	-	-	-	1	**1**	-
248	-	1	1	-	-	2	**4**	-
276	-	-	-	-	-	1	**1**	-
289	-	-	-	-	-	1	**1**	-
291	-	-	-	-	-	1	**1**	-
297	-	1	-	-	-	1	**2**	-
329	-	1	-	-	-	1	**2**	-
330	-	1	-	-	-	-	**1**	-
Totals	**12**	**148**	**14**	**90**	**1**	**367**	**632**	

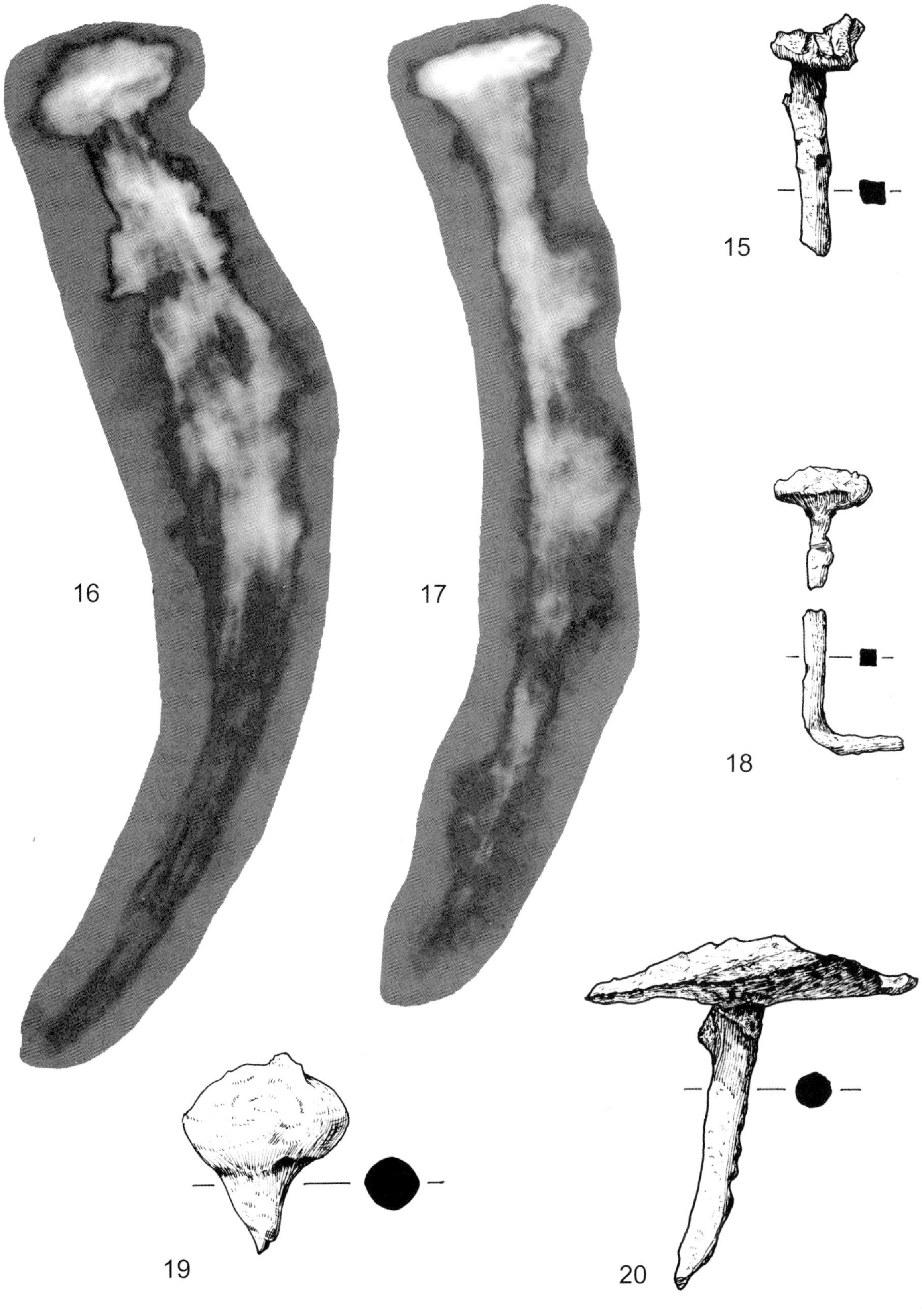

Fig. 2.94 Iron nails. Scale 1:1 (except no. 19 at 2:1)

square-sectioned tapering shank. Length as extant 177mm. Context [**544**]. Find no. 10,039.

Fe 17 Type 1A, almost complete. Round flat head and square-sectioned tapering shank. Length as extant 160mm. Context [**544**]. Find no. 10,043.

Fe 18 Type 7, almost complete. Tack with round, slightly pyramidal head and short square-sectioned tapering shank. Length as extant 22mm. Context [**285**]. Another four similar tacks were recovered from this context.

Fe 19 Type 10, complete. Hobnail with domed head and very short square-sectioned shank. Length 15mm. Context [**309**].

Fe 20 Nail, incomplete. Large flat (?)round head and square-sectioned shank. Length as extant 61mm. Context [**540**].

(ii) Post-Medieval

Iron objects: A total of 67 iron objects and object fragments as well as 187 nails were recovered from post-Roman deposits on the site. The identified objects were clearly associated with the clom and later stone buildings and all were items typical of a rural dwelling. The assemblage included horse equipment, parts of agricultural tools, milking pail fragments, numerous bar, strap and rod fragments and even a corkscrew. Once recorded, these objects were then discarded on site. A list of items by context is housed with the main site archive. As might be expected, the greatest numbers of items came from the farmyard surfaces [32] and [79], although interestingly [123] and [129], the fills beneath the pitched stone floor of the clom building, both yielded slightly more nails than the average (10 and 11 respectively), hinting at earlier structures in this area. The disturbance [49] cut through this floor also contained 10 nails and fragments forming two objects. The 'garden soil' [157] at the south end of the site, which was otherwise prolific in pottery, yielded a total of only nine nails and no iron objects.

One object, which was retained, was a sub-triangular block of iron, of uncertain date and function (Fe 21). It came from the fill [239] of an oval hollow [240], one of a series of features apparently connected with metalworking. Peter Crew comments as follows:

Fe 21 (Fig. 2.95). A sub-triangular block of iron, length 890mm, thickness 29mm, weight 824 grammes. This was originally interpreted as an anvil, supported by the observation that the edges and corners of the upper working face were rounded and damaged from use and that the slight depression of this working face was probably caused by repeated hammering. In fact the block is heavily corroded, with all its surfaces swollen and distorted by the corrosion products, and all the features observed would be consistent with the forging and refining of a small bloom. In any case, if this block had been an anvil, its most likely mode of use would have been with the narrow end set into a block of wood, with the wider end being the working surface. On balance, this block seems likely to be a piece of raw material, probably a cake of steel from which pieces would have been cut for tipping chisels. It is not clear if the triangular shape is deliberate or if it simply reflects the shape and nature of the bloom from which it was refined. Context [**239**].

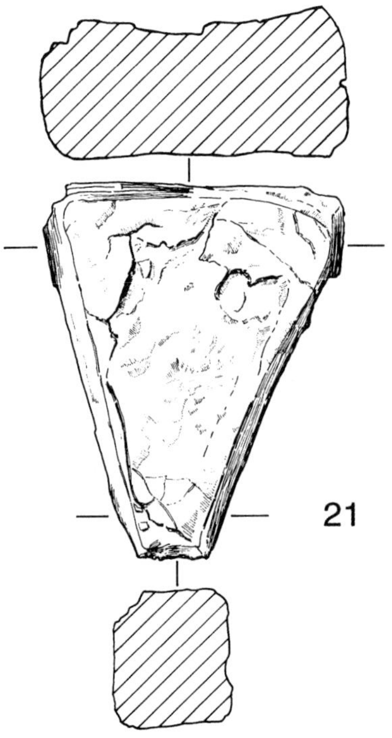

Fig. 2.95 Sub-triangular block of iron from [239]. Scale 1:2

(I) Iron slags and other metalworking residues (by Peter Crew)

The slags fall into three groups. Table 2.13 lists those which have the appearance of modern cinder, a mixed amorphous vesicular slag without secondary corrosion products. This material is typical of secondary smithing slags produced with the use of a coke or coal fuel, and it all derives from later contexts.

The second group (Table 2.14) comprises small fragments of fuel ash slag, which are light vesicular irregular globules. Slag of this type can form under a wide variety of high temperature conditions and it does not necessarily derive from metalworking. All the samples of this type from Pumsaint come from Roman contexts.

Some of the slags, however, have a quite different character to the cinder and fuel ash slags (Table 2.15). In all cases the slag is coated with heavy secondary corrosion products, but in cut section they are dense black slag with a fayalitic structure. All these slags are probably related to bloomery iron-smithing. Most of them derive from Roman contexts and the few pieces in more recent contexts are probably residual. The slag cake from [375] is a typical plano-convex

Table 2.13 Cinder and other forge slags

Context no	Pieces	Weight (gms)	Comments
u/s	3	105	forge cinder
u/s	4	90	heavily corroded forge cinder
38	3	300	forge slag, coke fuel, slightly magnetic, amorphous mixed material
71	1	20	forge cinder
121	1	22	forge cinder
129	1	19	forge cinder
131	2	89	vitreous vesicular slag, ? forge cinder
174	1	84	smithing hearth slag, ?from use of coke fuel
226	3	892	forge cinder
248	1	20	low density forge cinder
285	1	12	forge cinder
291	5	127	forge cinder
338	1	11	forge cinder
Totals	27	1791	

Table 2.14 Fuel ash slags

Context no	Pieces	Weight (gms)	Comments
233	1	12	prill of fuel ash slag
357	1	5	low density fuel ash slag
377	1	5	tiny fragment fuel ash slag, clean surfaces, reddish tinge
378	1	14	low density fuel ash slag
403	1	18	low density frothy fuel ash slag
487	1	14	low density fuel ash slag with charcoal fragment
Totals	6	68	

Table 2.15 Bloomery iron working and other slags

Context no	Pieces	Weight (gms)	Comments
224	2	166	amorphous slag, heavily corroded, strong accretions
224	1	26	black glassy vitreous slag
234	7	200	irregular amorphous slag
248	1	20	flow of dense tap-like slag, 5mm thick, very abraded
284	1	1312	large, irregular smithing hearth slag, heavy concretion products trapping charcoal fragments, non magnetic
297	1	376	part of large smithing hearth bottom, large angular pieces stone trapped in lower corrosion products, frequent round and angular stone in upper surface, slag only visible through heavy concretion
297	1	157	friable frothy amorphous slag, vesicular, glassy in section
309	1	23	amorphous slag
316	3	389	amorphous dense vesicular glassy slag, black glassy external surface, charcoal trapped in heavy corrosion products
375	1	31	fluxed lining slag, typical irregular concretion
375	2	266	dense irregular plano-convex slag cake, slightly magnetic, smithing hearth bottom
471	1	31	amorphous slag, heavy secondary corrosion products
Totals	22	2997	

smithing hearth bottom. The other smithing cakes from [284] and [297] are considerably larger and would imply heavy smithing, perhaps using a very slaggy raw material or involving a lot of fluxing and welding. The small run of fluid slag from [248] is a typical bloomery slag, though a piece as small as this could easily form in a smithing hearth and is not necessarily indicative of smelting. This fragment is from a recent context, but it is heavily abraded and probably residual.

Several of the contexts yielding more than 100 grammes of slag probably represent rubbish dumping during the Roman period; this could account for the significant concentrations recorded in some of the fills [234, 284 and 375] of the clay-lined cistern and the sump at its base, from just one fill [316] of the 'late' pit [323], and from the infill [224] across the demolition deposits in the large pit [245]. The material in the fill [297] of post-Roman gully [298], and the fill [291] of the robber trench [292] above the Roman stone tower, could either be residual or reflect rather later activity. The material from the makeup [38] beneath the flooring in Room 1 must represent much more recent activity.

From [375] came an unusual find which is probably related to the iron-smithing.

Fe 22 (Fig. 2.96). A sub-circular collar, with two irregular concretions of slag on one side. It is coated in heavy corrosion products, in which some charcoal is trapped. It is non-magnetic, but it is clear from the broken section that the collar was made of iron strip, *c.* 22 × 4mm in section, which has been completely mineralised. The collar had an external diameter of *c.* 55mm and the hole is *c.* 47mm diameter. A possible interpretation of this collar is that it was set into the clay wall of a smithing hearth to protect the end of the blowing hole. Tuyères (*sensu stricto* as pipes) have been found rarely on British ironworking sites and the normal method for directing air into a furnace or a smithing hearth is simply a hole through the clay wall of the structure. The inside face of the hole is in the hottest part of the hearth and it is very liable to damage from either melting or attempts to move slag blocking the hole. In this case the iron collar would have been luted onto the inside face of the clay hearth, perhaps so that it projected slightly into the hearth. The two concretions of slag are consistent with this interpretation and would have derived from the clay wall melting and becoming attached to the edges of the collar. One edge of the upper slag concretion has a straight fracture where it would have broken from the clay wall. As far as is known, this is the only example yet recognised from Britain.

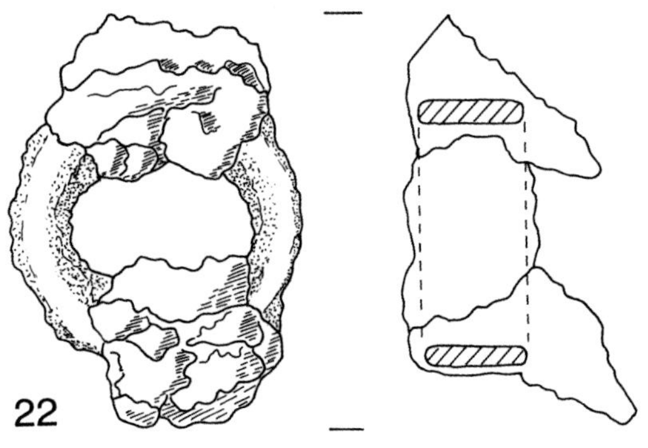

Fig. 2.96 Iron collar from [375] with slag accretions: front view and composite section. Scale 1:2

(J) Lead (by Peter Crew)
Five fragments of scrap lead, none of which merits detailed description, and part of a plano-convex cake were recovered from the site (Table 2.16). The latter, from [224] is of some interest, though it is not known if it is some type of ore or a waste product from lead working. SEM examination by Dr Tim Young, University of Cardiff, has shown the core to consist of lead oxide and XRF examination by Dr Justine Bayley, Ancient Monuments Laboratory, has shown the yellow efflorescence to contain lead with a trace of antimony.

All this material probably derived from clearance operations or rubbish disposal during the Roman period; this would account for its presence in a demolition horizon [411] post-dating the phase 4 structures, the layers [460 and 224] above the burnt timber structure, two fills [308 and 375] of the clay-lined cistern, and a fill [350] of one of the 'late' pits [351].

(K) Furnace or hearth lining (by Peter Crew)
A total of 18 fragments of furnace or hearth lining were recovered from the site, mostly from late Roman contexts (Table 2.17), though none was actually *in situ*. The character of all these pieces is rather similar and they could have derived from the same feature. None of the pieces has any superficial indications of which metal may have been involved. The majority came from fills [269; 300; 345 and 378] within the 'late' pits [302; 302; 351 and 410 respectively], the rest from the clay lining [376] of the cistern and one of its fills [284], a fill [224] in the large pit [245], a demolition layer [342] post-dating the phase 4 buildings, and the fill [343] of a phase 5 trench [414]. None merited detailed description.

Table 2.16 Lead working residues

Context no	No of frags	Weight (gms)	Comments
224	1	369	2 joining fragments, about half of a plano-convex cake 107mm diameter, 21mm maximum thickness. Soft dense crystalline material with external yellow efflorescence, in which is a fragment of charcoal.
308	1	41	Scrap waste
350	1	811	Sheeting
375	1	9	Scrap waste
411	1	95	Scrap waste
460	1	82	Scrap waste

Table 2.17 Furnace lining material

Context no	Pieces	Weight (gms)	Comments
224	2	10	soft red oxidised fabric, interior heavily vitrified, black glassy, with angular unreacted grog
234	1	9	
269	5	52	soft red oxidised fabric, black vitrifaction, traces of organic temper
284	1	38	grey reduced fabric, large angular stone grog, frothy irregular vitrification, maximum 35mm thick
300	2	32	red oxidised fabric, smooth vitrified surface, brown patches, heavy secondary corrosion
342	1	20	red oxidised fabric, smooth black vitrified surface, 20mm thick
343	1	21	black de-vitrified surface, no curvature
345	1	4	black de-vitrified surface, with fragment of fuel ash slag stuck to surface
373	2	41	with cinder attached
376	1	4	red oxidised fabric, black vitrified surface
378	1	15	red oxidised fabric, vitrification degraded
Total	18	246	

(L) Building materials (by Dee Brennan)

(i) Ceramic building materials

Roman: The total number of fragments of Roman brick and tile recovered was 303, only a representative sample of which was kept (Table 2.18). The discarded material (273 fragments) comprised mainly indeterminate lumps, either too small to identify with certainty or without measurable dimensions of length, width or thickness. Of the sample kept not one piece had a complete measurable length or width dimension but in most cases the complete thickness or depth could be measured.

This brick and tile came from a variety of contexts, all probably associated with clearance operations or rubbish disposal during the Roman period. By far the largest concentration (38.7%) came from the fills of just one of the 'late' pits [302], closely followed by that (24.2%) from one of the fills [224] in the large pit [245] at the south end of the site. A lesser concentration (14.2%) was recovered from the successive fills of the clay-lined cistern [376] and the sump at its base. Elsewhere, only the odd fragment was recorded, as in the top fill [233] of the large pit [410], the fill [303] of one of the phase 4 building trenches [304], and three of the fills of the other 'late' pits [328 and 351]. The remaining pieces came from post-Roman contexts, where they were clearly residual.

Brick fragments are more numerous than tile. None displays diagnostic features to indicate their use on the site. There are no finger impressed markings or animal paw prints, both common features on Roman building materials. The thinnest is 35mm, and the thickest 65mm. Most are between 45mm and 50mm thick.

Very little tile was recovered and the kept pieces are not very informative. One fragment from [224] is possibly part of a *tegula* as there is some evidence to suggest a flange on its upper face; it is however unusually narrow. One small fragment from [269] is identified as tile, probably *tegula*, on its thickness (20mm) alone. There are no other distinguishing features. One small angled fragment from [329] suggests a flange, and is therefore possibly a *tegula*. Two joining fragments from [329] are also possibly part of a different *tegula*.

The **fabric** varies in degree of colour and hardness, ranging from orange through to dark red; in very hard-fired examples a grey core is usual. All are tempered with frequent sandstone gravels, most usually red but sometimes grey/black in hard-fired pieces. The gravels are rounded and range in size from 2mm across up to 20mm or more. Sparse white quartzite grits are usual. The underside and edges of brick and tile are usually peppered with abundant gravels.

An analysis of the fabric would indicate a single source of manufacture although that source is not yet identified. There is a similarity in fabric to building material found on sites within Carmarthen where a 'local' source is suggested but not yet proven (James 2003, 348).

Post-Medieval: A total of 86 fragments of post-medieval ceramic building material were recovered from the site. Of that total 33 were sections of drainage pipe and 53 were fragments of brick. The material was discarded once its presence was recorded, apart from three fragments of brick which were kept for comparison. The largest number of brick fragments (34.9%) came from the farmyard surface [79], although the disturbance [44] cut through the pitched stone floor of the clom building also produced a high concentration of pipe fragments (12.8%).

(ii) Mortar

Traces of mortar were noted in a number of contexts, but in only five cases were the fragments considered large enough or sufficiently interesting to be worth collection. None merited detailed description. Three pieces from the mixed ashy fill [237] of a possible

Table 2.18 Distribution of Roman building material

Context no	Total found	Brick no	Brick thickness	Tile
CLAY-LINED CISTERN 376				
234	5	1	50mm	-
283	3	1	50mm	-
284	8	2	40mm, 50mm	-
285	14	2	45mm, 40mm	-
286	2	1	45mm	-
308	6	1	40mm	-
375	5	1	45mm	-
PIT 302				
269	57	2	50mm	1 tegula ?
300	6	1	55mm	-
301	34	2	?mm	-
PIT 328				
309	1	1	65mm	-
PIT 351				
345	1	1	50mm	-
350	4	Indeterminate		
PIT 245, CONTAINING BURNT TIMBER BUILDING				
224	73	1	58mm	1 tegula ?
OTHER ROMAN CONTEXTS				
233	1	1	35mm	-
303	3	Indeterminate		
POST-ROMAN CONTEXTS				
157	28			-
161	1	1	45mm	-
248	4	4	?mm	-
297	1	1	42mm	-
329	44	1	50mm	2 tegulae ?
359	1	Indeterminate		
461	1	Indeterminate		
Total	303	26		4 ?

metalworking hearth [238], may perhaps be a limestone or chalk flux, rather than mortar proper. Single fragments also came from the fill [235] of a post-pit beneath the 19th-century construction trench [74] and a layer [282] across the top of the clay-lined cistern [376].

(iii) Daub / low-fired clay

One fragment of daub or low-fired clay was recorded, from a fill [357] in the clay-lined cistern [376]. It did not merit detailed description.

(M) Flint (by R. N. E. Barton)

Eight pieces of flint were recovered from the excavation, from a variety of contexts, none of them *in situ*. Three came from Roman or probable Roman contexts [269; 303; 432], four from post-Roman horizons, predating the clom building [188; 248 (2); 291], and one was unstratified. The pieces from contexts 303 and 432 were burnt, probably natural, fragments, which did not merit detailed description.

CATALOGUE (Fig. 2.97)

(all measurements relate to the percussive axis of the artefact)

Fl 1 Backed bladelet.
10 × 11 × 03mm. Proximal fragment of bladelet of light brown stained flint in sharp condition, with continuous, direct semi-abrupt to abrupt retouch down the left margin. The break appears to be ancient. Backed bladelets such as this occur in the British late upper palaeolithic and mesolithic (Barton 1992). The nearest finds of comparable type are from Gwernvale (Powys) [beneath a chambered tomb] and from a number of caves in Pembrokeshire. Unstratified.

Fl 2 Flake scraper.
31 × 22 × 05mm. Steep-sided, ogival scraper of mottled grey-brown flint in mint condition, with continuous, direct semi-abrupt retouch around most of the perimeter. Has a thick rolled cortex and probably derives from a waterworn cobble. Undiagnostic, but probably post-palaeolithic. Context **[248]**.

Fl 3 Opposed platform core.
29 × 25 × 14mm. Flake core with two opposed platforms on a small, waterworn flint cobble. The back of the core is cortical. Incipient cones on one platform of the core indicate use as a hard hammer. Surface shows signs of burning. Undiagnostic, but likely to be bronze age, given the size of the material and nature of the flake scars. Context **[291]**.

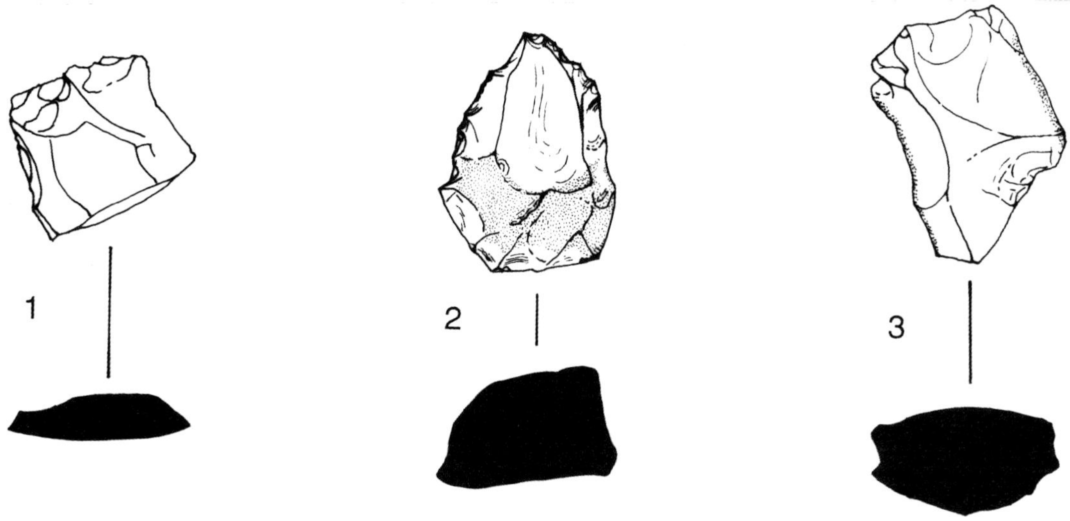

Fig. 2.97 Flints. Scale 1:1

Fl 4 Flake (not illustrated).
14 × 12 × 04mm. Hard hammer struck flake of grey speckled, lustrous flint, with a punctiform butt. The flake is partly cortical. On the left margin towards the proximal end are signs of direct, semi-abrupt retouch. The position of this retouch does not conform with that of a typical tool. This is likely to be the result of edge damage incurred spontaneously during debitage, use-wear, or some form of post-depositional contact. Undiagnostic. Context [**188**].

Fl 5 Flake (not illustrated).
15 × 17 × 04mm. Flake of grey flint, broken laterally, with a lustrous surface appearance. Unretouched with modern nick at distal end. The condition of the edges suggest some rolling. Undiagnostic. Context [**248**].

Fl 6 Flake (not illustrated).
24 × 17 × 05mm. Flake of brown stained flint in fresh condition, with cortical butt. Both margins are concave and are characterised by irregular nicks and notches. The irregularity of edge damage and the thinness of the flake suggest that some of the latter are the result of post-depositional processes than use. There is also some damage at the distal end. Undiagnostic. Context [**269**].

(N) Animal bone from Roman contexts (by Su Johnson)

99 fragments of bone were recovered from contexts which are likely to be Roman on stratigraphic grounds (Table 2.19). They were more or less evenly divided between contexts associated with the well/cistern [559/376] and those sealing the burnt timber building in pit [245]. All the bone comprised calcined material, mostly of very small size. Its condition did not allow definite identification of species, nor could most of it be identified to anatomical part. Of the fragments which could be so identified, most were from ribs and vertebrae, though fragments of pelvis, distal tibia, mandible and astragalus were also present. Context [540] produced three fragments of juvenile metatarsus of a size which suggests cattle (*Bos*). Whether this latter has any bearing on the function of the building must remain entirely speculative.

(O) Timbers from the burnt structure (by Jennifer Hillam)

13 samples from the burnt timber structure were examined at the Sheffield Dendrochronology Laboratory. The study aimed firstly to identify the wood type used in the structure, and secondly to assess the suitability of the samples for tree-ring dating, dependent upon the condition and stability of the timbers and on the number of annual growth rings they contained. Had any of the samples proved suitable, the study was to be extended to try to provide a date for the structure by dendrochronology. In the event, the samples were not entirely suitable, and recommendations were therefore made as to which might best be submitted for radiocarbon dating.

Methods

The samples were carefully unwrapped. Where sections through two planks had been sampled, it was often only possible to keep the larger of the planks intact. The samples were snapped to obtain a clean break. This provided a cross-section on which the annual rings could be isolated and the wood types identified. Where the samples remained intact, a rough ring count was made. A note was also made of the approximate dimensions of the cross-section, the average ring width and the orientation of the rings. This provided information about the timbers and the trees from which they had originated.

Table 2.19 Distribution of Roman animal bones

Context no	No of fragments	Weight (gms)	Identifiable fragments (all calcined)
WELL 559 / CISTERN 376			
284	2	1.2	-
285	2	1.6	-
308	3	3.6	? astragalus, ? sheep size
357	10	7.0	2 rib fragments
375	12	11.5	1 mandible fragment; 1 rib fragment; 1 vertebral fragment; ? 1 cranial fragment
377	1	0.3	-
471	16	6.8	1 rib fragment; 3 vertebral fragments (med size)
487	5	1.7	2 rib fragments; 1 vertebral fragment; 1 distal tibia fragment (med)
PIT 245, CONTAINING BURNT TIMBER BUILDING			
224	2	1.2	1 vertebral fragment
460	10	6.8	1 rib fragment 1 pelvis fragment (med size)
540	36	24.5	1 vertebral fragment; 3 metatarsal fragments (juvenile; cattle size)

The Timbers

All the samples were identified as oak (*Quercus* spp). No trace of sapwood was found. Full details of each sample, including their context numbers, are given in Table 2.20. In some cases, the samples were too broken for a ring count to be made. This was particularly true for the thin planks, for example, T 6 and T 7. All the planks had been tangentially split or cut from the tree. This allows smaller and generally younger trees to be used than those that would be needed to produce a radial plank of the same size. Unfortunately it also means that samples from such planks are less useful for dendrochronology than those from radial planks.

The planks contained between 20 and 60 rings, but could have come from trees with many more rings. Sample T 8, for example, with cross-sectional dimensions of about 435mm by 25mm, contained 30–40 rings, but it had no sapwood and could have many rings missing at the inside. (The 95% confidence limits for the number of sapwood rings from British oak trees over 30 years old is 10–55 rings - see Hillam *et al.* 1987 for further details.)

The four post-like timbers (T 5, 15, 22, 23) were converted in different ways. T 15 and T 23 were sections of timber, probably hewn from halved or quartered trunks and cut into shape. T 5 was probably a halved section of a smaller and younger tree. It contained 45 rings and may just have had the sapwood rings missing. The tree would therefore have been less than 100 years old when felled. The remaining post, T 22, was a whole stem with probably just the sapwood missing. It had 20 rings and was probably less than 50 years old when felled.

Although it was often only possible to calculate an approximate growth rate, these varied from sample to sample. The average ring width of sample T 5 was 2.23mm, that of T 19 was 1.37mm, and that of T 20 was 3.1mm. The sample which contained the maximum number of rings, T 21 plank 2, had extremely narrow rings at the centre (so narrow that they could not be measured with any accuracy) and much wider ones at the outside of the sample.

Dendrochronology

It was apparent from an initial examination of the samples that both the condition of the wood and the number of rings, or lack of them, was going to hinder, if not prevent, tree-ring dating. The thin planks proved to be unsuitable for tree-ring dating because they were broken in too many places and a complete ring sequence could not be obtained. The thicker timbers were more substantial and where there appeared to be more than 30 rings, an attempt was made to measure the rings. Two or three sets of measurements were made on each sample to check for accuracy. This is not usually necessary but it seemed a necessary precaution because many of the samples contained cracks. Some of the cracks were obvious but there were also less obvious hairline cracks which could have invalidated the accuracy of the ring sequence. Such cracks did in fact make it impossible to produce identical sets of measurements for two of the better samples, T 23 and T 21 plank 2.

When the samples which were broken, or those with less than 30 rings were rejected, two samples

Table 2.20 Details of tree-ring samples. Sketches are not to scale; some of the dimensions are very approximate

No	Context	function	No of rings	Av ring width (mm)	sketch	dimensions (mm)	comments
T 5	531	post	45	2.2		180×80	measured; contaminated by roots
T 6	518	plank	?	?		260×15	
T 7	519/532	vertical plank	?			395×20	broken
T 8	520/521	plank	30–40	?		435×25	broken
T 9	520	plank	20–25	2–3		230×25	
T 11	520	plank	30–40	3.0		210×20	
T 12	521	plank	30–35	2.0		245×35	
T 15	542	post	28	2.1	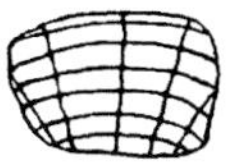	95×60	
T 19	Not contexted		43	1.37		140×45	
T 20	520/521	plank	29	3.1		170×20	
T 21	546	plank 1	33	2.1		125×35	
	546	plank 2	50–60	?	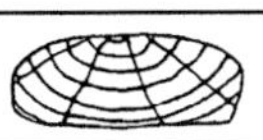	165×100	rings too narrow at centre
T 22	533	post/ upright	20	2.5		90×65	
T 23	555	post/ upright	30–40	1–2	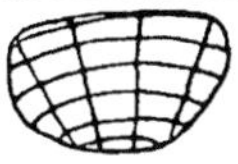	145×75	

Fig. 2.21 Ring width data

Pumsaint, timber T 5 (average of 3 sets of measurements)										
year	ring widths (0.02mm)									
1	79	79	124	125	129	161	147	139	134	130
11	158	120	130	108	156	130	114	144	165	135
21	117	96	120	111	86	94	107	68	52	75
31	110	73	120	111	86	94	107	68	52	75
41	72	92	94	84	94					
Pumsaint, timber T 19 (average of 2 sets of measurements)										
year	ring widths (0.02mm)									
1	60	79	60	61	68	73	88	58	42	50
11	91	85	50	93	110	61	82	62	63	84
21	77	56	77	60	50	69	47	49	79	59
31	59	93	72	43	61	50	55	87	56	88
41	68	82	79							

remained. Three sets of measurements from sample T 5 resulted in a ring sequence of 45 years, whilst two sets of measurements from T 19 gave 43 years (Table 2.21). At Sheffield, 50 rings is usually taken as the minimum number of rings required for reliable dating (see also Hillam *et al.* 1987). Exceptions are where there are several timbers with 30–50 rings from a tight context, but even then it is necessary to have a few samples with longer ring sequences (Hillam 1998).

Comparison of the two Pumsaint sequences showed some similarities between them, but not enough to indicate that they were definitely contemporary. When they were compared with dated reference chronologies, no consistent results were obtained. It is likely that the shortness of the ring sequences will prevent them being dated by dendrochronology in the future.

Recommendations for radiocarbon sampling

From an assessment of the samples the most suitable for radiocarbon dating is likely to be post T 22, because it is a young stem containing 20 rings with probably only the sapwood missing. The radiocarbon result should therefore be a close reflection of the date when the timber was felled. The same should be true of post T 5, although this sample is badly contaminated with root intrusions, and the radiocarbon laboratory would need to take this into account. Posts T 15 and T 23, with 28 and 30–40 rings respectively, would also be suitable, as would the wider outer rings of T 21 plank 2. The orientation of the rings suggest that these timbers are less likely to be from the inner part of a large tree than are some of the larger planks.

Conclusion

No tree-ring dates were obtained for the oak timbers, although it was possible to provide information about the type of timber used, and the trees from which it had been converted. The condition of the burnt timber and the insufficient number of annual rings both contributed to the failure of dating, although the latter was probably the decisive factor. Although no dating was obtained for the Pumsaint timbers, the study demonstrated that charcoal can potentially be used for dendrochronology.

(P) Radiocarbon determinations from the burnt timber building (by Quentin Dresser)

Method

The Radiocarbon measurements for the Pumsaint series were made at the Department of Geology, University of Wales, Cardiff. All ^{14}C ages are referenced to 1950 (NBS standard) and were corrected for isotopic fractionation. Calibrations and statistical tests were done using the Radiocarbon Calibration Program Rev. 3.0 produced by the University of Washington, Quaternary Isotope Laboratory (Stuiver and Reimer 1993, 215–230). The decadal dataset was used (Stuiver and Becker 1993, 35–65). A laboratory error multiplier of 1.25 was used and sample age ranges as follows:

CAR-1251. 10 years. Oak charcoal – outer 10–15 rings of charred post with *c.* 20 annual rings (T22).

CAR-1252. 20 years. Oak charcoal – outer 20 rings of charred post with *c.* 45 annual rings (T5).

CAR-1253. 30 years. Oak charcoal – random lumps from charred post with *c.* 28 annual rings (T15).

CAR-1254. 20 years. Oak charcoal – outer 20 rings of charred post with *c.* 30–40 annual rings (T23).

Results

The calibrated results are given in Table 2.22 as a 2 standard deviation range, providing a 95% confidence interval.

Initial examination of the Pumsaint results indicates that they are not statistically the same at the 95% confidence level, whereas all the samples probably come from the same structure and may thus be thought to be contemporary in use, if not in origin. During the pre-treatment of the samples, it was noted that CAR-1252 was abnormally heavily penetrated by rootlets and while these were removed, there could have been a residue of very fine rootlets in the sample. This sample is the youngest, which may confirm the presence of younger material. If this sample is excluded from calculations, the remaining three are statistically the same at the 95% confidence level, and the pooled mean Radiocarbon age is 1738±46, which calibrates to AD 177 – AD 419 (95% confidence level; intercepts at 266, 276, 330). The best estimate in round numbers for the date of construction is AD 180 – AD 420. However, there remains the possibility that some non-contemporary wood was used in the construction of the building, in which case the construction will be younger by an indeterminate amount, unlikely to exceed 100 years.

Table 2.22 Radiocarbon calibrations

Designation	Lab. No.	^{14}C Age	Calibrated range (2 σ)	Intercepts
Sample T 22 Context 533	CAR-1251	1780±60 BP	Cal AD 73–423	250, 310
Sample T 5 Context 531	CAR-1252	1530±60 BP	Cal AD 394–638	540
Sample T 15 Context 542	CAR-1253	1630±60 BP	Cal AD 250–607	420
Sample T 23 Context 555	CAR-1254	1830±70 BP	Cal AD 7–415	190, 230

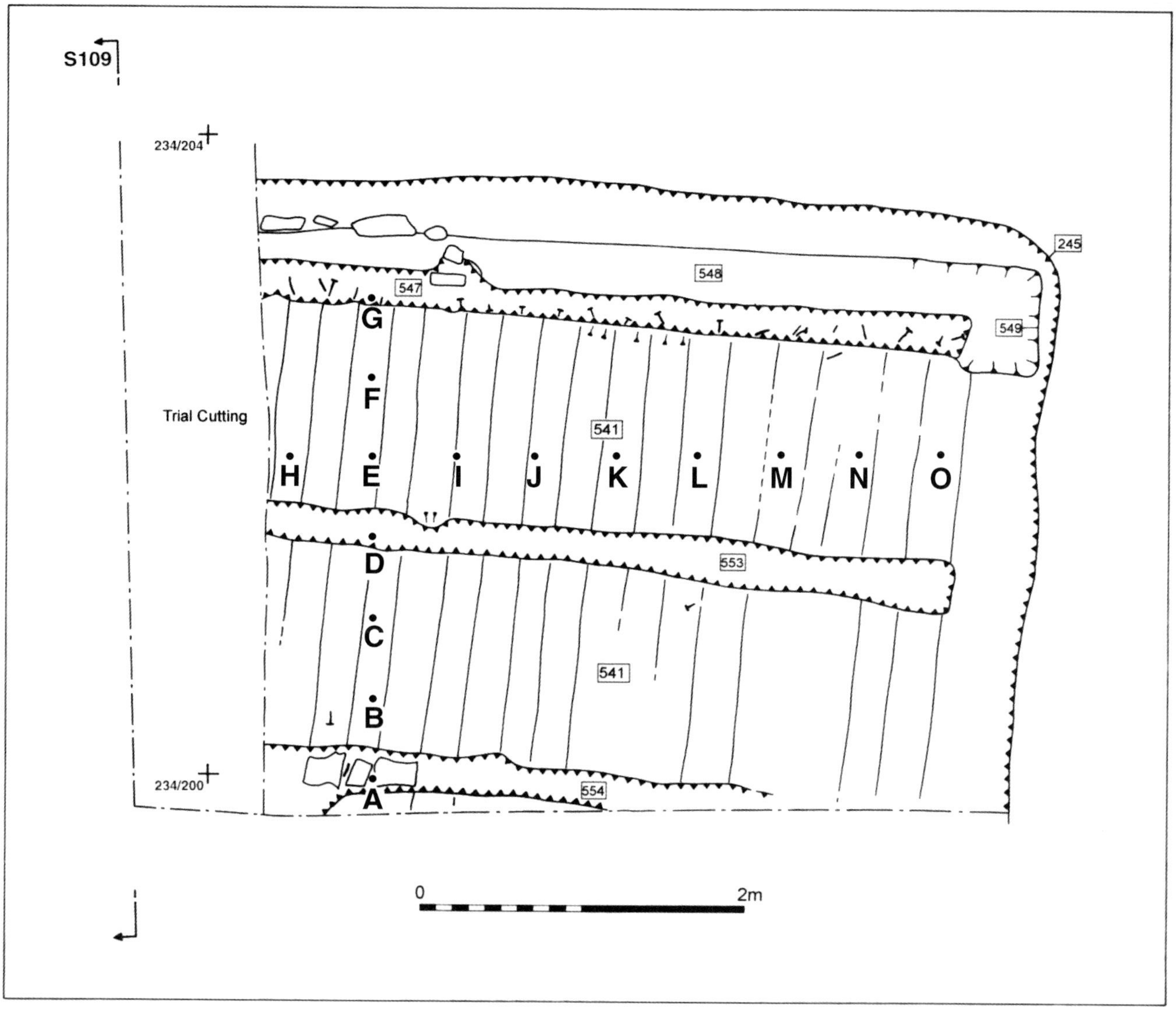

Fig. 2.98 Location plan of soil samples

(Q) Magnetic susceptibility and phosphate studies of sediments from the burnt timber building (by John Crowther)

Introduction

Sediments from what appeared to be a cellar of a burnt timber building at Pumsaint were analysed[1] in the hope that they might provide some insight into the function of the cellar and into the way in which it was ultimately destroyed (by fire). Samples were taken on a 0.50m grid across the whole of the cellar floor. This report presents the results of a pilot investigation undertaken on samples from two intersecting transect lines from the grid: one running parallel with the northern wall (grid line 202 N), and the other cutting north-south through what is thought to be the central part of the cellar (235 E). Although some useful findings emerged from the results, it was felt unlikely that much more would be achieved by analysing the full set of samples. At each of the 15 locations (identified by code letters A–O in Fig. 2.98), separate samples were taken from the thin minerogenic layers immediately above (context [534]) and below (context [543]) the burnt plank flooring. In addition, samples of the 'natural' sediments underlying the planking were taken at two points (24 and 25; see Fig. 2.42) along the western section wall (*i.e.* in the central part of the cellar), with samples being taken at depths of 5, 10, 15 and 20cm below context [543]. Unfortunately, no control samples were taken from the natural sediments outside the building. Previous studies of nearby exposures do, however, provide some basis for comparison (Crowther 1987). A single bulked sample of charcoal fragments from contexts [534] and [543] was also analysed.

[1] The laboratory work was undertaken by David Weston and Ian Clewes.

Attention focused upon two properties:

Phosphate-P: Phosphates, which are present in all organic matter (including bones, excreta and plant material), are relatively insoluble in a mineralised form and tend to become 'fixed' in sediments and soils. Phosphate enrichment is therefore often indicative of areas where there has been a concentrated input, and subsequent mineralisation through decomposition processes, of organic matter (Proudfoot 1976; Hamond 1983; Crowther 1997). Total phosphate (phosphate-P) was determined on the fine earth (*i.e.* <2 mm) fraction by alkaline oxidation with NaOBr using the method described by Dick and Tabatabai (1977).

Magnetic susceptibility: Low frequency magnetic susceptibility (χ) reflects the concentration of magnetic forms of iron oxide (*e.g.* maghaemite) in sediments, which is largely dependent upon the presence of iron and a period(s) of alternating reduction-oxidation conditions that favour the formation of magnetic minerals. Heating of sediments, particularly when this occurs under reducing, followed by oxidising conditions, can cause marked χ enhancement (Tite and Mullins 1971). Spatial variations in χ are therefore potentially valuable in identifying those locations where burning has taken place *in situ*, though care has to be taken to ensure that the variations in χ are largely attributable to differences in degree of enhancement, rather than underlying variations in the amount of iron present. In the present study this has been established by determining the maximum potential susceptibility (χ_{max}) and percentage fractional conversion (χ_{conv}) of a subset of 20 samples which are representative of the range of χ values encountered at the site. All the samples were air-dried and analysis undertaken on the fine earth fraction. χ was determined using a Bartington MS1 magnetic susceptibility meter. Determinations of χ_{max} were made by heating samples at 650°C in reducing, followed by oxidising conditions. The method broadly follows that of Tite and Mullins (1971), except that household flour was mixed with the soils and lids placed on the crucibles to create the reducing environment (after Graham and Scollar 1976). χ_{conv} is a measure of the extent to which the potential susceptibility has been achieved in the original soil, *viz*: $(\chi/\chi_{max}) \times 100.0$ (Tite 1972; Scollar *et al.* 1990). The statistical analysis undertaken and the interpretation of the results follow the principles discussed by Crowther and Barker (1995).

A qualitative assessment was made of the amount of charcoal and gravel present in the samples. In addition, loss-on-ignition (LOI) at 375°C for 16 hours (Ball 1964) was determined for the subset of 20 samples in order to provide an estimate of the organic matter concentration, and texture and pH were determined for a representative sample from each of the two contexts and the underlying sediments.

Results and discussion

It should be noted that both the minerogenic layers are very thin. Incorporation of some material from underlying or overlying deposits is therefore inevitable, either as a result of disturbance processes associated with the utilisation and subsequent destruction of the building, or difficulties in sampling. As a consequence, caution must be exercised in interpreting the results.

General character of the sediments: The building is located on a terrace of unconsolidated river gravels and sands. Contexts [534] and [543] are clearly distinguishable from the underlying sediments in that they contain much higher proportions of silt and clay (details in footnotes of Tables 2.23 and 2.24). The similarity of the texture of these contexts suggests that they may have a common origin. Thus, if context [534] represents a layer of fine deposits that had accumulated on the floor of the cellar prior to destruction, then context [543] could well comprise material that had fallen through gaps in the planking whilst the cellar was in use.

Although there are quite wide variations in χ_{max} (range, 488–3390 × 10^{-8} SI kg^{-1}), stepwise multiple regression analysis identified χ_{conv} as by far the dominant factor affecting χ – the r_2 value being 0.932 (*i.e.* explained variation of 93.2%). This finding demonstrates that the observed variability in χ (range, 14.8–1710 × 10^{-8} SI kg^{-1}) is largely attributable to differences in the degree of magnetic susceptibility enhancement, as might be associated with burning, rather than natural variations in the potential susceptibility of the sediments (see discussion in Crowther and Barker 1995). Natural background concentrations of phosphate-P in the terrace deposits are *c.* 2.50–3.00 mg g^{-1} (Crowther 1986). These levels are relatively high for such a coarse-textured sediment, and are presumed to reflect a high proportion of shale fragments (*cf.* quartz grains) in the matrix. Variations in LOI (range, 0.850–29.9%) appear to relate closely to charcoal content.

Layer above plank flooring [534]: If it is assumed that this layer represents fine deposits that had accumulated on the cellar floor prior to the destruction of the building, then the presence of moderate amounts of gravel in 8 of the 15 samples (Table 2.23) may be taken to indicate locations where there has been greatest incorporation of coarser sediments. This is presumably the result of the collapse of the walls of the cellar, which are cut in the natural terrace deposits. The amounts of charcoal (and LOI data) are difficult to interpret since they are probably largely derived from the burnt plank flooring and/or burnt timber that fell onto the floor during the conflagration. Apart from sample A, which is clearly anomalous, the χ values are high (>400 × 10–8 SI kg^{-1}) – a feature which is wholly consistent with the *in-situ* burning of the plank

Table 2.23 Analytical data for samples from the floor of cellar (locations shown in Fig. 2.98)

Sample location	Gravel content†	Charcoal content†	LOI (%)	χ (10^{-8} SI kg^{-1})	χ_{max} (10^{-8} SI kg^{-1})	χ_{conv} (%)	Phosphate-P (mg g^{-1})
Layer (Context 534) above flooring							
A	*	*	7.99	60.2	2170	2.77	5.93
B[1]	**	**	nd	877	nd	nd	4.45
C	**	**	nd	786	nd	nd	3.94
D	*	***	nd	642	nd	nd	6.40
E	**	*	7.68	1210	3060	39.6	3.59
F	*	**	13.2	1710	2800	61.1	3.69
G	**	***	nd	491	nd	nd	5.66
H	**	**	11.8	972	2260	43.0	4.80
I	*	**	nd	747	nd	nd	5.89
J	*	***	nd	753	nd	nd	7.51
K	*	***	nd	774	nd	nd	7.51
L	*	***	29.9	439	2440	18.0	5.36
M	**	**	12.6	1050	3390	31.0	4.28
N	**	**	11.6	1070	2660	40.2	3.08
O	**	*	9.48	680	2710	25.1	4.07
Layer (Context 543) below flooring							
A	**	**	nd	58.0	nd	nd	8.64
B[2]	**	**	13.4	271	2440	11.1	11.9
C	**	**	13.0	328	2260	14.5	10.8
D	**	**	nd	206	nd	nd	7.14
E	**	*	4.21	340	1780	19.1	5.79
F	***	*	nd	260	nd	nd	7.18
G	***	*	nd	73.7	nd	nd	4.40
H	**	*	nd	189	nd	nd	6.35
I	**	**	nd	285	nd	nd	4.94
J	***	*	3.20	39.0	1080	3.61	2.49
K	***	*	2.27	35.9	1100	3.26	1.12
L	***	**	nd	41.3	nd	nd	1.83
M	***	*	2.34	960	1750	54.9	3.73
N	***	**	4.37	1300	2660	48.9	2.78
O	***	*	2.11	57.8	1430	4.04	1.48

† Qualitative assessment: High (***), Moderate (**), Low (*)
1 Particle size (<2 mm fraction): 51% sand, 37% silt, 12% clay. pH (1:2.5, water): 7.2
2 Particle size (<2 mm fraction): 49% sand, 36% silt, 15% clay. pH (1:2.5, water): 7.2

flooring. Extremely high χ figures (>1000 × 10^{-8} SI kg^{-1}) were recorded close to the intersection of the two transect lines (samples E and F) and in the north-eastern corner (M and N). Such high values are suggestive of *in-situ* burning under strongly reducing conditions, as would have been the case if burning had occurred beneath collapsed rubble from the cellar walls. It should be noted that the charcoal in the samples does not contribute directly to the observed enhancement – the bulk sample of charcoal analysed had a χ value of only 75 × 10^{-8} SI kg^{-1}.

Phosphate-P concentrations range from 3.08–7.51 mg g^{-1} (Table 2.23). Unfortunately, because of the nature of the context, there is no background 'control' against which these data may be judged, and it is therefore difficult to assess whether the higher values recorded do actually represent significant levels of enrichment. What does emerge from the results, however, is that higher phosphate-P concentrations are found in samples that are charcoal rich. Thus, the three samples with phosphate-P concentrations of >6.0 mg g^{-1} all have high levels of charcoal, whereas all samples with values of <5.0 mg g^{-1} contain only low or moderate amounts. Since the concentration of phosphate-P recorded in the bulk sample of charcoal (5.17 mg g^{-1}) is not exceptionally high, the charcoal itself is clearly not the direct cause of the elevated values. It seems reasonable to assume therefore that the observed relationship is in part attributable to phosphate derived from the burnt wood that has become fixed within the mineral matrix. Under these circumstances, any original pattern in the phosphate-P distribution of across context [534] is likely to be masked by the effects of burning, and it would clearly be unwise to make any inferences from the results obtained.

Layer below plank flooring [543]: In general the samples from this layer are more gravelly, but contain less charcoal, than context [534] (Table 2.23). Unfortunately, these findings could simply be an artefact of sampling, in that this context may have been difficult to separate from the underlying terrace deposit, while being more easily separated from the overlying burnt timber. In contrast, both the χ and phosphate-P data are potentially significant. The χ values (range, 35.9–1300 × 10^{-8} SI kg^{-1}) are generally lower than in context [534]. This is consistent with the lower temperatures likely to have been experienced beneath the flooring during the fire. The low values (58.0 and 73.7 × 10^{-8} SI kg^{-1}) recorded for samples A and G, taken from beneath two of the beam slots along transect 235 E, may perhaps indicate that these beams did not actually burn through to the base. Notably high levels of enhancement (960 and 1300 × 10^{-8} SI kg^{-1}) were recorded, however, for samples M and N in the north-eastern corner of the cellar. This is in keeping with the results from context [534], thereby adding further support to the idea that this part of the cellar was subject to particularly intense heating under reducing conditions, as might have occurred following the collapse of a mixture of large burning timbers and terrace deposits from the cellar walls.

Phosphate-P concentrations within context [543] are extremely variable (range, 1.12–11.9 mg g^{-1}). What is especially significant is that samples A, B and C, from the central part of the building, have very high phosphate-P concentrations (8.64, 11.9 and 10.8 mg g^{-1}, respectively), yet contain only moderate amounts of charcoal. This provides a clear indication of phosphate-P enrichment in this part of the cellar, though its cause remains obscure. Since similar enrichment is not evident in the layer immediately above the plank flooring, this raises the interesting possibility that the cellar had more than one phase of use. Thus, originally the cellar may have been used for a purpose that led to an increase in phosphate-P in the minerogenic layer immediately below the flooring in the central part of the cellar, either as a result of particulate material falling through gaps in the planking or liquid soaking through. However, by the time the building was destroyed, no clear traces of this former activity were detectable in the thin minerogenic deposit that had accumulated on the plank flooring. Whether this represents a change to a different use, or simply the abandonment of the cellar, is uncertain.

'Natural' sediments beneath the building: The terrace sediments beneath the plank flooring show little evidence of χ enhancement (Table 2.24), suggesting that they were not significantly affected by the fire. Perhaps more interesting is the fact that, despite being located in the central part of the cellar, the sediments sampled do not show any signs of phosphate-P enrichment (range, 1.79–3.96 mg g^{-1}). In fact, phosphate-P concentrations tend to increase, rather than decrease, with depth. The implication of this is that the enrichment that occurred in context [543] was not transmitted downwards to any significant extent into the underlying natural. The source of phosphate in the cellar would therefore seem more likely to be particulate, rather than liquid, in origin.

Table 2.24 Samples of natural sediments beneath building

Sample depth (cm)	Gravel content†	Charcoal content†	LOI (%)	χ (10^{-8} SI kg^{-1})	χ_{max} (10^{-8} SI kg^{-1})	χ_{conv} (%)	Phosphate-P (mg g^{-1})
Location 24							
5	***	*	nd	20.3	nd	nd	2.54
10$	***	*	nd	16.2	nd	nd	2.27
15	***	*	nd	14.8	nd	nd	2.52
20	***	*	nd	15.7	nd	nd	3.96
Location 25							
5	***	*	0.850	16.4	488	3.36	1.79
10	***	*	1.05	18.9	820	2.30	2.64
15	***	*	1.00	16.7	833	2.00	3.90
20	***	*	1.16	18.0	981	1.83	3.14

† Qualitative assessment: High (***), Moderate (**), Low (*)

$ Particle size (<2 mm fraction): 87% sand, 10% silt, 3% clay. pH (1:2.5, water): 6.1

Conclusions

1. Magnetic susceptibility data appear to be useful in elucidating the way in which the floor of the cellar was affected by fire, though this may add little to what is perhaps already clear from field evidence.
2. Phosphate-P concentrations in context [534] (above the plank flooring) are relatively low, and appear to vary directly with charcoal content. Although the charcoal itself does not have a particularly high phosphate concentration, it seems likely that the enrichment is derived, at least in part, from the burning of the wood.
3. Context [543], in contrast, shows quite clear evidence of phosphate enrichment in the central part of the cellar. Analysis of the remaining samples would have enabled the area of enrichment to be defined more precisely, but it seemed unlikely that this would provide any additional insight into the function of the cellar.

(R) Pollen analysis from the pre-rampart and rampart area (by Astrid E. Caseldine)

Very few pollen studies have been carried out on Roman sites in Wales and most pollen diagrams from peat deposits lack radiocarbon dates for this period, thereby making it difficult to assess the effects of Romanisation on the landscape (Caseldine 1990; 1991). This work was undertaken with the aim of determining the effect of the establishment of the Roman fort at Pumsaint on the surrounding environment.

During excavation of the northern defences four monoliths were removed from sections for the analysis of pollen to ascertain the nature of the local environment prior to the construction of the fort and to give some indication of the environment during occupation. Monoliths 1 and 2 were from the area between the stone wall and the original rampart (the latter is shown on Fig. 2.8; the former was taken from the northern end of the trial trench across the defences, before the back face of the stone rampart was fully exposed). Monoliths 3 and 4 were from the original turf-clay rampart (for location see Figs 2.7 and 2.8). Unfortunately pollen preservation in monolith 3 was so poor that, after initial examination, further analysis was abandoned.

The sequence of deposits recorded in the three sections examined was as follows:

Monolith 1

0–12/13cm	Silty clay [380] ?redeposited turf-clay rampart
12/13–17/18cm	Turf horizon
17/18–22/23cm	Trampled horizon [385] ?redeposited
22/23–30cm	Subsoil

Monolith 2

0–20/21cm	Silty clay [380] ?redeposited turf-clay rampart
20/21–23/24cm	Turf horizon
23/24–30cm	Trampled horizon [385] ?redeposited

Monolith 4

0–21/22cm	Turf-clay rampart [381]. Possible turf lines *c.* 7cm and 16cm.
21/22–33cm	Trampled ground surface [383]
33–50cm	Subsoil

Methods

The samples were prepared following standard procedures outlined in Moore *et al.* (1991), including treatment with hydrofluoric acid and micro-sieving because of the high minerogenic content. A pollen count of 300 total land pollen (TLP) was achieved at all levels. Nomenclature is based on Bennett (1994) and Bennett *et al.* (1994).

Zonation

Local pollen assemblage zones have been identified for the lower levels of each of the diagrams (Figs 2.99 to 2.101). The upper levels, from dumped and redeposited material, have not been zoned.

Monolith 1 (Fig. 2.99)

LPAZ P1.1 This zone is characterised by high arboreal values. *Corylus avellana* type values are between 45–55% TLP and *Alnus* reaches over 20% TLP. Poaceae pollen is less than 20% TLP. *Calluna* is present in small amounts. *Plantago lanceolata* is present as are other herbaceous taxa, particularly Ranunculaceae. Spores are quite well represented. *Polypodium* is more than 10% TLP.

LPAZ P1.2 This zone is distinguished by a decline in arboreal pollen, notably *Corylus avellana* type pollen from over 45% TLP to less than 25% TLP, and a marked increase in Poaceae pollen initially to more than 35% TLP and ultimately 50% TLP. *Alnus* decreases slightly while *Betula* shows a slight increase. Herbaceous pollen is more frequent with an increase in *Plantago lanceolata*. Lactuceae is well represented. Cerealia type pollen occurs. *Calluna* is less frequent. Spores are less abundant.

LPAZ P1.3 The zone is characterised by an increase in arboreal pollen, largely *Corylus avellana* type and a slight decrease in Poaceae pollen. Herbaceous pollen is less frequent. Spores are quite frequent.

?Redeposited rampart [380]. Arboreal values are slightly lower than in P1.3. Poaceae values are *c.* 35–40% TLP. Herb pollen is quite well represented including *Plantago lanceolata*, Lactuceae and Ranunculaceae. Low amounts of *Calluna* are present.

Monolith 2 (Fig. 2.100)

LPAZ P2.1 Arboreal values are generally low, with the exception of *Corylus avellana* type. Poaceae and Lactuceae values are high. *Plantago lanceolata* and Cerealia type pollen are recorded.

LPAZ P2.2 This zone is characterised by a slight

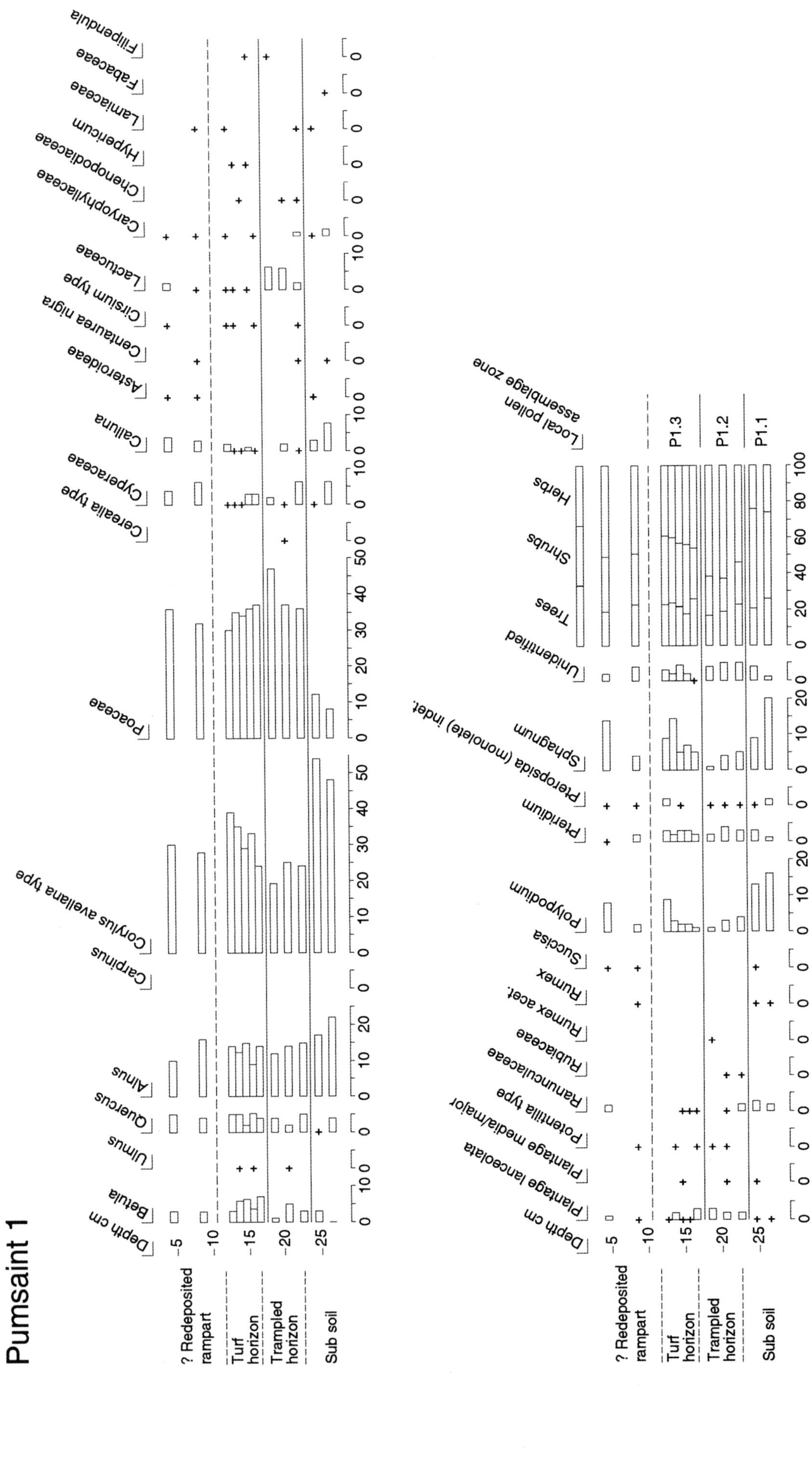

Fig. 2.99 Pollen analytical data – Pumsaint 1

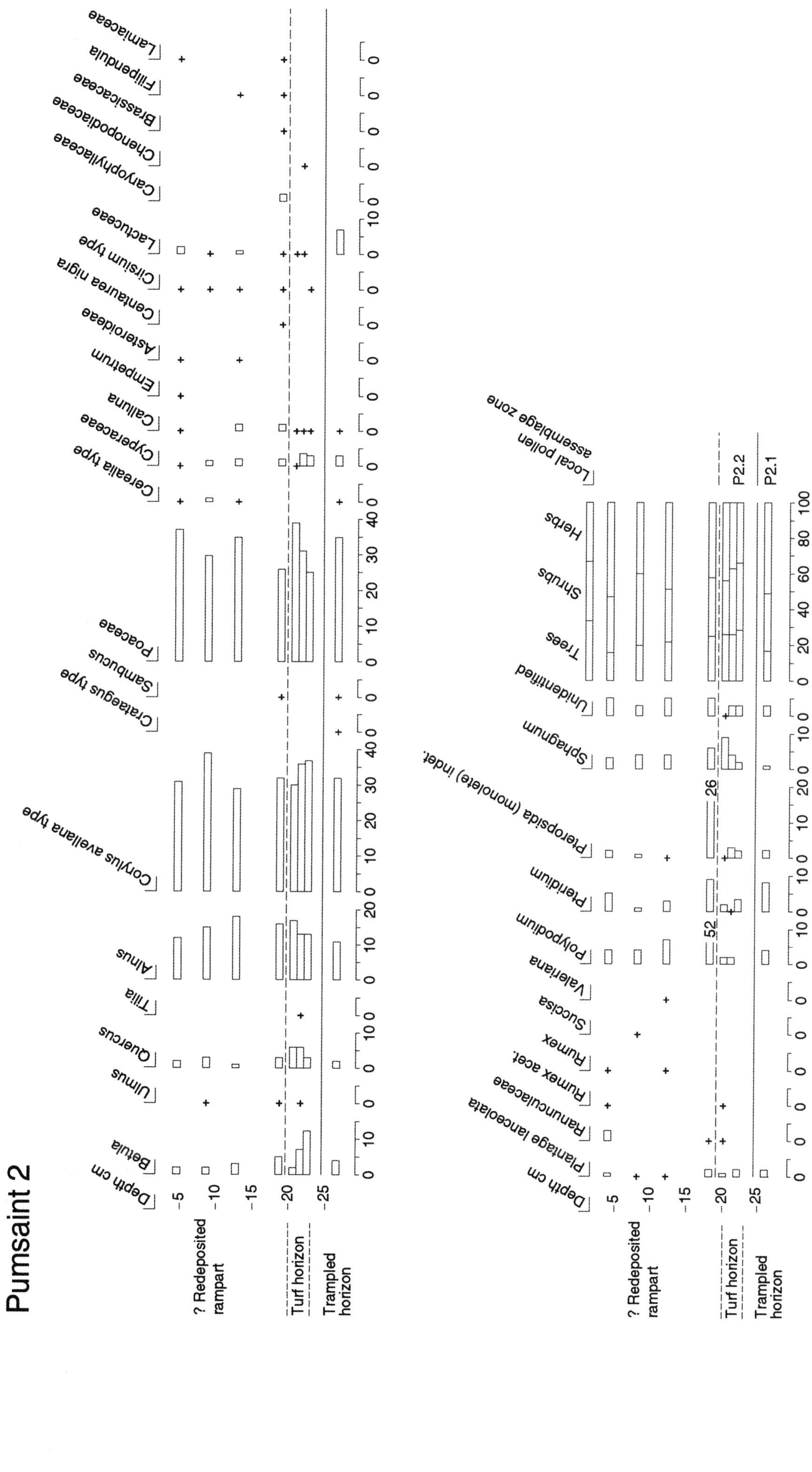

Fig. 2.100 Pollen analytical data – Pumsaint 2

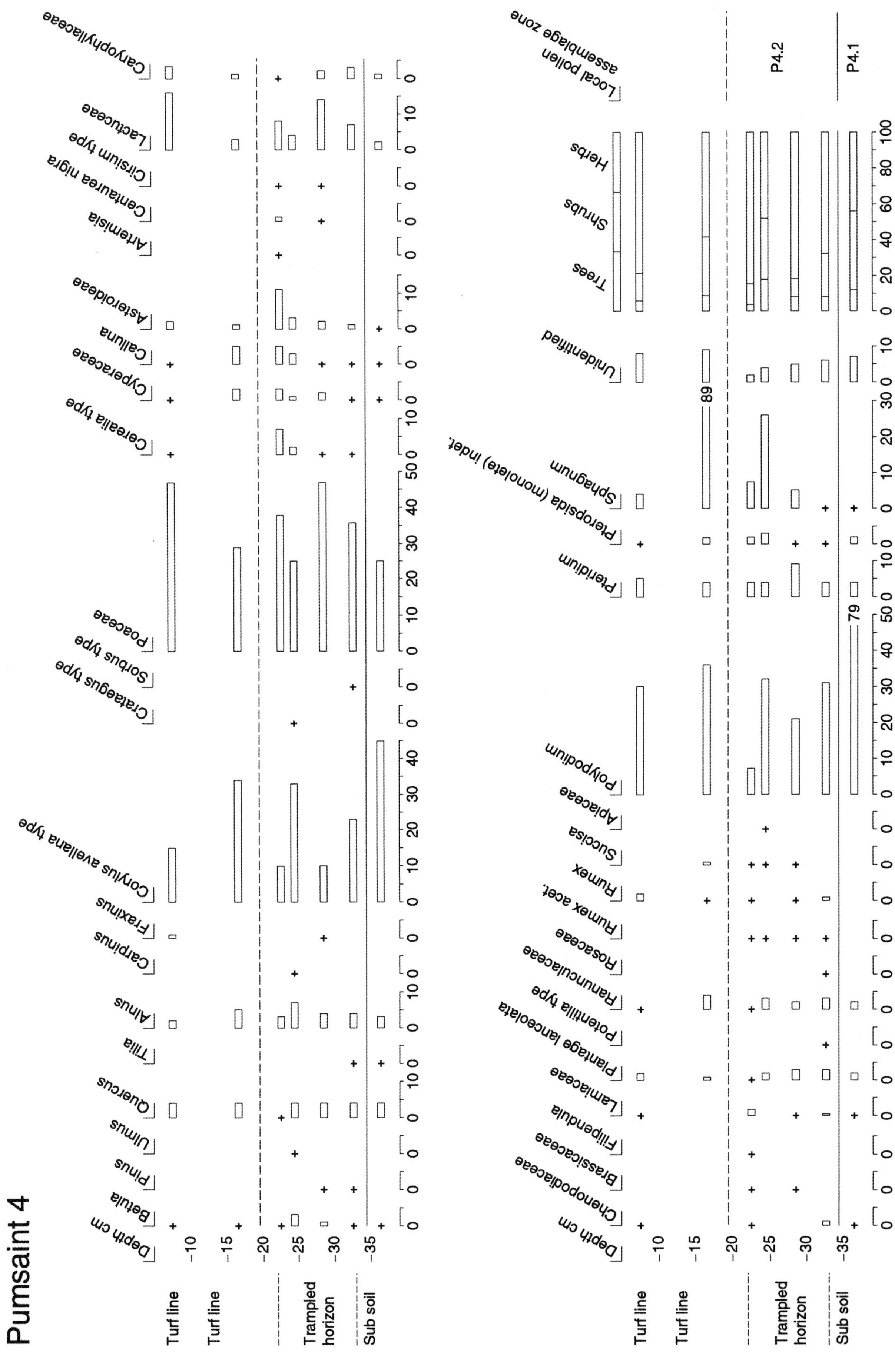

Fig. 2.101 Pollen analytical data – Pumsaint 4

increase in arboreal and a decrease in herbaceous pollen. Cerealia type pollen is absent.

?Redeposited rampart [380] Arboreal values are generally marginally lower than in P2.2. Poaceae is consistently *c.* 35% TLP. Cerealia type pollen is present and pollen of *Plantago lanceolata*, Lactuceae and Ranunculaceae is quite frequent. *Calluna* values exceed 1% TLP.

Monolith 4 (Fig. 2.101)

LPAZ P4.1 The earliest zone is marked by high arboreal values, largely *Corylus avellana* type *c.* 45% TLP. Poaceae values are *c.* 25% TLP and herbaceous pollen, including *Plantago lanceolata*, is present in low amounts. *Polypodium* spores are abundant.

LPAZ P4.2 The beginning of this zone is marked by a rise in Poaceae pollen and a decline in *Corylus avellana* type. Herbaceous pollen is much better represented than in the previous zone. Lactuceae reaches over 10% TLP as does Asteroideae at the end of the zone. Cerealia type pollen is more than 5% TLP at the end of the zone.

Turf lines. Arboreal pollen other than *Corylus avellana* type is poorly represented. Poaceae pollen is reasonably abundant and *Plantago lanceolata*, Lactuceae and Ranunculaceae pollen quite frequent. Cerealia type pollen is present.

Discussion

There are a number of problems associated with the interpretation of soil pollen and with samples from archaeological sites where the normal models of pollen deposition do not necessarily apply. In the case of the former there are problems of differential preservation, vertical mixing by the soil fauna and translocation downwards through the profile (Moore *et al.* 1991). However, it is to a large extent because of this downward movement that it is possible to determine pollen stratification. In the case of anthropogeneous deposits there are additional considerations which need to be taken into account in terms of the origin of the pollen and the deposit such as whether the pollen accumulated independently of the deposit (Faegri and Iversen 1989). Although some of the Pumsaint pollen was indeterminable the percentage was not too high. Stratification was also evident in the lower layers allowing zones to be identified, even where there was possibly some redeposition in monoliths 1 and 2 (trampled horizon [385]). A greater problem is the interpretation of the rampart material discussed below.

The earliest evidence is reflected in zones P1.1 and P4.1 and indicates the dominance of hazel scrub with some alder and oak woodland but also that already some clearance had taken place. Higher *Corylus avellana* type and *Alnus* and lower Poaceae values in P1.1 than P4.1 suggest the location of the former monolith was closer to woodland. Given the location of the fort on a terrace above the confluence of the Afon Twrch with the Afon Cothi, the *Alnus* pollen is possibly derived from alder growing on the wetter floodplains while the *Quercus* pollen may have come from oak growing on the valley sides. The herbaceous taxa indicate clearance and the presence of *Plantago lanceolata* and Ranunculaceae pollen together with the absence of cereal type pollen tends to suggest the nature of the farming activity was pastoral rather than arable. The high *Corylus avellana* type values are consistent with hazel scrub recolonising an abandoned area following earlier clearance. A marked expansion in Poaceae and anthropogenic indicator species and a decline in arboreal pollen in P1.2 and P4.2 demonstrates further clearance activity. The evidence suggests that it was largely local hazel scrub that was cleared. In contrast to the previous zone there is evidence for cereal cultivation as well as pastoral activity. The former is particularly clearly indicated in the final pollen level from P4.2 where cereal pollen values reach *c.* 7% and Asteroideae, which include cornfield type weeds such as *Chrysanthemum segetum*, reach over 10%. The evidence is much stronger at P4 than P1 and tends to suggest cultivation in the immediate area of P4 but not at P1. However, this zone coincides with a layer interpreted as an *in-situ* trampled horizon at P4 but possibly a mixture of pre-rampart soil and erosion off the rampart at P1, which may account for the slight differences in pollen representation. This zone is also represented at the base of monolith 2 (zone P2.1). At P4 oscillations in the *Corylus avellana* type curve during P4.2 indicate an episode with some re-colonisation by hazel scrub. This is accompanied by a slight increase in *Alnus*, *Quercus* and *Betula*. At P4 immediately above the trampled horizon are the clay blocks of the rampart, which suggests that the rampart was built directly upon an area which shortly before had been under cultivation. Given the nature of the site, above the flood-plain and relatively flat, it would have been a relatively favourable area for cultivation in the iron age.

The final zone is represented in monoliths 1 (P1.3) and 2 (P2.2) and is characterised by a slight increase in arboreal pollen, particularly *Corylus avellana* type and *Betula*, the absence of Cerealia type pollen and a marked reduction in Lactuceae pollen. Poaceae levels remain quite high. The absence of cereal type pollen is consistent with a change from cultivation at the site following construction of the fort and the development of grassland, resulting in a well-established turf horizon in this area of the fort in the period after the construction of the original rampart and prior to the construction of the stone defences. The remaining levels from P1 and P2 represent pollen from ?redeposited rampart material and are less easy to interpret. Essentially they show little difference to the zones below at P1 and P2, particularly P1.2 and P2.1 which coincide with the trampled horizon. Values for arboreal and Poaceae pollen are similar and cereal type pollen is present as well as slightly higher values for Lactuceae, though not as high as in P1.2 and P2.1, suggesting that at least some of the pollen incorporated in these deposits may be derived from the trampled horizon as well as from the contemporary environment. Similarly the pollen from the two traces of turf lines in P4 shows similarities with zone P4.2, which coincides with the trampled horizon, in terms of the occurrence of cereal type pollen, Lactuceae and other taxa. This may mean that turf from the immediate area was used to make at least part of the rampart. In addition the cereal type pollen and Lactuceae evident in P1 and P2 may be derived from either the trampled horizon represented in P4.2, from eroded turf material from the rampart itself or from both. Alternatively, the pollen in the upper levels from P1 and P2 may be totally or partially derived from the lower horizons as a result of activity by soil fauna.

Comparisons with other evidence

The nearest diagrams from the Gwenlais Roman leats (see section 3.3.3 below) differ in having lower arboreal pollen and higher Ericaceae pollen, reflecting the upland environment. However, as at Pumsaint the strongest representation of arboreal taxa is that of *Corylus avellana* type followed by *Alnus* and *Betula*. The data from the buried soil at Gwenlais is tentatively assigned to the Roman period while the lowermost pollen assemblage zone in the two leat profiles may belong to the medieval period. In terms of the representation of Poaceae pollen and arboreal pollen there is a greater similarity between the pollen evidence from Pumsaint and that from the lowermost pollen zone from the two leats than that from the buried soil, which may indicate that the pollen spectra in the very base of the leats began accumulating in the Roman period. However, the pollen data are from very different areas and therefore may simply reflect similar land use at different times.

The absence of radiocarbon dates makes any correlation with pollen evidence from Talley (Butler 1984), *c.* 5 miles to the south, highly speculative. It is proposed that the effects of Romanisation may be reflected in pollen zone UT-2 at Talley but equally they could be represented in a later zone. There is, however, some dated pollen evidence from both Tregaron (Turner 1964; 1965), *c.* 14 miles to the north, and Cefn Gwernffrwd (Chambers 1982a), *c.* 7 miles to the north-east. At Tregaron an extensive clearance episode, largely pastoral, is dated from about 400 BC until AD 1200. This follows a period when *Quercus* was first replaced by *Corylus* which in turn appears to have been cleared during the early iron age, *i.e.* at the beginning of the extensive clearance episode. At Cefn Gwernffrwd there is some evidence for the establishment of local *Corylus* scrub for a period during the iron age after a period of extensive upland heath development *c.* 2755 BP. It is suggested that clearance of this scrub occurred during the late iron age or Romano-British period. Similarly there is evidence from the Brecon Beacons (Chambers 1982b; 1983a) for the clearance of hazel scrub during the iron age and for the removal of woodland cover during the later iron age in the Waun-Fignen-Felen area (Smith and Cloutman 1988). The evidence from Pumsaint for the clearance of hazel scrub during the iron age/Romano-British times is therefore consistent with the evidence from these other sites in mid and south Wales.

Finally, there is evidence for clearance prior to the construction of a Roman marching camp in the Fforest Fawr area (Crampton and Webley 1964) and some limited pollen evidence from turves from the Roman fort at Pen Llystyn (Dimbleby 1968) in north Wales. Although the evidence is slight, one of the turves contained cereal type pollen indicating pre-Roman cultivation in the area, as at Pumsaint.

(S) Charred plant remains from the burnt timber structure (by Astrid E. Caseldine)

Bulk samples[1] were taken from immediately above and below the floor of the burnt timber structure during the excavations at Pumsaint. The purpose of the sampling was to see if the deposits contained any charred plant material other than wood charcoal, which might help to ascertain the function of the structure.

The samples were taken on a grid basis and six samples from each of the two layers were selected for initial processing. The samples were processed by Dyfed Archaeological Trust. The flots and residues recovered were sub-sampled (Table 2.25) and sorted for plant remains using a Wild M5 stereomicroscope.

Very few charred plant macrofossils were present (Table 2.26). A hazelnut fragment, a seed of *Galium* sp. (bedstraw, cleavers) and a grass rhizome were recorded in samples from above the floor. A *Carex* sp. (sedge) nutlet was present in one of the samples from below the floor. Apart from the latter possibly indicating damp conditions locally, the remains do not help to determine the function of the structure.

[1] I would like to thank C.J. Barrow for sorting the samples.

Table 2.25 Proportions of samples sorted

Context [534]	**Flot**	**Residue**	**Fine Residue**
234/200	1/32 of 525ml	1/32 of 550ml 1/32 of 625ml	1/128 of 1000ml
234/202	1/16 of 550ml	1/16 of 300ml	1/128 of 600ml
235/201	1/16 of 400ml	1/64 of 1000ml 1/32 of 550ml	1/128 of 900ml
236/200	1/32 of 600ml 1/32 of 825ml	1/64 of 1200ml 1/32 of 800ml	1/128 of 1350ml
236/202	1/16 of 300ml 1/32 of 600ml	1/32 of 600ml 1/32 of 575ml	1/128 of 1150ml
237/201	1/16 of 400ml		1/64 of 700ml
Context [543]			
234/200	1/8 of 150ml	1/16 of 400ml	1/64 of 550ml
234/202	1/8 of 125ml	1/32 of 1150ml	1/64 of 700ml
235/201	1/8 of 120ml	1/64 of 1200ml	1/64 of 600ml
236/200	1/8 of 150ml (All sorted (10ml)	1/32 of 1100ml 1/64 of 900ml	1/128 of 850ml
236/202	All sorted (5ml) All sorted (<5ml)	1/16 of 575ml All sorted (700ml)	1/32 of 220ml
237/201	1/8 of 75ml	1/64 of 1000ml	1/32 of 300ml

Table 2.26 Plant macrofossil remains

	Context [534]			**Context [543]**
	234/200	236/202	237/201	236/200
Corylus avellana L. frag.	-	1	-	-
Galium sp.	1	-	-	-
Carex sp. - trigonous	-	-	-	1
Poaceae rhizome frag.	-	-	1	-

Table 2.27 Concordance of contexts and associated finds

Context no	Description	Plan nos	Sect. nos	Finds	Cat nos
1	modern concrete floor in Room 1				
2	concrete flooring in Room 1 underlying 1				
3	concrete threshold in NW door of Room 1	2.68			
4	concrete spread in Room 2 to N of partition 6	2.68			
5	spread of cobbles in NE corner of Room 2	2.68			
6	E-W timber partition towards N end of Room 2	2.68			
7	concrete spread to N of cross-passage in Room 2				
8	scattered shale flagstones S of partition 6 in Room 2	2.68			
9	fragment of cobble flooring N of W end of cross-passage in Room 2	2.68			
10	concrete floor of cross-passage in Room 2				
11	partition of shale slabs S of cross-passage in Room 2	2.68			
12	area of cobble flooring S of partition 11 in Room 2	2.68			
13	fill at S end of Room 2	2.68		PR pot late C18+; PM glass; PM Fe; drainage pipe	
14	concrete in cut 15 in Room 3				
15	cut towards NW of Room 3, (?) for generator base				
16	cobbled floor in Room 3	2.68		PR pot late C18+; PM glass; PM Fe	
17	concrete floor E of partition 26 in Room 4				
18	fragmentary concrete floor W of partition 26 in Room 4				
19	scattered shale flagstones underlying 18	2.68		PM glass	
20	strip of concrete against N wall in W part of Room 4	2.68			
21	threshold to doorway on W of Room 4	2.68			
22	threshold to doorway on S of Room 4	2.68			
23	drain cut through floor 24 in Room 1	2.68			
24	cobbling beneath concrete 2 at N end of Room 1	2.68			
25	turf and topsoil along western edge of trench			PR pot C18+; clay pipe *c.*1660; PM glass	
26	N-S timber partition in Room 4	2.68			
27	sandy concrete floor beneath 17 in E end of Room 4	2.68			
28	two shale flagstones W of stalls in Room 1, below 1	2.68			
29	concrete post-packings in Room 1	2.68			
30	makeup beneath 4 and 5 at N end of Room 2	2.68		1957 halfpenny; PR pot late C18+; PM glass; PM Fe	Co 4
31	makeup between partition 6 and cross-passage in Room 2	2.68		PR pot late C18+; clay pipe *c.*1700; PM glass; PM Fe	
32	recent farmyard surfaces			1917, 1921 pennies; PR pot late C18+; clay pipes 1660-1800; R gl bottle, bead; PM glass; PM Fe	Co 5, 6; LP 8; Gl 13
33	shale partition like 11, N of cross-passage in Room 2	2.68			
34	fill beneath 10 along cross-passage in Room 2	2.68		PR pot late C18+; PM Fe	
35	threshold in E doorway of Room 2	2.68			
36	infill beneath western floors in Room 4 (19, 20, 27)			PR pot late C18+; clay pipe *c.*1700; PM glass; PM Fe	LP 32; Gl 20, 21, 22
37	(?)beaten earth floor along S 1.5m of Room 1	2.68		PR pot late C18+; PM glass; PM Fe	
38	infill beneath flooring and stalls in Room 1	2.68		PR pot late C18+; clay pipe *c.*1850; PM glass; PM Fe; cinder 300g	
39	concrete overlying pitched stone floor 41 at N end of Room 2 area	2.64		PR pot late C18+; clay pipe *c.*1850; PM Fe	
40	mortar overlying pitched stone floor 41 S of 39 in Room 2 area	2.64			
41	pitched stone flooring of clom building in area of Room 2	2.62		PR pot late C18+; PM Fe	
42	post-holes cut through 39/40/41, probably for features relating to stone building	2.68			
43	disturbance cut through 39	2.62		PR pot late C18+; PM Fe	
44	disturbance cut through 39	2.62		drainage pipe	
45	disturbance cut through 39	2.62			
46	disturbance cut through 39	2.62			

Context no	Description	Plan nos	Sect. nos	Finds	Cat nos
47	disturbance cut through 41	2.62			
48	disturbance cut through 41	2.62		PM Fe; drainage pipe	
49	disturbance cut through 41	2.62		PR pot late C18+; PM Fe	
50	disturbance cut through 41	2.62		PM Fe	
51	disturbance cut through 41	2.62			
52	infill beneath 38 in Room 1			PR pot late C18+; clay pipes 1700-1900; PM Fe	CTP 10, 11
53	infill to S of N wall 63 of clom building in area of Room 1			PR pot late C18+; PM glass; PM Fe	
54	infill across N wall 63 of clom building (=63)			PR pot late C18+; PM glass	
55	trench for plastic drainpipe W from Room 3	2.68			
56	trench for ceramic and iron drainpipe W from Room 4	2.68			
57	N-S trench for plastic water pipe serving Room 1	2.68		PR pot late C18+; PM Fe; drainage pipe	
58	trench for ceramic drain W from Room 2	2.68			
59	wall butted to SW corner of stone building	2.68			
60	post-pipe relating to gate attached to 59, in post-pit 105	2.68			
61	post-pipe in post-pit 66	2.68			
62	post-pipe in post-pit 187	2.62 2.68			
63	N wall of clom building	2.62		PR pot C17+; PM glass	
64	concrete floor to S of 63 in area of Room 1 (=39)				
65	fill of post-pit 66			PR pot C18+	
66	post-pit in farmyard area relating to stone building, a pair with 187	2.68			
67	disturbance cut through 41, sealed by 40	2.62		PR pot late C18+; clay pipe *c.*1800; drainage pipe	
68	gap in 41, possibly partition, sealed by 39	2.62			
69	wall on W, N and E of Room 1	2.68			
70	fill of construction trench 71 for wall 69			PR pot C18; clay pipe *c.*1660; R gl bead; PM Fe	Gl 16
71	construction trench for wall 69	2.62		PR pot C18+; cinder	
72	wall on N, S, E and W of Room 2	2.68		PR pot C18; PM glass; drainage pipe, brick	
73	fill of construction trench 74 for wall 72			clay pipes 1700-1880; PR pot late C18+; PM glass; PM Fe; Fe slag	CTP12; Gl 23
74	construction trench for wall 72, present on W, N and E sides only	2.62			
75	S wall of clom building, underlying S wall 72 of Room 2	2.62			
76	wall on W, S and E of Rooms 3 and 4	2.68		PR pot C18+; PM Fe; Fe slag	
77	cross-wall dividing Rooms 3 and 4	2.68			
78	pitched stone flooring beneath concrete floor 64 (=41)	2.62			
79	farmyard surface relating to stone building	2.68		PR pot late C18+; clay pipes 1700-1850; PM glass; PM Fe; brick	LP 9, 30
80	masonry platform (?) for stair, on W of stone building	2.68			
81	post-hole cut through 41 (bottom of 42?)				
82	post-hole and disturbance cut through 41	2.62		PR pot C18; clay pipe *c.*1850; PM Fe; Fe slag	
83	surviving post in pit 85				
84	fill of pit 85			PR pot late C17+ (joins 92); PM glass	LP 22
85	disturbance at N of area of Room 1	2.62			
86	mortar patches at N of area of Room 1	2.62			
87	patch of pitched stone flooring in area of Room 1	2.62			
88	E-W line of pitched stone flooring S of (?)partition 89	2.62			
89	E-W partition parallel with and N of 63	2.62			
90	part of mortar spread adjoining N side of 63	2.62			
91	pitched stone flooring to N of 63	2.62			
92	infill to N of 63 in area of Room 1, underlying 52 and 85-91 inclusive			PR pot late C18+ (joins 84); clay pipes 1600-1700; PM glass; PM Fe; Ae ring	LP 19, 22; Ae 2
93	timber post in post-pit 95	2.68			

Context no	Description	Plan nos	Sect. nos	Finds	Cat nos
94	fill of post-pit 95			PR pot C16+; PM glass	Gl 19
95	post-pit cutting 100 in farmyard area	2.68			
96	disturbance cut through 41	2.62		PR pot C18+; PM Fe; brick	
97	infill beneath 16 in Room 3			PR pot late C18+; clay pipes 1600-1700; PM glass; PM Fe; brick	
98	foundation cut for cross-wall 77	2.62			
99	foundation trench for walls 76	2.62		PR pot late C18+; PM Fe	
100	fill of pit 101			PR pot C16+; PM Fe; brick	
101	irregular pit in farmyard area	2.68			
102	fill of robber trench 103			PR pot late C18+; clay pipes 1660-1730; PM glass; PM Fe	
103	robber trench for wall built in trench 107	2.62			
104	fill of post-pit 105			PR pot late C18+; clay pipes 1660-1730; Fe slag; drainage pipe, brick	LP 10, 31; CTP 7
105	post-pit containing gate-post 60	2.62			
106	fill of 107			PR pot late C18+	LP 11, 12
107	(?)construction trench for earlier S wall in area of Rooms 3/4	2.62			
108	fill of 109	2.62			
109	post-pit cut through pitched stone flooring 87 in area of Room 1	2.62			
110	fill of 111				
111	truncated feature on W side of area of Room 1	2.62			
112	post-pipe in post-pit 114				
113	fill of 114				
114	post-pit at W entrance to Room 3	2.62			
115	post-impression in post-pit 117				
116	fill of 117				
117	post-pit at W entrance to Room 3	2.62			
118	post-pipe in post-pit 120				
119	fill of 120				
120	post-pit at W entrance to Room 3	2.62			
121	spread towards E of area of Room 3	2.62		PR pot late C18+; clay pipe *c.*1660; PM glass; PM Fe; cinder; brick	LP 3
122	spread towards W of area of Room 3, above surface 128			PR pot late C18+; clay pipes 1660-1730; PM glass	LP 27; CTP 9
123	infill beneath pitched stone floor 41 in area of Room 2 (=126)			George II or III penny; PR pot late C18+; clay pipes 1750+; PM glass; Ae button; PM Fe; drainage pipe, brick	Co 7
124	foundation cut for wall 59	2.62			
125	pebble surface at S end of site	2.62		PR pot C18; clay pipe 1660-1730; PM Fe	LP 13
126	infill beneath pitched stone floor 78 in area of Room 1 (=123)				
127	patches of pitched stone flooring at SW of area of Room 2	2.60		1773 penny	Co 8
128	rough flooring in area of Room 3	2.62		PR pot late C18+; clay pipes 1660-1730; R gl bottle; PM glass	
129	clay surface underlying 123 and 126	2.60		1775, 1730 pennies, PM coin Co 11; PR pot C18; clay pipes 1660-1880; PM glass; PM Fe; cinder; drainage pipe	Co 9, 10, 11
130	fill of post-medieval cellar 132			PR pot C18; clay pipe *c.*1660	
131	fill of post-medieval cellar 132			samian; PR pot late C18+; clay pipes 1600-1900; PM glass; PM Fe; cinder; brick	
132	walling of post-medieval cellar	2.62			
133	N-S scatter of stones below 92	2.60			
134	fill of 135, below 92				
135	post-hole in area of Room 1	2.60			

Context no	Description	Plan nos	Sect. nos	Finds	Cat nos
136	E-W scatter of stones below 92	2.60			
137	stake-hole filled with 92	2.60			
138	hollow filled with 92	2.60			
139	hollow filled with 92 and rotted timber	2.60			
140	hollow filled with 92	2.60			
141	oval hollow filled with 92 and traces of rotted timber	2.60			
142	E-W trench filled with 92	2.60			
143	stake-hole filled with 92	2.60			
144	timber post in post-pit 159, beneath 92				
145	fill of 146, below 92				
146	cut against N side of clom wall 63	2.60			
147	stake-hole filled with 92	2.60			
148	fill of 149, below 92				
149	oval feature in area of Room 1	2.60			
150	stake-hole filled with 92	2.60			
151	stake-hole filled with 92	2.60			
152	fill of 153			PM glass; PM Fe	
153	foundation trench for clom wall 63 (on N only)	2.60			
154	fill of 155, below 129				
155	irregular hollow in area of Room 2	2.53 2.57			
156	foundation trench for S wall 75 of clom building			samian; R pot; PR pot C18+	CP 1
157	extensive spread of garden soil; uppermost fill of timber building pit 245		2.42	pennies 1730, undated; samian; R pot; PR pot late C18+; clay pipes 1660-1880; R gl bottle; PM glass; R Ag finger ring; R Fe nails; PM Fe; R brick	Co 10, 11; S 1; CP 2-4; LP 1, 4, 6-8, 14-16, 20, 23-26, 29; CTP 2, 3, 4, 6; Gl 8; Ag 1
158	fill of 159			clay pipe *c.*1800-50	
159	post-hole containing timber 144				
160	stake-hole filled with 92	2.60			
161	fill of 162			PR pot late C18+; R brick	LP 17
162	rectilinear pit in area of Room 3	2.53			
163	walling running into trench from W at S end	2.62			
164	fill of 165, below 92				
165	small post-hole in area of Room 1	2.60			
166	fill of 167, below 92				
167	hollow in area of Room 1, cutting 164	2.60			
168	fill of 169, below 92				
169	hollow in area of Room 1, cutting 164 and 166	2.60			
170	fill of 171, below 92				
171	hollow in area of Room 1	2.60			
172	fill of 173, below 92				
173	hollow in area of Room 1	2.60			
174	fill of 175, below 133			PR pot C18; clay pipes *c.*1660-80; cinder	LP 2, 19
175	hollow in area of Room 1	2.60			
176	fill of 177, below 92				
177	oval hollow in area of Room 1	2.60			
178	blackened patch in area of Room 3	2.53			
179	stone packing of E-W drain 221 in area of Room 3	2.62		PR pot C18+; clay pipes *c.*1680-1780; PM glass; brick	CTP 8
180	probable farmyard surface relating to clom building (?=223)	2.62		samian; PR pot C18+; clay pipes *c.*1680; brick	
181	scatter of stones set into 129 to E of area of Room 2	2.60			
182	fill of 183, below 92			PR pot C17+; clay pipe *c.*1680	
183	oval hollow in area of Room 1	2.60			
184	fill of 185, below 92				
185	oval hollow in area of Room 1	2.60			
186	fill of 187			PR pot C17+; clay pipe *c.*1660-90; R Fe nail; brick	CTP 1

Context no	Description	Plan nos	Sect. nos	Finds	Cat nos
187	post-pit in farmyard areas relating to both clom and stone buildings, containing post 62, a pair with 66 and 212	2.62			
188	area of gravel below drain 221 towards W of trench	2.53		PR pot C17+; clay pipes *c.*1660-1730; PM glass; PM Fe; flint	CTP 5; Fl 4
189	fill of 190, below 123				
190	post-pit in area of Room 2	2.60			
191	fill of 192, below 123				
192	post-pit in area of Room 2	2.60			
193	fill of 194, truncated by cut 201				
194	post-pit filled by 193	2.53			
195	fill of 196, truncated by cut 201				
196	post-pit filled by 195	2.53			
197	fill of 198, truncated by cut 201			PR pot late C18+	
198	post-pit filled by 197	2.53			
199	fill of 200, truncated by cut 201				
200	post-pit filled by 199	2.53			
201	levelling cut running S from S side of N wall of clom building	2.60			
202	number never allocated				
203	fill of 204, below 123				
204	hollow in area of Room 2	2.60			
205	fill of post-impression 206				
206	post-impression in pit 210				
207	fill of post-impression 208				
208	post-impression in pit 217				
209	fill around post-impression 206				
210	post-pit beneath stone-wall construction trench 74, towards N of W wall of Room 2	2.60			
211	fill of post-pit 212, cut by later post-pit 66			PR pot C17+	
212	post-pit in farmyard area relating to clom building, a pair with 187	2.62			
213	void within post-impression 214				
214	post-impression in pit 219				
215	clay-bonded walling across N end of cellar 132	2.62			
216	fill of post-pit 217				
217	post-pit in farmyard area relating to clom building, a pair with 219	2.62			
218	fill of post-pit 219				
219	post-pit in farmyard area relating to clom building, a pair with 217	2.62			
220	gravel spread adjoining drain 221 in eastern part of trench	2.53		PM coin Co 12; PR pot C18+; clay pipes *c.*1660-80; PM glass	Co 12; LP 28
221	E-W drain in area of Room 3, lined with stones 179	2.62			
222	cut for insertion of post-medieval cellar wall 132	2.60			
223	gravel spread recognised in 1987 trial trenches but unclear beyond (?=180)	2.53		PR pot C17+	LP 21
224	fill in top of timber building pit 245, below 157		2.42	samian; R pot (1 joins 460); R gl bottle, vessel, window, bead; R Fe nails; Fe slag 192g; Pb slag 369g; furnace lining; R brick, R tile, bone	S4; CP 5-9; Gl 4, 9, 14
225	mixed layer in area of Room 1, below 92				
226	reddish clay across S end of area of Room 2, below 129	2.57		PR pot late C18+; cinder 892g	
227	ashy fill of square hollow 228			PR pot late C18+; PM Fe	
228	square (?)metalworking hollow cut into 226	2.53 2.57			
229	ashy fill of small oval hollow 230				
230	small oval hollow cut into 226	2.53 2.57			
231	fill of post-pit 232				
232	post-pit cutting corner of timber building pit 245	2.53			

Context no	Description	Plan nos	Sect. nos	Finds	Cat nos
233	uppermost fill or subsidence into large ?late-Roman pit 410	2.53		R Fe nails; fuel ash slag; R brick	
234	uppermost fill or subsidence into clay-lined cistern 376		2.21	samian; R gl beads; R Fe nails; Fe slag 200g; furnace lining; R brick	Gl 11, 12
235	fill of post-pit 236			PR pot late C18+; mortar	
236	post-pit beneath stone-wall construction trench 74, towards N of W wall of Room 2	2.60			
237	mixed ashy fill of (?)possible hearth 238			PR pot C18; clay pipe *c.*1660; PM Fe; mortar	
238	(?)possible metalworking hearth; replaced 355	2.53 2.57			
239	fill of oval hollow 240, below 226			Fe ?anvil	Fe 21
240	small oval hollow cut into 359	2.53 2.57			
241	fill of (?)post-hole 242				
242	(?)post-hole, cut into 359	2.53 2.57			
243	spread in hollow 244, above 188, cut by drain 212				
244	shallow elongated hollow, to N of timber building pit 245 on W of trench	2.53			
245	cut of pit containing burnt timber building				
246	fill of (?)post-pit 247			R pot (joins 378)	CP 10
247	(?)post-pit, truncated on S by cellar cut 222	2.53			
248	fill of linear gully 254		2.55	samian; PR pot medieval; R gl bottle; R Fe nails; cinder; Fe slag; R brick; flint	LP 5; Fl 2, 5
249	patch of cobbles to E of cellar cut 222	2.53			
250	fill of pit 251, below 129				
251	subrectangular pit cutting E edge of clay-lined cistern 376, postdating 258	2.53			
252	fill of (?)post-pit 253				
253	(?)post-pit cutting turf-clay rampart 381	2.53			
254	N-S V-cut linear gully running from near clay-lined cistern 376 to E of timber building pit 245	2.53	2.55		
255	fill of post-pit 256, below 129				
256	post-pit, to E of cellar cut 222	2.53			
257	fill of elongated hollow 258 and oval hollow 260				
258	elongated hollow to E of clay-lined cistern 376	2.53			
259	stake-hole in base of elongated hollow 258	2.53			
260	oval hollow ovelapping elongated hollow 258 on S	2.53			
261	fill of (?)post-hole 262				
262	(?)post-hole overlapping elongated hollow 258 on N	2.53			
263	fill of hollow 264				
264	hollow to S of elongated hollow 258	2.53			
265	fill of post-impression 266				
266	post-impression in post-pit 268				
267	fill of post-pit 268		2.7		
268	post-pit running into W section near N of intervallum road 434	2.68	2.7		
269	fill of 'late' pit 302		2.52	samian; R pot; R Fe nails; furnace lining; R brick, R tile; flint	CP 11; Fl 6
270	fill of small post-hole 271				
271	small post-hole to W of cellar cut 222	2.53			
272	fill of pit 275				
273	fill of pit 274, below 225			PR pot C16+	LP 18
274	pit in area of Room 1	2.53			
275	pit near W section, W of 'late' pits	2.68			
276	fill of post-impression 277, below 225			R Fe nail	
277	post-impression in area of Room 1	2.53			
278	fill of gully 279		2.55		
279	N-S gully near E edge of site	2.53	2.55		
280	fill of pit 281			clay pipe *c.*1660	
281	pit cutting N end of foundation trench 304	2.53			

Context no	Description	Plan nos	Sect. nos	Finds	Cat nos
282	fill in top of clay-lined cistern 376		2.21	mortar	
283	fill in top of clay-lined cistern 376		2.21	R pot; R gl vessel; R Fe nails; R brick	CP 12
284	fill in clay-lined cistern 376		2.21	samian; R pot; R gl vessel, bottle; R Fe object; R Fe nails; Fe slag 1312g; furnace lining; R brick, bone	CP 13-16; Gl 1; Fe 11
285	fill in clay-lined cistern 376		2.21	R Ag coin; samian; R pot; R gl bead; R Fe nails; cinder; R brick, bone	Co 1; CP 17-18; Gl 10; Fe 18
286	fill in top of clay-lined cistern 376		2.21	R pot; R brick	CP 19
287	fill of post-impression 288			PR pot C17+; PM glass	
288	post-impression in post-pit 290				
289	fill of post-pit 290			PM glass; R Fe nail	
290	post-pit towards SW corner of W extension of trench	2.53			
291	fill of robber trench 292			samian; R pot; PR pot C17+; R Fe nail; cinder 127g; flint	CP 20; Fl 3
292	robber trench above W stone guardchamber of fort				
293	fill of oval hollow 294				
294	oval hollow, possible (?)grave	2.53			
295	fill of foundation trench 296		2.33	samian	
296	N-S (internal) foundation trench of E phase 4 building	2.10	2.33		
297	fill of broad shallow gully 298		2.55	R pot; Ae slag; R Fe nails; Fe slag 533g; R brick	CP 21
298	broad shallow gully running NE-SW across E part of site	2.53	2.55		
299	fill of 'late' pit 323		2.52	R gl bottle; R Fe nails	Gl 7
300	fill of 'late' pit 302		2.52	samian; R pot; R Fe nails; Fe slag; furnace lining; R brick	CP 22-23
301	fill of 'late' pit 302		2.52	samian; R pot; R gl bottle; R Fe nails; R brick	CP 24
302	'late' pit		2.52		
303	fill of foundation trench 304		2.32	R Fe object; R Fe nail; R brick/tile; flint	
304	N-S (external) foundation trench of W phase 4 building	2.10	2.32		
305	fill of oval hollow 306				
306	oval hollow, possible (?)grave	2.53			
307	charcoal lens in clay-lined cistern 376		2.21		
308	fill in clay-lined cistern 376		2.21	R pot (1 may join 375); R gl vessel/bottle; R Fe nails; Pb scrap; R brick, bone	CP 25-26
309	fill of 'late' pit 328		2.52	R pot (joins 327); R Fe lorica; R Fe nails; Fe slag; R brick	CP 27; Fe 3, 19
310	fill of 'late' pit 302		2.52		
311	modern post-pit			PR pot C17+; PM glass	
312	modern post-pit				
313	fill of gully 315				
314	upright stones in gully 315				
315	shallow gully to SE of clay-lined cistern 376, cutting its upper fill 284; possible (?)grave	2.53			
316	fill of 'late' pit 323		2.52	samian; R Fe lamp hanger; R Fe nails; Fe slag 389g	Fe 5
317	fill of 'late' pit 323		2.52		
318	fill of 'late' pit 323		2.52		
319	fill of 'late' pit 323		2.52		
320	fill of 'late' pit 323		2.52		
321	fill of post-pit or trench 322				
322	post-pit or trench running into E section, cutting pit 535 of first phase timber gate	2.9			
323	'late' pit		2.52		
324	stake-hole near N end of gully 279				
325	fill of post-impression 326				
326	post-impression	2.53	2.33		
327	fill of 'late' pit 328		2.52	R pot (joins 309); R Fe nails	CP 27
328	'late' pit		2.52		

Context no	Description	Plan nos	Sect. nos	Finds	Cat nos
329	tumble to N of stone rampart		2.7	R brick, R tile; R Fe nails	poss
330	fill of robber trench 331		2.7	samian; PR pot late C17+; R gl bottle; PM glass; R Fe nail	
331	robber trench above stone rampart 382		2.7		
332	fill of stake-hole 333				
333	stake-hole near N end of foundation trench 304	2.53			
334	fill of (?)oval hollow 335				
335	(?)oval hollow, (?)possible grave	2.53			
336	fill of shallow trench 337				
337	shallow trench within area of second-phase stone tower				
338	stone spread associated with possible (?)metalworking hearths	2.57		cinder	
339	fill of 'late' pit 323		2.52	R Fe nails	
340	fill of 'late' pit 323		2.52	R Ae coin; R gemstone; R Fe lamp	Co 2; Gm 1; Fe 4
341	fill of 'late' pit 323		2.52	R Fe nails	
342	layer intermediate between phase 4 and 5 buildings	2.39		R pot (joins 481); R Fe nails; furnace lining	CP 28
343	fill of foundation trench 414		2.32	R gl bottle; Ae offcut; R Fe nails; furnace lining	Ae 3
344	fill of 'late' pit 351		2.52		
345	fill of 'late' pit 351		2.52	samian; R pot; R gl bottle; R Fe nails; furnace lining; R brick	CP 29-30
346	fill of 'late' pit 351		2.52		
347	fill of 'late' pit 351		2.52		
348	fill of 'late' pit 351		2.52		
349	fill of 'late' pit 351		2.52		
350	fill of 'late' pit 351		2.52	samian; R pot; R Fe nails; Fe slag; Pb sheeting; R brick/tile	CP 31
351	'late' pit		2.52		
352	fill of gully 353		2.52		
353	shallow gully cutting uppermost fill 344 of pit 351		2.52		
354	fill of (?)possible hearth 355	2.53 2.57		Ae ?fitting	Ae 1
355	(?)possible metalworking hearth	2.53 2.57			
356	fill of clay-lined cistern 376		2.21	samian	
357	fill of clay-lined cistern 376		2.21	samian; R pot; R gl bottle; R Fe nails; fuel ash slag; daub/low fired clay; bone	CP 32-33
358	fill of clay-lined cistern 376		2.21		
359	disturbed horizon across E end of Roman structures	2.57		R gl vessel/bottle; R brick/tile	
360	fill of (?)oval hollow 361			Ae slag	
361	(?)oval hollow, (?)possible grave	2.53			
362	fill of hollow 365		2.7		
363	fill of hollow 365		2.7		
364	fill of hollow 365		2.7	R Fe nail	
365	elongated E-W hollow above foundation trench 557		2.7		
366	fill of foundation trench 367		2.33	R Fe nail	
367	E-W (external) foundation trench of E phase 4 building	2.10	2.33		
368	post-impression in 367	2.10			
369	post-impression in 367	2.10			
370	post-impression in 367	2.10			
371	post-impression in 367	2.10			
372	post-impression in 367	2.10			
373	fill of foundation trench 374		2.33	samian; R pot; R Fe nails; Fe slag; furnace lining	CP 34-35
374	N-S (external) foundation trench of E phase 4 building	2.10	2.33		

Context no	Description	Plan nos	Sect. nos	Finds	Cat nos
375	fill of sump in clay-lined cistern 376		2.22	samian; R pot (1 may join 308); R gl vessel (joins 377), bottle, vessel/bottle; R Fe objects; R Fe nails; Fe slag 297g; Pb scrap; R brick; bone	CP 36-37; Gl 2; Fe 7-9, 12-14
376	clay lining of cistern		2.21 2.22	samian; furnace lining	
377	fill in large pit 560, beneath clay-lined cistern 376		2.21 2.22	samian; R pot (may join 487); R gl vessels (1 joins 375); fuel ash slag; bone	S 2; CP 38; Gl 2
378	fill of large pit 410		2.8	R pot (joins 246); R Fe nails; fuel ash slag; furnace lining	CP 10
379	secondary clay dump to rear of turf-clay rampart 381	2.9 2.20	2.8		
380	secondary infill between turf-clay rampart 381 and stone rampart 382/496	2.20	2.7 2.8	samian	
381	first-phase turf-clay rampart	2.9	2.7 2.8		
382	second-phase stone rampart	2.17	2.7 2.8		
383	trample layer beneath turf-clay rampart		2.7 2.8		
384	trample layer behind turf-clay rampart		2.8		
385	turf horizon and erosion off front face of turf-clay rampart		2.8		
386	fill of foundation trench 387				
387	E-W construction trench for rubble/clay foundations 499		2.7		
388	fill of foundation trench 389		2.32	samian; R gl vessel (joins 478)	Gl 3
389	N-S (external) foundation trench of phase 5 building	2.10	2.32		
390	post-impression in 374	2.10			
391	post-impression in 374	2.10			
392	post-impression in 374	2.10			
393	stake-hole immediately E of 'late' pit 302	2.53			
394	erosion off turf-clay rampart across N part of intervallum road 434	2.28	2.7		
395	fill of possible post-pit 396		2.12		
396	possible post-pit, cut through clay rampart	2.53	2.12		
397	fill of post-pit 398		2.7		
398	modern post-pit in W section		2.7		
399	fill of post-pit 400		2.7		
400	modern post-pit in W section	2.68	2.7		
401	fill of large pit 410		2.8		
402	fill of large pit 410		2.8		
403	fill of large pit 410		2.8	R pot; Fuel ash slag	CP 39
404	fill of large pit 410		2.8		
405	fill of large pit 410		2.8		
406	fill of large pit 410		2.8		
407	fill of large pit 410		2.8		
408	fill of large pit 410		2.8		
409	fill of large pit 410		2.8		
410	large pit, 'late' Roman?		2.8		
411	demolition layer post-dating phase 4 buildings	2.39		R gl bottle; R Fe nails; Pb scrap	
412	fill of foundation trench 413		2.12		
413	E-W foundation trench of phase 5 building		2.12		
414	Right-angled (external) foundation trench of phase 5 building	2.10	2.32		
415	fill of drain 416		2.7		
416	shallow E-W drain at back of turf-clay rampart	2.9 2.28			
417	secondary gravel spread along N side of intervallum road 434	2.28	2.7		

Context no	Description	Plan nos	Sect. nos	Finds	Cat nos
418	secondary clay dump behind turf-clay rampart, (?)probably the same as 379		2.8		
419	trample horizon above pit 410, (?) probably the same as 384		2.8		
420	demolition layer intermediate between phase 4 and 5 buildings	2.36		R Fe nails	
421	fill of drain 422	2.28	2.7		
422	shallow E-W drain at back of turf-clay rampart 381	2.9 2.28	2.7		
423	fill of post-impression 424				
424	post-impression in trench 296	2.10			
425	gravel makeup of N-S road	2.29			
426	fill of drain 427		2.7		
427	shallow E-W drain at back of turf-clay rampart 381	2.9 2.28	2.7		
428	fill of post-impression 429				
429	post-impression in 431	2.53			
430	post-packing and fill in 431				
431	post-pit immediately W of 'late' pit 302	2.53			
432	fill of foundation trench 433		2.33	R gl bottle; R Fe nails; flint	
433	E-W (internal) foundation trench of E phase 4 building	2.10	2.33		
434	gravel makeup of E-W intervallum road, west of post-medieval cellar	2.10	2.7		
435	post-impression in 414	2.10			
436	post-impression in 414	2.10			
437	post-impression in 414	2.10			
438	post-impression in 414	2.10			
439	post-impression in 414	2.10			
440	post-impression in 414	2.10			
441	post-impression in 414	2.10			
442	clay layer inside stone tower				
443	fill of truncated feature 444				
444	truncated feature of uncertain extent, to W of cellar cut 222				
445	fill of truncated feature 446				
446	truncated feature perhaps associated with phase 4 buildings, cut by 'late' pit 323 to W				
447	fill of foundation trench 451				
448	secondary infill to E of turf-clay rampart 381, south of stone tower	2.20		Ae slag	
449	fill of foundation trench 450				
450	E-W (internal) foundation trench of E phase 4 building	2.10			
451	N-S (internal) foundation trench of W phase 4 building	2.10			
452	clay layer beneath N-S road 425				
453	stake-hole cutting 454				
454	clay layer beneath 452 and N-S road 425				
455	(?)demolition layer intermediate between phase 3 and 4 buildings	2.36			
456	(?)demolition deposit post-dating phase 4 buildings				
457	fill of rectangular hollow 458				
458	rectangular hollow post-dating phase 4 buildings	2.39			
459	(?)demolition layer intermediate between phase 3 and 4 buildings	2.36			
460	demolition deposits in timber building pit 245		2.42	samian; R pot (1 joins 224) R gl bottle, bead; R Fe weapons; R Fe nails; Pb scrap; bone	S 3; CP 9, 40-45; Gl 17; Fe 1, 2
461	fill of (?)oval hollow 462			R brick/tile	
462	(?)oval hollow, (?)possible grave	2.53			
463	fill of possible tree pit 464		2.7		
464	possible tree pit, seen only in W section		2.7		
465	gravel makeup of E-W intervallum road, east of post-medieval cellar				
466	silty horizon above 472	2.36			
467	fill of pit 468		2.7		

Context no	Description	Plan nos	Sect. nos	Finds	Cat nos
468	early pit pre-dating intervallum road		2.7		
469	fill in large pit 560, above stone-lined well		2.21		
470	fill in large pit 560, immediately above stone-lining 559		2.21 2.22		
471	fill of projected upward extension of stone-lined well 559		2.22	samian; R pot (1 joins 487); R gl bottle, gaming piece; R Fe nails; Fe slag; bone	CP 46-51; Gl 5, 18
472	silty-clay spread intermediate between phase 3 and 4 buildings	2.36		samian	
473	fill of foundation trench 474		2.32		
474	E-W (internal) foundation trench of E phase 4 building	2.10	2.32		
475	fill of post-pit 477				
476	fill and possible packing in post-pit 477				
477	post-pit post-dating phase 4 buildings, on line of phase 2 N wall 480	2.39			
478	silt horizon (?) above fill 479		2.33	R gl vessel (joins 388)	Gl 3
479	fill of foundation trench 480		2.33		
480	right-angled foundation trench of phase 2 building	2.10	2.33		
481	fill of foundation trench 482		2.33	R pot (joins 342)	CP 28
482	E-W (external) foundation trench of phase 3 building	2.10	2.33		
483	fill of foundation trench 484				
484	N-S (internal) foundation trench of phase 3 building	2.10	2.33		
485	fill of butt-ended feature 486		2.33		
486	butt-ended (internal) foundation trench of E phase 4 building	2.10	2.33		
487	lower fill of stone-lined well 559		2.22	samian; R pot (1 joins 471, 1 may join 377); R gl vessel, bottle; R Fe nails; fuel ash slag; bone	CP 51-55; Gl 6
488	fill of foundation trench 489				
489	N-S (external) foundation trench of phase 3 building	2.10	2.33		
490	fill of foundation trench 491				
491	E-W (internal) foundation trench of phase 2 building	2.10	2.33		
492	fill of foundation trench / (?)drain 493		2.32		
493	possible N-S foundation trench of phase 1 building or drain(?)	2.10	2.32		
494	rubble/clay foundations for stone tower, infilling 495	2.17			
495	construction trench for rubble/clay tower foundations 494	2.9 2.17			
496	rubble/clay foundations for wing wall 498, infilling 497	2.17	2.16		
497	construction trench for rubble/clay wing wall foundations 496		2.16		
498	wing wall of stone tower	2.17	2.16		
499	rubble/clay foundations for stone rampart 382, infilling 387	2.17	2.19		
500	silt horizon (?) above fill 501				
501	fill of shallow gully/foundation trench 502				
502	possible E-W foundation trench for phase 1 building or (?) shallow drain	2.10			
503	erosion deposit N of stone wall above ditch fill 504		2.16		
504	fill of fort ditch 508		2.16		
505	fill of fort ditch 508		2.16		
506	natural silting in fort ditch 508		2.16		
507	primary fill of fort ditch 508		2.16		
508	first-phase fort ditch		2.16		
509	redeposited gravel from behind planking in timber building pit 245		2.42	R gl bead	Gl 15
510	fill of sausage-shaped features 511		2.33		
511	irregular sausage-shaped features apparently cut by phase 2 buildings	2.34	2.33		
512	fill of fort ditch 508	2.9	2.16		
513	secondary backfill above post fill 514		2.15		
514	fill of post-impression 515		2.15		
515	post-impression of first-phase gate tower in post-pit 517	2.9	2.15		
516	fill of post-pit 517 around post-impression 515		2.15		

Context no	Description	Plan nos	Sect. nos	Finds	Cat nos
517	post-pit at NW corner of first-phase gate tower	2.9	2.15		
518	upper carbonised plank along N side of burnt timber structure	2.43		timber	T 6
519	middle carbonised plank along N side of burnt timber structure	2.43		timber (joins 532)	T 7
520	upper carbonised plank along E side of burnt timber structure	2.43		timber (2 join 521)	T 8, 9, 11, 20
521	middle carbonised plank along E side of burnt timber structure	2.43		timber (2 join 520)	T 8, 12, 20
522	timber upright on N side of burnt timber structure	2.43 2.45			
523	fill of post-impression 524		2.15		
524	post-impression of first-phase gate tower in post-pit 538	2.9	2.15		
525	timber corduroy traces beneath first-phase rampart 381				
526	(?)gravel flooring inside stone tower				
527	carbonised timber from superstructure of burnt timber building	2.45		R Fe nails	
528	carbonised timber plank from superstructure of burnt timber building	2.45			
529	carbonised timber plank/beam from superstructure of burnt timber building	2.45			
530	carbonised plank associated with N wall of burnt timber building	2.45			
531	carbonised vertical post on N side of burnt timber building	2.45		timber; C14 sample	T 5
532	lower carbonised plank along N side of burnt timber building	2.43		timber (joins 519)	T 7
533	carbonised timber upright on E side of burnt timber building	2.45		timber; C14 sample	T 22
534	silty deposit across plank floor 541				
535	fill of post-pit 539				
536	fill of fort ditch 508	2.9			
537	fill of post-pit 538, around post-impression 524		2.15		
538	post-pit at SW corner of first-phase gate tower	2.9	2.15		
539	post-pit at SE corner of first-phase gate tower	2.9			
540	destruction horizon from burnt timber building in pit 245	2.43	2.42	R Ae coin; R Fe objects; R Fe nails; bone	Co 3; Fe 6, 10, 15, 20
541	carbonised timber planks forming floor of burnt timber building	2.47	2.42		
542	part of fallen post or beam from burnt timber building	2.45		Timber; C14 sample	T 15
543	layer beneath plank floor 541				
544	iron nails along N side of plank flooring 541			R Fe nails	Fe 16, 17
545	carbonised NE corner post of burnt timber building	2.45			
546	carbonised timber from superstructure of burnt timber building	2.45		timber	T 21
547	carbonised sleeper beam along N side of burnt timber building	2.47	2.42		
548	fill of construction trench 549 around sleeper beam 547	2.47	2.42		
549	construction trench along N side of burnt timber building	2.47	2.42		
550	fill of sleeper beam slot 553		2.42		
551	fill of sleeper beam slot 554		2.42		
552	lower carbonised plank on E side of burnt timber building	2.43			
553	construction trench for sleeper beam 550	2.47	2.42		
554	construction trench for sleeper beam 551	2.47	2.42		
555	carbonised timber beam from superstructure of burnt timber building	2.45		timber	T 23
556	fill of foundation trench 557		2.7		
557	E-W (external) foundation trench of W phase 4 building	2.10	2.7		
558	stone collapse inside stone-lined well 559		2.22		
559	stone-lined well and infill in pit 560	2.25	2.22		
560	large pit containing stone-lined well 559 and clay-lined cistern 376	2.9 2.25	2.21 2.22		
561	E-W (internal) foundation trench of phase 3 building	2.10			

2.2.5 Discussion of Roman levels

(A) Broad phases and chronology

The 1989 excavations provided a valuable section through the northern part of the fort at Pumsaint, to complement that recovered from the earlier excavations in 1972–73 (Jones and Little 1973; 1974). Coherent sequences were recorded for each part of the site (Fig. 2.102), the key features of which included:

A two-phase defensive sequence and gate;
A two-phase well/cistern;
Four or five successive phases of building, variously related to a gravel street;
A burnt timber structure set in a 2.00m-deep hole.

While these individual elements were reasonably clear, it was far less easy to establish a more general sequence or phasing for the site as a whole for two main reasons: first, because several of the critical stratigraphic relationships, such as those between the defences and the well/cistern, had been truncated by later activity; second, because vital floor levels and occupation horizons, which could have been used to establish or clarify the building sequence, were absent from within the timber structures. Enough key relationships did survive, however, to provide a framework for correlating different areas:

(a) while the first three building phases respected the gravel road, the phase 4 and 5 construction trenches progressively encroached across its line, thereby making it redundant;
(b) the phase 5 building trenches cut into the clay lining of the cistern, making it likely that the latter had ceased to function and was partly infilled;
(c) the 'late' pits to the west of the phase 5 building cut into both the phase 4 trenches and the gravel road;
(d) the large pit housing the burnt timber structure at the south end of the site had truncated the foundation trenches relating to all the earlier timber structures.

From such details, and the wider perspective of previous excavations, the following broad sequence emerges:

Phase A: limited pre-Roman activity, represented by some traces of pits or possible root cavities where shrubs or bushes had been grubbed up prior to the construction of the fort; the environmental evidence recovered by Astrid Caseldine also suggests some pre-Roman cultivation in the vicinity of the site.

Phase B: the construction of the fort, including the first-phase defences and the timber gate tower, the well to its rear, and the gravel road flanked by the first in a series of internal timber buildings; the phase 1 structure was successively replaced by phases 2 and 3, all of them respecting the line of the road.

Phase C: the next significant phase is likely to be associated with the decision to construct the stone wall and its associated stone tower or gate; following Jones and Little, this is most likely to be a consequence of the decision to reduce the fort to fortlet status, and would more than likely have involved other significant internal changes, perhaps including the conversion of the well to a clay-lined cistern, and the encroachment of the phase 4 buildings onto the defunct gravel road as space was at a premium. There is no absolute stratigraphic proof to correlate these separate events, but this seems the best hypothesis on the available evidence. During this phase, the cistern would seem to have gone out of use as a consequence of the construction of the phase 5 building, which further encroached onto the gravel road.

Phase D: after a significant interval of time, based on the radiocarbon dates, a large timber structure was constructed in a deep pit at the south end of the site; it clearly still reflected the alignment of the earlier buildings and the north-south street, suggesting that some elements of the fortlet had been retained in the interval. There is no clear evidence for the status of the defences during this period, though they might well have been maintained intact, even if the internal buildings had long been demolished; their remains may be represented by the material used to seal the destruction horizon and the burnt timbers of the late structure.

Phase E: the destruction of the late structure and the infilling of the associated pit with demolition debris. A considerable interval appears to have passed thereafter, before the later medieval and post-medieval activity. During this period, the stone wall was progressively robbed down to its foundations, together with most of the stone (gate) tower.

Discussion: This provides a reasonably coherent picture and conforms with the known complexity of the sequence elsewhere in the fort. The details could only be confirmed or extended by further excavation in the open area to the west, where preservation was shown to be so much better by the 1987 trial excavations (Burnham and Burnham 1987b).

The absolute chronology of the 1989 sequence is inevitably dependent upon the excavated finds, the quantities of which varied over the different parts of the site. Finds were especially abundant in the various feature fills, but surprisingly scarce in the critical contexts associated with the defences and the successive phases of timber building. Since there were only two legible coins, the bulk of the dating has to be derived from the pottery sequences, the glass being less susceptible to precision. Much ultimately depends upon the samian assemblage, though this was generally poorly preserved, with few diagnostic pieces in the most critical stratigraphic contexts. Problems are further compounded by the fact that the floor levels and occupation horizons associated with the buildings have not survived, leaving much

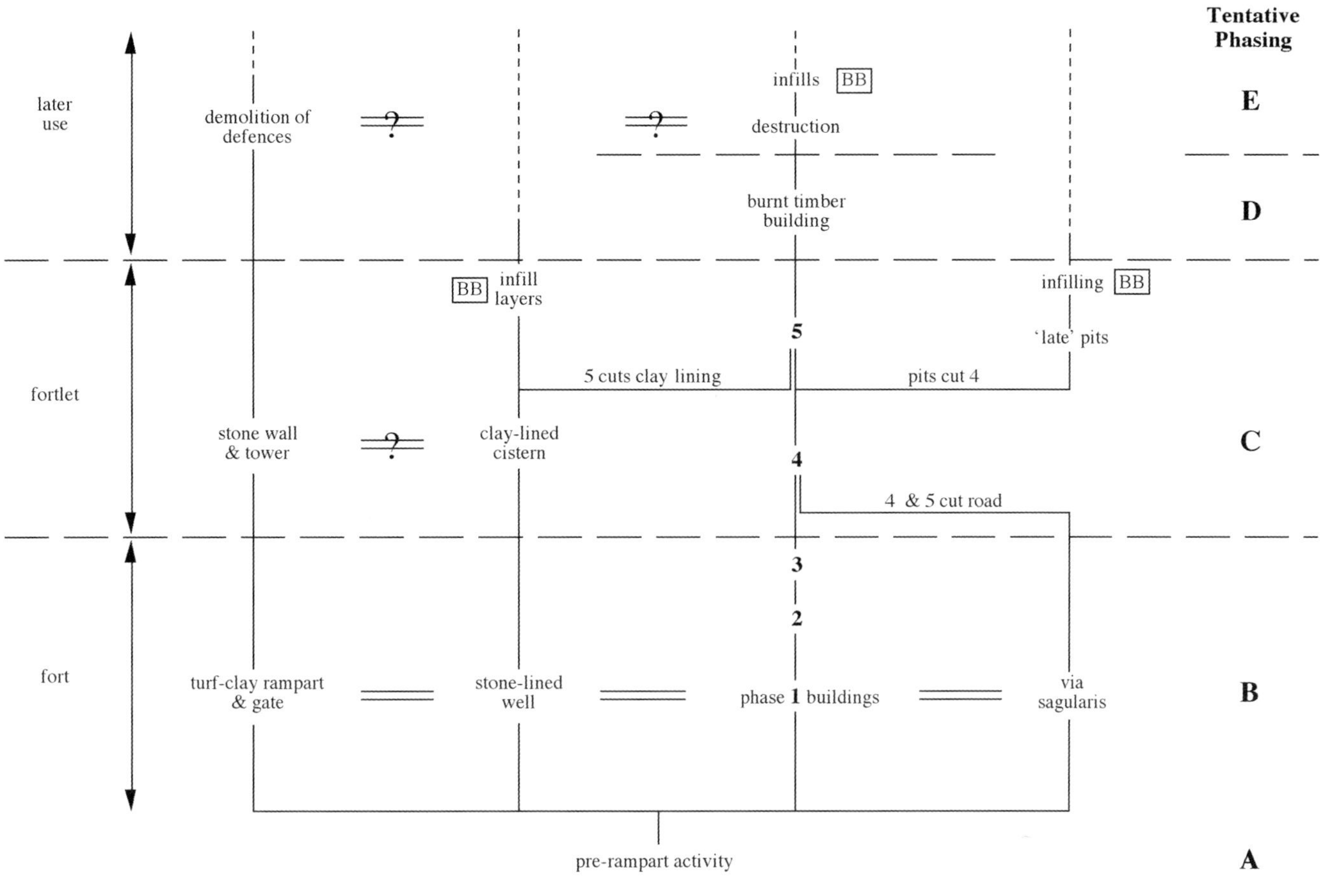

Fig. 2.102 Simplified site sequence for 1989 excavations

of the material in potentially secondary contexts.

There can be very little doubt from the samian that the 1989 site was occupied from the mid AD 70s, well in line with the picture derived from the earlier excavations and the main Flavian thrust into west Wales (Davies 1981). Surprisingly, however, the assemblage as a whole is almost exclusively South Gaulish in origin, except for a single Central Gaulish piece from Les Martres, which must surely suggest that occupation on the 1989 site did not extend far into the first decade of the 2nd century. The coarse pottery, which is overwhelmingly Flavian to Trajanic, would seem to be broadly in line with this assessment, provided we accept that the Black-burnished wares are at the early end of their respective date ranges. This would mean that the overall site sequence, including the fort/fortlet transition and all five phases of building, would have to be accommodated within a 30–year period extending from *c.* AD 75 to 105. If so it would contrast with the pottery from the 1972 excavations, which would seem to indicate that activity continued there into the early Hadrianic era, prior to the fortlet being abandoned by the mid 120s.

What remains difficult on the 1989 site, however, is to refine the chronology of the individual phases, not least those relating to the defences and the internal buildings. This is principally because of the paucity of diagnostic pieces in critical stratigraphic contexts and the loss of the associated floor levels. Certainly, there is nothing to date the fort/fortlet transition, which Webster has now tentatively assigned to *c.* AD 100 on other grounds. Neither of the first two phases of building was associated with any pottery; phase 3 produced a single sherd of probable Flavian date; and phase 4 was associated with material of Flavian-Trajanic date. None of this is sufficiently sensitive, however, as the basis for a more refined chronology. Far more material was recovered from the successive fills of the well/cistern, the 'late' pits and the large pit at the south end of the site, but this was so overwhelmingly of mixed Flavian-Trajanic date, that it can do little more than confirm the generally late date of the phase 5 building.

The discrepancy between the samian assemblages from the 1972 and 1989 sites raises a number of problems, given that both areas are known to have been incorporated within the fortlet. In particular, we would either have to accept that the 1989 site was effectively not used to any significant extent for 10 to 15 years, while other areas were more extensively built up, or else question the nature of the 1989 assemblage. A matter of particular concern is the loss of the floor levels associated with the buildings, because it has effectively removed material which would have been contemporary with the structures, while leaving the finds in the foundation trenches which could be residual. This would have the effect of lowering their overall date and, thereby, the demise of the 1989 site. The samian evidence might tell against this, but this presumes that this part of the fortlet was just as likely as other areas to receive Central Gaulish material, when it might have been functionally or socially different. Some such distinction is certainly suggested by the high percentage of jars in the 1989 assemblage, when compared with the 1972 material where bowls were much more common. Given the different structural histories of the 1972 and 1989 sites, some caution seems necessary.

There might also be a tendency to give less weight to potentially later material where it survives, irrespective of the absence of Central Gaulish material. In particular, it is interesting to note that the Black-burnished wares generally all come from late stratigraphic contexts, with none at all being found in the foundation trenches; thus Black-burnished wares were found in (i) the fills of the cistern (but not the well), (ii) the fills of the 'late' pits probably associated with phase 5 onwards, and (iii) the fills overlying the destruction horizon of the burnt timber building. These latter also produced a Hartshill-Mancetter mortarium and a Martin-Kilcher Dressel 20 amphora, both dating to the 2nd century, which clearly indicates that sufficient later material was around to become incorporated into these horizons. This must sound a note of caution, given the truncation of critical levels which might have been more informative. Irrespective of this, the relatively late appearance of Black-burnished wares in the stratigraphic sequence is in accord with what we would expect, and provides a useful marker in otherwise difficult circumstances.

(B) The 1989 site in context

The character and extent of the fort were established in part from the 1972 excavations on the granary and associated buildings and in part from trenches designed to locate the four sides of the defences (see Fig. 2.1). Those on the north, west and south were fairly convincing in section or description, while that on the east was dependent on two shallow ditches located in the cricket field. Jones and Little accepted that the fort faced east, with the central range including the granary at its south end, the headquarters in the centre and the two phases of building at its north (the 1973 trench), while their barracks, workshops and other structures were assigned to the *retentura* of the fort. The road between the granary and *retentura* was convincingly shown to link up with that along the west side of the 1973 trench and buildings, and, thereby, the gate identified in the northern defences. This latter was seen to be an exceptional additional gate, over and above the main gate which lay to the east in line with the predicted course of the *via principalis*. The need for six gates was seen as being a reflection of the fort's role in supervising the mines and its additional functions. Elements of the fortlet remained far less well known, though traces of its stone wall and ditches were said to have been located east of the main road through the village, while Jones and Little argued that the subsidiary north gate was blocked at this stage.

The 1989 site has done much to corroborate aspects of this sequence, while at the same time adding significant new detail to the overall picture. The confirmation of the two-phase defences, the site of the gateway, and the line of the north-south road located in 1973, all helps to reinforce earlier conclusions about the general alignment of the main fort. It also demonstrates that the 1989 buildings should lie in the same spatial relationship to the central range as did those examined to the west of the granary in 1972. Both sets of buildings (with the exception of the western structure in phase 4) would also seem to have occupied a strip of land *c.* 10.00m wide, very probably sandwiched between the two north-south roads which are thought to extend across the width of the fort between the 1972 and 1989 sites; that to the east is well established, of course, on both sites, but its western counterpart remains to be proven outside the confines of the 1972 excavation. If accepted this would pose a problem for the western of the two phase 4 structures, since if correctly identified, it would have overlain the postulated north-south street along the western side of the 1989 site. No such encroachment was identified on the 1972 site, however, which might cast some doubt on the interpretation of the 1989 structure pending a clearer view of its principal elements.

Far more of the defences was available for excavation in 1989 than had been possible in the 1973 trenches. For the most part the sequence was as expected, though some of the details were less than easy to correlate between the two sites east and west of the gateway. The first-phase clay and turf defences proved to be surprisingly wide on the 1989 site, to judge by comparable evidence elsewhere (Jones 1975), but this may well reflect a deliberate thickening of the rampart as it approached a gate terminal as was evidently the case at Elginhaugh (Hanson 1987). The presence of the timber gate had been predicted by the 1973 excavations, and the 1989 site duly confirmed three of the four large post-pits designed to house the upright timbers supporting the gate tower on the west. These had been subsequently truncated or removed when the stone wall was inserted in front of the rampart, along the line of the infilled fort ditch. One unresolved problem is the reconciliation of the four-post gate tower on the 1989 site and the palisade trench with several post-holes seen on the 1973 site (Fig. 2.103); as currently

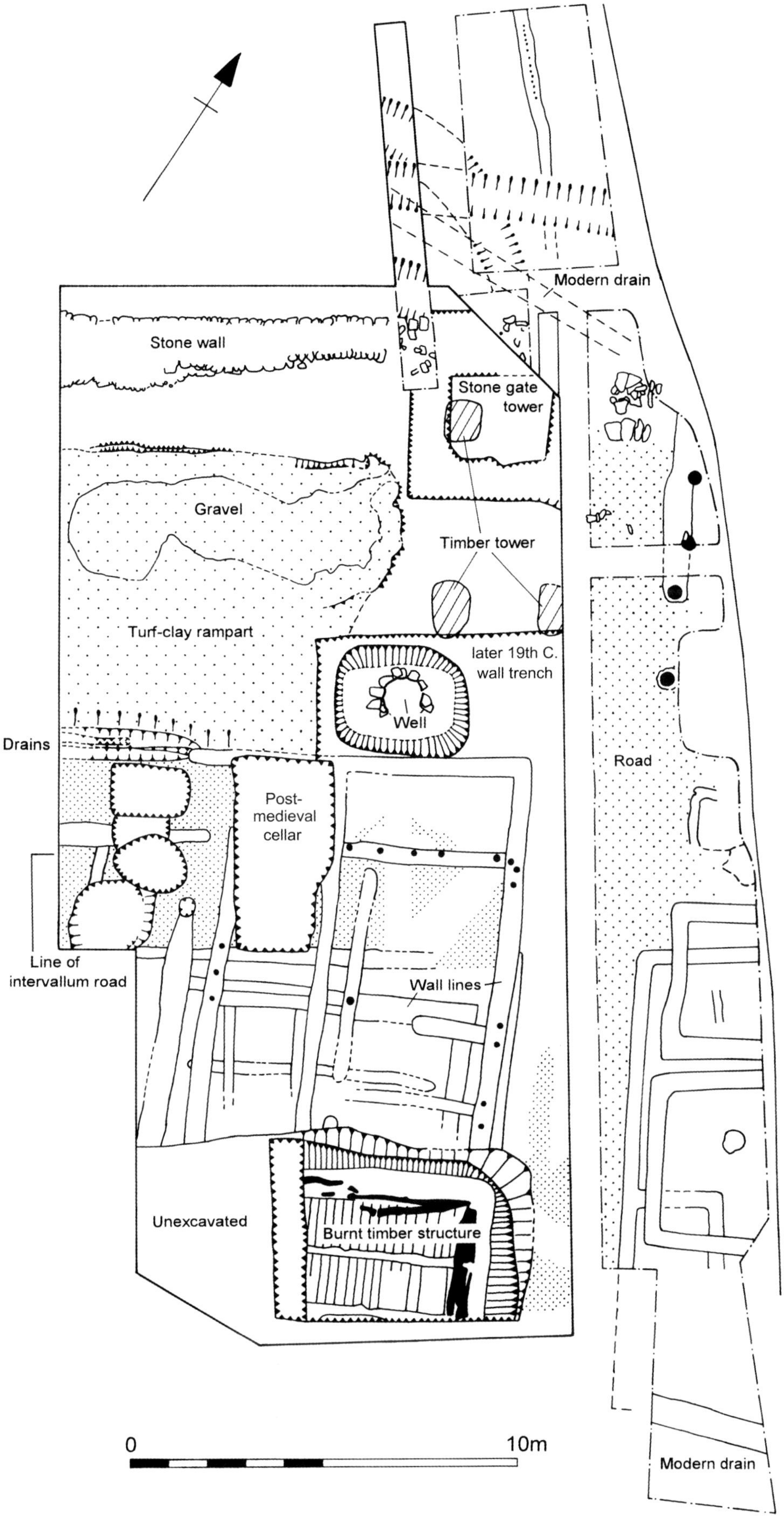

Fig. 2.103 Composite of 1973 and 1989 excavations side by side

envisaged, this would be unlike anything known from conventional gateways as defined by Manning and Scott (1979). One possibility is that the 1973 site did not reveal enough of the east side of the gate passage to identify the core structural elements, and that the palisade trench in part concealed the robbed-out foundations of a stone tower paralleling that found in 1989; if so the timber posts might relate to a structure flanking the gate tower rather than being directly connected with it. None of this can now be easily checked.

The discovery of the robbed-out remains of a stone tower also raises problems, in that Jones and Little claimed that the first-phase gate was blocked when the stone fortlet was constructed, partly because the fortlet ditch was seen to extend across the approach area to the north. This cannot now be readily re-examined, but one has to wonder whether the 'wall' shown in the 1972 trench might represent the foundations of a stop-block extending across a revamped gate between a pair of stone towers. Moreover, recent work on the line of the Roman road from Llandovery to Pumsaint (see section 5.6 below), coupled with extensive geophysical survey on the eastern side of the fort (see section 2.6 below), must also raise doubts as to whether the fort ever had six gates. In particular, it suggests that the main road by-passed the site on the west, with the important implication that the fort itself ought to have faced west (*pace* Jones and Little). This would mean that the north-south road extending across the excavated areas ought to represent the fort's *via principalis*; if so, the 1973/1989 gate can hardly be a subsidiary entrance, which would make it unlikely that it was ever blocked during the lifetime of the military occupation.

Behind the defences, the discovery of a well/cistern represents a significant addition to our understanding of the site. Most forts had some sort of water supply, whether in the form of wells or cisterns, water-collection tanks, or a piped supply provided by a leat from a nearby stream (Johnson 1983, 202–10; Stephens 1985). The Pumsaint example forms a welcome addition to the list, demonstrating by its location just inside one of the main gates, the primary importance which was here attached to the provision of a fresh water supply. Though its interpretation is not without its problems, it clearly began life as a simple well, almost certainly stone-lined throughout, the upper part of which had been robbed out, leaving the original construction backfill in place, prior to the well shaft itself being infilled with successive dumps of material. In this respect it is identical with a 3rd-century well known from the Bays Meadow site at Droitwich (Hurst, forthcoming). Its subsequent conversion into a clay-lined cistern finds no immediate parallels in the literature, however, hence the difficulty of reconciling some aspects of the stratigraphic sequence. Once it went out of use, probably prior to the construction of the phase 5 building, water would have had to be obtained from alternative sources, perhaps via the leat which would have been needed to supply the postulated bath-house (see section 2.6.3 below).

The 1989 site provided evidence for at least four, and possibly five, phases of timber buildings. The main four were all aligned north-south, like their 1972 counterparts, while the less certain phase 1 structure may have been aligned east-west; in this it would conform to the alignment of Jones and Little's period 1A structures which they considered were aligned *per strigas* rather than the later *per scamna* arrangement. Not enough was either visible or excavated on either site to confirm this arrangement, however, and more excavation is perhaps needed before it is accepted as fact. The other four phases appear to be a mixture of fairly typical barracks (1989, phases 2–4; 1972, period 1B), workshops with furnaces (1972, periods 1C-E) and one unidentified structure (1989, phase 5); the precise identification of the 1989 examples is discussed in the next section. Those at the north end originally conformed to the line of the gravel road or *via sagularis*, but the later phases 4 and 5 clearly encroached upon, and eventually cut right across it. This may well be a reflection of increasing pressure on space, no doubt necessitated by the conversion to a fortlet. Such rapid and repeated changes in the internal arrangements at Pumsaint, over a period of 30 years or so, is no longer seen as unusual for auxiliary forts, as the excavated evidence from Loughor and elsewhere attests (Marvell and Owen-John 1997, 228), though we are not much closer to understanding what this means in terms of local and provincial strategies, nor for the character of the associated garrisons.

Nothing on the 1989 site has helped to clarify the nature of the transition from fort to fortlet, in part because the two phases of defences coincided at this point, and in part because potentially critical stratigraphic links between the defensive modifications and the different phases of internal buildings had been truncated by later activity. To some extent this is offset by the recovery of new information about the eastern side of the fort/fortlet and the associated settlement, as a result of recent geophysical work, but its interpretation is not without its associated problems, the implications of which are best discussed in the appropriate place (see section 2.6.3 below). What is clear, however, from the combined evidence of the pottery from the 1972–3 and 1989 sites is that military activity associated with the fortlet did not outlast the first couple of decades of the 2nd century, after which occupation seems to have continued only at locations south of the river. Normal practice at other military sites would certainly have involved the demolition of the internal buildings, while the defences would also have been slighted to avoid their use by any potential hostile group, provided that the state had no further need for the enclosure. At Pumsaint, what happened to the stone defences remains uncertain, however, because the fortlet might have been worth maintaining as a convenient enclosure, close to the mines, for intermittent use as need arose. Beyond the 17th-century pottery in the robber trench associated with the wall, there is little to help us date the process of its demolition, which may in any case have occurred in several stages well into the post-Roman era.

The possibility that the enclosure was maintained in military hands gains some credence from the discovery of the enigmatic burnt timber structure, which clearly retained the alignment of the military phases and continued to respect the line of the main north-south street across the site. Dating as it probably does to some time in the 3rd or 4th century, this structure could as easily be civilian as military, but its location inside a former military compound inclines one to the view that it is more likely military or official, occupying land still in government hands. In this respect, it is worth noting similar late phases of activity and re-occupation at other military sites across Wales (*e.g.* Brecon, Castell Collen, Neath), which must remind us that we know all too little about the processes of control during this critical era, despite recent analysis of the complex evidence (Davies 1991a). Just how long the structure was maintained remains unclear on current evidence, as does its function, but with its destruction, Roman activity on the site (whether military or civilian) was effectively ended. Such issues will need further discussion in section 6.2 at the end of this volume.

(C) The internal buildings

Four certain phases of internal structure and a possible fifth were recorded in 1989, excluding the burnt timber structure discussed in the next section. This complexity is paralleled by at least five phases of building on the 1972 site in the hotel car-park, though only two were noted on the site to the east in 1973. On the 1989 site phases 1 to 3 are the least easy to discuss in detail because so little of their plan was recovered during excavation, but where the later phases 4 and 5 had increasingly encroached across the *via sagularis*, more can obviously be said with confidence about their overall character and design.

Given the earlier discussion of the fort's layout, phases 2 to 5 ought to lie in the *praetentura* of the fort, immediately west of the main central range, in an area traditionally given over to barracks, stables or storebuildings (Johnson 1983, 33–5). Such buildings have been the subject of an extensive recent analysis by Davison (1989), to which one can turn for a wealth of comparative data and statistical information. For the Flavian-Trajanic era, in particular, the evidence so collated indicates that the majority of barracks fell within the size range 30–50m in length by 5–12m in width (*ibid.*, 6, 79–80, figs 2.3 and 2.4). At Pumsaint barracks of average width within this range could easily have been accommodated in the available space (*c.* 10.00m) on both the 1972 and 1989 sites, though a positive attribution remains dependent upon an analysis of the individual plans. The situation is slightly complicated, however, by the fact that the excavated area encompasses that part of the buildings traditionally given over to the officer's quarters; so unless the internal arrangement in part perpetuated the longitudinal sub-division of the men's quarters further south, as for example at Elginhaugh (Hanson 1987), Flavian Strageath (Frere and Wilkes 1989), and Hofheim Period 1 (Nuber 1983), certainty of identification may be impossible. For the sake of argument this latter has been assumed in what follows, but only further work will elucidate the reality.

Phase 1 (see Fig. 2.31): this is the most problematical of the structural phases to interpret with any certainty, in part because it does not conform to the north-south alignment of the other phases and in part because neither its overall length nor breadth could be established. As excavated, its minimum dimensions proved to be 8.00m east-west by 6.00m north-south, with no obvious internal partitions, in contrast with most of the later structures and those of the earliest phase on the 1972 site. In their absence and that of any finds there is little to indicate a positive function, though one might have expected a barrack in this location to correspond with the primary structure on the 1972 site.

Phase 2 (see Fig. 2.31): though only the northern 4.00m of this building was examined on the 1989 site, enough survived to establish that its east-west dimension measured *c.* 8.00m. There was also one associated internal feature, extending east-west, centred at *c.* 2.00m south of the external north wall. The possibility that this was the first in a series of partitions, whether at regular or irregular intervals, could not be established, which makes a positive identification difficult. Regularly spaced partitions might have pointed to a granary (*cf.* Manning 1975), but this is unlikely in this location, given the existence of an example in the central range beneath the later stone one (Jones and Little 1973, 12). A more likely alternative is that it represents the north end of an officer's range in a barrack block, though the narrow space across the width of the building would seem odd, unless there was originally a north-south partition as well; this could well have been cut away by later building trenches. If so, it would have extended down the centre of the building, thereby perpetuating the longitudinal division of the men's quarters further south, as in a barrack of Davison Type B (1989, 5, fig. A); in this it would conform with one of the phase 4 structures.

Phase 3 (see Fig. 2.35): even less of this structure was available for excavation, but again enough survived to establish that its east-west dimension measured *c.* 6.40m. It was divided into two, with a narrow section to the west, no more than *c.* 1.50m wide, and a wider portion to its east measuring *c.* 4.80m across. The narrower western part was also sub-divided towards the south end of the excavated area to produce a room 2.70m by 1.50m. Again, this could be the officer's end of a barrack, with the partition perhaps perpetuating the longitudinal division of the men's quarters into a narrow *arma* area to the west and a wider *papilio* area to the east; but, if so, the building would fall very much at the extreme limits of the range for the known proportions of *arma* to *papilio* in auxiliary barracks (Davison 1989, 14–15, fig. 12). Though this might advise caution, a similar, if less extreme, sub-division seems to have existed in the period 1B barrack on the 1972 site, which was itself no more than *c.* 6.00m wide (see Fig. 2.2). The period 1C workshop was also relatively narrow at *c.* 6.50m, but this in itself is insufficient to posit an alternative identification.

Phase 4 (see Fig. 2.37): of the two adjacent structures assigned to this phase, the eastern one was the more informative in terms of its plan. It was basically 7.60 to 7.70m wide and at least 8.40m long, with a roughly central north-south partition apparently made up of separate lengths of foundation trench. Several other partitions running east-west created irregular spaces on either side. The presence of the central partition, potentially perpetuating the longitudinal division of the men's quarters further south, is highly suggestive that this is a barrack block of Davison Type B (1989, 5, fig. A). Moreover, such a division of the men's quarters into *arma* and *papilio* on a ratio of 1:1 would conform with a trend which became increasingly common from the Flavian period onwards, especially in barracks without a veranda (Davidson 1989, 15, fig. 12). The irregularity of the east-west partitions confirms that this part of the building represents the officer's quarters.

Not enough of the western building was excavated for a definitive discussion of its character and function, but it could well have been a 'partner' to its eastern neighbour. What survives of its internal partitioning is less convincingly 'central', however, because it would involve a restricted amount of space for the *papiliones* in the east range. There is also the problem discussed in the preceding section that no parallel structure was recognised on the 1972 site, where a north-south street had clearly existed; as such the interpretation of the western building must remain in doubt.

Phase 5 (see Fig. 2.40): this was a long narrow building, 5.00m wide east-west by at least 10.30m north-south, with no recognisable internal partitions. It is, therefore, much less easy to interpret as a barrack, but beyond that there is little else to indicate its function. Although the width would be suitable for a stable, in line with Davison's Building Type O (1989, 151–52), there were none of the characteristic internal features now so convincingly known from other sites in the form of drains and cesspits (for the evidence at Wallsend, in particular, see Hodgson 1999, fig. 15). The alternative, that it represents a store shed or some such structure, seems more likely.

The nature of the superstructures and construction details of the different buildings has not survived well on the 1989 site. In all probability all of them were of post-in-trench construction, even if traces of the original timber uprights were only recognised with confidence in the trenches associated with phases 4 and 5. Those of phase 4 were noted as distinct soil discolorations at several points, indicating an original spacing at regular intervals of 0.70 to 0.80m. Above ground level no floor levels had survived, which was somewhat surprising given the abundant evidence for clay floors recorded in previous excavations in 1972–3 and 1987. Nor was there much to indicate the nature of the superstructure and the roofing for certain. Very little wattle and daub was seen, beyond the much disturbed material in the upper fills above the burnt timber building; only one fragment of window glass was found, and this not certainly associated with any of the structures; very little tile was recovered from the site, so a roofing of shingles or thatch seems the most likely. Oak shingles are certainly attested at Vindolanda, both as surviving examples and in one of the letters from a late 1st- to early 2nd-century context (Birley 2002, 69 and 92).

In summary, at least two (and possibly three) of the phases on the 1989 site are likely to have been barracks (phases 2, 3? and 4), one may well have been a storebuilding or shed (5), while the nature of another remains uncertain (1), though a barrack remains a possibility. The presence of barrack-type structures in the early phases is hardly surprising, given both the site's location within the fort and the 1972 parallels. There was no evidence, however, for any hearths or furnaces to match the later workshops on the 1972 site.

(D) The burnt timber structure

The discovery of a large timber structure at the base of a 2.00m-deep hole was somewhat unexpected in the context of an auxiliary fort or fortlet. As such it deserves more detailed treatment. It had clearly been cut through all the previous building lines, and so represented the latest Roman activity at the south end of the site. Since no clear relationship could be established with the 'late' pits and other features at the north end of the site, there was no way of ascertaining whether the fortlet rampart was still standing when the structure was in use.

Only about a quarter of the structure was excavated, to judge from the size of the pit and the angle of the infill deposits, which would mean that the original must have measured *c.* 8.50m by 6.50m overall. It had clearly burnt down in an extensive conflagration at the end of its life, but whether deliberately or unintentionally cannot now be determined. Much of what has survived would seem to be derived from a subterranean structure, apart from a jumble of timbers from the horizon above its plank floor, which may well have fallen or collapsed from the superstructure. Above this, none of the surviving structural timbers extended up to ground level and no post-holes were found relating to any above-ground features. This makes it difficult to reconstruct the original with any certainty beyond the basal, burnt levels, and even here the fire has destroyed vital evidence both for the measurement of individual timbers and for the nature of the associated woodworking techniques. Useful details can be gleaned, however, from the wealth of well-preserved timber structures which are now known from a variety of military and civilian sites, in part from waterlogged contexts and in part from destruction horizons, among them Valkenburg (Glasbergen 1972), Colchester (Crummy 1992; 1997), Carlisle (McCarthy 1991), and London (Milne 1985; Dillon 1989).

The Pumsaint structure had clearly been constructed upon at least three, and probably originally five, horizontal beams, each *c.* 0.20m in cross-section, which had been dug into the natural gravel at the base of the pit (see Fig. 2.47). They were all aligned roughly east-west, with no obvious trace of any corresponding north-south timbers across their ends. The use of such large, squared timbers is not uncommon in military contexts, usually but not exclusively at ground level, primarily to act as sill-beams into which the various timbers of the superstructure were secured. This is certainly the case with the various military buildings at Valkenburg (Glasbergen 1972), but can also be seen in some of the 2nd century structures at Castle Street, Carlisle (McCarthy 1991, 56–7, fig. 41). In subterranean contexts, similar sill-beams provided a foundation for a building in Southwark (Dillon 1989, 229, fig. 1), while the burial chamber at Folly Lane, Verulamium, also employed horizontal sill-beams, in part set into shallow slots, in part laid directly on the base of its pit (Niblett 1999, 33, fig. 15). In the case of Southwark the beams were up to 6.00m in length, but even longer timbers have been recorded in the London quays (Milne 1985), well in line with the overall dimensions of the Pumsaint structure.

In the case of Pumsaint the beams were also designed as joists to support a floor of closely-set planks, laid horizontally across its narrow dimension from north to south. To judge from the fact that each of these planks had a single large iron spike at its northern end, but not obviously elsewhere, it would seem likely that all of them had been secured in place only at their extremities, suggesting that they originally extended right across the building over a distance of *c.* 6.50m. An oak plank floor also formed an integral feature of the subterranean structure at Southwark, though here the individual planks extended lengthwise along the building, and rested on a series of joists which were themselves dovetailed into the sill beams (Dillon 1989, 229); even here the planks at the eastern end measured 5.50m in length, not much short of those at Pumsaint. Such measurements have obvious implications for the ready availability of good lengths of timber from the neighbouring woodlands, which would have been felled and utilised without the need for any seasoning (Hanson 1978).

Excavation demonstrated that the side walls of the Pumsaint structure had been made from a series of timber uprights placed at intervals along the north and east sides of the surviving remains (see Fig. 2.45). These apparently lay at intervals of *c.* 1.50 to 1.60m, which might indicate a possible arrangement of 6 by 5 along the exterior sides. Not enough survived to show how they had been attached to the underlying sill-beams, but a simple mortise and tenon joint would seem the most likely, as was the case in the Southwark example already quoted (Dillon 1989, 231). An alternative possibility is suggested by the evidence from Folly Lane, Verulamium, where the uprights seem to have been recessed into the sides of the sill beams, no doubt with some sort of wooden peg to secure them in place (Niblett 1999, 34). A similar arrangement was employed in one of the structures at Castle Street, Carlisle, but in this example the uprights were set in their own post-pits, which was clearly not the case at Pumsaint (McCarthy 1991, 58). Without clear evidence for the nature of the superstructure, little can be said about how such uprights were secured at their upper ends, let alone whether they extended to the full height of the building.

Once in place, these uprights had served as a framework against which the external cladding of the structure had rested. At Pumsaint this comprised a series of planks laid horizontally edge to edge, one on top of the other (see Fig. 2.43). Up to four courses had survived in places, though there was no clear indication as to how they had been fastened, if at all, to the main uprights. Iron nails had certainly been used to secure the planks in one of the burial chambers at Stanway, Colchester (Crummy 1992, 2), but if this had been the case at Pumsaint there were no obvious nails or spikes at the appropriate points in the surviving timbers. The same would seem to have been true at Southwark, where again there were no discernible means of attaching the planks to the uprights (Dillon 1989, 229). In such circumstances they must either have been secured by wooden pegs or else held in place by the infill which had been backfilled behind them as the structure rose upwards.

The excavated remains at Pumsaint represent no more than the subterranean elements of what may have been a larger building. The exact means of access remains unclear, but presumably there was either an internal stair linking with a ground floor level, or else an external flight of steps leading to a door in one of the sides. Above the surviving remains, however, reconstruction of the building's superstructure is a matter of speculation, as is the question of whether it was of one or more storeys. In particular, it remains unclear whether the timber uprights forming the sides of the cellar extended as structural members to the full height of the building, or whether there were extra uprights at ground level, designed to carry the external walls of the structure and its roof. The former was clearly the case with the timbers surrounding a substantial pit inside the fortress at Chester (Strickland, *pers. comm.*), but the situation at Pumsaint is less clear because the dimensions of the timbers remain unknown and so their weight-bearing capacity cannot be determined. Had there been extra uprights to carry the main walls, however, some trace of their post-pits might have been expected to have survived around the exterior, unless these had been incorporated into the backfill around the cellar and thereby lost when it all subsided during the conflagration.

Such uncertainties compound the problem of finding either parallels for the building or satisfactory explanations for its function. The situation is further exacerbated by the fact that nothing was found in association with the plank flooring of the structure, nor with the overlying destruction horizon, except for a small number of calcined bone fragments, various iron nails and spikes (all structural), and a possible coin (Co 3). On analysis (see section 2.2.4(N) above), the bone proved to include three fragments of juvenile metatarsus of a size consistent with cattle (*Bos*), though this is hardly sufficient in itself to draw any firm conclusions. Samples of the sediments located both above and below the flooring were also analysed, in the hopes of providing some insights into the question of function (see section 2.2.4(Q) above), but beyond recording a high phosphate concentration in the underlying deposit, most notably towards the central part of the structure, these too proved inconclusive. Given such negative evidence, our only available recourse is to seek suitable parallels with other, primarily subterranean, structures elsewhere in Britain and the other western provinces.

Fortunately for our purposes a variety of such structures have now been identified in sufficient contexts, both military and civilian, that we can begin to narrow down some of the options. Perhaps the most distinctive of those found in the military sphere are the structures identified as tanning pits. One such, outside the fort at Brithdir, proved to have contained a wooden tank, up to 5.60m long, 3.00m wide and 1.80m deep, which had apparently been laid on a raft of cross timbers supported by five horizontal sleeper beams, themselves set into a thick clay deposit; its side walls were also supported by a clay-bonded stone lining to ensure that the structure remained watertight (White 1978, 41–3, fig. 11). A similar tank with a clay and timber lining has also been recorded from the area outside the presumed fort at Carmarthen, set in a rectangular pit *c.* 7.70m long, 4.90m wide and 1.80m deep (James 1992, 8, fig. 3; 2003, 37–8). Though there are some superficial similarities with elements of the Pumsaint structure, these rapidly disappear when it is remembered that the latter has no evidence for any sort of clay lining, either across its base where the plank floor rested directly on the underlying gravel, or behind its side walls where the infill comprised clean natural gravel. As such, the Pumsaint example could not have held water and so would not have been suitable for use either as a tanning pit or as a cistern.

Potentially more interesting in a military context is a large timber structure located in a walled compound inside the fortress at Chester (Strickland, *pers. comm.*). This comprised a large rectangular pit, at least 15.5m long, 4.80m wide and *c.* 1.80m deep, which had been partly rock-cut into the sandstone bedrock. Along the length of this feature pairs of substantial timber uprights had been set into specially designed emplacements, themselves partly rock-cut, at regular intervals of *c.* 1.50m; they may also have extended across the narrower ends, though the evidence is less certain. By contrast with Pumsaint, however, there was no trace of any planking across its floor, nor between the various timber emplacements. Undoubtedly these remains represent the weight-bearing members of a substantial, possibly timber-framed, building, which is thought to date within the period *c.* AD 220 to *c.* 330. Numerous suggestions have been made for the function of this structure and its associated pit, including some connection with tanning or metalworking, or usage as a possible abattoir, granary, cistern, latrine, wine store, service pit (for wheeled vehicles) or saw pit, though few of these can be made to satisfy the surviving remains. For our purposes, however, the question of function may be less important than the presence of such a large subterranean feature (and its associated building) within an explicitly military context, at a date not inconsistent with that of the Pumsaint example, since this might lend weight to the view that we are dealing with a military rather than a civilian structure.

Structures with subterranean elements have, nevertheless, been recorded in civilian contexts. The most distinctive, perhaps, is a class of early burial chamber, well represented by the recent publication of the material from Folly Lane, Verulamium (Niblett 1999). This comprised a substantial pit, 6.90m square and 2.90m deep, the vertical sides of which had been

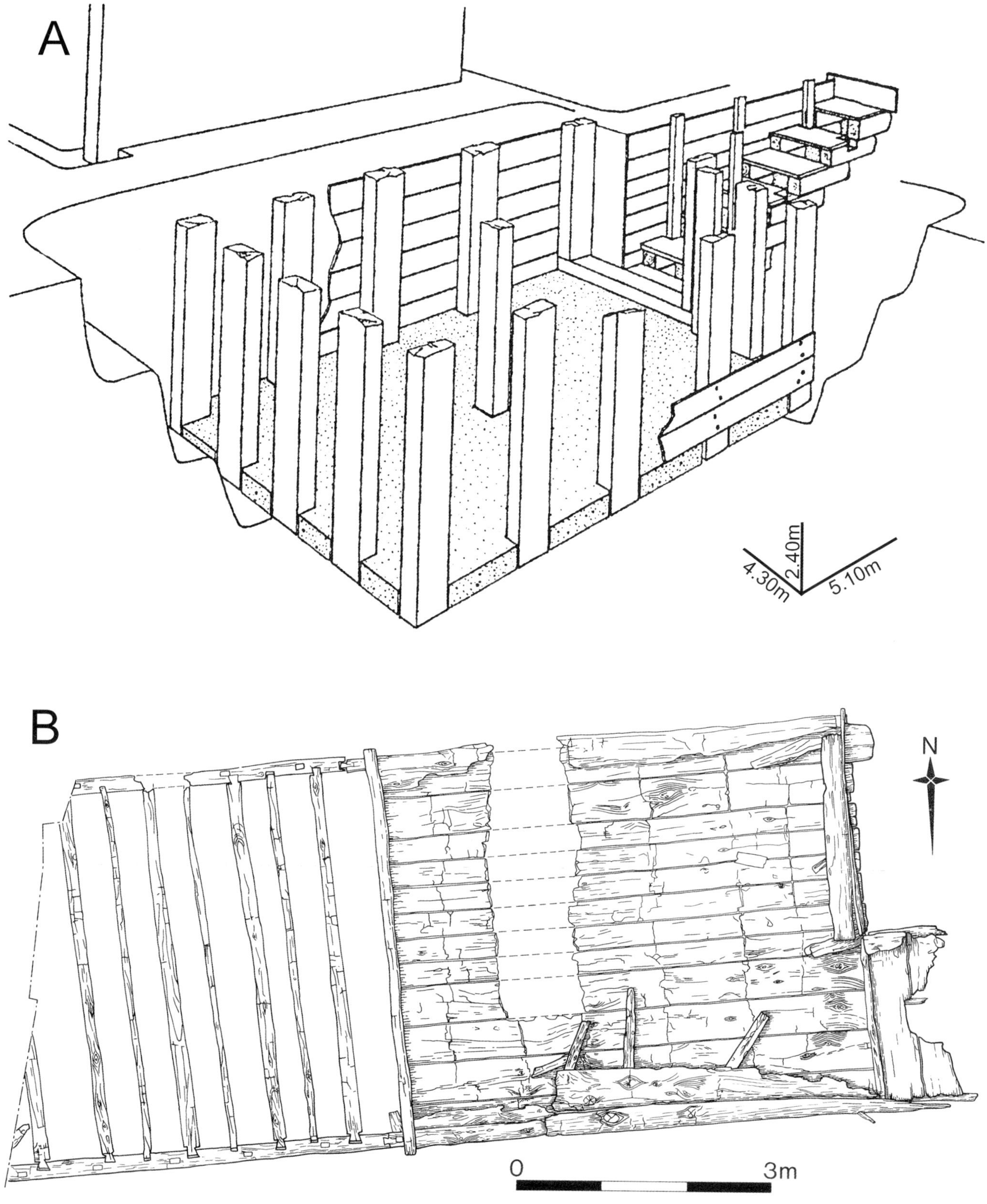

Fig. 2.104 Timber buildings from (A) Mont Beuvray (after Vitali and Zvald 1999, fig. 23) and (B) Southwark (after Dillon 1989, fig. 1)

revetted by a complex timber structure, supported on a series of horizontal sill-beams; in this case, however, there would seem to have been some sort of roof structure at about ground level, apparently with a mound on top (*ibid.*, 29–44, figs 21–2). Similar, if simpler, timber-lined burial chambers have been identified at Stanway, south-west of Colchester (Crummy 1992, 2–3; 1997, 22–8) and under the Lexden Tumulus (Foster 1986), while examples are not unknown on the continent (*cf.* Niblett 1999, 396–400). At Stanway, the largest of the chambers (in enclosure 3) proved to be 4.00m long, 3.40m wide and just over 1.00m deep, with walls of timber planking held in place by nails. Despite some superficial similarities in the structural details and the carpentry between such structures and the Pumsaint example, there is absolutely nothing to suggest that it served a similar function; in particular, the dearth of any significant finds associated with the latter contrasts markedly with the rich assemblages of material now known from the interior of the burial chambers, all of it intimately derived from the accompanying funerary ritual and ceremony.

A more general group of subterranean structures, often found in association with domestic or commercial properties, seems to have served as cellars. These are especially common in civilian contexts on the continent, but an increasing number of examples are also known in Britain (Perring 1989). More often than not they were of stone construction, as was the case at such sites as Alesia (Bénard and Mangin 1994, 42–5, figs 11–12), Bliesbruck-Reinheim (Petit and Schaub 1995, 30–5, 69–79, and associated figs), Mâlain (Roussel 1994, 66–7, fig. 18), Schwarzenacker (Gilles 1994, 279–80, Abb. 26) and Verulamium (Frere 1983, 14), but good examples in timber are afforded by a cellar of Augustan date at Mont Beuvray (Vitali and Zvald 1999, 39–40, fig. 23) and by a 3rd- to 4th-century structure in *Insula* XIV at Verulamium (Frere 1972, 103–8, fig. 22). That at Mont Beuvray, which was especially well preserved in the wake of a fire, proved to be 5.50m long by 4.00m wide, with a framework of upright timbers set into a large pit dug to a depth of *c.* 2.40m (Fig. 2.104A). These uprights supported a cladding of timber planks, very reminiscent of the Pumsaint example, though in this case they were held in place by nails. Access from ground level was afforded by an external set of wooden steps, which led down to a door on the west side. The example at Verulamium, though less well preserved, was somewhat larger at 10.36 by 4.34m, and thus far more comparable in size with Pumsaint. It too had a framework of heavy timber uprights, set in their own separate post-holes, as well as a plank floor resting on timber joists. Such cellars could obviously have served a variety of functions over their lifetime, but their potential for providing a cool, stable environment, ideal for the storage of perishable goods, might suggest a plausible parallel for Pumsaint, even if context and date are not directly comparable.

In this context the building from Southwark is of particular interest, in part because its exceptional state of preservation affords valuable parallels, in part because it too may have served as some sort of storebuilding or warehouse (Dillon 1989, fig. 1). Dating to the 2nd century, it comprised the lower portions of a collapsed timber structure, at least 11.00m long by 4.75m wide, resting upon a foundation of squared oak sill-beams and floor joists at right angles, which had been laid in the base of a large construction pit at least 0.80m deep (Fig. 2.104B). The sill-beams had supported a framework of timber uprights, against which had been laid the oak planks forming the external cladding of the building, much like the Pumsaint structure. Oak planks had also been laid across the joists to provide a strong floor, access to which was gained down a ramp in the south-east corner. Although there was no conclusive evidence as to the function of this structure, its location alongside the Thames foreshore, coupled with a floor capable of supporting heavy weights and an attempt to create a cool environment by part sinking it into the ground, lends itself to the idea that it may have been a warehouse forming part of a dockside installation. Once again an emphasis on storage might suggest a parallel for Pumsaint, even though their respective contexts and dates are very different.

In summary, much of the preceding discussion about the character and function of the Pumsaint structure must remain inherently inconclusive, in the absence of definitive parallels. There would seem to be little doubt, however, that it represents the remains of a subterranean cellar, probably attached to a larger building with at least one storey above ground. On balance, given its location inside a former military establishment, a military or official function seems the most likely, but whether this should be associated with the collection and storage of foodstuffs or the like, perhaps in relation to taxation, must remain a matter of pure speculation.

2.2.6 Discussion of the post-Roman levels

The main features of the post-Roman levels on the site were the successive clom (clay-walled) and stone buildings, although there were a number of apparently earlier features. The clom building can probably be dated to the earlier 19th century, while the overlying stone building, which survived until demolished in advance of the excavation, appears to have been erected in three sections between the mid and the late 19th century. The first portion constructed was probably the southern extension (Rooms 3/4), which is likely to have been added to the clom building while it was still standing. Later, the clom building itself was replaced by Room 2, with a new northern wall slightly further to the south, and, finally, a northern extension (Room 1) was added as well. A number of features in the yard area to the west, including a post-medieval cellar immediately to the west of the clom building, could also be related to the different buildings with varying degrees of confidence.

Numerous features clearly pre-dated the known clom and stone building episodes, but for the most part, they were very difficult to sequence or date except on stratigraphic grounds, particularly at the earlier end of the range. Good markers were provided by layers [129], [225] and the associated yards to the west, where a series of features were sealed, while at a higher level, [92] and [123/126] were also useful defining horizons. Where finds were present, the TPQs obtained have been of considerable help in constructing some parts of the suggested sequence (Table 2.28).

(A) The nature of the pre-building activity and its dating

The activity pre-dating the clom building is likely to span much of the period from the end of Roman occupation to the early 19th century, when that building would appear to have been erected. The traces are however fairly sparse before the 17th century (though this may reflect a general lack of archaeologically visible material rather than a lack of activity), while after this date there are copious hints of fairly concentrated usage of what is likely to have been a site on the frontage of the road through the incipient village. The pub next door to the site may well have been in existence in some form for much of this later period, though it is first definitely attested in the later 18th century, at which period it may have been a focus for droving activities.

The pre-building features can be divided into three groups (see Figs 2.53 and 2.60): (i) a series which, on stratigraphic grounds, supported by very limited dating evidence, appear to be pre-17th century; (ii) features suggesting scattered activity in the 17th and 18th centuries, pre-dating any definite traces of structures; and (iii) a number of contexts which appear to be the fragmentary remains of earlier, probably 18th-century, structures on the site of the later buildings. While many features were scrappy and not readily intelligible, the more coherent ones in each group are considered here.

(i) The following features are likely to be pre-17th century, though there was little clear dating evidence.

(a) A broad shallow gully running obliquely across the area later occupied by the southern end of the clom building. This alignment does not appear to tally with either the Roman plan or that of the later village, and its date, purpose and context remain obscure in the absence of any finds.

(b) Six shallow oval hollows, aligned roughly north-east to south-west, broadly parallel with the end walls of the later clom and stone buildings. While these features are not clearly associated with either of these, they may represent the very tenuous remnants of earlier structures of which no trace otherwise survived, although they do not combine happily into any single plan. They may alternatively represent shallow or truncated graves, although there were no obvious traces of human remains and little evidence for any episode of truncation. Again, there was no useful dating evidence for these features.

(c) A V-shaped gully running north-west to south-east down the length of the area later occupied by the clom building, and swinging away to the east at the south end of the site. This gully was very shallow at the northern end, becoming deeper and wider as it went further south. Its northern portion was again compatible with the alignment of the later buildings, though the portion to the south of the later southern clom wall was not. The finds from the fill of this gully were all of residual Roman material, apart from one sherd of medieval pottery.

(d) one small pit towards the north of the west side of Room 1, which yielded a sherd dated to the 16th century onwards.

(ii) A number of further features, somewhat better dated, suggested scattered activity in the 17th and 18th centuries, pre-dating any definite traces of structures. These were:

(a) The infilled robber trench overlying the Roman stone tower and the wall to its west, which yielded pottery dating to the 17th century onwards. This dating may reflect an episode of activity, possibly levelling, on the site near the beginning of the main sequence of post-medieval occupation.

(b) One feature from the area of the later yards, which yielded a fragment of 17th-century clay pipe.

(c) Numerous features sealed by surfaces [225] at the north end and [129] in the central area. Few of these were readily intelligible, and few yielded any dating evidence. Four post-holes, forming a rough line, lay in the area between the later northern clom wall and the foundation trench for the north wall of the stone-built Room 2; these may have been traces of an earlier building phase or boundary. One yielded pottery of the later 18th century onwards, but this may be intrusive. These post-holes had been truncated by a cut alongside the inner face of the northern clom wall, probably intended to help level the floor of that building or a predecessor.

To these might be added the fact that, throughout the period from the 17th to the 19th century, there is a marked rise in the number of clay tobacco pipes on the site, which would suggest an increase in its importance for drovers (see section 2.2.4(D) above), and by implication, some degree of structural development.

Table 2.28 TPQ dating for post-Roman contexts

Source key: P=pottery, CTP=clay pipe, Co=coin, Gl=glass

Context no	TPQ	Source	Description
13	late C19-C20	Gl	fill at S end of Room 2
16	"	Gl	cobbled floor in Room 3
25	C18+	P	turf and topsoil along W edge of trench
30	1957	Co	makeup beneath 4 and 5 at N end of Room 2
31	late C19	Gl	makeup between partition 6 and cross-passage in Room 2
32	1921	Co	recent farmyard surfaces
34	late C18+	P	fill beneath 10 along cross-passage in Room 2
36	late C19+	Gl	infill beneath western floors in Room 4
37	"	Gl	(?)beaten earth floor along S 1.5m of Room 1
38	*c.*1850	CTP	infill beneath flooring and stalls in Room 1
39	"	CTP	concrete overlying pitched stone flooring 41
41	late C18+	P	pitched stone floor of clom building in area of Room 2
43	"	P	disturbance cut through 39
49	"	P	disturbance cut through 41
52	*c.*1900	CTP	infill beneath 38 in Room 1
53	late C18+	P	infill S of N clom wall 63 in area of Room 1
54	"	P	infill across N clom wall 63 in area of Room 1
57	"	P	N-S trench for plastic water pipe serving Room 1
63	C17+	P	N wall of clom building
65	C18+	P	fill of post-pit 66 in farmyard area
67	*c.*1800	CTP	disturbance cut through 41, sealed by 40
70	C18	P	fill of trench 71 for wall 69 of Room 1
71	C18+	P	construction trench for wall 69 of Room 1
72	C19	Gl	wall of Room 2
73	*c.*1850	CTP	fill of trench 74 for wall 72 of Room 2
76	C18+	P	wall on W, S and E of Rooms 3 and 4
79	*c.*1850	CTP	farmyard surface relating to stone building
82	"	CTP	disturbance cut through 41
84	late C17+	P	fill of disturbance 85 at N of Room 1 NB Sherd joins 92
92	late C18+	P	infill N of clom wall 63 in area of Room 1 NB Sherds join 84 and 174
94	mid-late C18	Gl	fill of post-pit 95 cutting 100 in farmyard area
96	C18+	P	disturbance cut through 41
97	late C18+	P	infill beneath floor 16 in Room 3
99	"	P	foundation trench for wall 76 of Rooms 3 and 4
100	C16+	P	fill of pit 101 in farmyard area
102	late C18+	P	fill of robber trench 103 at S end of site
104	"	P	fill of post-pit 105 in farmyard area
106	"	P	fill of 107, trench for earlier S wall?
121	"	P	spread towards E of area of Room 3
122	"	P	spread towards W of area of Room 3
123	"	P	infill beneath pitched stone floor 41 in area of Room 2
125	C18	P	pebble surface at S end of site
127	1773	Co	patches of pitched stone flooring at SW of area of Room 2
128	late C18+	P	rough flooring in area of Room 3
129	1880 1775	CTP Co	clay surface beneath 123 and 126 NB The date is archaeologically unacceptable. Other dates from context are C18th. Pipe fragment must come from fill of a disturbance.
130	C18	P	fill of post-medieval cellar 132
131	1900	CTP	fill of post-medieval cellar 132
156	C18+	P	foundation trench for S wall 75 of clom building
157	1880	CTP	garden soil, uppermost fill in timber building pit 245
158	*c.*1850	CTP	fill of post-hole below 92 in area of Room 1
161	late C18+	P	fill of pit 162 in area of Room 3
174	C18	P	fill of hollow below 92 in area of room 1 NB sherd joins 92
179	1780	CTP	Stone packing of E-W drain 221 in area of Room 3
180	C18+	P	probable farmyard surface relating to clom building
182	C17+	P	fill of oval hollow 183, below 92 in area of Room 1
186	"	P	fill of 187, post-pit in both farmyard areas
188	"	P	gravel below drain 221 towards W of trench

Context no	TPQ	Source	Description
197	late C18+	P	fill of post-pit 198, truncated by cut 201
211	C17+	P	fill of post-pit 212 in farmyard area relating to clom building
220	C18+	P	gravel spread adjoining drain 221 in E part of trench
223	C17+	P	gravel spread on W of trench, =180
226	late C18+	P	reddish clay across S end of area of Room 2, below 129
227	"	P	ashy fill of square hollow 228
235	"	P	fill of post-pit 236 beneath stone-wall construction trench 74, towards N of W wall of Room 2
237	C18	P	mixed ashy fill of (?)hearth 238
248	medieval	P	fill of N-S linear gully 254
273	C16+	P	fill of pit 274 below 225 in area of Room 1
280	*c.*1660	CTP	fill of pit 281 cutting N end of foundation trench 304
287	C17+	P	fill of post impression 288 in pit 290, in SW corner of W extension of trench
291	C17+	P	fill of robber trench 292 above stone guardchamber
311	early C18	Gl	modern post-pit
330	late C17+	P	fill of robber trench 331 above stone rampart

(iii) The following features represent increasing evidence for buildings or structures dating to the 18th century. These were much easier to see and more clearly datable than the earlier features, despite their considerably fragmented state.

(a) Two irregular gravelled areas towards the southern end of the site, probably originally parts of the same surface. The 18th-century or later material from the eastern area is likely to provide a date for the spread as a whole. While it lay at a lower level than the clom building, its northern edge appeared to respect the line of the southern clom wall. A concentration of stake-holes forming a rough line, with a post-hole at their eastern end, may have defined a fence, partition or wattle wall *c.* 4.00m south of the clom wall; this alignment lay just inside the northern edge of the pit which contained the Roman burnt timber building.

(b) Features dug into the topmost infill of the Roman pit south of the gravel surface. These included an elongated pit, with pottery dating to the late 18th century onwards, and an undated gully, running roughly north-south near the eastern edge of the excavation, the fill of which was cut by the post-hole at the eastern end of the line of stake-holes discussed above.

(c) Hearths and hollows apparently associated with metalworking activity. These fell into two groups, separated by the clay spread [226] which contained 18th-century material. This spread, despite pre-dating the south wall of the clom building, the foundations of which had ultimately been cut into it, did not appear to extend to the south of it. The earlier group of features itself appeared to consist of two successive phases of activity, the later of which was dated by 18th-century pottery in the fill of one of the features, while finds from the later group ran up to the later 18th century or beyond.

(d) Features at the north end cut into [225]. While the construction trench for the northern clom wall had been dug into this layer, it was not otherwise dated. The features all lay to the north of the clom wall and were all sealed by an overlying layer [92]. They were so numerous that they must surely relate to a number of successive internal arrangements, if not to a succession of actual structures. While no trace was seen of any external wall line, due to the depth of the foundation trenches relating to the later stone building, it seems likely from the hints contained in the internal layout that any structure in this area would have been of similar dimensions to this later building. The only dating evidence for the features came from two of a group of three pits, towards the middle of the northern part of the area, one of which yielded 17th- and another 18th-century material. A number of pairs of post-holes 2.00m apart, one containing the remains of timbers, hint that this area may have contained stalling for paired cattle.

(e) Features lying stratigraphically between [129], the layer filling the cut to the south of the northern clom wall, and [123/126], the makeup of the pitched stone floor which indisputably belonged to the clom building. It is not altogether clear whether the features at this level relate to the clom building itself or to a predecessor of the same extent. The layer [129] was noticeably clayey in consistency, suggesting that it may represent clay flooring. It contained a very mixed range of finds including mainly 18th-century material, in particular a penny of 1775, while a clay pipe fragment of *c.* 1880 must be intrusive, since the overlying stone building was in position by then. The various features at this level included a spread of stone flooring towards the south-west corner, probably a repair, containing a coin of 1773.

(f) The post-medieval cellar. This had been cut off from any direct stratigraphic relationship with the building to its east by the construction trenches of the later 19th-century stone-built Room 2, which appeared partially to truncate whatever had been the original structural arrangement on its east. This would suggest that the post-medieval cellar was either contemporary with or earlier than the main clom building, and had gone out of use by the time the stone building was erected, though, puzzlingly, finds from the infill continued down to 1900. It was not clear how the post-medieval cellar had been roofed, whether it had supported a superstructure, or exactly how it was accessed, though a ladder at the south end or on the east is the most likely. Its function is unknown. It could possibly have served as a dairy, where milk products could be kept cool during processing, though there are few if any obvious parallels.

(B) The clom building

Despite the evidence discussed in the preceding section for activity of some sort on the site from the 17th century onwards, an estate map of 1803 (NLW Dolaucothi 2) shows no building in this position. It is possible that only stone structures were included, but the map as a whole does appear to be a fairly accurate survey, and a number of farm buildings are shown adjacent to the mansion, which lies 1 km to the east of the site. The village lies on the western edge of the land belonging to this home farm. The pub to the south of our site was clearly already present by this date, and is shown running north-south, supporting the conclusion reached by Darlington and Noake in their vernacular survey (1989) that the north-south section now forming the north-western part of the main building was the earliest, and had been truncated to allow the addition of the more spacious and stylish southern portion. They dated the internal roof timbering of the earlier section on stylistic grounds to the later 18th century. By the 1820s or 30s, another map (NLW Dolaucothi 20018–P24451, rather sketchy and not definitively dated), shows a building running north-south on our plot to the north of the pub. The latter is itself shown running east-west and labelled the 'Pumsaint Inn', under which name it also appears in various estate records down to the early 1870s, when it becomes the 'Dolau Cothi Arms'. A date before 1833 is likely for this map, since the fine stone coach house and stables opposite the pub, which were built at that date, are not shown.

The clom building as excavated consisted of two lengths of walling, each 0.85m wide, running roughly north-east to south-west, 10.3 to 10.6m apart, flanking the north and south edges of a carefully-laid pitched stone floor (see Fig. 2.62). The construction trench of the southern wall yielded a fragment of 18th-century or later pottery. No trace was seen of any wall to the east or west. The area to the east of the east wall of the later stone building lay outside the excavation, within the area previously examined by Jones and Little, while on the west, the floor certainly did not appear to have continued to the west of the western wall of that building.

The presence of the post-medieval cellar on this side (whether of this date or earlier) probably limited the area available. There is room for a wall 0.85m thick, like those to north and south, in the area cut away by the later foundation trench, although the section past the post-medieval cellar may not have been all that stable. It is likely from the pattern on the floor that the eastern side of the building lay roughly along the later line; if the southern portion of the later stone structure, as seems probable, was built as an extension to the clom building, this would lend further weight to this hypothesis.

The northern and southern walls are not precisely parallel to one another, with the southern approximating more closely to the alignment of the later stone buildings, though at a slight skew to it, while the northern heads away rather more towards the north-east. The line dividing the pitched stone floor along its length suggests that the east and west walls of the clom building may also have been slightly off the later alignment, compensating for the skew on the northern and southern ones to some degree, though still with some variation from a perfectly squared plan. The western edge of the stair base [80] at the south-west of the building does, however, run exactly parallel with the division in the flooring, suggesting that the western wall of the building, at any rate, also followed this alignment. If this were the case, however, the line of the wall past the post-medieval cellar becomes problematic. It is inviting to see the outer face of this wall as running from the substantial westernmost stone of the southern stretch, suggestive of a quoin, along the stone facing on the easternmost side of the post-medieval cellar, but a wall strictly parallel with the pattern on the floor, taken from this same quoin, would overlap into the post-medieval cellar space, which would require the cellar to have gone out of use by this time. The wall on this side may well have been somewhat unstable in any case, even if it did follow the stone facing. Assuming that the suggested positions for the east and west walls are at least broadly correct, the building would have had an internal width of at least 4.60m and probably rather more.

Although primarily of clom construction, the building had footings containing a higher proportion of stone to provide support for the remainder of the structure, and some protection against splashing at the vulnerable foot of the walls. The area of southern Ceredigion and northern Carmarthenshire is one of only two in Wales (the other being the Lleyn Peninsula) where clom buildings survive in any numbers, and it is likely that the distribution was always limited, since the technique depends on the ready local availability of suitable clays. Wiliam (1988, 14–24) discusses the use of clom in the construction of cottages, together with the types of openings, flooring and roofing commonly found in such buildings. A stone footing was quite typical, while above this the clom material, made by puddling clay (usually including a proportion of small stones) and adding straw and even dung, was laid onto the walls in courses about 0.60m thick, often between wooden shuttering. The different walls of a building were not necessarily bonded to one another. The completed structure would usually be given a surface treatment of limewash, which would have been renewed at least once a year, to help keep the weather out. The single-storey cottages Wiliam discusses typically had walls *c.* 0.60m thick, and between 2.00 and 2.70m high; our building clearly had slightly thicker walls than this and may have supported a low second storey, which would have been normal for a farm building of the period.

There is little evidence for the doors and windows typically used at the time when clom buildings were originally built, since these features have been modified in most surviving examples. Written sources suggest that windows in cottages were normally either absent altogether, or very small, although gratings composed of interwoven sticks were sometimes used, while doorways were often closed by separate wattle hurdles, rather than hinged doors. Roofing was typically of thatch, though both the thatching material and the supporting timberwork varied depending on what was locally available. In the Dyfed area two forms of roof truss were attested: the scarfed cruck and the more ordinary A-frame. While the scarfed cruck was popular for clom buildings, because the timbers continued down into the wall, reducing the pressure from the roof upon its fabric, there is no clear evidence from our building as to which form of truss was employed. The thatch could be supported using a variety of methods, although often a number of closely-set thin poles were laid across the purlins, their lower ends sometimes being covered by an additional course of clom at the top of the wall. Wickerwork was sometimes used, or the thin poles might be laid horizontally like extra purlins instead of vertically. On this foundation would be set an underthatch, usually of gorse, over which the main thatch would be laid. Apart from the poorest properties, which used heather or bracken, this was usually of wheat straw; hanks of this would be knotted at one end and pushed into the underthatch with a special tool known as a *tobren*. The thatch was additionally pinned into place along the ridge, eaves and around the gable ends with straw ropes, although these were inclined to need replacing at frequent intervals.

The flooring of Wiliam's cottages (1988, 20) was normally of mud, which, properly laid, produced a durable, hard and reasonably clean surface. The mix used might consist of four or five barrowfuls of earth and one of good manure, with a bucket and a half of lime. This material would be broken up and sieved, and any stones removed, then laid wet and rammed down to form a layer 100 to 130mm deep; when the desired thickness had been reached, it would be stamped down by somebody wearing a pair of special flat wooden clogs. The floor would take several weeks to dry fully. Repairs were easily made by wetting and re-smoothing the material, and a polish could sometimes be achieved on the surface. The deposit [129], discussed in sections 2.2.3(Cii) and 2.2.6(Aiii) above, may well be the remains of such a floor, although it is uncertain whether it belonged to this clom building or to a predecessor of the same extent.

It is the pitched stone floor with its distinctive design which provides the main evidence for the function of the clom building. This floor is not far off flat, with a slight dip, averaging only 20–30mm, into the division in the pattern running the length of the building, and a fall of *c.* 100mm from the north end to the south. While the longitudinal division in the pattern finds no obvious parallels in domestic architecture, it is clearly very similar to that recorded by Wiliam (1982, 201–2, fig. 31, pls 26 and 27) in a cow-house at Hafod Adams, Glynceiriog (Fig. 2.105). In this example, measuring 9.50 by 5.00m internally, the cattle were tethered in pairs facing across the width of the building, with a passage to allow feeding along the length of one wall beyond the front panel of the stalls. The division in the pattern of the cobbling ran behind the beasts, in the position in which a channel to facilitate mucking out is often found. The animals would enter the building through a central door on this side, and a cobbled area was often provided on the outside as well, to minimise wear; these areas can sometimes be seen nowadays sitting 'high and dry' in a farmyard whose level has been much reduced by constant removal of muck. The building was lit by a pair of unglazed windows, provided with wooden bars arranged in a chevron pattern and solid hinged shutters, evenly spaced to either side of the doorway.

While this particular example provides the closest parallel for the arrangement of the floor (although the cobbles run in the opposite directions to ours), our building would not appear to duplicate the internal arrangements at Hafod Adams in other respects. It is worth noting, however, that stalling could often be secured to the ceiling, where it would leave little recoverable archaeological trace. The disturbances in the pitched stone floor certainly do not seem to suggest the presence of a feeding passage, unless the apparently worn area, perhaps hinting at a doorway, at the south-west corner points in this direction. A similar plan without a feeding passage is still quite common, and generally rather earlier in date (Wiliam 1982, 196–7); even so, the intervals

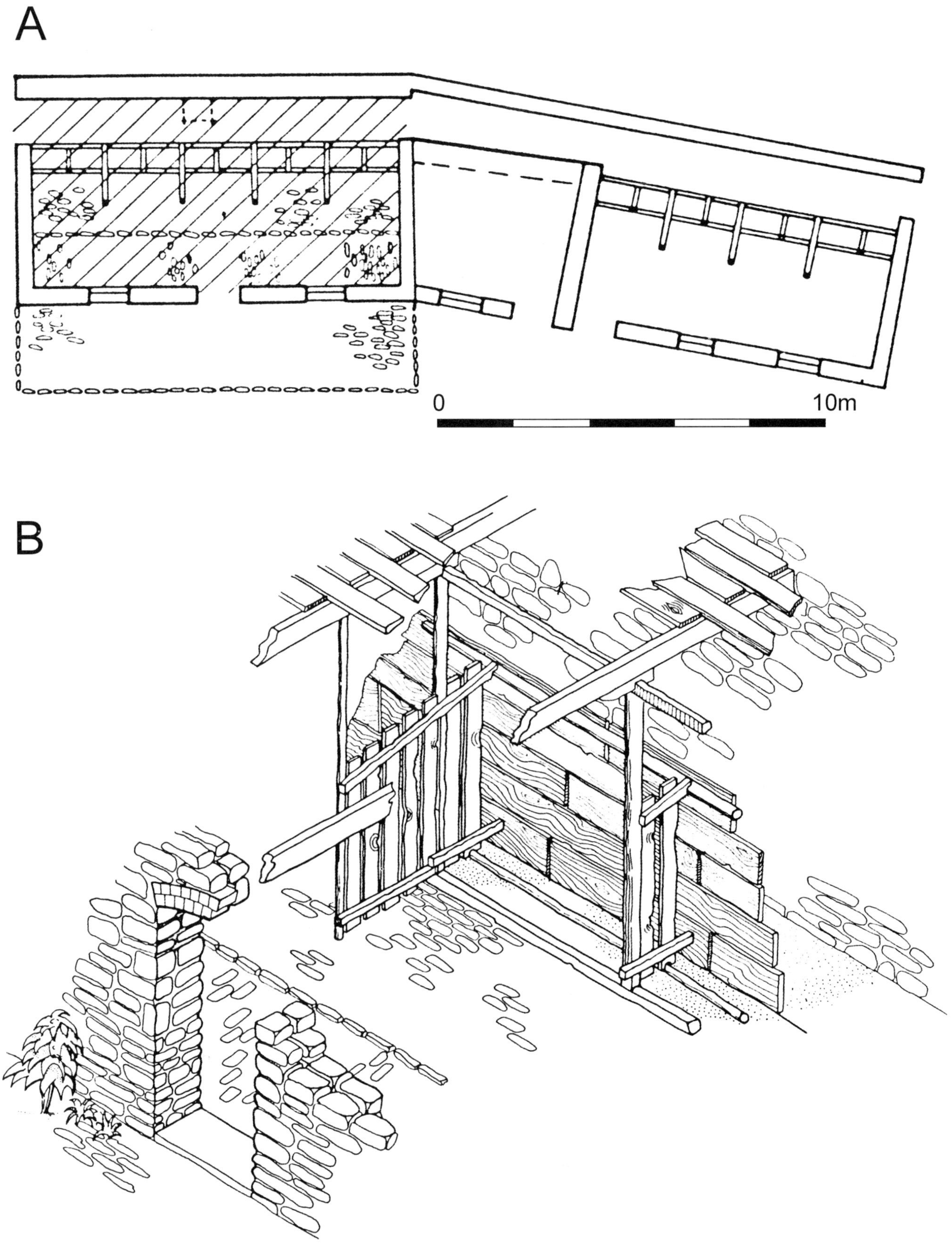

Fig. 2.105 Hafod Adams cowshed: (A) Plan and (B) Exploded sketch (after Wiliam 1982, fig. 31)

between the disturbances in our building cannot be made to fit the 1.90 to 2.00m module required for paired cattle stalls.

The distance from [44] to [68] (the most obvious contenders for the remains of stall partitions) is 1.35m which would, however, fall happily within the range of widths for single horses; the overall flatness of the floor might also suggest a stable rather than a cow-house (Wiliam 1982, 172). An interval of 1.35m between stall partitions can readily be applied to the disturbances in the northern half of the building (and hypothesised beneath the later stone wall), but further south the features suggest an interval nearer to 1.50 or 1.60m, which would still be acceptable. It is however possible that the layout at this end was originally different, perhaps incorporating one or more loose-boxes, and was altered, perhaps to provide larger stalls, at the time the concrete floor (which does not seem to relate to the earlier set of stalling) was inserted into the northern half. Loose-boxes allowed more space for the animal, which was not tethered, and were often used for carriage or hackney horses, which did not necessarily get the same amount of exercise as a working horse; they could also be used for delivery of foals or for sick animals. When present, they were often incorporated into the same stable as individual stalls (Wiliam 1982, 166–8; 1986, 130–2). There is room for (and features suggestive of) three stalls of the larger size between the edge of the concrete flooring and the southern end of the building; it is not altogether clear what use the concreted area was put to, although it may have been intended for cattle, which were increasing in importance relative to arable cultivation (represented by horses) as the 19th century progressed (Wiliam 1986, 83). Most of the stables recorded in the Clwyd sample had been converted to cow-houses (Wiliam 1982, 161–3).

Stables with stalls at right angles to a long wall were widespread on Wiliam's Clwyd farms (1982, 164–6). While most of them post-dated the Tithe survey of 1839, the earliest example, probably of this type despite later alterations, was at Hendre, Clocaenog, which bore a tablet of 1738. This is the most frequently found pattern throughout Wales as a whole, particularly on larger farms, and appears normally to have been intended for between four and six horses (Wiliam 1986, 130). For the most part, they still appear to date back no farther than the early 19th century, when the use of horses for farming was gaining ground over the earlier preference for oxen. Examples dated earlier than this were more usually of the other common pattern, where the beasts were stalled facing a gable end of the building, an arrangement which might well reflect a stage when horses were first brought into buildings designed for oxen (*ibid.*, 128). Wiliam (1982, 169; 1986, 135) reviews the range of agricultural literature regarding the size of stable stalls; horses got bigger during the first half of the 19th century, with earlier writers advocating a width of 1.20m per beast, and later ones 1.50 or even 1.70m. Partitions in stables were generally longer than those in cow-houses, up to 3.00m, with the 2.40m or so suggested by [68] falling well within the norm, and were also higher, so that horses could not see one another. If our building is indeed a stable, it might potentially relate to a fair-sized farm; alternatively, it may have had some function in relation to the pub next door. If this was the case, it may well have been the predecessor of the stabling (for six horses) provided across the road in the fine stone-built coach-house of 1833; the provision by about mid-century of the dairy in the southern extension provides a further pointer to a change of use for the building.

A stable would generally be somewhat better lit and ventilated than a cow-house, and would have a higher ceiling (Wiliam 1986, 132–135). There would usually be a loft, reached by an external staircase, which might be used as a fodder store, a granary, or, as was often the case in west Wales right up to the 20th century, as sleeping accommodation for unmarried male servants. Often there would be some arrangement whereby fodder could be dropped through openings in the floor straight into racks beneath. The fodder in the loft would help to keep the stable warm, while the animals in turn would provide a fair amount of warmth to the loft. Furthermore, if staff slept there they would be on hand quickly to deal with any problem, such as fire, theft or sickness, threatening the valuable horses.

It is not entirely clear where the access to any loft in our building would have been, and it could have been by a simple ladder, though this would have made the movement of fodder rather difficult. A further metalled surface [128], overlying those discussed earlier, adjoined the southern wall of the clom building but did not extend as far as any of the walls of the later stone structure. It contained a drain, which ran southwards from a point close to the south-eastern corner of the clom building for *c.* 1.10m, before turning westwards parallel with the southern clom wall, and which could be traced passing beneath the foundation trench for the west side of the later stone building. There would have been room between this drain and the wall for a stair base, although no evidence of one survived. It is not clear whether the metalled surface and drain related to an area of yard, or whether they represent flooring within a possible earlier structure at this south end. The only evidence for such a structure is the presence of a pair of robber trenches, both following roughly the same east-west line beneath the later stone-built Room 4, and both containing pottery dating to the late 18th century onwards.

The stone footing [80], to the south-west of the clom building, clearly intended for an external stair, does not lie entirely parallel with the later stone buildings and appears to align better with the postulated western wall of the clom building. If it does indeed relate to this building, it might provide further evidence for the earlier southern extension suggested by the robber trenches, though it is impossible to tell whether the entrance served was in the clom building or the extension. It is likely, however, that if not the same stair, then at least one on the same footing, was used to gain access to the upper floor, visible in the photograph of *c.* 1900 (see Fig. 2.71), of the mid 19th-century stone-built southern extension (Rooms 3 and 4) and possibly also, through it, to a loft above the stone-built Room 2. If the stair was in place before the demolition of the clom building, there could clearly have been no ground-floor door at this corner by this stage. The northern edge of this platform is only 0.50m from the edge of the cut for the post-medieval cellar, which may suggest that the latter had gone out of use by the time it was built.

(C) The 19th-century stone building sequence

(i) The southern portion (Rooms 3 and 4)

The piecemeal erection of the stone-built complex (see Fig. 2.68), which was demolished immediately before the excavation, appears to have begun in the mid 19th century, with the construction of the southern portion of the later building (for this and other details see the vernacular report by Darlington and Noake in section 2.2.3(N) above). This would seem to have been added as a southern extension to the clom building, while the latter was still standing. The photo of *c.* 1900 (see Fig. 2.71) also makes it clear that this extension had an upper floor, which was probably an original feature, and what appears to be a slated roof. By the 1950s, however, it had been reduced to a single storey (G. and D. Morgan, *pers. comm.*). As seen in 1989, it had a loft floor at eaves level, though the loft was not entered during the survey prior to demolition, as the vernacular report makes clear; the way into it by this date appears to have been through the pitching door high in the stone-built gable on its northern side. It is not entirely clear whether this wall was built as part of the extension or whether it was erected later when the central portion of the range (Room 2) replaced the clom building.

While the clom building appears from its pitched stone floor to have faced primarily east, and indeed its successor had its main emphasis on this side, the southern extension had its main entrances on the west. Prior to the 1970s, the road through the village ran some distance from our building, in front of the cottages opposite. The photograph of *c.* 1900 shows a stone boundary wall running alongside the road and enclosing an area to the east of the building, with an access gate adjoining the corner of the pub. It is likely, therefore, that stock could be corralled in the area to the east of the building, while the area to its west could be accessed independently. The arrangement may also have been intended to protect the cart and the other tools and equipment commonly stored in a cartshed (Wiliam 1986, 142–3) from the light fingers of passers-by on the road (the unusual presence of what appears to be a fastening for a door may point the same way; Wiliam 1982, 235). The evidence would seem to suggest that the area to the west of the building was roughly metalled from a fairly early stage, though it is not clear how the post-medieval cellar related to the other features in the area.

While the building was clearly originally of two storeys, and had since been reduced, it is not altogether readily apparent whether or not its ground plan had also been altered. The wall dividing the main body of the cartshed from the dairy area was not integral to the structure of the building, although the positioning of the 'arrowslit' ventilator on the east side centrally to the dairy portion does strongly suggest that the arrangement was planned from the outset. It is also clear from the photograph of *c.* 1900 that the door and window on the southern elevation, serving the dairy area, were already in position by this time, which might further indicate that the division of the dairy area into a larger, western and a smaller, eastern portion was also, if not original, relatively early. Some later improvements are likely, and it is possible that the generator for which a concrete standing was provided in Room 3 was linked with the need to power cooling and sterilising equipment in line with the demands of such regulations as the Milk and Dairies Order 1926 (Harvey 1970, 178). By 1989, the whole range had been supplied with electricity, and there was a piped water supply to the dairy and the northern extension; these facilities may well have been installed at the same time as other work in the early 1960s. Drains, leading west across the yard, had also been provided from the dairy and (at a late stage) Room 3; that from the dairy had a trap outside the building, again in accordance with hygiene regulations. The only feature noted in the dairy area at the time of the vernacular survey was a slate bench running along the northern wall of the western room.

The presence of a separate dairy appears somewhat unusual, to judge from Wiliam's comments (1982, 223–4; 1986, 24–5). More normally, dairying work, consisting of cooling and separating the milk and making butter and cheese from the cream, would have been carried out in the farmhouse; cheeses could be stored in lofts above the dairying area. The presence of a separate dairy in this case may indicate either that the farming enterprise was carried out by someone living at a distance, or that it was more convenient for those involved in running the pub to ensure that dairy products were kept well away from the hostelry operation. Whatever the case, the system had been well thought out, since the pigsties were conveniently located a short distance to the south-west of the dairy. Here buttermilk and whey, the waste products of butter- and cheese-making, would have been used to fatten pigs, which, with relatively little effort, would yield a variety of useful products.

Pigs were not normally kept in the farmyard before the 18th century (Brunskill 1982, 75), although it is possible that in Wales the change was later still. Some of the famous corbelled pigsties of south Wales (Wiliam 1986, 26–9) do date to the late 18th century, but the type represented at Pumsaint lies clearly in the 19th century. The typical 19th-century Welsh pigsty, of what is known as the 'in and out' type (*ibid.*, 29, fig. 20), consisted of a rectangular building divided into two halves across its shorter dimension, with walled yards of roughly similar size to the resulting rooms attached to one side. Access from room to yard would have been through a low (pig-sized) doorway. Typically the roof would be low, and often single or unequally pitched, although a fairly common variant had a hen-house, with separate access, forming an upper floor above the sty. The pigsties at Pumsaint, dated on architectural grounds to the mid to late 19th century (section 2.2.3 above), were converted in the 1950s into public conveniences, so that some of the original detail has been lost. They were a little unusual, in that the open yards were appended to either end of the building, which ran basically north-south, rather than along the long side. Each yard had an access gap on the east next to the respective corners of the building. The roof was unusually high, with a one and a quarter pitch, although there was no obvious evidence of there having ever been a hen-house in the extra height. The doorways from sty to yard were towards the west side, where the pitch of the roof came lowest. It is tempting to see the construction of the pigsties as part of the same plan as, and probably contemporary with, the redevelopment of the main range incorporating the new cartshed and dairy extension. The adaptation of the clom building to cater for some cattle (if this was indeed the case) may also have been part of the same overall scheme.

(ii) The central portion (Room 2)

Slightly later in the 19th century, the clom building itself was demolished and replaced with a new structure in stone. The vernacular report suggests that, on architectural grounds, this may have taken place in about the 1860s or 1870s; this coincides with evidence from estate records (NLW Dolaucothi 3411–3412), which certainly indicate a good deal of stone construction in the early 1870s, though it is not possible to identify this individual building. The stone building was slightly shorter at the northern end than its predecessor, but probably much the same width, giving overall dimensions of 9.60m by 5.30m internally. The southern wall of the clom building may already have been replaced in stone at the time the southern portion of the complex was built. The photograph from *c.* 1900 (see Fig. 2.71) suggests that this middle part of the building, while still basically single-storeyed, may at that time have been perhaps 0.30m or so higher than it was when seen in 1989, while the internal ledge and pitching door in the southern wall, together with the joist holes in the east and west walls, suggest that it may originally have had a loft. The loft was clearly removed at some stage, and the building as a whole would appear to have been lowered and re-roofed, perhaps at the same time as the portion to the south was reduced to a single storey, although the slates seen on the roof in 1989 may well be the original ones re-used.

The plan of the building is distinctive, incorporating three doorways on the east, the middle one of which lay opposite the single doorway on the west; smaller 'arrowslit' type ventilators on this side corresponded with the other two doorways. This layout is that of a well-known type of cow-house, designed so that the stalled cattle would stand along the length of the building in short parallel rows (Brunskill 1982, 60–66, figs 36–37; Wiliam 1982, 186–193, pl. 25; 1986, 92–5, figs 76–77). This basic type of design, used in the byre portion of longhouses, can be traced back to the later medieval period, while the earliest dated example of a free-standing building recorded in Wales, at Ty'n-y-celyn, Glynceiriog, bears a datestone of 1725. The basic type is illustrated (in the nose-to-nose variant) in the model farm plans of Waistell and Robinson, of 1827 and 1830 respectively (reproduced in Woodforde 1983, figs pp 53 and 23). By the later 19th century, it was declining in popularity in most areas (though there were wide regional variations) in favour of designs where the animals stood across the width of the building in a longer row or rows. In Carmarthenshire, however, some cow-houses of this general type were still being built at the end of the century (Wiliam 1986, 94–95).

While there was a long tradition of having some form of accommodation for milking cattle, particularly through the winter, provision before and even during the 18th century appears to have been patchy, to judge by the government's Agricultural Reports of 1794 (Wiliam 1986, 87–89). Throughout the 18th century, agricultural writers had debated the best way of utilising the manure from cattle, and the faction in favour of housing had gradually gained ground, although it was not until the mid 19th century onwards that much accommodation was provided for fatstock, and even then they were not always tethered in stalls like the milkers, but allowed to circulate in sheds or boxes. This development also in part reflected an increased emphasis on cattle as a result of the slump in corn prices which took place in the mid 1870s, when the market was flooded with cheap grain from the recently opened prairies of north America (Brunskill 1982, 29; Wiliam 1986, 83; Wade Martins 1991, 72–73). By the end of the century, however, the advent of refrigerated ships also allowed the import of cheaper meat, forcing the emphasis throughout Britain onto dairying. Fortunately the ever-growing cities and improved transport provided a ready market for liquid milk and milk products, for which there was a growing demand throughout the period (Harvey 1970, 177–78; Brunskill 1982, 29–30).

Our part of Carmarthenshire, however, had a rather longer tradition of milk production. Butter was collected together at Llandovery market and sent to Merthyr and Bristol as early as 1815, while skimmed milk cheese was also made for local use (Davies, quoted in Wiliam 1986, 82). Many parts of Wales, particularly the uplands, lacked the good pasture and winter fodder needed to fatten beef cattle, and these were usually sold and sent east with the drovers (*ibid.*). In view of this tradition, and the presence of the dairy in the southern extension, it seems most likely that the main emphasis of our building was on dairying.

Even for the milking cattle, winter fodder was always a problem. Hay was the most favoured, with oat straw as a good second choice. In poor seasons or marginal areas cattle might be given barley straw, gorse or bracken (Wiliam 1982, 180; 1986, 86–87). The limited supply always restricted the number of cattle which could be kept over the winter; even as late as 1896, farmers were complaining of the difficulty (Wiliam 1986, 91). Upland Wales seems to have been slow to benefit from developments such as the introduction of roots and advances in the manufacture of prepared cattle feed (Wiliam 1986, 72–73), which in much of Britain gradually removed this limitation and further encouraged the intensification of cattle rearing already noted. For bedding and litter, however, the bracken widely used in the uplands made better manure than ordinary straw, because of its high potash content (Wiliam 1982, 181; 1986, 92).

The basic design of cow-house described so far could appear in a number of forms, depending on how many cattle needed to be stalled, and on the ease of management desired (Fig. 2.106). The most basic, and probably earliest, examples needed only one door, through which animals, fodder, and manure would all have to pass. The animals would normally be tethered at right angles to the door with their heads against one, or sometimes, in the case of a larger building, both, gable walls, and would have to be fed from behind. In some longhouses, however, a single row of beasts would have faced, instead of a blank wall, the cross-passage between house and byre, from which they could be fed. An obvious development of this facility, where more cattle were kept, was to provide a second feeding passage in front of a second row of beasts, which would stand back-to-back with the first, giving the familiar plan with three doors on one side. The stalls in such a building could in fact be laid out with the cattle either facing each other across one central feeding passage, or back-to-back across a central mucking out passage;

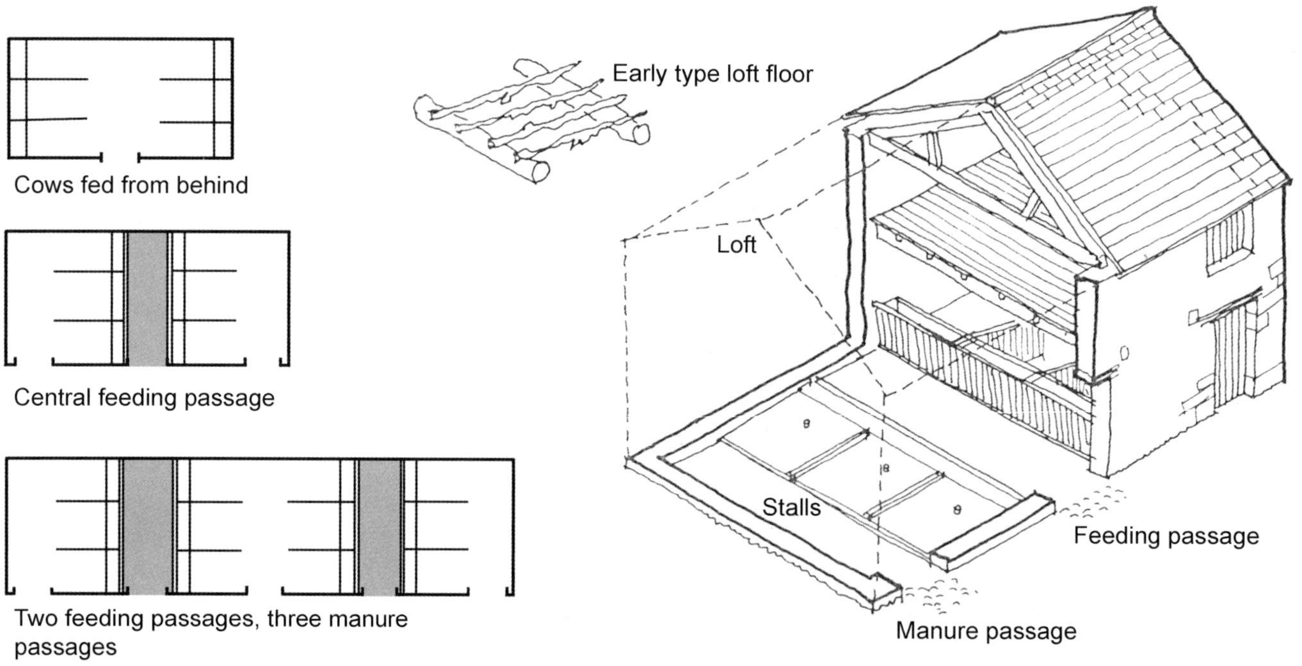

Fig. 2.106 Cowshed plans (after Brunskill 1982, fig. 36)

while the first arrangement proved initially more popular, considerations of animal health (it was considered undesirable for the beasts to breathe over one another) tended to lead to its gradual abandonment in favour of the second. In Wiliam's Clwyd sample (1982, 188) just over half of the cow-houses of this general type identified were designed for only a single row of cattle, involving for the most part only one door, while designs with three doors accounted for a further third of the examples noted, with nose-to-nose plans roughly twice as common as back-to-back ones with additional feeding passages.

Most early cow-houses were lofted, the loft being used for the storage of the fodder. Lofts providing less than 2.50m of headroom below were common (Wiliam 1986, 105), and although it is no longer possible to check the measurement, the loft originally provided in our building is unlikely to have been any higher than this. Loft flooring could often be quite rough. As the 19th century progressed, there was increased emphasis on the benefits of light and ventilation for the housed cattle (Wiliam 1982, 219–22); previously it had been thought that too much light made the beasts restless and encouraged them to fatten. This change of thinking may well be behind the removal of the loft in our building and the conversion of the eastern doors, and the northern 'arrowslit' ventilator on the west, into windows at some point. The doors had probably always been of two parts, enabling the upper part to open separately, but the through passage arrangement was also very conducive to through draughts, and ventilation in the roof was preferred by the agricultural writers. Some at least of the alterations to our building may coincide with the construction of the northern extension in the 1890s.

The basic dimensions involved tended to become standardised as the 19th century progressed, in part as a result of the stipulations (regarding space per beast) of the Cattle Disease Prevention Act of 1866 and its successors (Harvey 1970, 134) and in part as deliberate breeding led to the development of larger beasts. In 1874, the small black Pembroke cattle were crossed with the broadly similar Anglesey and Lleyn black breed to give the Welsh Black, which became the standard animal over much of west Wales in the years that followed (Wiliam 1986, 83). By the later 19th century, the standard width appears to have been 0.90m to 1.00m per animal, with a standard length of 1.80m. Animals in earlier cow-houses were often tethered singly (Wiliam 1986, 101), and this practice continued in some Carmarthenshire long-houses even as late as 1894, though stalling in pairs was gradually adopted as the norm (Fig. 2.107). While the stalls found at the time our building was demolished consisted of two pairs, with a passage to the side, there is in fact room across its width for five beasts.

Numerous variations on the basic layout were possible, though that suggested in the vernacular report, involving an arrangement of three rows of cattle all facing north, does not seem to be paralleled in the published literature. There is certainly room at the north end of our building for a third row of beasts, which would presumably have had to face north with their heads against the wall; this arrangement might be paralleled by an example from Llwyn Dedwydd, Llangwm (Fig. 2.108A), although in that case there was no passage at the other end either, giving only two passages to three rows of cattle (Wiliam 1982, 308, fig. 33). A rather earlier, but more up-market example from Nerquis Hall, dated to 1804–13 and designed as part of a mixed range, (*ibid.*, 318, fig. 43) had two rows of cattle facing one another across a feeding passage, with a 'calf cote' (which in this case had a separate entrance) behind one of them. Wiliam's closest parallel to our design (1986, 94–5, fig. 77) comes, however, from a longhouse at Nant-y-ffin, Llandeilo (a mere eight miles, as the crow flies, from our site), recorded by the Royal Commission on Land in Wales and Monmouthshire in 1896 (Fig. 2.108B). In this case, the cross-passage forms one of the feeding walks, in the way already described; two rows of five cattle, all stalled individually, stand back-to-back, with a manure passage between them, while a further area beyond a second feeding walk appears to be used as a pen. A similar stalling arrangement was in use at Lan a couple of miles to its south (*ibid.*), though here the area beyond the second feeding passage was used as a barn (Fig. 2.108C). Back-to-back plans of this basic type appear to be particularly prevalent and long-lived in the Carmarthenshire area, and some complex variants were built. In most cases, however, where three rows of cattle were accommodated, the third row was stalled the opposite way round to the second, and a further passage provided; this would give a plan with four doors, which preserved the distinction between feeding passages and manure passages.

While our building clearly belongs to a readily recognisable type, therefore, and contains sufficient space to stall 10 or possibly 15 beasts, there is no clear evidence to show what the original layout of the stalling was. There is no trace of any really coherent flooring until after the eastern doorways had been blocked, while the only set of stalling recovered is the north-facing set, comprising two pairs, to the north of the cross passage, which survived before demolition in 1989. It would seem surprising, however, if such a spacious unit was built solely to house the four beasts attested by these stalls. Examination of the ground levels recorded after removal of the most recent concrete surfaces is unfortunately rather inconclusive, though it does provide some possible pointers. South of the cross-passage, there is a drop of about 70mm towards the southernmost passage, while the cross-

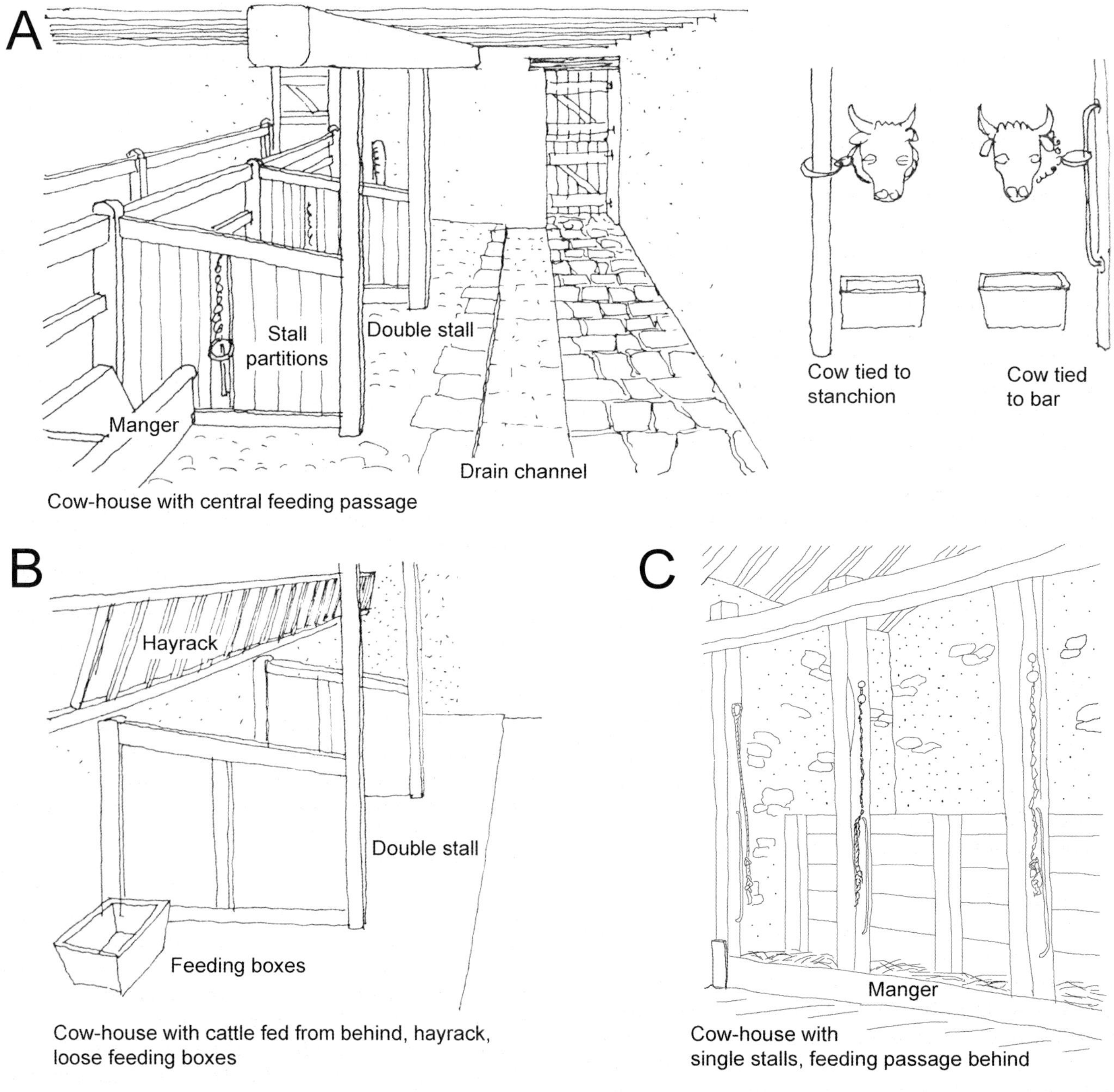

Fig. 2.107 Cowshed partitions (A and B after Brunskill 1982, fig. 37; C after Wiliam 1986, fig. 84))

passage itself has a drop of 160mm towards the western door. The area to the north of the cross-passage, where the stalls survived, was somewhat damaged during demolition, but appears to have a drop of about 120mm southwards towards the cross-passage; its northern half is virtually level, however, and the drop is particularly pronounced at the south-west corner, just north of the west door. To the north of the northern passage, there is again a drop of about 60mm southwards towards the passage.

It is tempting to wonder whether the stalling in the building was originally designed with two rows of beasts, possibly tethered singly, nose-to-nose across the central cross-passage, with either a third row or a pen occupying the northernmost area. Virtually no evidence for such a layout survives, however, unless the post-hole [48], immediately to the south of the shale partition bordering the southern side of the cross-passage, was part of such a scheme. The shale or slate partitions on either side of the cross-passage might themselves have formed part of such a set of stalls (Wiliam 1986, 102 and fig. 87). The actual stalling arrangements need not have left any archaeological trace; indeed Wiliam (1986, fig. 84) gives an example from Cil-hir Uchaf, Talog, Carmarthenshire, where each cow was tethered to one of a set of evenly-spaced uprights set into horizontal members running along the floor and the ceiling (see Fig. 2.107C). Other

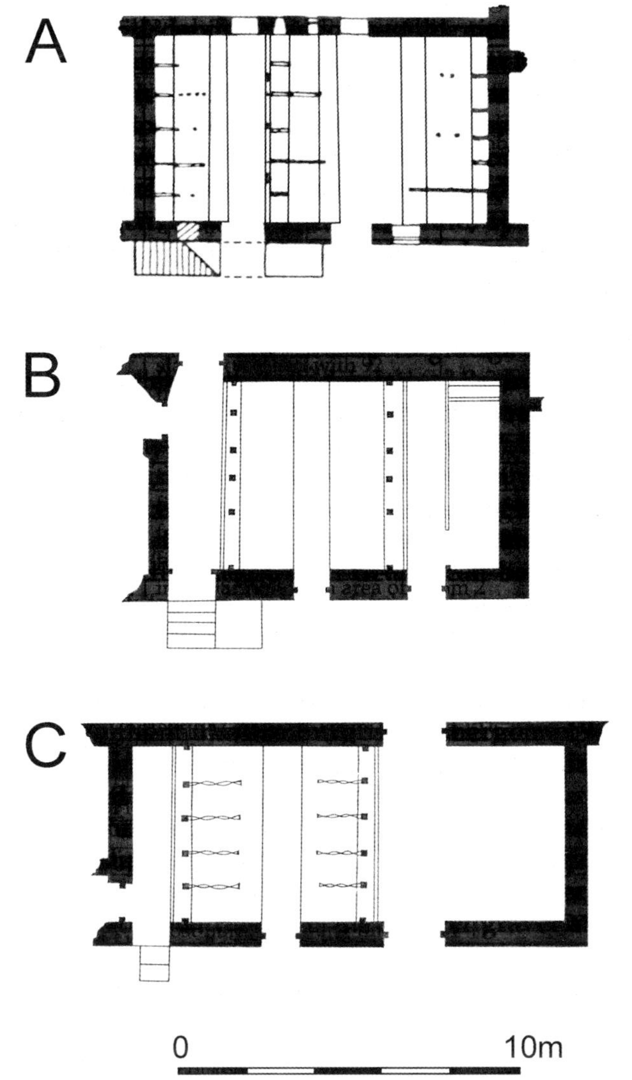

Fig. 2.108 (A) Llwyn Dedwydd (after Wiliam 1982, fig. 33), (B) Nant-y-ffin (after Wiliam 1986, fig. 77) and (C) Lan (after Wiliam 1986, fig. 77)

variants involved an upper horizontal at a lower level (*cf.* Wiliam 1986, fig. 86). These would be set a little way back from the front of the stall area, leaving a space between the lower beam and the front partition which could be used as a manger.

It is uncertain how long any such arrangement would have lasted. The construction, probably in the 1890s, of what appears to be further cattle accommodation in a northern extension of the building (Room 1) may have provided the impetus for a change of plan, though the two pairs of stalls (facing south) for which there is evidence in this northern building, of very similar dimensions and broadly similar construction to those (facing north) which survived in our building in 1989, may not be the earliest. The eastern doors of Room 2 are unlikely to have been blocked until some time afterwards, since Room 1 was built with two doors on the east (and one on the west); the blocking and conversion to windows of all five eastern doors appears relatively uniform and is likely to have been carried out as a single operation. It is of course possible that the two eastern doorways of Room 1 were built already blocked to preserve the appearance of this side of the range, but the provision of the window between them would tend to militate against this idea.

It is however possible that the northern 'arrowslit' ventilator in Room 2 was converted to a window, if not at this stage, then not much later, the new stalls (dated in the vernacular report to the late 19th or early 20th century) constructed, and the loft floor removed; all measures which would have 'modernised' the accommodation according to the thinking of the time. This would have meant that fodder would either have to be fetched down through the pitching door from the upper floor of the southern extension, which would make it awkward to have a row of beasts stalled in the southern part of the room, or would have to be brought from elsewhere. The initial plan of the northern section (Room 1) suggests that fodder may at that stage have been brought into the building from the east, but clearly once the doors were blocked, a western source would have made more sense. There seems, according to local information, to have been a Dutch barn in the north-west corner of the farmyard, where fodder could have been stored. This survived up until some time in the 1960s or '70s, though its date of construction is unclear. It is shown on the 1:2,500 OS map and traces of it may be represented by the large pits [398] and [400] in the western section of the excavation at this end. Dutch barns were not commonly built in Wales before the second half of the 19th century, and were usually of timber posts and corrugated iron (Wiliam 1986, 112–15).

The northern part of Room 2 appears to have been used as a pen or pens for quite some time, if not from the outset. The exact arrangement is not entirely clear; the area certainly seems to have been demarcated latterly by the partition [6], which appears to have post-holes defining doorways at either end. The easternmost of these posts, however, projects slightly across the northernmost eastern doorway, and while this may not have been enough to cause a problem, it may indicate that the partition in this form post-dates the blocking of the door. An earlier partition may have extended between the disturbances [44] and [46], visible in the underlying pitched stone floor of the clom building; arrangements secured to ground-sills and/or to the loft are of course quite feasible. At a later stage again the two round uprights [42a] and [42b] appear to have divided the pen into two, and drains were provided in each part. There are traces of pitched stone flooring in the eastern section, while the western had been concreted, probably in the

1960s. The eastern pen would therefore have been about 2.10m square, while the western measured roughly 2.10m north-south by 3.10m. The dimensions and position of this area compare quite well with the pen at Nant-y-ffin, Llandeilo (see Fig 2.108B), which was just under 2.00m wide and about 4.60m long (the width of the building). There was no sign, however, that that example had been sub-divided. Pens of this type are most likely to have been used for calves, some at least of which would have been kept to ensure the continuity of the milking herd. While Wiliam (1982, 203–9; 1986, 106–9) states that the received wisdom of the time was that calves should not be kept in the same building as the cattle, and could cite few if any instances of it, this certainly was the case in the northern extension (Room 1), where the last farmer to use the shed could remember the area to the west of the stalls being so employed.

After the eastern doors had been blocked and converted to windows (sadly the photo of *c.* 1900 (see Fig. 2.71) does not show this area clearly), the northern cross-passage and the area to the west of the stalls were given a flagstone floor; this would have provided access for feeding and to the pens. This flooring may well originally have been more extensive. The paired stalls must have been in position by this time at the latest; the slope on the floor towards their south-west corner suggests that they were well used. The wear is likely to be the result both of the passage of stock and of regular mucking out. There is little sign of much other change to the building in the first half of the 20th century, apart from the apparent lowering and replacing of the roof, which may well have taken place at the same time the southern extension was reduced to a single storey, some time before the 1950s.

Otherwise the main developments were the concreting of the western pen, the cross-passage and much of the area around the stalls, said by Darlington and Noake to have been in the early 1960s, although it may well have been rather earlier. The rebuilding of the outer skin of the western wall, which had obviously failed, probably because of the presence of the adjacent post-medieval cellar, in concrete blocks, does however probably date to about this period.

(iii) The northern portion (Room 1)

A further stone-built unit was added to the north end of the range, probably in the 1890s, according to the vernacular report, to judge from the architectural features. As already indicated, this was provided with two doors on the east, rather wider than those of Room 2 to the south, with a window between them, and one door on the west, at the northern end. A further small window lay on the west, opposite the southern of the eastern doors. The building does not appear ever to have been other than single storeyed, and had a small skylight halfway up the roof on the east side, towards the south end. The two eastern doors were later blocked, and converted into windows, in the same way, and probably at the same time, as those of Room 2 to the south.

Despite the spacious and well-lit design, the internal fittings are clearly intended for cattle rather than horses; there is no sign that the structure was ever anything other than a cow-house. It would probably have been state-of-the-art when built, conforming to the latest ideas discussed in the preceding section about light and space for the beasts, although the original stalls do not seem to have left any trace. There are traces of a set of two paired timber stalls adjoining the eastern wall and facing south; these would appear to be of very similar design to those which survived (at least in part) in Room 2, and it is likely that they were installed at a similar date. Once again there is room for a fifth beast on the western side, but to judge from the pitched stone flooring, which continues right across the northern passage and up this western space, this standing too, like the westernmost one in Room 2, was used as a passage rather than a stall. The arrangement of the building is rather unusual, in having two passages for one row of cattle; Wiliam recorded no examples of the type (1982, 190; 1986, 94), unless the variant where the through passage in a longhouse was used for feeding, is counted. Once again, however, there is a parallel in Robinson's model farm plan of 1830 (reproduced in Woodforde 1983, fig. p. 23), where the single-row unit is marked 'sick cattle'.

It is clear from the plan that the cattle must have faced south, making the southern passage a feeding walk and the northern one the manure passage. Unless there was access round the stalls from the west from the outset, fodder must have been brought in on the east of the building, though it is likely that the Dutch barn on the north-west of the yard provided the main source once the eastern doors had been blocked. The cattle too must have been corralled in the western yard by this stage.

This part of the building seems to have remained in regular use the longest. At some stage, the wooden stalls appear to have been concreted around, though the work was rather more sophisticated than that carried out in Room 2, and at a later stage again, probably the early 1960s, according to the memory of the last farmer, the building was completely remodelled with concrete and metal stalls, still facing south, against the western wall and a passage on the east. Although there are some signs of modernisation in the dairy section of the building, there are no obvious traces of any mechanical milking equipment, although it is likely that something of the sort was used, at least in conjunction with this final remodelling of Room 1. The installation of running water for drinking troughs, and probably also mains electricity, is likely to date from this time. The work may well have been in response to the introduction of the Farm Improvement Scheme in 1957 (Harvey 1970, 219), an initiative which provided advice and grants to farmers for the modernisation of farm buildings. Room 2 was clearly out of regular use as a cow-house by the time all this work was done; it is not altogether clear when Room 1 too fell into disuse, but it is likely to have been some time in the later 1960s or 1970s, when the building became a general outhouse for the pub.

2.3 EVALUATION WORK AT PUMSAINT, 1993

The small-scale excavations discussed here were undertaken in response to an application by the National Trust for Scheduled Monument Consent for proposed works within the fort at Pumsaint. This application had been necessitated by pressure from the local community for the restoration of the public conveniences which had been demolished in the wake of the 1989 excavations. The proposal involved a single-storey building, measuring *c.* 7.00 by 4.00m, with a connection to the water and drainage services in the highway.

The site chosen lay immediately north of the 19th-century coach house on the eastern side of the A482, immediately opposite the Dolaucothi Arms Hotel (Fig. 2.109). This placed it in the centre of the Roman fort, where one would expect the remains of the original headquarters building and the projected line of the defences of the second-phase fortlet defined by Jones and Little (1974), and made it likely that the planned building would involve damage to the archaeological horizons. For this reason, Cadw advised the National Trust that they could not proceed further with the application until small-scale excavation had been undertaken to define, if possible, the top of Roman archaeology and to relate this to ground levels within the site and to the adjacent

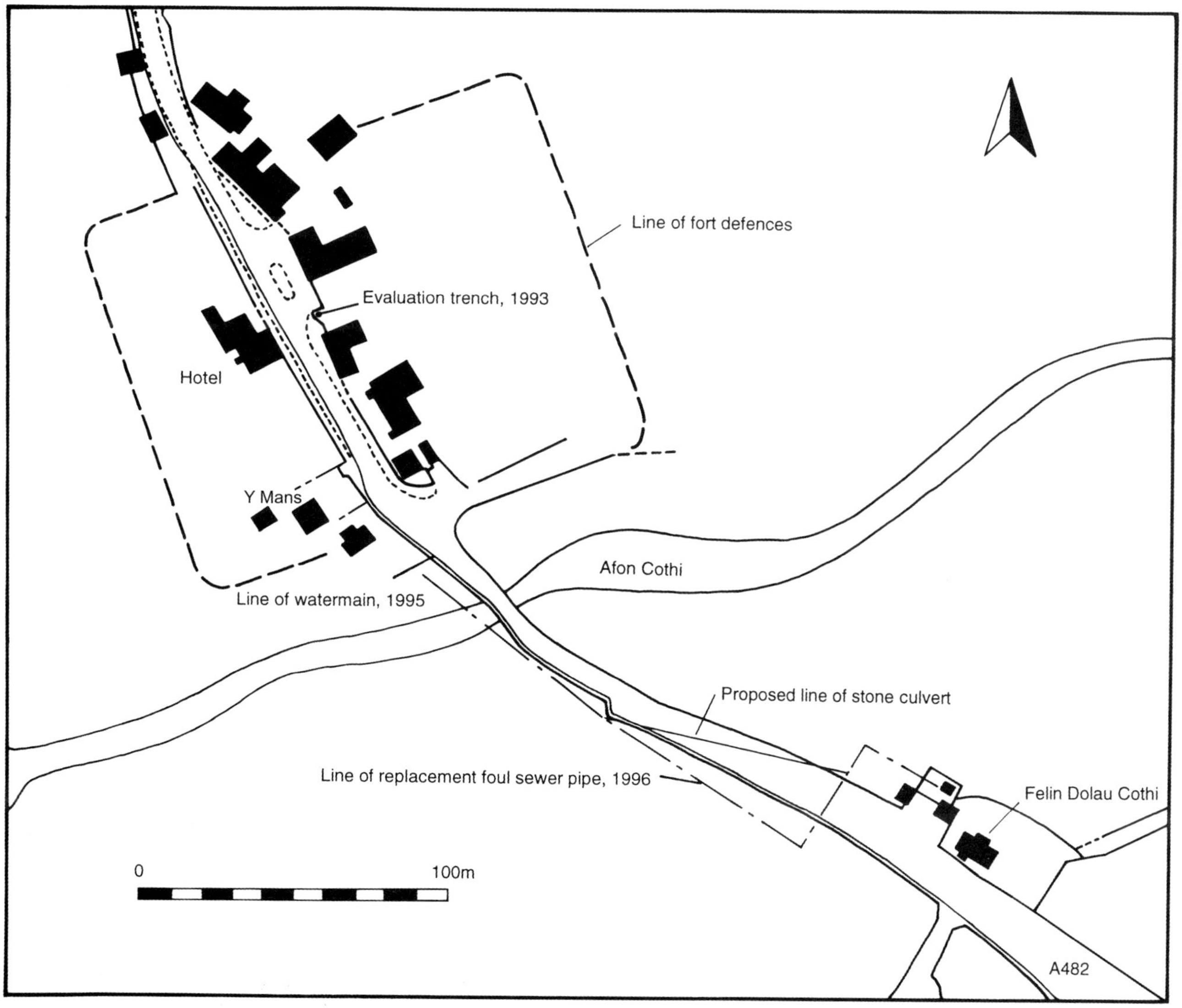

Fig. 2.109 Location plan of 1993 trial work and 1995/96 watching briefs

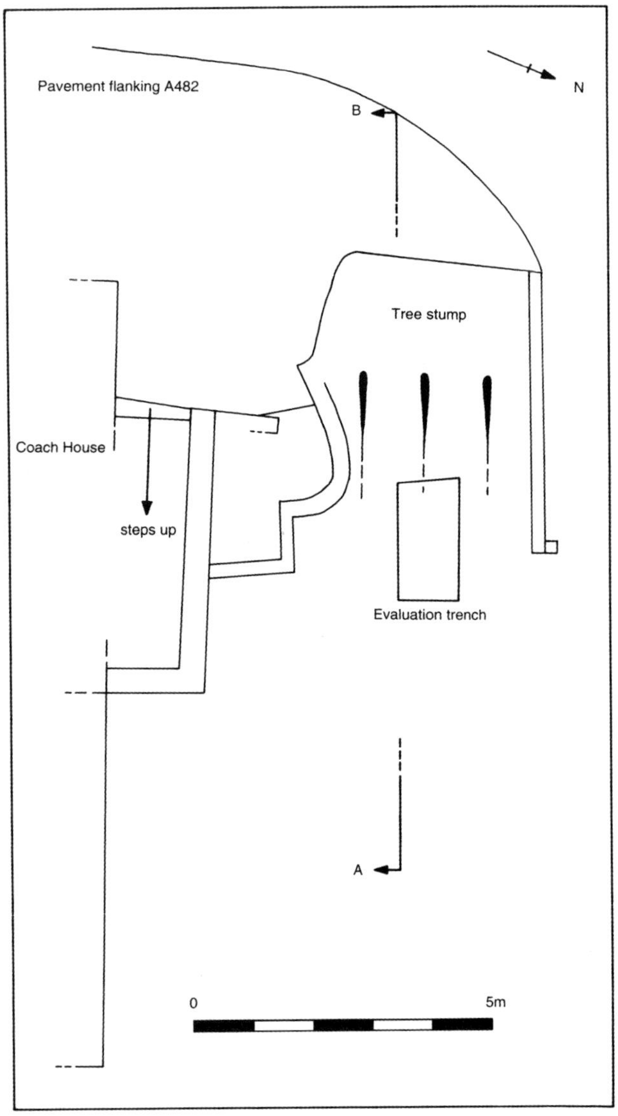

Fig. 2.110 Plan of 1993 trial work

pavement. This work was undertaken by Barry and Helen Burnham in January 1993.

Within the site, a 2.00 by 1.00m trench was excavated to the east of the surviving tree stump (Fig. 2.110), into the build-up which had accumulated at this point to a depth of *c.* 1.00m above the level of the pavement (131.70m OD). The surviving deposits above context [5] proved to be somewhat uninformative in archaeological terms, and to be exclusively associated with relatively modern finds. These contexts were recorded as follows:

1 a mixed upper surface incorporating breeze blocks, redeposited cobbles and a thick spread of sawdust.
2 a brownish-black loam or 'garden soil' (10 YR 2/2), with 10% small stones.
3 a spread of medium to large stones extending into the south section which appeared to have been laid on the underlying horizon, but with no obvious structure.
4 a brownish-black loam or 'garden soil' (10 YR 3/1) very similar to [2], with 30% small and medium stones which increased in density towards the base of the context. This layer had been much disturbed by tree roots.

Below context [4], the picture changed, with the discovery of a distinctive surface of rounded pebbles, medium-sized stones and gravels [5], set in an olive-brown matrix of loamy coarse sand (2.5 Y 4/3). This intact archaeological horizon, sealed beneath 'garden soils', was very reminiscent of the sequences identified to the west of the A482 in 1987 and 1989, where such a transition was a very good indicator of the top of Roman archaeology (see section 2.2.1 above). For this reason, no attempt was made to excavate into this horizon and so no diagnostic finds, Roman or otherwise, were recovered to corroborate the hypothesis. The context did, however, contain traces of the charcoal and degraded daub which are so characteristic of the Roman levels elsewhere.

Careful levelling of the surface of context [5] demonstrated that it lay at a height of 131.67 to 131.70m OD, much the same as the kerb of the adjacent pavement (Fig. 2.111). This is somewhat

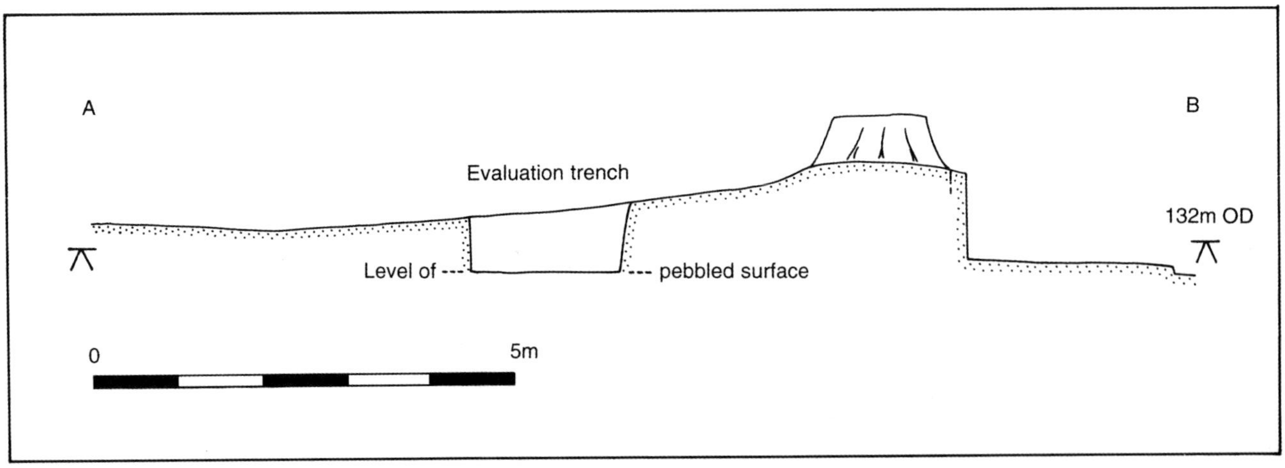

Fig. 2.111 Section of 1993 trial work

higher than the comparable Roman levels recorded across the road to the west, but one has to make allowances for the slope of the original ground surface and for the fact that the Roman levels closest to the west side of the A482 had been extensively truncated. The discovery of an intact, possibly Roman, horizon had inevitable implications for the development proposed in the Scheduled Monument Consent application. As a consequence, no action was taken and the site has been reinstated with a wall fronting the pavement.

In so small an area, a definitive interpretation of the evidence is impossible, but the most plausible hypothesis is that we are dealing with a yard or road surface. While we cannot entirely discount the suggestion that this was related to the adjacent post-medieval building, the site does lie close to the projected line of the defences of the second-phase fortlet (see Fig. 2.1), suggesting the possibility that it represents the remains of an *intervallum* road running to the rear of the rampart structure. Only larger-scale excavation could hope to clarify this point for certain.

2.4 WATCHING BRIEF AT PUMSAINT, 1995

The watching brief discussed here was undertaken by the Dyfed Archaeological Trust in September 1995, during the laying of a watermain by Welsh Water along the line of the A482 through Pumsaint. Staff at the Trust had first become aware of this work in May 1995, by which time operations had progressed to within *c.* 100m of the north end of the village. In view of the known archaeological potential of the site, the Trust made contact with the engineer in charge of the project in order to obtain details of the works. Welsh Water subsequently agreed to a watching brief through the village, to ensure a record of any features of archaeological significance. The work was directed by Ken Murphy, on whose report this account is based (Murphy 1995).

Within the village the trench for the main was excavated in the modern road, *c.* 2.00m to the west of centre, beyond which, to north and south, it ran into the roadside verges to the west (Fig. 2.109). The trench was 1.10m deep and *c.* 0.35m wide in the road, shallower in the verges. It was machine dug, with the spoil deposited into a lorry for off-site disposal. Feeder trenches to houses were *c.* 0.50m deep. Only a short length of trench was excavated prior to the laying of the hard-core cushion, insertion of the pipe and backfilling with hard-core. Because of these working methods and the narrowness of the trench, it was not possible to examine the sides closely, nor to make detailed drawings. It was equally impossible to examine spoil as it was dug out, hence the lack of artefacts recovered during the watching brief.

For much of its course through the village the excavation of the watermain trench lay within modern makeup, laid as part of road-widening operations in the mid 1970s. Throughout much of its length, therefore, the following stratigraphic sequence was recorded:

1 0 – 0.30m. Modern tarmac.
2 0.30 – 0.50m. Coarse modern hard-core.
3 0.50 – 0.70m. Fine- to medium-sized brown and orange-brown compact gravel.
4 0.70 – 1.00m. Brown and orange-brown gravel with patches of orange-brown silty loam, possibly a buried soil.

At the southern end of the village, however, opposite Y Mans, more complex and interesting archaeological deposits were revealed in the trench cutting (Fig. 2.112). A substantial foundation trench [1], containing cobbles and rounded boulders packed in a white-grey clay, was butted on its north side by a stone-free orange-brown clay [2] and on its south by a possible ditch or gully [3] with a loose gravel fill. To the south of this, what appears to be a clay-filled trench [6] was overlain by a single course of massive masonry. Beyond this and an area disturbed by an old pipe trench, a layer of very compact gravel [7] may well represent a Roman road surface.

The interpretation of these horizons is not without its problems, not least because conditions for archaeological recording were far from ideal. There can be no doubt, however, that the trench as a whole ran obliquely through an area known to be occupied by the north gate of the fort, the postulated central range of buildings, and the proposed line of the two-phase south defences (*cf.* Jones and Little 1973; 1974; see also section 2.2.1 above). Indeed, the watermain actually passed through a 1970s excavation trench on the site of the north gate. At this end also, the modern road is known to overlie the line of one of the Roman

Fig. 2.112 Section of 1995 work adjacent to Y Mans

roads through the fort, though nothing was actually discernible in section, as the lower levels of the modern makeup, any possible Roman levels, and the subsoil all consisted of the same compact brown or orange-brown gravel.

Opposite the hotel car park and Y Mans, the trench must have cut through or skirted the east side of the granary excavated in 1972, across an area which ought to lie on the projected line of the second-phase defences (see Fig. 2.1). Their eastern side, represented by a stone wall, was certainly recorded in Trench 7, east of the A482, while what was taken to be the south-east angle was reconstructed from the curving traces of a possible *intervallum* road, cutting obliquely across the demolished south-east corner of the granary (see Fig. 2.2). Such a sequence could account for some of the remains identified in 1995, especially the substantial foundation trench [1], with its distinctive fill, and the stone-free clay [2] to the north, both of which are reminiscent of the second-phase stone wall and its associated clay backing as recorded in 1989. This would presume, of course, that we are dealing with the corner of the fortlet. The only other alternative is that the various features represent part of the granary itself, the full extent of which was not established in 1972, or part of a neighbouring structure.

If the former interpretation were accepted, however, the south defences of the fortlet would clearly have to follow a different line from those of the first-phase fort, which were otherwise located in 1972, *c.* 20 to 30m to the south in Trench 6. While this would not be impossible, it would be difficult to reconcile such evidence with the section in Trench 6, which certainly seems to record a classic two-phase defensive sequence, including a wall of ashlar blocks set in a foundation trench with a clay backing to the rear. Confirmation of this line is also provided by the remains of the gravel surface [7], which must surely represent the *intervallum* road known elsewhere in the circuit of the fort. Any resolution of these issues must await opportunities for further excavation at the critical points, though this seems unlikely in the immediate future.

2.5 WATCHING BRIEF AT PUMSAINT, 1996

The watching brief discussed here was undertaken by the Dyfed Archaeological Trust in February 1996, during the replacement of an existing foul sewer pipe to the south-east of the village of Pumsaint. The work was commissioned by Dinefwr Borough Council, on the advice of the National Trust, to ensure a record of any features of archaeological significance. The replacement pipeline crossed the river Cothi, its associated flood-plain and relict river terraces, running roughly parallel to the A482, which it traversed in the vicinity of Felin Dolau Cothi (Fig. 2.109). The work was carried out by Belinda Allen and Ken Murphy. This account is based on the report by Allen (1996).

The works involved the cutting of a machine trench, *c.* 260m long, on average 2.00m wide and 2.00 to 3.00m deep. It was excavated in contiguous sections, with those of interest being left open for detailed archaeological recording, prior to the instalment of the box frames. The depth of the narrow trenches restricted access, but both sections were examined and recorded throughout the work; the south-west facing section proved the more informative. Spoil and areas of stripped topsoil were also regularly scanned for archaeological evidence.

The majority of the deposits encountered were of natural origin, following much the same alluvial sequence throughout. This included: (i) a combination of buff and red/brown clay loams of varying thickness, underlying the topsoil; (ii) a band of river gravels, averaging 1.20m thick, composed predominantly of Silurian shale in line with the local solid geology; (iii) at the base, an homogeneous blue/grey silty clay alluvium, at least 1.00m thick. Throughout, the water table lay at the interface of the gravels and the impervious clay alluvium.

Changes within the stratigraphy were noted at the far north-west end of the cutting, on the north side of the river. Here, the earliest feature, recorded in the south-east facing section, comprised a partially observed cut feature *c.* 2.50m deep and 2.00m wide, the upper sides and top of which had been truncated by later cuts. The base of this feature was not reached by the machine trench, but it was clear that it cut the river gravel deposit. Its infill comprised blue/grey silty clay, heavily leached and containing abundant river gravel. It had been truncated to the north-east by the old foul sewer pipe, and in the south-west end of the trench by a U-shaped cut, *c.* 1.50m wide and 1.20m deep; the fill of this latter feature comprised friable red/brown sandy loam, containing mixed angular and sub-angular stones of varying sizes. No archaeological material was recovered from the deposits in the section.

South of the river, a stone culvert was located at two points, one on either side of the A482. This had apparently been constructed in a substantial foundation trench, dug down to the top of the impervious silty clay, onto which the drystone barrel vaulting of shale slabs with an internal facing had been laid. The clay clearly served as the base of the drain, as no other form of lining was observed. Various shale slabs in the basal infill would seem to represent the subsequent collapse of the vaulting, intermixed with a highly organic matrix containing fragments of oak planking. A second, distinctive layer, above this, contained large shale slabs within an inorganic silty clay. The uppermost fill comprised a very friable sandy gritty loam, with subangular shale fragments and distinct tip lines of grit and angular gravel. This latter is interpreted as the backfill of a trench dug to house a modern storm drain.

These features add new detail to our knowledge of the Pumsaint area, though their interpretation is not without its problems. Particular interest necessarily attaches to the substantial cut features located at the far north-west end of the site, in view of the known proximity of the defences of the Roman fort. The infill of the secondary U-shaped feature, in particular, did not resemble any of the deposits seen elsewhere in the trench, nor those described by Burnham and Burnham (1987a) in the eroded river section some 30m away, to the north-east of the bridge across the river. According to the excavators, however, there were some characteristics in common with previously excavated deposits associated with the second-phase defences, when the existing ditches were backfilled to make way for a secondary stone wall (Jones and Little 1973). If the deposits exposed during the watching brief did relate to second-phase alterations, then the large feature could represent a more extensive first-phase circuit than hitherto postulated, though such evidence would be difficult to reconcile with the accepted evidence in Trench 6 along the southern edge of the garden of Y Mans (see section 2.2.1.above). Pending further excavation, this issue will not be resolved.

As far as the culvert is concerned, there is cartographic evidence (Tithe Map 1841) for a post-medieval mill with the same name as the extant buildings at Felin Dolau Cothi, east of the A482; a leat is also shown to run from the river Cothi in the north-east to the corn-mill in the south-west. Though the recently discovered culvert is not marked on any of the maps studied, its alignment and the preservation of wood in one of its fills strongly suggest an association with the said mill.

2.6 GEOPHYSICAL SURVEY ON THE EAST SIDE OF THE FORT, 1999

2.6.1 Introduction

Of the four sides of the fort at Pumsaint, that on the east is the least well-known. Two possible ditches, both rather shallow, were located in the cricket field during trial trenching in 1973, though neither looks particularly convincing in section (Jones and Little 1973, fig. 3). There is also the unresolved question of the extent of any associated settlement, which might have been expected to develop in the vicinity of the fort/fortlet, paralleling those at better known military sites elsewhere (Sommer 1984; Davies 1991b). Limited evidence in support of this had been forthcoming from the area south of the fort, in particular from the eroding river section east of the modern bridge, where the remains of various clay-floored buildings and pits have been recorded (Lewis and Jones 1969, 260; Burnham and Burnham 1987a). In this area, much has obviously been lost to river action, but just how much will never be known. The same is true along the west side of the fort, where a tributary of the Cothi, the Twrch, has even encroached upon the north-western defences.

To the north and east of the site, however, nothing of significance had been recorded prior to 1997, even where there was ample level ground on the gravel terrace north of the Cothi. In the wake of the geophysical work on the settlement area south of the river (see Chapter 5), a trial survey was undertaken in the western part of the cricket field in late 1997 and early 1998, as part of a student training exercise. Promising results were obtained, suggesting the presence of an extensive area of settlement, but without proving the line of the eastern defences with any certainty. Given the problems that this latter would pose for our understanding of the fort phase at Pumsaint, the chance was taken in August 1999 to carry out a much more extensive programme of geophysics, with finance provided by the National Trust. Significant new evidence has thereby been added to our knowledge of the Pumsaint fort and its environs, even if problems of interpretation remain.

2.6.2 Geophysical survey, 1999 (by David Jordan)

Magnetic and electrical resistivity survey was carried out under the direction of David Jordan of Terra Nova Ltd. across the whole of the cricket field, which now forms the southern part of OS parcel 7265 east of the modern village. The magnetic survey was carried out by David Jordan, and the resistivity by Helen Burnham and a team of Lampeter graduates. The aim of the survey was to confirm, if possible, the line of the fort's eastern defences and to map the variety of archaeological features which might be expected to lie both within and beyond its confines. It was also intended to detect any significant changes in the gravel terrace surface which might indicate the condition of preservation of the remains.

The magnetic survey was carried out using a Geoscan FM18 twin fluxgate magnetic gradiometer with readings taken at 0.25m along traverses spaced at 1.00m. **The resistivity survey** was carried out with a Geoscan RM15 resistance meter using a twin-electrode array. The survey was carried out twice using two mobile array widths, 0.50 and 0.90m, which has the effect of measuring to two different depth ranges. The narrower array is sensitive to *c.* 1.00m and the wider array to about 1.80m, although there is no absolute limit in either case.

All the data were automatically logged in the field

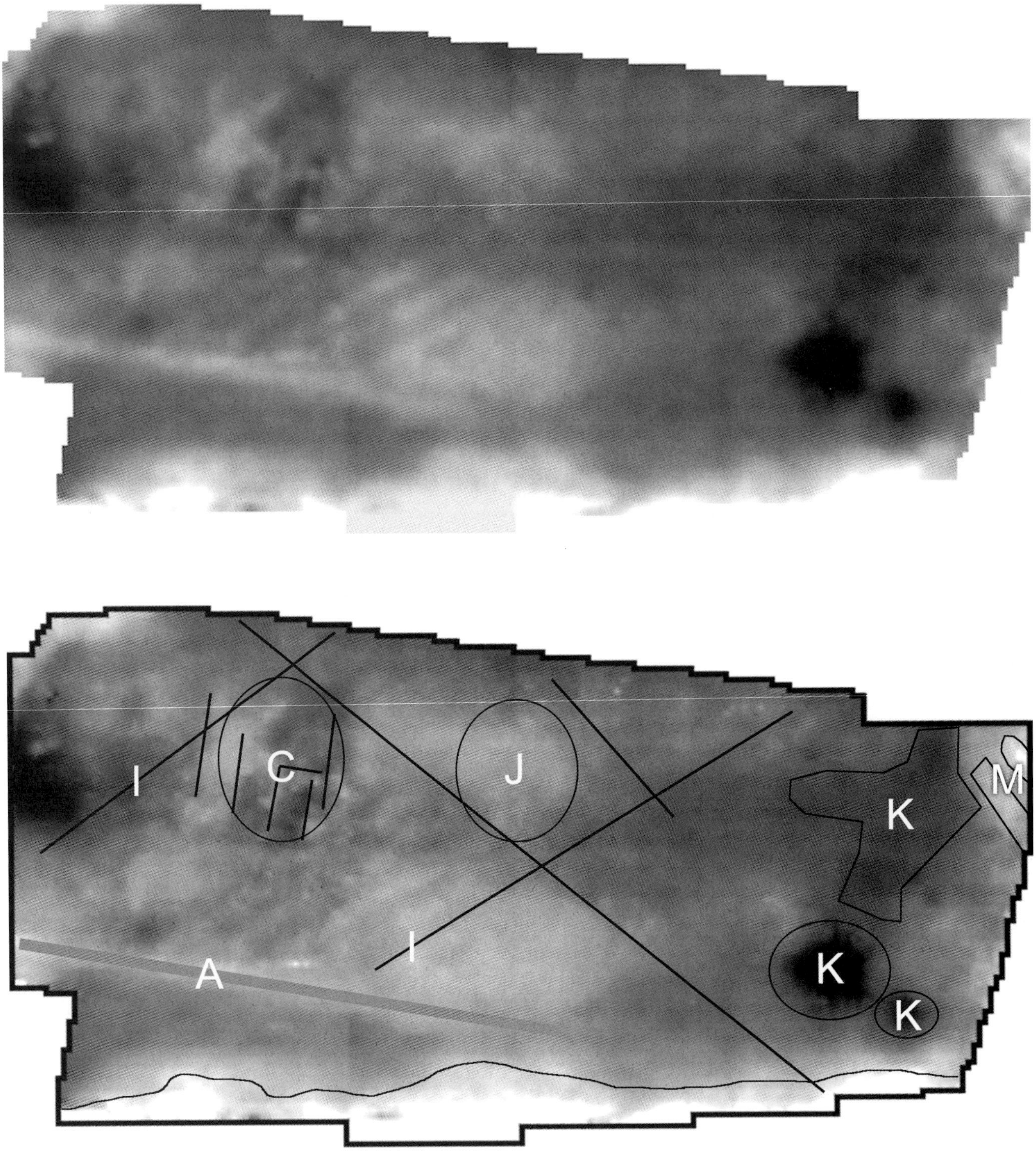

Fig. 2.113 Electrical resistance survey. Top: results – light greys indicate high resistance, dark greys lower resistance. Bottom – results with interpretative overlay

and processed by computer in the laboratory. The magnetic data were corrected to remove position errors and reduced to a common mean. Magnetic noise due to metal debris was removed and the data stretched with 5% saturation, which removes the highest and lowest peaks but greatly improves the contrast between the remaining 95% of the data points. The resistance data were corrected to a common mean and stretched in a similar way to the magnetic data and with a similar improvement in contrast.

The **results** are presented graphically, the electrical resistance in Fig. 2.113 and the magnetic in Fig. 2.114.

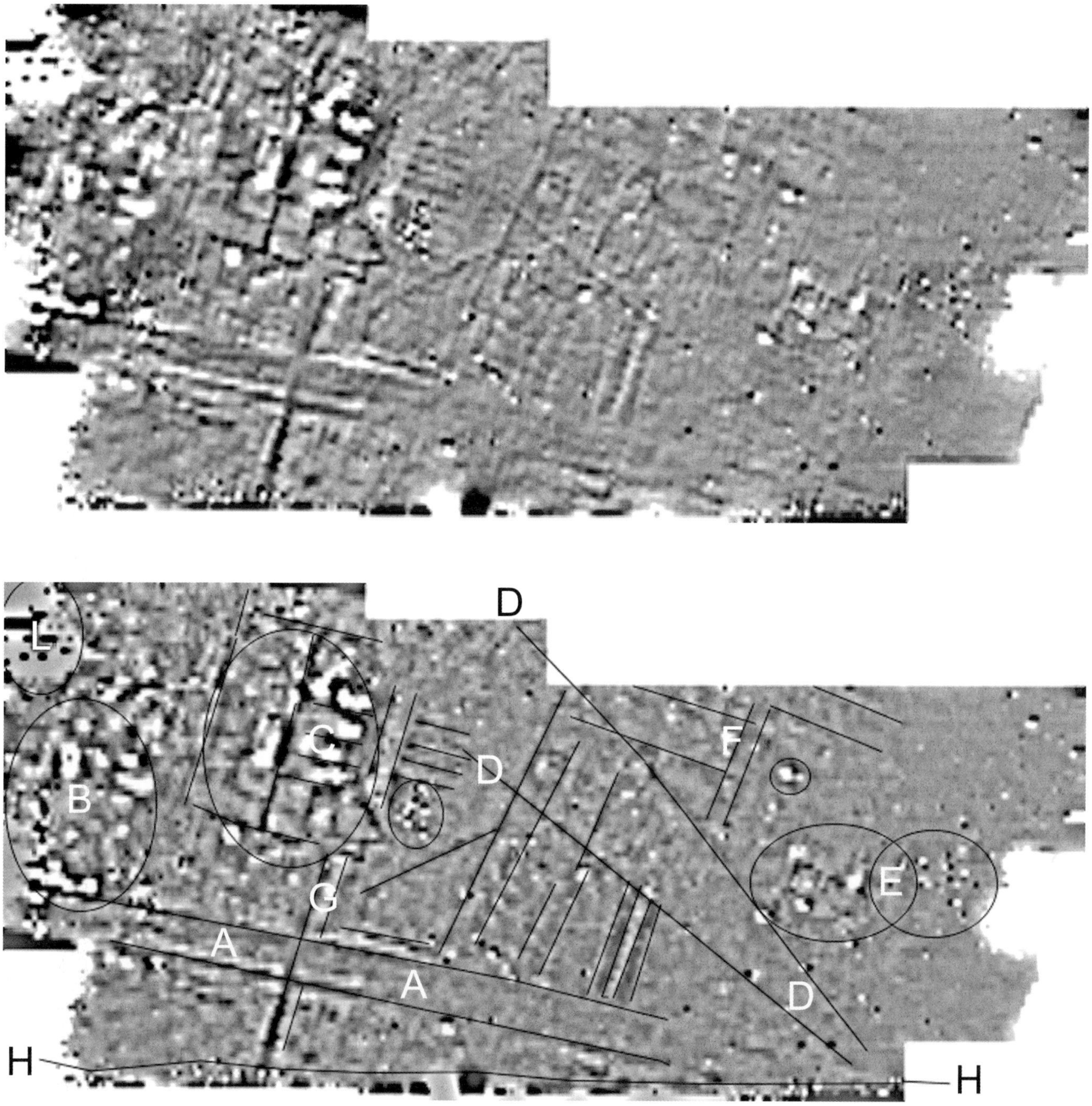

Fig. 2.114 Magnetometer survey. Top: results – high positive magnetic distortions are shown as light greys and high negative distortions as dark greys. Bottom – smoothed magnetometer results with interpretative overlay

Both surveys proved successful; the images show a range of archaeological and natural structures, although the magnetic survey has been more sensitive to the archaeology and the resistance survey to the superficial geology. As with most successful surveys, the images largely speak for themselves and this interpretation is meant only as a guide to the major structures. The whole area is crossed by archaeological features, except for the eastern quarter of the field, and we may presume that dense archaeological features are preserved over most of the field.

A major boundary structure (G) – perhaps a bank

and external ditch – appears partly to overlie a roadway (A), which may be of gravel construction, leading out into an area (F) containing abundant, regularly structured buildings. Some of the buildings along the boundary line (around C) appear to be built of stone but most, both inside and out, seem to be defined by ditches and were probably built of timber. Some strong magnetic anomalies (B) corresponding with areas of low resistance inside the fort may indicate floors surviving within deeper soils while other very local anomalies (E) may be due to Roman industrial refuse, perhaps in pits.

The resistance survey shows clearly that the site is crossed by later cultivation ridges (on alignment I) and by trenches (D) which probably carry modern pipes. Readings along the southern edge of the survey (H) are significantly disturbed on both surveys showing the magnetic effects of the boundary fence and resistance effects of soil disturbance.

The survey shows a number of higher resistance areas (J) which may correspond with ridges in the gravel surface. These are likely to have been eroded by cultivation and it is possible that only negative features survive. Lower resistance areas (K) correspond with areas of deeper, wetter soil where positive features are more likely to be preserved. The very strong magnetic anomaly to the north-west (L) is due to the metal swings of the play area. The strong positive resistance anomalies to the north-east (M) may be due to recent structures and possibly to the foundations of a building.

It appears that the site contains evidence for several construction phases. The magnetic results, in particular, suggest that boundary structure (G) may have been disturbed by the construction of building (C) while (G) appears partly to overlie roadway (A). The structures to the west of (G), however, seem denser and more magnetically disturbed which suggests that this may have been the focus of most activity even if further development occurred to the east.

2.6.3 Discussion

The two surveys clearly combine to give a useful impression of a site which, being scheduled and unthreatened by modern agriculture, is not susceptible to investigation by traditional excavation methods. One of the pressing questions on this side of the village remains the location of the fort's eastern defences, postulated otherwise only from limited trial trenching by Jones and Little. For this reason, the discovery of a substantial boundary structure (G), extending roughly north-north-west to south-south-east across the southern part of the field, assumes a measure of significance (Fig. 2.115), as this is the only feature in any way resembling a possible defensive line (perhaps a bank and ditch) anywhere in the survey area. At first glance it would seem reasonable to associate this with the eastern side of the fort, even though its line is obscured towards the northern part of the field by the anomalies associated with the large building (C). Such an interpretation is seemingly reinforced by the fact that it coincides fairly closely with the line of the defences as predicted from the known layout of the fort and from trial trenching.

There are problems with this interpretation, however, which are incapable of resolution on current evidence. One is the absence of any definable features relating to an east gate in the predicted central position or of any associated roadway leading out from such a gate. This might suggest that we should associate the visible roadway (A) with one of the main roads of the fort, exiting presumably via one of the principal gates, but if so it would have to be either the *via praetoria* or the *via decumana*, in which case it should be central to the defences on this side; this is patently not the case, unless we accept that our current reconstruction of the fort is incorrect. Traces of the original east gate might well be obscured, however, by the overlying building (C), while the absence of the road might well not be too significant if the fort actually faced west onto the Llandovery to Llanio road, which would thereby have been the natural focus for any related civilian developments (for a discussion see section 6.2.3(A) below).

A second problem concerns the fact that the boundary structure (G) appears partly to cut across the visible roadway in the magnetometer survey, though not apparently in the resistivity results, which would suggest a secondary date incompatible with what ought to have been a primary feature on the site. While there is a suggestion that more than one feature could be involved at this point, the reconciliation of the resistivity and magnetometry results remains a high priority. If a secondary date were proven, however, this would have significant consequences for our understanding of the eastern side of the fort, as it would involve relocating the defences to somewhere west of the current survey area, which is otherwise devoid of any appropriate features; such a relocation would effectively call into question the very nature and extent of the fort as currently reconstructed.

Leaving these problems aside for the moment, the visible roadway (A) looks very much like an extension of the road presumed to lie between the central headquarters building and the granary to the south, the line of which has recently been confirmed by a watching brief during work on the Coronation Hall (Page 2003). As such it would be tempting to see this as a consequence of the decision to reduce the fort in size to a fortlet, when it may well have been considered convenient to extend an existing road into an extramural area via a newly-created gate in the fortlet's defences. This would mean that many of the features recorded in the geophysical survey should represent the remains of timber and stone buildings (C and F)

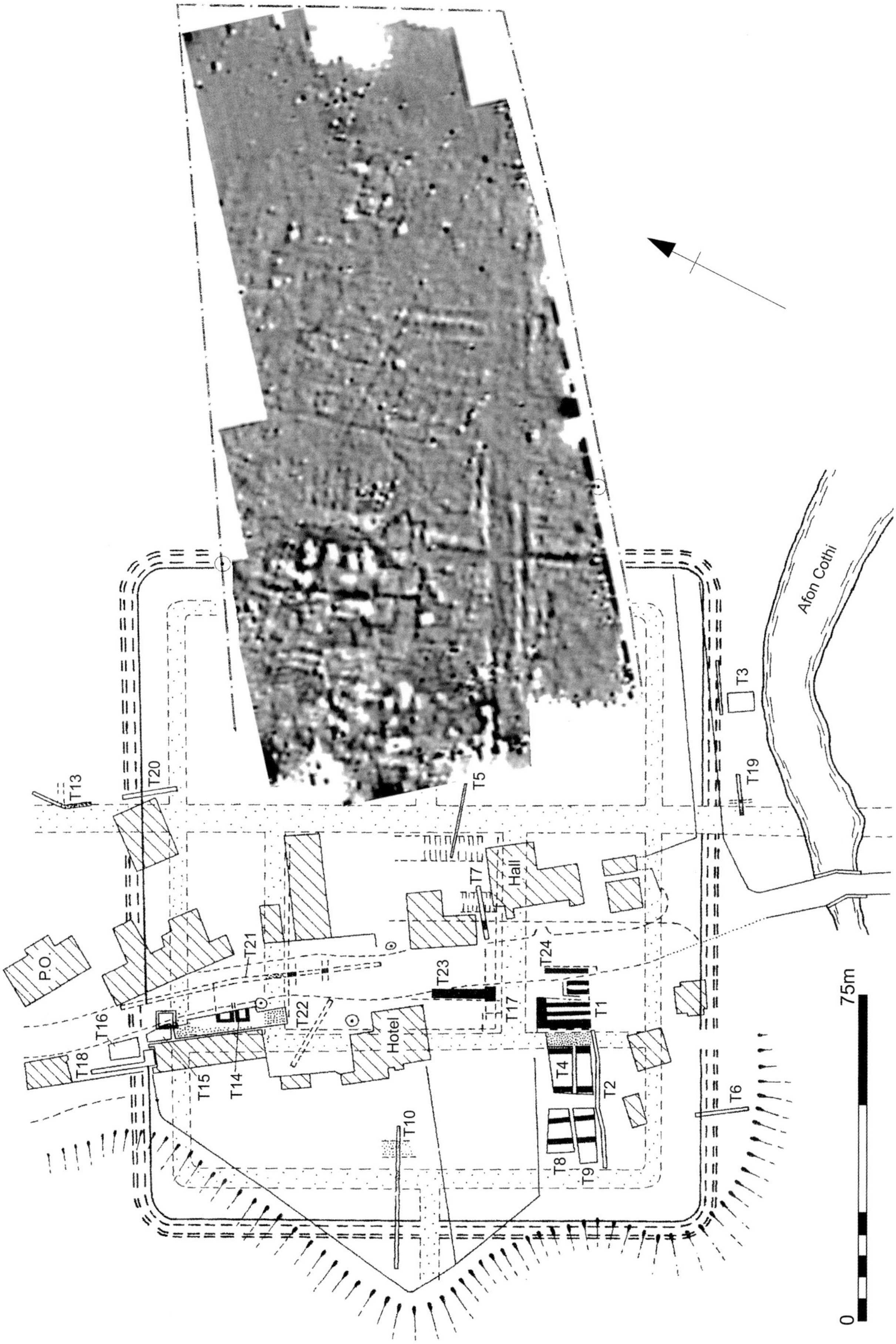

Fig. 2.115 Plan of the fort showing integrated results of excavation and geophysical survey

Fig. 2.116 Aerial photograph of building under cricket field, looking west (APDAT 95–90.29). Reproduced by permission of Cambria Archaeology

associated with the fortlet phase, since their layout is predicated upon the existence of the roadway. Hitherto, the lack of any significant development on this side of the fort might be a reflection of the overriding importance of the main road which by-passed the site along its western side.

Most of the recognisable buildings would seem to have been of timber construction (F), set inside plots defined by ditches, but one at least (C) was clearly a much larger, stone-built structure, apparently set into the ground to judge from the resistivity data. This latter building is also visible as a parch mark on a pair of aerial photographs, taken in August 1995, as part of Cadw's aerial monitoring programme of scheduled sites (Fig. 2.116). Although difficult to interpret, the linear arrangement of the rooms with several possible side chambers, taken alongside the high magnetic anomalies, might suggest a bath-house of characteristic row type, with associated plunge baths. The high readings might then indicate the survival of floors or underfloor heating arrangements. This would be a significant addition to our knowledge of the site, especially if it could be shown to be certainly associated with the fortlet phase; in view of this, any opportunity to assess the level of survival by excavation would naturally be welcome, despite the lack of any immediate threat.

The discovery of the later cultivation ridges (on alignment I) in the resistivity data, on an alignment which does not conform with that of the village, adds another fragment to our wider understanding of the evolving landscape of the Dolaucothi area, particularly as traces of such post-Roman, pre-modern ploughing regimes have also been located in work south of the river (see section 5.4 below).

Much obviously remains speculative in the absence of excavation, and the picture presented here can do no more than attempt to reconcile the postulated configuration of the fort/fortlet sequence with the new evidence derived from geophysical survey. Some complexities cannot be resolved by geophysics alone, however; for the future, an extended programme of survey using ground-penetrating radar and electromagnetic techniques might usefully add to the picture, as would a campaign of carefully located trial trenches across the cricket field.

3 SURVEY AND EXCAVATIONS ON THE LEAT SYSTEMS, 1990 AND 1993

3.1 INTRODUCTION

The existence of at least two major leat systems supplying water to the gold mines at Dolaucothi has, not surprisingly, been a source of considerable interest in the literature in view of their presumed Roman date (see Fig. 1.2); the Cothi leat has been known since the 18th century at least (Jones *et al.* 1960), while the so-called Annell leat was first recorded only in 1969 (Lewis and Jones 1969). Both are seen as supplying water to a series of tanks and reservoirs above the mine workings, with the water being used apparently in the processes of hushing, fire-setting and quenching, hydraulicing and ore-processing (*cf.* Bird 1984; 2001). Despite the potential importance of the Annell system, however, its surviving remains have never been published in detail, with the result that questions have been raised in recent years about its exact character, date and function, prompted in the main by the discovery of the Gwenlais leat in 1988 (Bick 1989). The publication of this volume, therefore, provides a useful opportunity to review the current state of our knowledge about the Annell-Gwenlais system, alongside the detailed results of two separate, small-scale investigations, undertaken by the authors in 1990 and 1993 (Burnham *et al.* 1993; Burnham and Burnham 1994). The chance is also taken to publish an appendix of some recent aerial photographs of the Nant Dâr leat, a subsidiary supply recorded, but not published in any detail by Lewis and Jones, on the hillslopes above the main Cothi leat.

3.2 THE ANNELL-GWENLAIS SYSTEM

The presumed traces of the so-called Annell leat are best discussed in two sections - those in the upper reaches of the river Annell, on the flanks of Cefn y Bryn, and those further down the river, extending along the flanks of Allt Cwmhenog down to the mines (*cf.* Lewis and Jones 1969, 254 and 266; Jones and Lewis 1971a, 292–4). In the upper reaches of the Annell (Fig. 3.1), the leat was first encountered as a series of shallow channels in the vicinity of the so-called 'maintenance hut' at SN 707 426, with which they were thought to be associated (Hook 1971). They were said to be tapping an extensive area of marshy ground from which the river itself ultimately drew its water. From this point the leat was seen to descend along the northern side of the valley, which here forms a steep-sided gorge, to a prominent series of earthworks now protected as a scheduled ancient monument at SN 699 426. This section was said to comprise at least four parallel gullies, descending steeply across the contours of the slope onto the lower-lying and marshy ground on the col above Llwyn y Ceiliog farm. At the time of the original survey, it was thought that one of the gullies may have been aiming to cross the col into the Cothi valley, though no specific surface remains were adduced in support of this hypothesis and the idea was subsequently dropped in later publications.

Beyond this, the middle and lower reaches of the leat along the flanks of Allt Cwmgerwyn and Allt Cwmhenog proved to be very difficult to trace, having supposedly been destroyed in the former area by a forestry plantation and in the latter by extensive ploughing. The apparent exception was a well-preserved stretch, now also protected as a scheduled ancient monument, which was located at SN 674 405 on Gwarnoethle farm. It was said to comprise a series of three parallel leats, the upper one of which was seen to diverge uphill to tap a local spring. Excavation across these features certainly seemed to confirm their identification as silt-choked channels with traces of a watertight clay lining (Lewis 1977, fig. 6). Beyond this point, the line of at least two of the leats is said to have been seen on aerial photographs, swinging around the spur of Allt Cwmhenog in the direction of the mines and the large tank (G) at the head of the Allt Cwmhenog pit (SN 667 404), though the relevant references have never been published; a possible line has also been recorded around the spur of the hill during geophysical work by Cardiff's (Mining) Field Centre (Alwyn Annels, *pers. comm.*).

Following the initial surveys in 1969, little new work was done to extend our understanding of this important feature. The exception in this context was

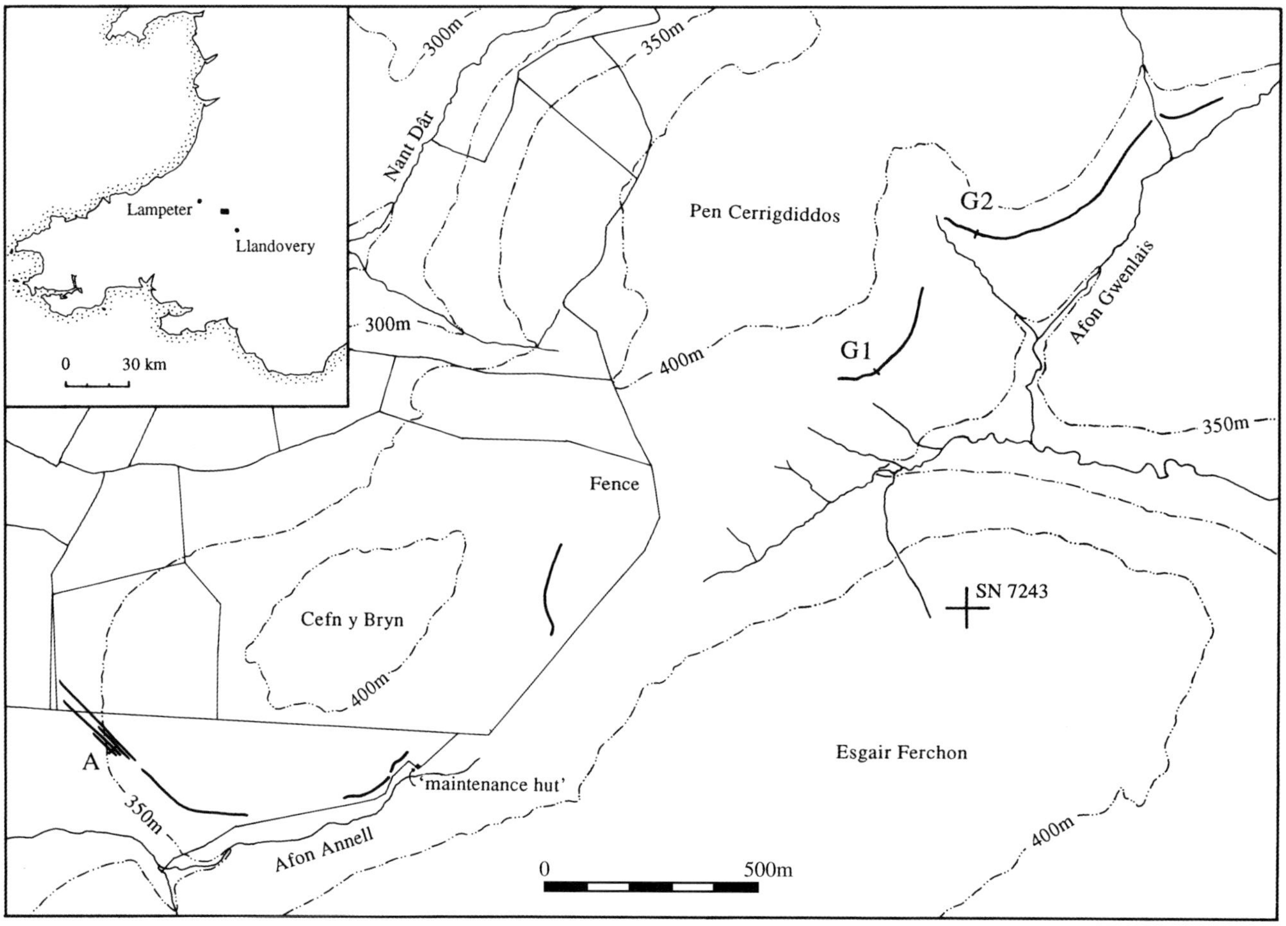

Fig. 3.1 Location plan of Upper Annell and Gwenlais system

the field survey carried out by David Bick (1989) in the upper reaches of the rivers Annell and Gwenlais (Fig. 3.1), which not only questioned the interpretation of the parallel gullies near the start of the Annell (SN 699 426) but also drew attention to a further, hitherto unrecognised leat, apparently extending *c.* 3km beyond the postulated end of the Annell, in order to tap the headwaters of the river Gwenlais. This latter proved to be surprisingly well preserved on the sloping ground closest to the Gwenlais, where there was clear evidence both for an earthen bank on its downhill side and for several rock-cut sections, but its course was less easy to trace in detail where it had to traverse the lower-lying and boggy ground between the Gwenlais and the headwaters of the Annell. As a result, any presumed relationship between the Gwenlais leat and the upper reaches of so-called Annell system could not be established for certain, despite a very strong suspicion that they were actually one and the same feature.

Bick's discovery of the Gwenlais leat inevitably raised several interesting questions about the nature of the so-called Annell leat, the inter-relationship of the surviving elements of both systems, and the destination of the Gwenlais leat, the details of which have been a matter of some debate (Bick 1989; Jones and Maude 1991; Bick and Boon 1993; Bick 1996). The discovery also provided a context for a renewed, if limited, campaign of survey and excavation, which was undertaken during 1990 and 1993.

3.3 THE ANNELL AND GWENLAIS SECTIONS, 1990

Given the importance of the Gwenlais leat and the need to ensure its protection as a scheduled ancient monument, as well as the obvious question marks over the upper reaches of the Annell, the authors were asked by Cadw to undertake small-scale excavations across both systems in order to establish their profile and character. This was duly carried out in July 1990, when two small sections were cut across the line of the Gwenlais leat, one of them in an area where it was clearly rock-cut, while a third, longer section was excavated across the two lowest gullies in the scheduled area, near the presumed start of the Annell; it had been hoped to extend this latter uphill across the rest of the features, but this proved impossible in the time available due to the inclement weather conditions. In the event, the sections across the Gwenlais proved especially rewarding, not least because they provided environmental evidence for recent landscape changes in this part of mid Wales.

3.3.1 The Annell Section, 1990

The surface archaeology of this scheduled ancient monument is represented by at least four parallel gullies which descend obliquely and at a relatively steep gradient across the prevailing contours of the local topography. They were first identified in 1969, when they were interpreted as being a series of parallel leats designed to help the flow of water down a particularly steep section of its course. This hypothesis has gained widespread acceptance in the literature on the gold-mines, despite the lack of any excavation, but recently it has been questioned by Bick (1989) in the wake of his survey of the upper Annell. He has argued instead that they resemble sunken cart-tracks or hollow-ways which are a common feature in the Welsh uplands. Indeed, there are other not-dissimilar features in the neighbourhood, in particular a rather straggling set of roughly parallel gullies which can be seen on the hillside opposite the central section of the Gwenlais leat.

The features themselves are not easy to interpret on surface remains alone. In section we are dealing with a series of at least four parallel V-shaped gullies or ditches, each at least 4.00m wide from crest to crest and *c.* 1.00m deep. This makes them surprisingly large for simple leats, unless they have been subsequently deepened either by human agency or by the flow of ground water during heavy rain. Equally problematical is the way in which the lowest of the gullies can be seen to terminate at its extremes in prominent butt ends, with little sign of continuity in either direction. Even the other three gullies are difficult to trace uphill beyond the modern trackway which cuts across their line, with the exception of a single wide hollow which seems to have been artificially enlarged by use as a track. The downhill end poses just as many problems because all the gullies can be seen to peter out at a point where the slope becomes less marked, while beyond this no obvious line can be traced across the rough ground to the north-west.

Given such uncertainty of interpretation, a trench 14.20m long was excavated across the two lowest gullies in an attempt to establish their profile and function. In the event this work has raised more problems than answers, because the two excavated gullies turned out to be so different in character (Fig. 3.2).

On removal of the covering layer of peaty turf [1], most of the surface within the length of the trench, with the exception of the bottoms of the two gullies, consisted of gleyed silty clays [2] and [7], divided into distinctive crumbly nodules *c.* 50–100mm in diameter which, in the case of [2], varied in colour between brownish-grey (7.5 YR 4/1) in their centres and brown (7.5 YR 4/4) around their iron-stained edges; this difference was less pronounced in the case of [7]. At the western end of the trench, on the downhill slope of the lower gully, there was very little thickness of [2] over a crumbly, shaly bedrock which was in places virtually degraded to clay. Towards the eastern end of the trench, on the uphill slope of the higher gully, the deposit represented by [2] and [7] was replaced by [6], a dark brown (10 YR 3/4) coarse sandy loam, which appeared to overlie a gravelly natural. It seems likely that [2], [6] and [7] are all of natural origin, the only irregularity in their spread being a slight thickening of the deposit on the crest of the ridge between the two gullies. Across this ridge, it overlay a fine orange-brown deposit which was assumed to represent the underlying natural at this point.

In the bottoms of the gullies, the deposits were more complex, and, interestingly, the two gullies excavated differed considerably from one another. The uphill gully contained a thin lens [5] of dark greyish-yellow (2.5 Y 4/2) loam beneath the turf in its centre; this in its turn overlay a deposit [4] of dark greyish-yellow (2.5 Y 4/2) sandy clay loam. Both [4] and [5] contained about 50% very small and small angular shale fragments. These two deposits are so similar that they are likely to reflect simply a differentiation in particle size caused by the passage of groundwater along the base of the gully; observation during the excavation indicated that this can be quite considerable. Beneath [4] and [5], which themselves contained lenses of iron-staining, was an iron-stained surface marking the top of a deposit of compacted gravel which was assumed to be natural in origin. The upper surface of this deposit displayed a distinctive cross-section, with a central, deeper, hollow flanked by a shallower one on either side.

The lower of the two gullies proved rather more difficult to resolve. At its lowest point, the silty clay deposit [2] blended into a deposit [3] of greyish-brown (10 YR 4/2) silty clay loam, containing a considerable admixture of roots. The bottom of the gully at this point was marked by clumps of rushes at intervals, indicating damp conditions, and, as with the upper gully, it is likely that this apparent differentiation simply reflects the effects of groundwater movement along the line of the gully. Beneath [2] and [3], the base of the hollow contained a substantial thickness [8] of compact dull yellowish-brown (10 YR 5/4) silty clay, containing about 50% of very small, small and medium-sized angular shale fragments; the concentration of these increased towards the bottom of the layer. Below [8], lay a dull yellowish-brown (10 YR 5/4) deposit [9], which

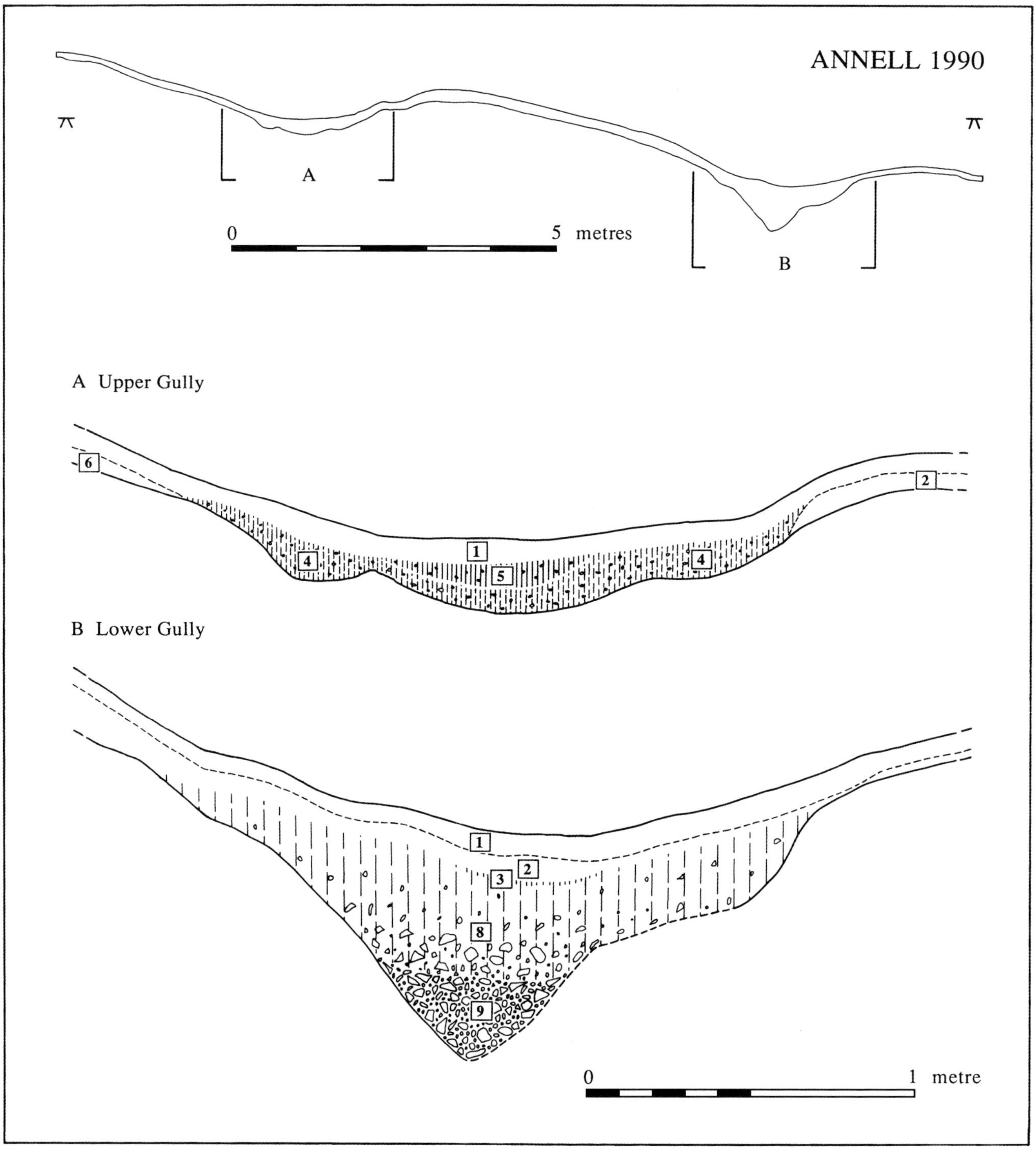

Fig. 3.2 Annell Section, 1990

contained about 95% very small, small and medium-sized angular shale fragments. It was difficult to isolate enough of the matrix of this deposit for analysis, but on excavation it gave the subjective impression of being a silty clay similar to [8]. While the cross-section gives the impression that this gully had a clear V-shaped profile, the position was not in fact so simple. The uphill (E) side of the feature was fairly clearly defined as a face of fine orange-yellow silty clay with pockets of fine pea-grits, from which [8] and [9] could be readily distinguished, but the downhill (W) side was not at all clear, and although the concentration of stone grew less towards the presumed edge, the profile drawn cannot confidently be assumed to reflect any genuine boundary to the feature. Work on this area was hampered by adverse weather conditions at the end of the excavation, and it was unfortunately not possible to devote further time to resolving these problems before backfilling. Nonetheless it seems plausible to suggest that the infill deposits [8] and [9] represent the products of natural weathering from the sides of the feature.

Discussion

Although the detailed interpretation of these features presents problems, neither of them seems to be all that convincing as the line of an ancient leat, especially when they are compared with the Gwenlais sections described below. Equally they do not compare well with the other published sections across the Annell leat itself, not least in their overall width and general profile (Jones and Lewis 1971a; Lewis 1977, fig. 6). The upper gully, for instance, has a very distinctive profile which cannot simply be explained as the frequent recutting of a leat line given the character of the infill horizons. Moreover it bears an almost uncanny resemblance to a pair of wheel or sled marks flanking a central hollow caused by the passage of either human or animal feet; this resemblance might be further reinforced by the distance between the two tracks which approximates too readily to the recognised norm of 4 ft 8.5 ins! This would seem to favour Bick's hypothesis discussed above, though one has to admit the possibility that such a track might have made use of an existing hollow/leat line, thereby widening and deepening (and potentially obliterating) it in the process.

This explanation cannot readily be applied to the V-shaped profile of the lower gully, even bearing in mind the possibility that its infill horizons may have been altered to some extent by the passage of carts/sleds. Unfortunately the excavated evidence does not allow a conclusive interpretation. If the feature was entirely artificial, it becomes difficult to find a satisfactory explanation for its shape and extent, its location, and its function; particularly puzzling is the manner in which it peters out at each end. On the other hand, if the feature is considered as a natural one, arising from irregularities in the underlying geology, then this could go some way to explaining the presence of the entire group of parallel features cutting obliquely across the hillside at this point. Such natural irregularities could have been exploited as a convenient line for a leat, but all trace of this could have been subsequently removed by the passage of carts and sleds in much the way described, together with erosion caused by groundwater.

On the basis of these results and the fact that only two of the actual gullies were sampled in 1990, it would clearly be unwise to draw too firm a set of conclusions for the whole series or to suggest any amendment to the scheduling of this part of the 'Annell leat'. For the moment it is enough to cast serious doubt on the received wisdom that the visible features represent a series of parallel leats designed to help the flow of water down a particularly steep section of its course. That said, however, even if all the gullies could be shown to be the result of later activity and/or natural, we would still not be in a position to disprove the suggestion that a leat had originally aimed for or crossed this sector of the hillside *en route* from its source to the gold-mines. This point has been reinforced recently by Bick (1996), who has drawn attention to the uppermost of the channels as a possible excavation along the back of a mineral lode, potentially served by a leat bringing water for hushing or other purposes.

3.3.2 The Gwenlais Sections, 1990 (Sites 1 and 2)

The line of this leat is particularly prominent in the upper reaches of its course, either as a rock-cut feature or as a narrow ledge with a low mound on the outer, downhill side. In places it is also well defined by a line of rushes clearly marking the presence of deeper, waterlogged conditions. In order to gain the clearest impression of these variations, sections were excavated at two points, one where the leat was visibly rock-cut, the other where the ledge and the associated bank were particularly well represented.

(A) Site 1 (Figs 3.3 and 3.4)

Prior to the beginning of the excavation it was clear that this part of the leat had been rock-cut, but there was no trace of any associated bank. Upon excavation the outer, downhill side of the leat was found to have been cut more or less along the bedding planes of the natural shale and had thereby preserved its original profile reasonably intact; on the uphill side, however, where it had been cut across the bedding planes, there had been a considerable amount of later erosion caused presumably by frost-shattering and weathering. It was possible nevertheless, to establish its basic profile with reasonable accuracy and to record the sequence of infill layers which post-dated its usage.

In profile, the leat proved to be roughly U-shaped, with steeply-sloping sides and a flat base. Across the top it must originally have measured *c.* 0.75m, but at the bottom it was found to have narrowed to *c.* 0.30m. Its effective depth must have been of the order of *c.* 0.35m though whether this bears any relationship to the original depth of the water cannot be established. There were no obvious traces of any tool marks on either the sides or the base.

Within the leat a sequence of five successive infills [2] to [6] was recorded below the modern peaty turf [1]. At the very bottom was a thin layer [6] composed almost entirely (95%) of small to medium-sized angular shale fragments set in a matrix of greyish-yellow-brown (10 YR 4/2) sandy clay. This was sealed by a somewhat thicker layer [5] of greyish-brown (7.5 YR 4/2) sandy clay with a lesser proportion (60%) of the angular shale fragments. Both these layers would seem to be the result of fairly rapid weathering of the

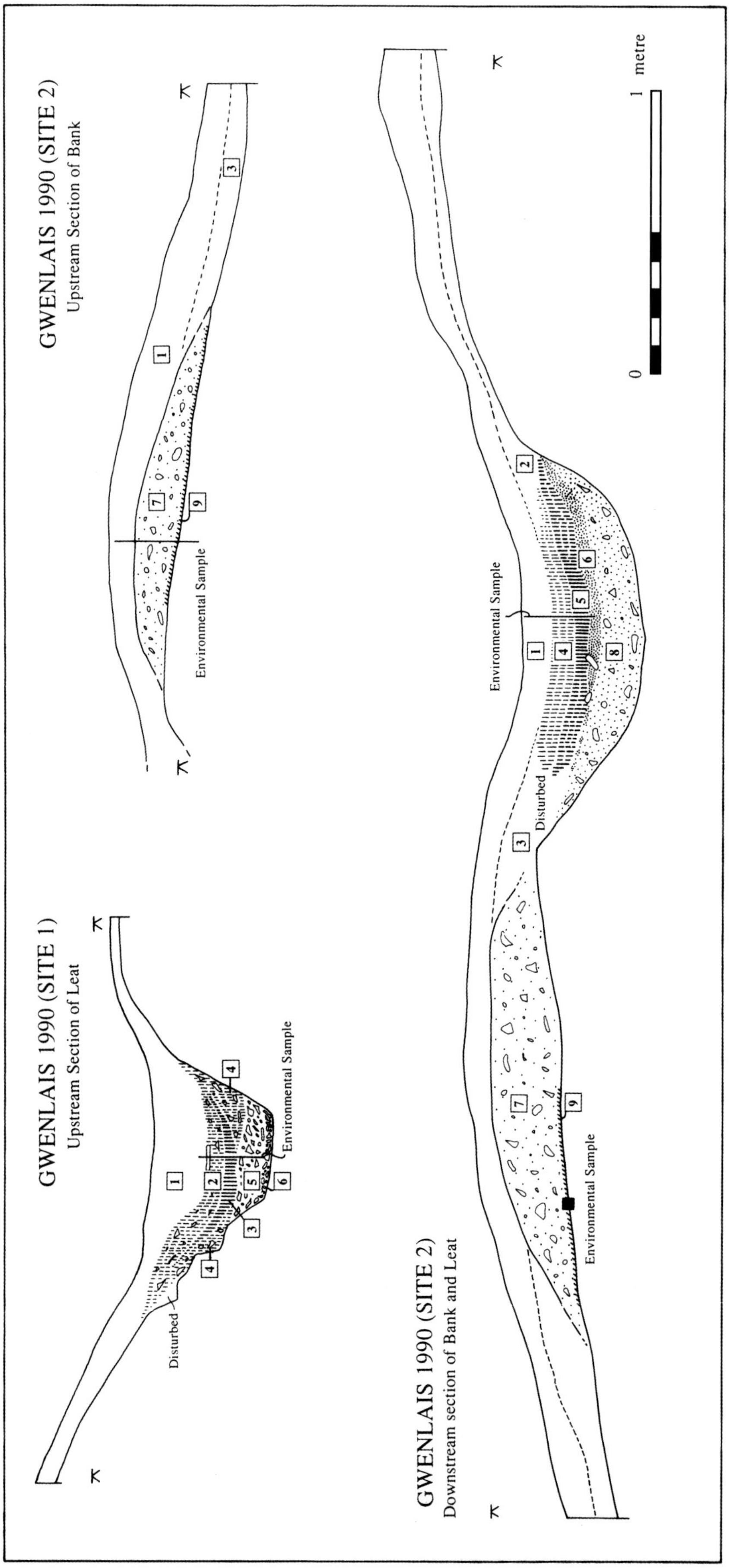

Fig. 3.3 Gwenlais sections, 1990

sides of the leat, particularly on the uphill side, in the period immediately after it had gone out of use. Some of the uppermost shale fragments in context [5] were almost clay-like in their consistency, a feature which became even more marked in the later infills [2] and [4]; this would seem to represent a progressive slowing down in the rate of deposition, during which time the shale became weathered and decomposed as a result of natural processes.

Above these two primary infills, and resting against the sides of the leat, was a secondary infill [4] of greyish-yellow-brown (10 YR 4/2) sandy clay loam with only a 30% admixture of small to medium-sized shale fragments, some of them much weathered. This was sealed by a black (7.5 YR 2/1) peat deposit [3] with the odd fragment of very weathered shale. A comparable peat horizon was also noted in the other section across the Gwenlais. The final infill [2] above this was a brownish-black (10 YR 3/3) sandy clay loam with a marked admixture of very weathered orange/yellow clay-like shale fragments. This too had a high peat content which was not always easy to differentiate from the underlying horizon. A column sample was taken through the uppermost deposits [1], [2], [3] and [5] in an attempt to recover material relating to the local environment at a time when the leat was out of use and already partially infilled.

Fig. 3.4 Rock-cut section of Gwenlais leat, looking north-east

(B) Site 2 (Fig. 3.3)
On the surface this section of the leat was very different in character and profile from that just described, comprising instead a broad, shallow ledge with an associated bank on the downhill side. These differences were duly confirmed by the excavations. The leat itself proved to be a broad, U-shaped feature measuring *c.* 1.40m across the top, with gently sloping sides and a rounded base. Clearly it had not been rock-cut at this point, a fact which would have made some form of watertight clay lining essential to its operation. From the available evidence the effective depth of the leat cannot have been much more than *c.* 0.35m.

The trench also cut across the associated bank [7] on the downhill side, which was shown to be between 1.50 and 1.70m wide and no more than 0.20–0.25m high. It was made up of a single dump of brown (10 YR 4/4) sandy clay loam with a high content (80%) of very small to small slightly abraded shale fragments, clearly derived from the associated leat. Of particular importance was the fact that this bank sealed a well-defined buried soil [9] across most of its width, which was sampled for evidence relating to the local environment immediately prior to the construction of the leat.

Within the leat a sequence of four successive infills [4] to [6] and [8] was recorded beneath the modern turf [1]. This proved to be very similar to that already described for the first section, despite the differences in profile and character. Across the base of the leat lay a spread [8] of bright yellowish-brown (10 YR 6/6) silty clay, up to *c.* 0.15m thick, with a 20% admixture of very small to small, slightly abraded shale fragments. This must represent a period of rapid infilling caused by the weathering of the sides of the leat, as well as the associated bank. This layer was sealed by a thin spread [6] of greyish-yellow-brown (10 YR 5/2) silty clay with virtually no stone content. At first this was thought to be a remnant of the original clay lining to the leat, because of its compacted upper surface, but it soon became apparent that this was little more than a natural iron pan formed at the junction of the overlying peat accumulation [5] with the lowest layers of natural infill [6] and [8].

As with site 1, these lowest fills were sealed by a brownish-black (10 YR 2/2) peaty deposit [5] with virtually no stone content, probably representing a phase of peat formation at a date some time after the leat had ceased to be used. Above this came a marked horizon [4] of greyish-brown (7.5 YR 4/2) silty clay loam containing very few stones and an admixture of peaty material. This in its turn was sealed by even later infill deposits [2] and [3] deriving either from the uphill side of the leat or from the bank on the downhill side. Again a column sample was taken through the principal deposits [1], [4], [5] and [6] in an attempt to recover evidence relating to the local environment.

Discussion

There can be no doubt from these two sections that we are dealing with a well-preserved leat which has hitherto escaped detection. This fact is somewhat surprising given all the work which has been done on the leat systems supplying the gold-mines, and clearly a very strong case exists for extending statutory protection to it as an ancient monument. Its discovery has inevitably raised the question of its relationship with the upper reaches of the so-called Annell leat in the vicinity of the 'maintenance hut', leaving aside for the moment the wider issue of any onward course towards the mines. While it has not proved easy to establish a precise link between the surviving stretches, principally because of the difficulty of tracing them across the lower-lying and boggy ground between the Gwenlais and the headwaters of the Annell, a *prima facie* case can be made for believing that they were originally part of the same, inter-linked system, designed to collect water not from the marshy area at the head of the Annell as was originally argued, but from the headwaters of the Gwenlais over 3 km away. If the case is accepted, however, it would raise a major question mark over the nomenclature of the so-called 'Annell leat', as Bick has consistently maintained, at least with reference to the surviving remains in its upper reaches, along the flanks of Cefn y Bryn.

A detailed evaluation of this hypothesis remains problematical, given the difficulties of the terrain outlined above, while the situation has not been helped by an episode of ploughing during 1995, which may well have damaged the most ephemeral traces of the upper Annell beyond recovery. It is worth recording, however, that Jones and Maude (1991) have recently described what seems to be a very similar stratigraphic sequence to that noted along the Gwenlais, including a phase of peat formation, in an unpublished section across the 'Annell' somewhere near its postulated source. This is all highly suggestive of a contemporary date for both systems, which emphasises the importance of the two radiocarbon dates recovered from the Annell section. The peat itself produced a date of 1175±15 bp, while an alkali extract yielded an age determination of 1085±20 bp, both of which are taken to indicate that by *c.* AD 800 the leat channel was already three-quarters infilled. Although these dates can be little more than a *terminus ante quem*, they seem to strengthen the case for a Roman date for the upper reaches of the so-called Annell leat and, by implication, the recently discovered Gwenlais leat. This has important implications for the results of the pollen analyses outlined below, which would seem to indicate significant changes in the local landscape spanning the period of use of the leat to the present day.

The belief that the upper Annell and Gwenlais leats are part of the same system has also raised questions about their relationship with the other postulated elements of the Annell leat lower down its course, to such an extent that Bick and Boon (1993) have even doubted the latter's existence (see section 3.4 below). Certainly it is difficult to see why such an extensive water system would have been deemed necessary, with its inherently steep rate of descent, when it would have been possible (and easier) for the engineers to tap the Annell lower down its course without any significant loss of water at the minehead. The alternative, that the upper Annell-Gwenlais leat represents an independent system, still leaves us with the difficult problem of explaining its original destination and function, which recent fieldwork has done little to resolve (Bick 1996). There remains a distinct possibility, however, that such a leat was actually aiming to cross the col into the Cothi valley, much as Lewis and Jones originally envisaged.

3.3.3 The environmental evidence (by Michael Walker)

This section describes the results of pollen analyses carried out on sediments collected during the 1990 excavations across the Gwenlais leat. These included (a) a 21cm monolith from each of sites 1 and 2, taken through the deposits forming the infill of the leat, and (b) three samples taken from the buried soil horizon beneath the upcast bank on site 2.

Laboratory methods

Standard laboratory procedures were employed to extract the pollen from the sediment matrix (Moore *et al.* 1991). Slides were counted on a Laborlux K microscope at x400 magnification, with critical identifications under oil at x1000. In some horizons, the pollen was badly deteriorated, the grains being either degraded or so heavily folded that identification was impossible. At no level, however, did the number of indeterminate grains exceed 10% of the pollen sum (300 land pollen). The results of the pollen analyses are shown on Fig. 3.5. Botanical nomenclature follows Stace (1991).

The Pollen Diagram (Fig. 3.5)

Although differences in detail are apparent between the two pollen profiles from the leat infills, there is a broad measure of agreement between the trends in the curves from the principal pollen taxa. Three local pollen assemblage zones (LPAZs) can be identified.

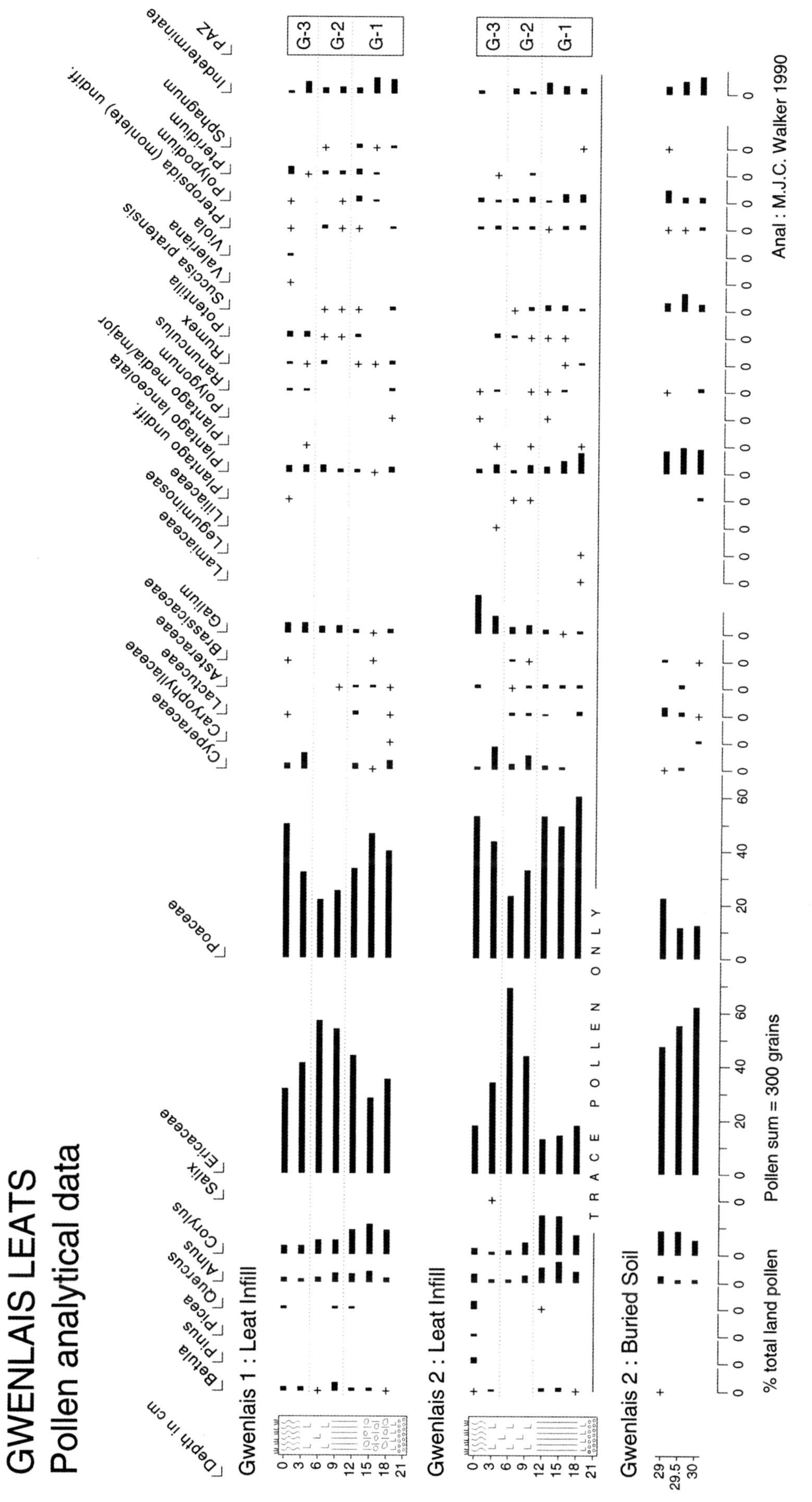

Fig. 3.5 Pollen analytical data from the Gwenlais leat

LPAZ G-1: The lowermost levels are characterised by high counts for Poaceae (grass), with significant representations of Ericaceae (heaths) and *Corylus* (hazel). Because of the often poor state of preservation of the Ericaceae grains, it was not possible to effect a consistent separation of this family into individual genera. However, the dominant species present was clearly *Calluna vulgaris* (heather/ling), although other genera (*Vaccinium, Erica*) were also recorded. *Alnus* (alder), *Plantago lanceolata* (ribwort plantain), *Succisa pratensis* (devilsbit scabious) and ferns (Pteropsida and *Polypodium*) are present in this biozone.

LPAZ G-2: A significant increase in Ericaceae (mainly *Calluna*) pollen is the distinctive feature of this part of the pollen diagram. Poaceae remains a major component of the pollen spectrum, although percentages are reduced by comparison with the previous zone. Counts for *Corylus* and *Alnus* are lower, while in site 2, values for *Plantago lanceolata* also decline whereas Cyperaceae (sedge) increases. In both profiles *Galium* (bedstraw) is more strongly represented in this zone.

LPAZ G-3: This zone spans the upper 5–6cm of the two profiles and is distinguished by a marked increase in Poaceae pollen and a complementary fall in values for Ericaceae. *Corylus* and *Alnus* counts are also generally lower. By contrast, frequencies of *Plantago lanceolata* and especially of *Galium* are higher than in G-2. Other taxa present in one or both profiles include *Potentilla* (tormentil), *Pteridium* (bracken), *Rumex* (sorrel), Pteropsida and *Polypodium.*

The buried soil: The pollen spectra from this horizon are dominated by Ericaceae grains (which again include large numbers of *Calluna vulgaris*), Poaceae, *Corylus*, *Plantago lanceolata* and *Succisa pratensis*. *Alnus* and Asteraceae (daisy) are also present, while relatively large numbers of fern spores (*Polypodium*) were recovered.

Interpretation

Two particular problems surround the interpretation of the pollen records from the Gwenlais leat: first the difficulty of establishing pollen source area, and second the possibility that the deposits contain varying quantities of secondary pollen. Ditch infills such as these typically contain a predominance of locally derived pollen, and hence vegetational reconstructions based on the pollen data are usually more likely to reflect local as opposed to regional vegetational patterns. However, the similarity between the two Gwenlais diagrams and those from peat and lake sites from elsewhere in the Welsh uplands (see below) suggests that the pollen record from the leats is a reasonable reflection of broader-scale vegetational developments.

The likelihood that secondary pollen is contained within the leat infills, however, is a more difficult problem to resolve, particularly as the materials which have accumulated in the leat may, in part, consist of inwash from former soils and therefore contain non-contemporaneous pollen. The occurrence of secondary grains in a pollen diagram is often reflected in sudden increases of particular pollen taxa; one interpretation of the sharp rise in the curve for Ericaceae in LPAZ G-2, for example, in both the Gwenlais profiles, is that this reflects the inwash of soils containing large numbers of Ericaceae pollen (soils, for instance, which are similar in age to the buried soil on Site 2). One way of tackling this problem is to monitor the numbers of deteriorated pollen (Lowe 1982; Lowe and Walker 1986), *i.e.* those grains showing evidence of exine damage which could have occurred during reworking of soils and other sediments. Significantly, no major variation in pollen deterioration was detected throughout the Gwenlais profiles. Many Ericaceae grains were crumpled and shrunken, but this could be attributed simply to desiccation during the drying out of the top 10cm or so of the profiles. Hence, although the extent of secondary influence in the pollen records cannot be quantified, it may be that redeposited grains constitute less of a problem in the interpretation of the Gwenlais pollen diagrams than might initially have been envisaged.

Environmental reconstruction

The earliest pollen sequence is that from the buried soil horizon beneath the upcast bank on Site 2, as this clearly relates to the period preceding the leat construction. The palaeoecological inferences which can be drawn from the pollen spectra from this soil horizon, however, are constrained by the problems which are encountered in all soil pollen analyses, most notably differential destruction of pollen grains, the relatively high frequencies of deteriorated pollen and the translocation of pollen grains through a soil profile as a result of pedogenic or faunal processes (*e.g.* Caseldine 1983; Anderson 1986; Walker 1989). This means that the pollen record recovered from a buried soil profile may not necessarily be a true reflection of the contemporaneous pollen rain falling on the former land surface. These difficulties notwithstanding, however, certain general inferences can be drawn about the vegetation cover of the area immediately preceding the construction of the leat which has been tentatively assigned to the Roman period. The data from the buried soil suggest an open heathland landscape, interspersed with grassland. Some wood and scrub (mainly *Corylus* and a limited amount of *Alnus*) were present in the vicinity, most probably on the more sheltered valley slopes to the west of the Gwenlais uplands. The possibility that some of the hillsides were given over to pasture may be indicated by the occurrence of pollen of *Plantago lanceolata*, although the relatively low counts for Poaceae pollen suggest that grazing was somewhat restricted. Certainly, the pollen spectra from the buried soil contrast quite sharply with those from the surface layers of the two leat infills. These show much higher counts for Poaceae and lower frequencies of Ericaceae pollen, reflecting the extensive use of the upland pasture at the present day and in the relatively recent past.

The heathland episode recorded in the pollen from the buried soil may well be the equivalent of that found in the profile from Cefn Gwernffrwd *c.* 5km to the north of the Gwenlais site. There, the evidence suggests that the expansion of *Calluna* moorland began after *c.* 2755 BP and lasted until *c.* 2220 BP, the phase of upland heath expansion being attributed to a possible relaxation in grazing pressure (Chambers 1982a).

The lowermost pollen assemblage zone in the two leat profiles (G-1) appears to reflect a landscape somewhat different from that recorded in the buried soil. Markedly lower Ericaceae values and a significant increase in Poaceae pollen indicate a reduction in heathland and more extensive grassland in the period after the leat had gone out of use. There is also a stronger representation of woody taxa in the pollen spectra, principally *Corylus* but also *Alnus* and *Betula*. This implies a degree of wood and scrub regeneration, although it seems unlikely that trees would have become established on the exposed and windswept upland plateaux. Some alder, hazel and perhaps also birch stands may have been found on valley floors and on more sheltered hillsides, however. There is no dating control on this grassland phase in the area, but there are indications elsewhere of pastoral activity in the mid-Wales uplands during the early medieval period and continuing into Norman times (Moore and Chater 1969; Chambers 1983b; Walker 1993). In the Gwernffrwd profile, for example, the upper part of the diagram (post Roman?) is characterised by a decline in *Calluna* and a significant increase in the Poaceae pollen (Chambers 1982a). The period from *c.* AD 700 to 1200 was one of increased warmth in both summer and winter (Lamb 1977; 1982) which would have enabled greater use to have been made of the potential offered by the higher slopes for sheep grazing. The grassland phase reflected in G-1 of the Gwenlais profiles may therefore be tentatively ascribed to this time period.

If this interpretation is correct, the subsequent increase in Ericaceae and decline in Poaceae pollen in the two Gwenlais profiles could reflect a further relaxation of grazing pressure, perhaps brought about by the increasingly severe climatic conditions of the Little Ice Age. Throughout north-west Europe, the markedly cooler summers and wetter winters and the significant increase in storms which characterised the period from the 14th to 18th centuries, resulted in widespread farm abandonment on the high moors and the retreat of the agricultural frontier from the upland fringes (Grove 1988; Bell and Walker 1992). A natural consequence would have been the replacement of former moorland pasture by heathland, and in a number of pollen diagrams from upland Wales, those parts of the profiles equated with the late medieval is characterised by a significant increase in Ericaceous pollen (Moore and Chater 1969; Walker 1993; Smith and Green 1995). The uppermost levels in the Gwernffrwd diagram, for example, show a rise in the curve for *Calluna* and falling values for Poaceae (Chambers 1982a). In the Coed Taf site, on the uplands above Merthyr Tydfil, the *Calluna* expansion in the upper part of the profile has been dated to 470±65 bp (Chambers 1983b). Hence the records from the Gwenlais sections tend to conform with the regional pattern.

The re-establishment of grassland around the Gwenlais catchment represented by zone G-3 of the pollen profiles may therefore be a relatively recent phenomenon. While this probably reflects changing moorland management practices, mainly in the increased use of upland pastures following the climatic improvement which began early in the 18th century (Manley 1974), the higher levels of atmospheric pollution which accompanied the onset of the Industrial Revolution may also have played a part in the expansion of grassland communities on the Welsh uplands (Chambers *et al.* 1979). The effects of recent afforestation are apparent in the occurrence of *Pinus* (pine) and *Picea* (spruce) pollen in the surface sample from Site 2, while many of the contemporary grassland species such as *Potentilla erecta* (tormentil), *Galium saxatile* (heathland bedstraw), *Plantago lanceolata*, and occasional stands of *Pteridium aquilinum* (bracken) are faithfully represented in the uppermost levels of both pollen diagrams. The pollen evidence suggests, therefore, that this type of upland pasture which is so characteristic of the rolling hills of mid-Wales has existed in its present form in the Gwenlais area for no more than 200 years.

3.4 TANK G AND THE ANNELL SECTION, 1993

In considering the function of their upper Annell-Gwenlais leat (see section 3.3.2 above), Bick and Boon (1993) have been tempted to doubt the very existence of the lower reaches of the Annell leat, arguing first that its conjectured route along the flanks of Allt Cwmgerwyn and Allt Cwmhenog would have involved an unnecessarily steep gradient when compared with many other leats, and second that the tank which is presumed to be its ultimate destination could have been filled naturally from the rising ground behind it, as was the case at Craig-y-Main in

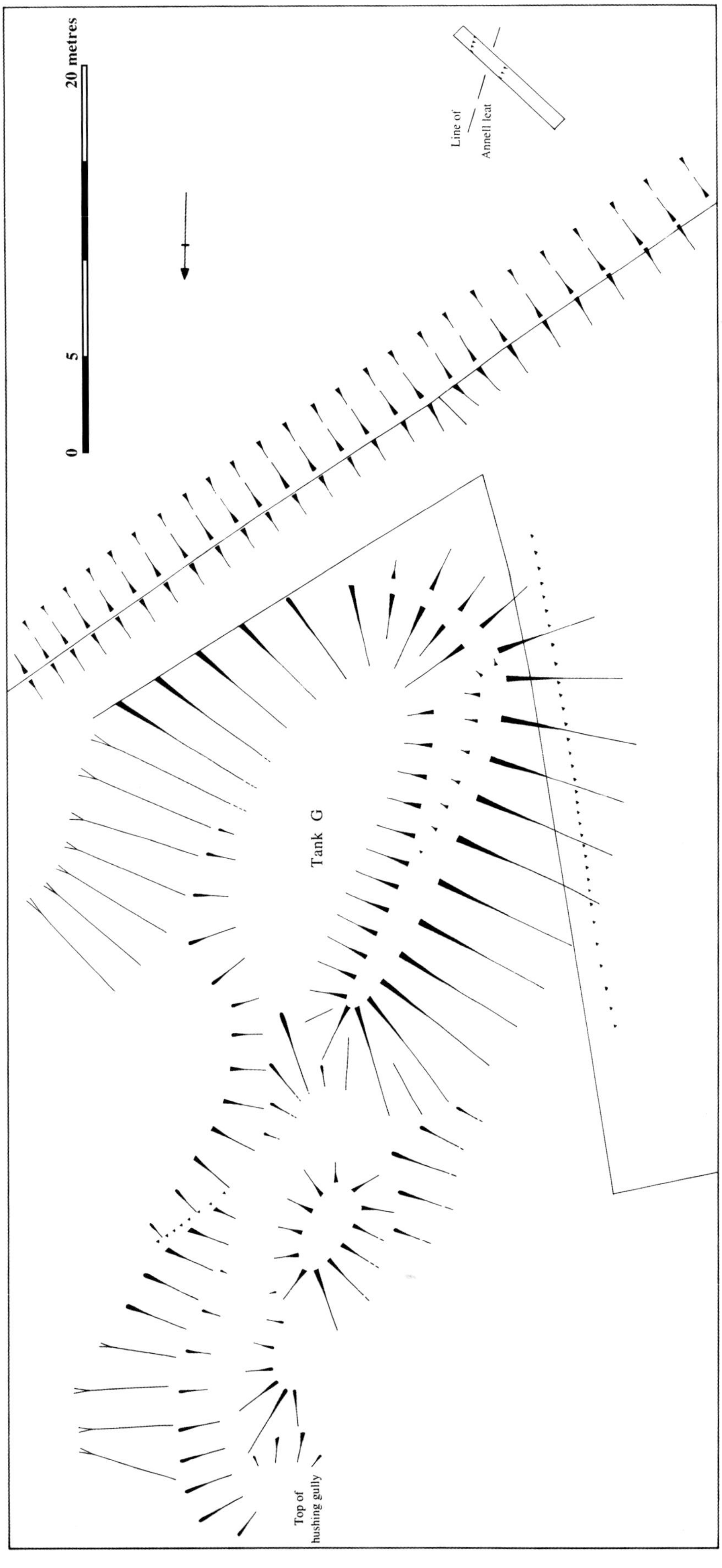

Fig. 3.6 Survey plan of Tank G

Powys (Bick 1991). Notwithstanding the difficulties of connecting all the surviving features from the upper reaches of the Annell/Gwenlais via Gwarnoethle to the top of the workings, the authors took the opportunity in 1993 to investigate the earthworks associated with the supposed termination of the Annell leat in the immediate vicinity of Tank G and to cut a section across its predicted line in the field immediately to the south of the scheduled area. In the event, this latter conclusively proved the existence of a water course heading directly for the known tank.

3.4.1 Tank G, 1993

The survey initially concentrated on defining the massive tank (G) which lay immediately inside the corner of the forestry plantation (Fig. 3.6). Its remains proved to be particularly well preserved on the downhill side, where a substantial outer bank still survived to a height of nearly 2.00m throughout most of its length. In the modern field outside the plantation, a pronounced plough scar parallel with the fence line made it clear that there had been some disturbance at this point from recent land improvements. On the southern and eastern, uphill sides the tank was less well preserved, partly because of the interference caused by the later trackway and partly because of the way in which natural hillwash had effectively smoothed its original profile, but enough survived to demonstrate that it must originally have measured some 27.00m north-south by at least 12.00m east-west to the top of the bank. At the northern end of the tank, a prominent gap in the encircling bank clearly represents the location of an original sluice or spillway which fed water downslope to a series of related features.

These features were not easy to survey in detail because of the thickness of the surrounding plantation and several collapsed trees, but even so a reasonably coherent picture emerged. Immediately below the sluice there appeared to be a second roughly rectangular area, with a low mound along part of its outer, downhill side, which has sometimes been interpreted as a second smaller tank below the main reservoir, though the surviving remains are somewhat less convincing. Beyond this, a shallow, sloping channel fed into the top of an irregular gully up to 5.00m wide, which plunged off steeply downhill towards the rim of the neighbouring opencast. No attempt was made to survey beyond the top of this gully. These features, along with other similar gullies further to the north, have long been identified as the clearest evidence for the use of hushing at Dolaucothi, whereby large quantities of water were released from a holding tank for the purpose of prospecting a hillside or for opencast development.

3.4.2 The Annell Section, 1993

Using the levels of the tank's outer bank and its interior as a relative guide, a trial section was cut across the predicted line of the leat in the heavily-ploughed field immediately outside the scheduled area. The upper end of the resulting trench encountered stony material which turned out to be part of the probable retaining bank of the leat, and a short extension up the slope soon located the rock-cut uphill face of a channel, confirming the presence of an artificial feeder running around the hill towards the tank (Fig. 3.7). The base of the channel as finally excavated was at roughly the level of the present top of the tank's retaining bank.

The uphill side of the channel was particularly well preserved, having been cut at an angle of about 45° into the bedrock which here comprised a loose shale bedded at right angles to the surface and running roughly up and down the slope. It had been excavated originally as part of a more extensive platform over 2.20m wide, with a slight downhill slope to its profile, onto which the remaining parts of the leat had subsequently been constructed. This took the form of several dumps of shaly material [4], [5], [6] and [9], most probably derived from the excavation of the platform itself, which together represent all that remains of the original retaining bank along the downhill side. The initial framework for this bank was clearly formed by two distinctive dumps of stone [5 and 9], both set in a matrix of dull yellowish-brown silty clay (10 YR 5/3 to 5/4), which had been laid across the centre of the platform and downslope from its outer edge respectively; these must have formed the inner and outer edges of the original bank. Between the two lay a spread of rather less stony material [6], which was intermixed with a brown silt loam containing little or no clay (10 YR 4/4); this is likely to have been derived from the original topsoil across the site.

The inner of the two stone dumps [5], already described, also provided a levelled surface across the inner side of the platform, which supported a compacted surface [3] of dull yellow-orange silty clay (10 YR 6/4) forming the base of the channel itself, up to 0.45m wide. This lay some 0.50m vertically below the lip of the rock cutting, although it is not clear just how much of this depth would have been available to hold water in the absence of so much of the original retaining bank. The outer side of the channel was clearly formed by another stony dump [4] with a pronounced admixture of dull yellowish-brown silty clay (10 YR 5/4); this must originally have formed a watertight lining on the downhill side, as well as being a substantial part of the structure of the retaining bank, most of which has been removed by later ploughing. It is likely that this side of the channel also rose at an angle of about 45°.

The outer side of the presumed bank had been

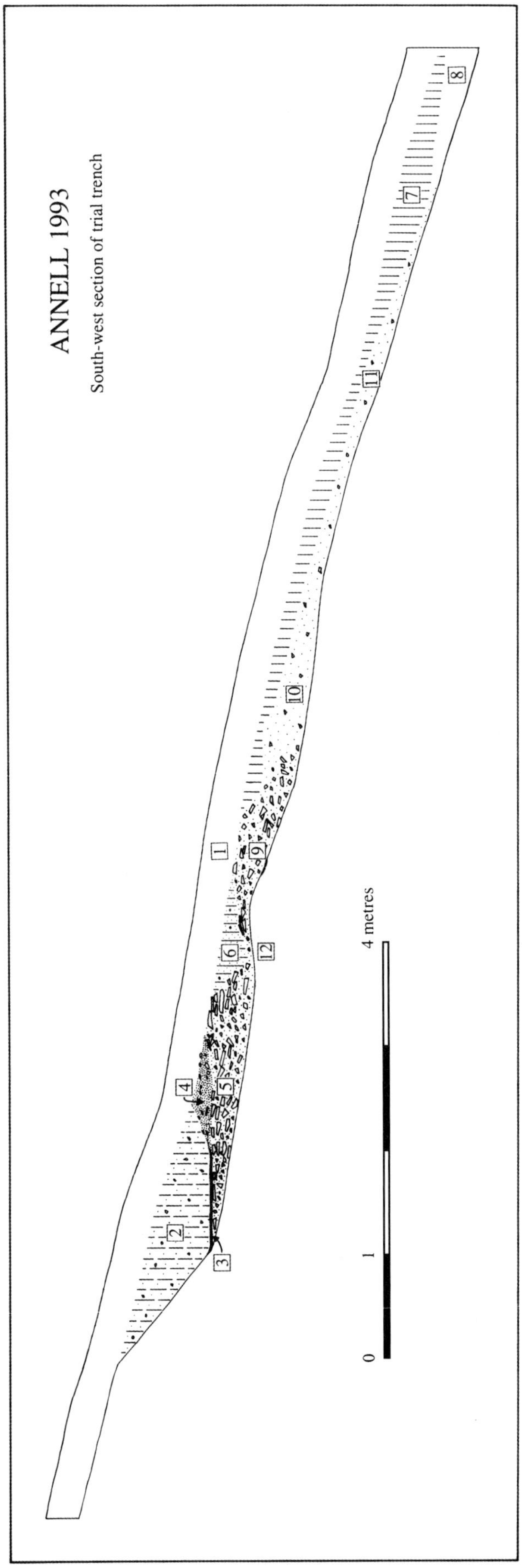

Fig. 3.7 Section across the predicted line of the Annell leat, 1993

noticeably affected by ploughing and worm action, to such an extent that the lower end of the stony dump [9] on this side merged quite gradually into the apparently natural soil [10], although there was a clear reduction in stone content at the approximate point of junction.

Discussion

Such evidence for a feeder leat in the immediate vicinity of Tank G and its associated hushing gullies has obvious implications for some of the rather negative views expressed by Bick and Boon (1993), not least by demonstrating conclusively that a water supply of sorts must have existed along the Annell side of Allt Cwmhenog. What still remains uncertain, however, is its relationship with the other evidence both at Gwarnoethle and in the upper reaches of the Annell-Gwenlais, which only further fieldwork can hope to resolve. On balance there seem to be plausible grounds for accepting a link with one of the three silt-choked channels at the former site, no more than 0.80 km away around the contour of the hill, even if the destination of the other two still awaits clarification. The jury must still be out, however, when it comes to the case for a direct link between any of the three channels at Gwarnoethle and the well-preserved leat system in the upper Annell-Gwenlais, more than 3.25 km away. At least one tapped a local spring, but either of the others could have tapped into the Annell itself, or linked in with a more extensive system to bring water from further afield.

3.5 CONCLUDING OVERVIEW

Given the limited nature of our excavations and the unresolved issues surrounding the nature and extent of the different leats it is difficult to draw absolute conclusions at this stage of research. Several points are worth emphasising, however: (i) there is increasingly little reason to doubt that the upper reaches of Lewis and Jones' 'Annell' leat and Bick's Gwenlais leat are one and the same; (ii) this combined leat can be traced with reasonable confidence as far as the area now scheduled as an ancient monument at SN 699 426 (Bick 1996), though none of the visible features here can be certainly associated with it; (iii) beyond this its destination remains uncertain; the jury remains out on any direct link with the postulated remains along the lower reaches of Lewis and Jones' Annell leat, while a route across the col into the Cothi valley remains unproven; (iv) there are significant question marks over the character and extent of any system along the lower reaches of the Annell valley, despite suggestive indicators of its existence.

Such debates necessarily emphasise two key points: (i) there is an urgent need for the detailed publication (and assessment) of all the available evidence relating to the so-called Annell leat, as has been done for the Cothi (Jones *et al.* 1960); and (ii) there is a case for a detailed programme of geophysical survey and small-scale excavation designed to trace the line of its various components from minehead to source and to clarify their relationship (if any) with the remains in the upper reaches of the Annell and Gwenlais. That a programme of small-scale excavation can be both highly productive and cost-effective has been amply proven by the 1990 and 1993 sections.

That such work is essential has been further reinforced by the publication of the two radiocarbon dates from what must surely now be seen as the lower end of the Gwenlais leat (said to be the upper reaches of the Annell when first published by Jones and Maude 1991). Their evidence that the leat channel was already three-quarters full by *c.* AD 800 (see section 3.3.2 above) clearly strengthens the case for a Roman date for this part of the system at least. It would also have enormous implications for our wider understanding of the history of Roman mining at Dolaucothi, as Jones and Maude have rightly emphasised, if a link between the isolated elements or an alternative destination could be conclusively demonstrated. This should be incentive enough for renewed research, not least because of the difficulties of dating the otherwise complex surface remains inside the mine area proper.

3.6 APPENDIX: Recent aerial photographic evidence for the Nant Dâr leat

The so-called 'Nant Dâr' leat was first identified by Jones and Lewis in the course of their 1968–69 fieldwork on the water supply systems to the mines. This fieldwork itself included a certain amount of analysis of aerial coverage, of a part of the valley which has been increasingly improved in the succeeding period, but this leat was only ever published in the form of sketchy descriptions and plans, which themselves vary somewhat (Jones 1971, 19–20; Jones and Lewis 1971a, fig. 1; Jones and Lewis 1971b, 6 and fig. 2; Lewis 1977, fig. 1).

It was said to tap the Nant Dâr, a tributary to the east of the Cothi, at a position which the plans suggest must be in the vicinity of SN 709 444. Jones (1971, 19) describes it as roughly following the 800ft (244m) contour, which, together with the two 1971 plans, is close enough to be confident that this is indeed the

Fig. 3.8 Aerial view of Cothi and Nant Dâr leats at Coed Maes-yr-Haidd (AP 955125–42). Crown Copyright: RCAHMW

same feature as that visible on the recent air photographs. The plan in Lewis 1977 appears to take a variant line at a higher level and does not seem to represent the route discussed in this appendix.

The set of photographs featured here was taken on 21st April 1995 by The Royal Commission on the Ancient and Historical Monuments of Wales (RCAHMW) as part of Cadw's aerial monitoring programme of scheduled sites, and was primarily intended as a record of the better known Cothi leat, which runs at a lower level some distance down the slope. The upper line is best seen on the unimproved north-facing slopes at Coed Maes-yr-Haidd (SN 694 433; Fig. 3.8) and Llwyn-y-Ceiliog (SN 688 428; Fig. 3.9), although there is also a hint of what may be its course above Briwnant (SN 677 415; Fig. 3.10). Unfortunately the area closer to the putative source of the leat was not included in the coverage, while the forestry plantation in the immediate vicinity of the mines conceals its possible lower course and ultimate destination.

At Coed Maes-yr-Haidd, this leat appears to lie between *c.* 270 and 260m OD (estimated from its relationship to local boundaries), while at Llwyn-y-Ceiliog, it is at roughly 260m OD. The stretch at Briwnant (if this is indeed part of it) lies between 260 and 250m OD. This leat survives today as a somewhat slighter field feature than the main Cothi leat.

By extrapolation along the contour lines, the leat above Coed Maes-yr-Haidd would meet the stream flowing westward down from Rhiw Garegog somewhere about SN 702 433. This would itself have provided a useful source of water, but Jones and Lewis clearly considered that the leat continued around the next bluff to the north, where the land has now been considerably improved, and into the valley of the Nant Dâr, which it would probably have tapped somewhere near SN 709 444. The exact spot would have depended on the gradient of the leat, but the maximum volume of water could presumably have been captured using a shallow gradient, giving a tapping point as far down the course of the stream as possible.

Assuming that the feature visible at Briwnant is a continuation of the same leat, and not a trackway connected with the old farmstead at the farther end of the long field, and assuming that the leat continued at much the same level from here into the mine area, it would have arrived at a convenient height for the

Fig. 3.9 Aerial view of Cothi and Nant Dâr leats above Llwyn-y-Ceiliog (AP 955124–62). Crown Copyright: RCAHMW

Fig. 3.10 Aerial view of Cothi leat and possible course of Nant Dâr leat above Briwnant (AP 955124–56). Crown Copyright: RCAHMW

exploitation of the Cwmhenog pit, which was also served by the so-called Annell leat. It is however possible that the Nant Dâr leat took a steeper course across the slope of the hill to join the main Cothi leat, and even if this was not its original intention, it might have been so modified at a secondary stage. There are hints of such a line on the Briwnant photograph (Fig. 3.10), although at least part of this is a trackway today. The well-worn track towards the upper left of this photograph provides a salutary warning of how easy it can be to confuse the old trackways of the area with portions of the leats, unless the latter conform clearly to the contour lines.

There is still considerable scope for investigation of this leat on the ground, not least at its presumed extremities. This is particularly important because, apart from in the immediate vicinity of the mines, these do not enjoy the statutory protection which now covers the stretches at Coed Maes-yr-Haidd and Llwyn-y-Ceiliog. In particular, any clearance of the area of forestry by the mines would provide a crucial opportunity, which should not be missed, to establish whether there is indeed a continuation of its line as surmised at its lower end, and if so, to examine its destination in greater detail.

4 EXCAVATIONS IN THE VICINITY OF THE CARREG PUMSAINT, 1991–1993

4.1 INTRODUCTION

While considerable effort has been directed at elucidating the nature of the surface workings and water supply systems, the related question of ore-processing has received far less attention. Lewis and Jones (1969), for instance, could only cite a number of crushing stones and quantities of crushed quartz in the vicinity of Ogofau Lodge, together with several recognisable stepped washing tables cut into the local shale, most notably in association with Tank C. The identification, therefore, of the possible remains of a water-driven mill complex in 1982, thought at the time to date on technological grounds to the medieval or early modern period (Austin and Burnham 1984), inevitably raised some interesting questions about our overall understanding of the early phases of exploitation at Dolaucothi, not least the extent to which all the visible surface remains should be attributed to the Roman period. Such a discovery, while not denying the importance of Roman mining, even on a large scale, necessarily warranted further investigations, hence the excavations undertaken by the authors over three successive seasons from 1991 to 1993 in the area immediately north of the presumed mill site, where characteristic processing wastes might well be located.

4.2 THE 1982 SURVEY

This work focused on a previously unsurveyed triangle of land bounded on the south by the lip of the main opencast (the Ogofau Pit), on the east by the unmetalled track running past the entrance to the so-called Mitchell Adit, and on the north by the road heading up the Cothi valley (see Figs 1.1 and 4.1). The land here slopes steeply from south to north towards a relatively flat area in which, on a low mound, stands the Carreg Pumsaint or 'Five Saints Stone' (Fig. 4.2). At its western end lies a far larger, tree-covered mound, traditionally identified in the literature and on the Ordnance Survey maps as a 'motte' (Nash-Williams 1952, fig. 3). Prior to 1982, only one other archaeological feature had been recognised in print, namely Tank J, a roughly square depression lying at the southern limit of the survey area immediately adjacent to the unmetalled track (Lewis and Jones 1969, 256).

In the course of the survey, a complex series of earthworks was recorded, apparently terraced into the hillside overlooking the level area and the Carreg Pumsaint. The most prominent of these was an elongated depression (Tank Y), extending over c. 40m, the outer edge of which had been characteristically reinforced by a bank, presumably to aid in the retention of large quantities of water. Along its length, several shallow breaks may represent the remains of possible sluice gates. Immediately above this lay a second levelled feature, less certainly identifiable as a tank (Tank/Leat Z). Both would seem to have been supplied with water from their eastern ends, though the original arrangement has been somewhat obscured by the modern trackway and later mining activity. A pair of shallow gullies extending up the hillside to the south east of the track might, in part, be connected with this, especially in view of their possible relationship with Tank P identified by Lewis (1977, fig. 2).

At the western end of these two features, several well-defined channels swung away to the north-west to focus on a steeply-sloping ramp, the western edge of which was formed by a small horseshoe-shaped opencast of later date. No features were recognised along the length of this ramp, except at the very foot of the slope, where there were the clear traces of a stone-lined channel and, immediately to the east, the remains of a 7.00 by 4.50m platform apparently terraced into the slope. These would seem to mark the position of some sort of structure. Several other features on the hillside, most notably the four squarish hollows (U to X), proved less easy to interpret, though originally they may have been part of a larger complex.

While full confirmation of the character of the site would have required large-scale excavation, enough survived to suggest that the combined features could represent the remains of a water-powered mill, most likely connected with ore-crushing and gold extrac-

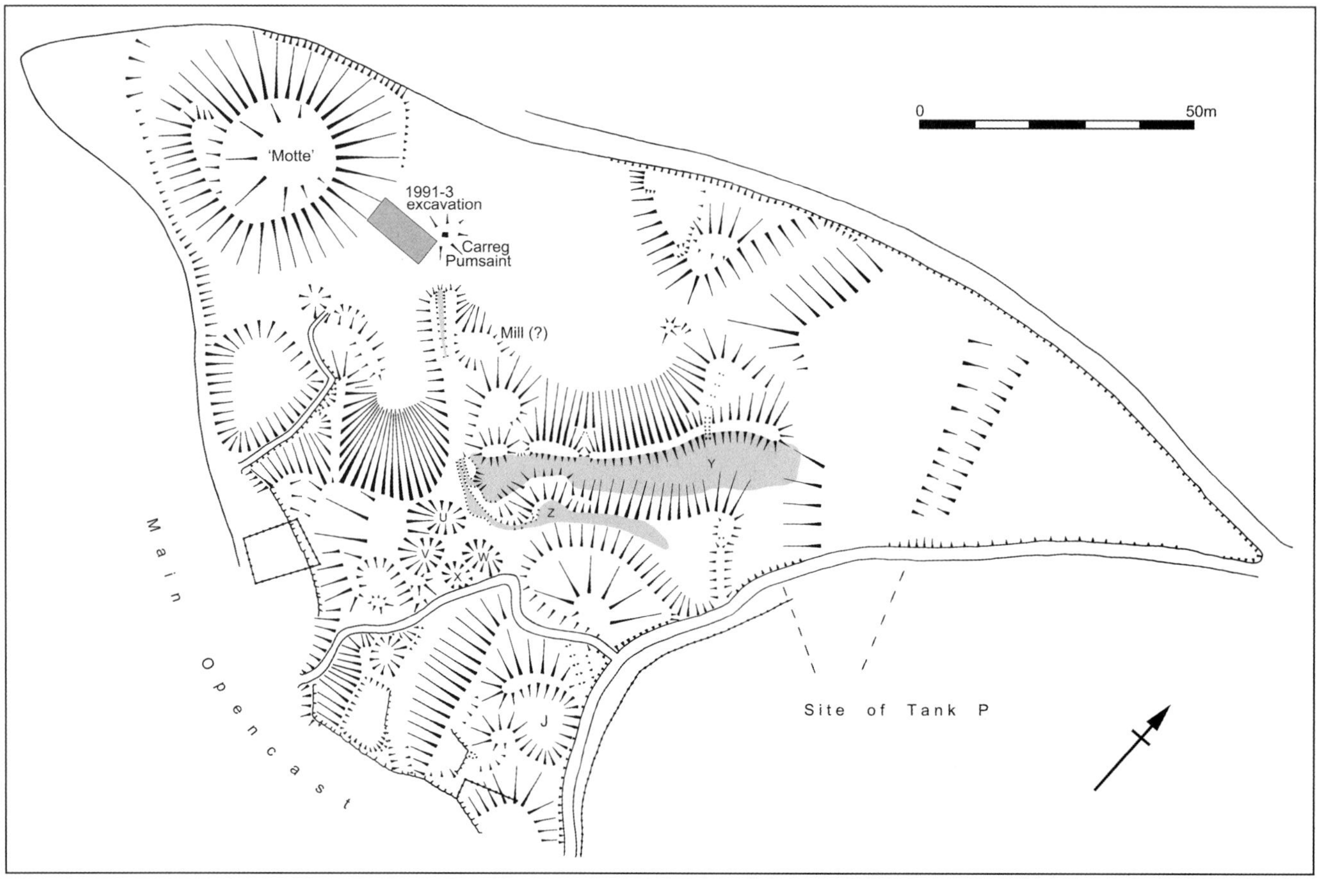

Fig. 4.1 1982 mill survey and location of 1991–93 excavations

Fig. 4.2 The Carreg Pumsaint

tion, in which the crushing machinery would have been driven by a waterwheel set alongside the mill and served with water from the tank complex on the hillside above. Drawing upon comparative evidence from the tin industry in Devon and Cornwall, where water-powered crushing and smelting sites are common (*cf.* Austin *et al.* 1989; Gerrard 2000), it was further suggested that the crushing machinery comprised a series of vertical hammers, operated by a system of trip levers or cams attached to the extended axle-drum of the waterwheel. These hammers would have pounded the ore on a mortar stone, itself set on the ground inside some sort of wooden box or frame; in the process the stone became indented and had to be turned as its successive faces became too worn. The occurrence of what is obviously one such mortar stone in the nearby Carreg Pumsaint, no more than 5.00m away from the presumed mill site, obviously lent further credence to the postulated interpretation of the site, as did the well-known representations of similar stamp mills in the pages of the 16th-century writer Agricola, involving both dry and wet stamping (1556). Additional processing activities would have been carried out in the flat area below the mill, hence the spreads of crushed quartz known from ploughing in the fields to the north.

Establishing the date of such a complex was obviously difficult without excavation, though three possibilities were considered at the time of the survey, *viz*. Roman, medieval or early modern, or mid 19th century. On balance, however, it was felt that a convincing *prima facie* case could be made for a hitherto unsuspected phase of mining at Dolaucothi at some time in the period *c*. 1200 to 1700. This was based on three principal lines of argument: first, the identification of the Carreg Pumsaint as a mortar stone of a type characteristic of the medieval or early modern tin industry in Devon and Cornwall; second, the apparent absence of any concrete evidence for the use of a cam-operated or trip-hammer technology during the Roman period; and third, the postulated relationship of the mill complex and its associated processing wastes with the presumed quarry ditch of the nearby mound traditionally identified as a Norman motte. Such arguments were necessarily based on a series of probabilities, the details of which could only be tested by excavation.

4.3 THE EXCAVATIONS

4.3.1 Location and aims (Fig. 4.1)

In order to examine some of the ideas generated by the survey work, a small-scale trial excavation was undertaken for two weeks in September 1984. This involved a single trench, 10.00m long by 2.50m wide, designed to sample the surviving archaeology in an area close to the site of the Carreg Pumsaint at the base of the 'motte'. In the event, the increasing complexity of the stratigraphy soon made it clear that a larger-scale excavation would be necessary to elucidate the full site sequence (Austin and Burnham 1985). In the face of other excavation commitments this did not prove possible before 1991, when the decision was taken to extend the original trench towards the modern Cothi valley road, in order to examine an area totalling 12.00 by 5.00m. This proved to be the first of three successive seasons through to 1993, the preliminary results of which were published in a series of interim reports and articles (Burnham and Burnham 1991c; 1992; 1993; Burnham 1995; 1997).

The renewed excavations of 1991–93 had a number of clearly defined objectives:

(i) the recovery of a clear sequence through the successive processing wastes indicated by the work in 1984, but over a wider area;
(ii) the examination of any related or underlying features, not least any possible motte ditch;
(iii) the investigation of the structure of the so-called motte itself;
(iv) the recovery of any associated dating evidence.

In the event the excavations produced some unexpected results, which raise intriguing questions both for our knowledge of ancient processing technologies and for our understanding of the character of Roman and later mining at Dolaucothi.

4.3.2 The site sequence (1991–93)

Post-excavation analysis of the stratigraphic sequence has indicated three broad phases in the makeup of the site, the first two of which may well have overlapped, with considerable implications for the character of the mining and ore-processing activities carried out in the vicinity:

Phase 1: a phase of extensive dumping across the site, represented by the core of the mound associated with the Carreg Pumsaint in the north-east corner, a dump of shale and sandy clay loam extending along the eastern side of the trench, and several successive shale spreads which merged together to form the basal core of the 'motte' dominating the site's western side. Late in this process, a 'built-up' path was imposed across the shale spreads at the base of the 'motte', connected perhaps with the decision to create a conical spoil heap. These various horizons defined a convenient hollow across the centre of the site. Several shallow hollows were also cut into the dump along the eastern side.

Phase 2: a phase of activity indicative of ore-processing in the immediate vicinity of the site, represented by a complex sequence of interleaved fine clay slimes and coarsely-crushed quartz and shale horizons which had progressively infilled the hollow across the centre of the site, and by an elongated depression or pit with its associated fills which had cut into the sequence of shallow hollows along the eastern side. These were all overlain and sealed by a pronounced turf horizon, representing a period of significant stability across the site.

Phase 3: a series of at least three separate road surfaces and other miscellaneous features post-dating the turf horizon, which were clearly associated with finds of 19th- and 20th-century date.

4.3.3 Phase 1 (Figs 4.3 to 4.10)

(A) The underlying topography

With the removal of the different horizons associated with ore-crushing and processing, the way was open for an examination of the successive deposits extending across the central and eastern parts of the site, their stratigraphic relationship with the material forming the structure of the so-called motte to the west, and the underlying topography. In the event, this did not prove as easy as had been hoped, partly because of the stratigraphic problem of separating the successive stone spreads across the centre of the site, where they were increasingly found to merge together without any discernible intervening horizons, and partly because of the safety problems imposed by the increasing depth of the excavations. As a consequence, exploration of the lowest deposits had to be confined to small-scale trenches aimed at investigating their critical stratigraphic relationships.

This ensured that the 'natural' was not encountered at any point in the excavation, which makes it impossible to gauge either the original depth of the dumped deposits or the underlying configuration of the site. All that can be said with any certainty is that, whatever the original character of the open area north of the mill complex, it was clearly substantially remodelled by extensive dumping, much of it composed of barren mining debris devoid of the characteristic quartz or black shale in which the gold is located. One possibility, which remains incapable of proof for the present, is that the area began life as an exploratory working or opencast, hence its convenience as a landfill site for later mining waste.

(B) The topography at the eastern end of the site

Discussion of the successive episodes of dumping can best begin at the eastern end of the site (Figs 4.3 to 4.6), where the early topography was dominated by two distinctive deposits, one forming the sloping core of the mound [172] associated with the Carreg Pumsaint in the north-east corner, the other a dump of shale and sandy clay loam [151] extending across the eastern side of the trench. The former [172], which could only be examined in plan over a very limited area, was found to be composed entirely of a mass of angular shale fragments of varying sizes, apparently sloping away towards the south and west beneath all the later horizons. As such, it was clearly the earliest stratigraphic feature in this part of the site. In an effort to pursue its profile and to establish more about its overall extent, a narrow trench was cut through the overlying dump [151], parallel with the main north-south section, but the mound proved to be sloping off at such a steep angle that work had to be abandoned before any significant new details could emerge.

The overlying horizon [151], which was exposed right across the eastern end of the site, proved to be an extensive dump of small to medium-sized angular shale fragments, intermixed with a distinctive yellowish-brown sandy clay loam (10 YR 5/6). In section at the north-eastern corner, it was clearly shown to ride up onto the core of the mound [172] associated with the Carreg Pumsaint, where it attained a maximum thickness of *c.* 0.55m. Elsewhere its full depth could not be established, though the evidence suggests that it might originally have formed part of a more prominent ridge extending along the eastern flank of the site, the top of which was extensively transformed by the excavation of a succession of hollows from a higher level [152; 165; 167; 169; 171]. To the west, the surviving remnants of this ridge were seen to be sloping away beneath the first [163] in a series of substantial stone spreads, which ultimately merged with the shale core [49] of the 'motte' dominating the western side of the site.

In an attempt to establish more information about the extent and profile of this ridge, a narrow trench was excavated to a maximum depth of 0.70m through the stone spread [163], extending from east to west across the central part of the site (see section projected on Fig. 4.5). This demonstrated that the ridge sloped progressively downwards and westwards to a point some 4.50m from the main east section, where it petered out altogether across yet another mass of vacuous shale, the limits of which could not be traced in the area available. To the west, this shale merged imperceptibly with the earliest recognisable horizon [174] forming the core of the so-called motte, the separate existence of which was indicated by a thin lens of silty clay [173] trending upwards and westwards across its surface for *c.* 1.50m. Beyond this point, it too merged imperceptibly with the overlying stone spread [163].

(C) The topography of the central hollow

In the hollow thus formed across the central part of the site, excavation revealed a complex sequence of dumped deposits, the details of which were established both in plan and section (Figs 4.5 to 4.9). This included the substantial stone spread [163] already mentioned, a series of stony deposits [159 to 162] which clearly intervened stratigraphically between it and two discrete layers of shale and frost shatter [157 and 158] extending across the northern and southern sides of the hollow respectively, and two later stone spreads [154 and 156] both of which had clearly been dumped against a 'built-up' area [155] apparently reinforcing the base of the 'motte' (see section 4.3.3(D) below). This latter [49] proved to dominate the western half of the site, its profile rising increasingly steeply up to and beyond the west section.

The most extensive of these successive deposits also proved to be the lowest in the stratigraphic sequence [163]. It comprised a spread of medium to large angular shale fragments, exposed as a coherent horizon right across the central part of the site (Figs 4.6 and 4.7). To the east, it clearly rode up onto the ridge of shale and sandy clay loam [151] already described, but to the west, where it merged with the sloping profile of the 'motte' beneath the 'built-up' area, its stratigraphic relationships could not be defined in plan. The situation became clearer, however, when a narrow section was cut through it, extending from east to west across the centre of the site (Fig. 4.5). This demonstrated that it increased in thickness downslope from the crest of the ridge, to a

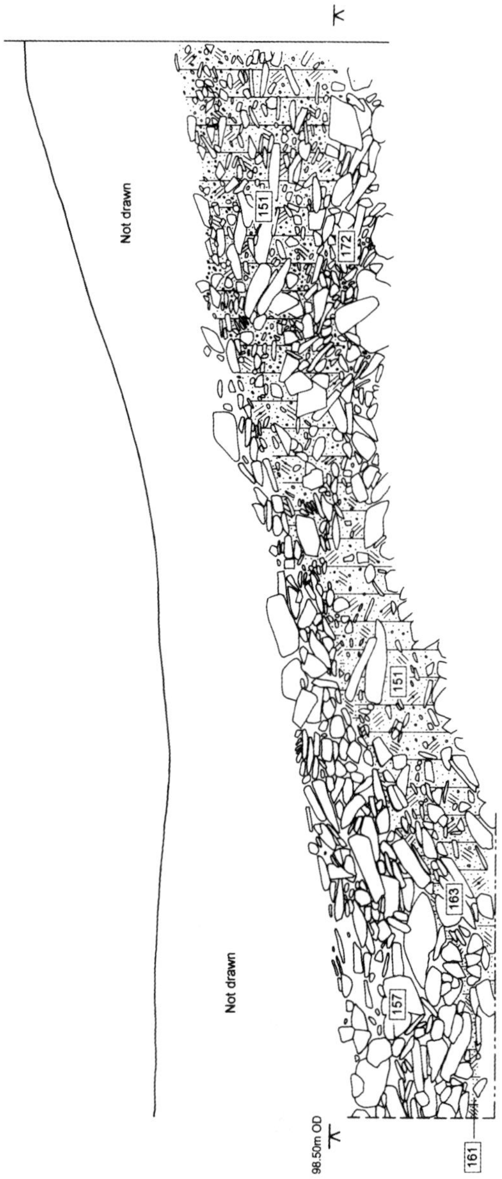

Fig. 4.3 North section (east end only)

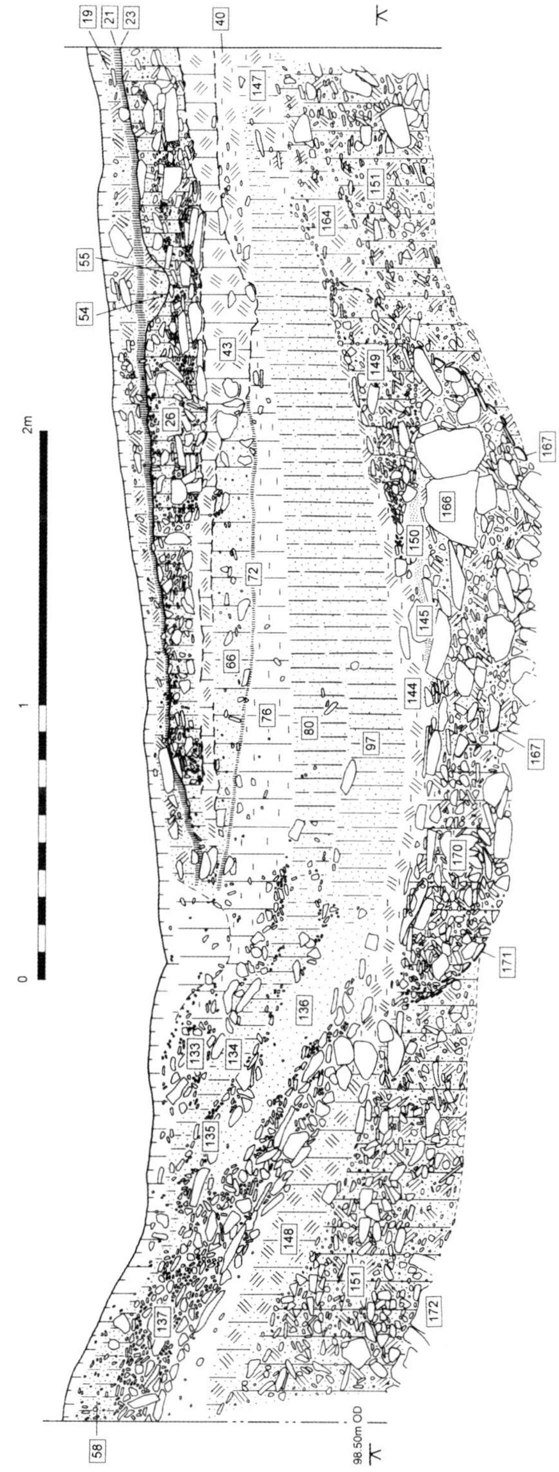

Fig. 4.4 East section

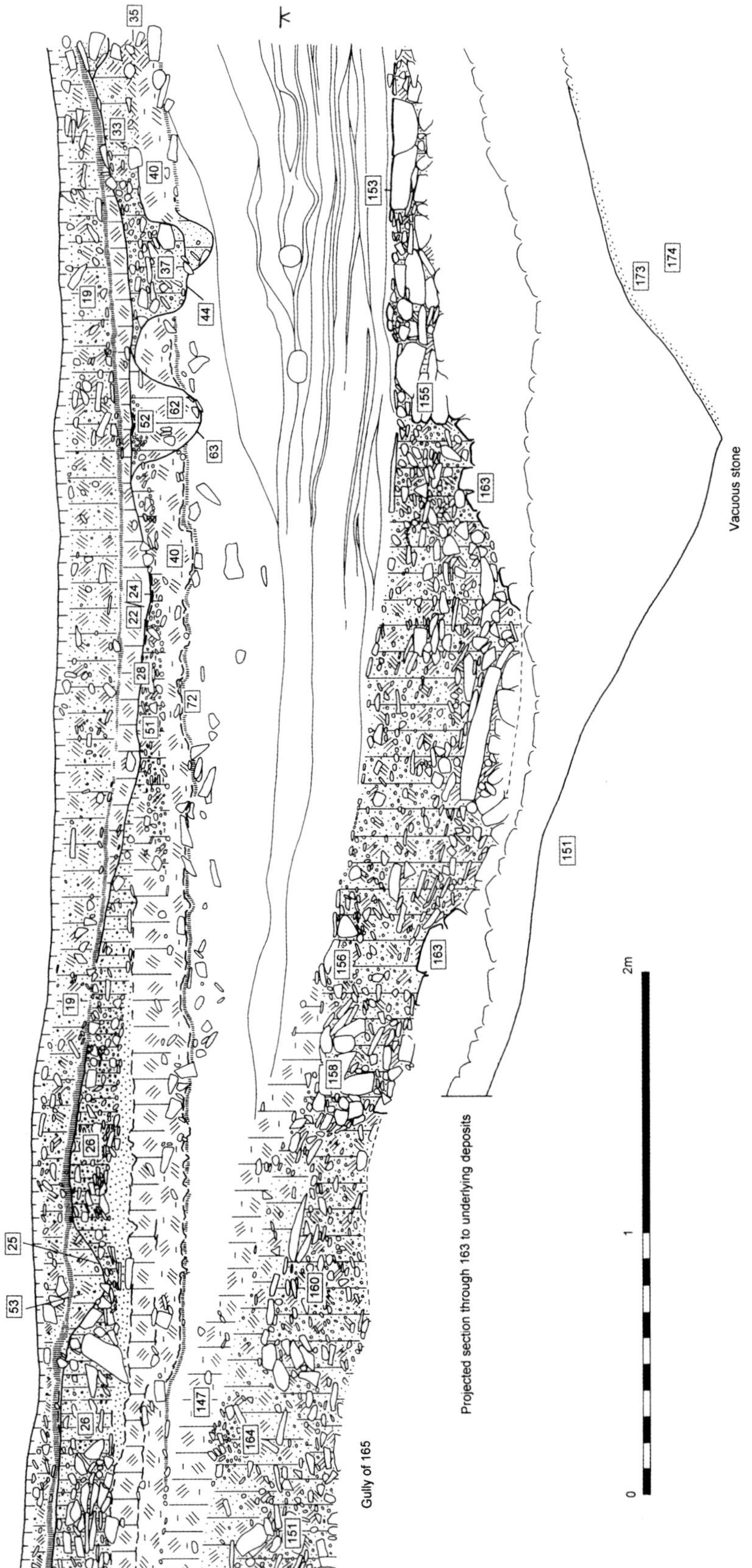

Fig. 4.5A South section: eastern half; For a detail of the central part including the processing horizons see Fig. 4.11

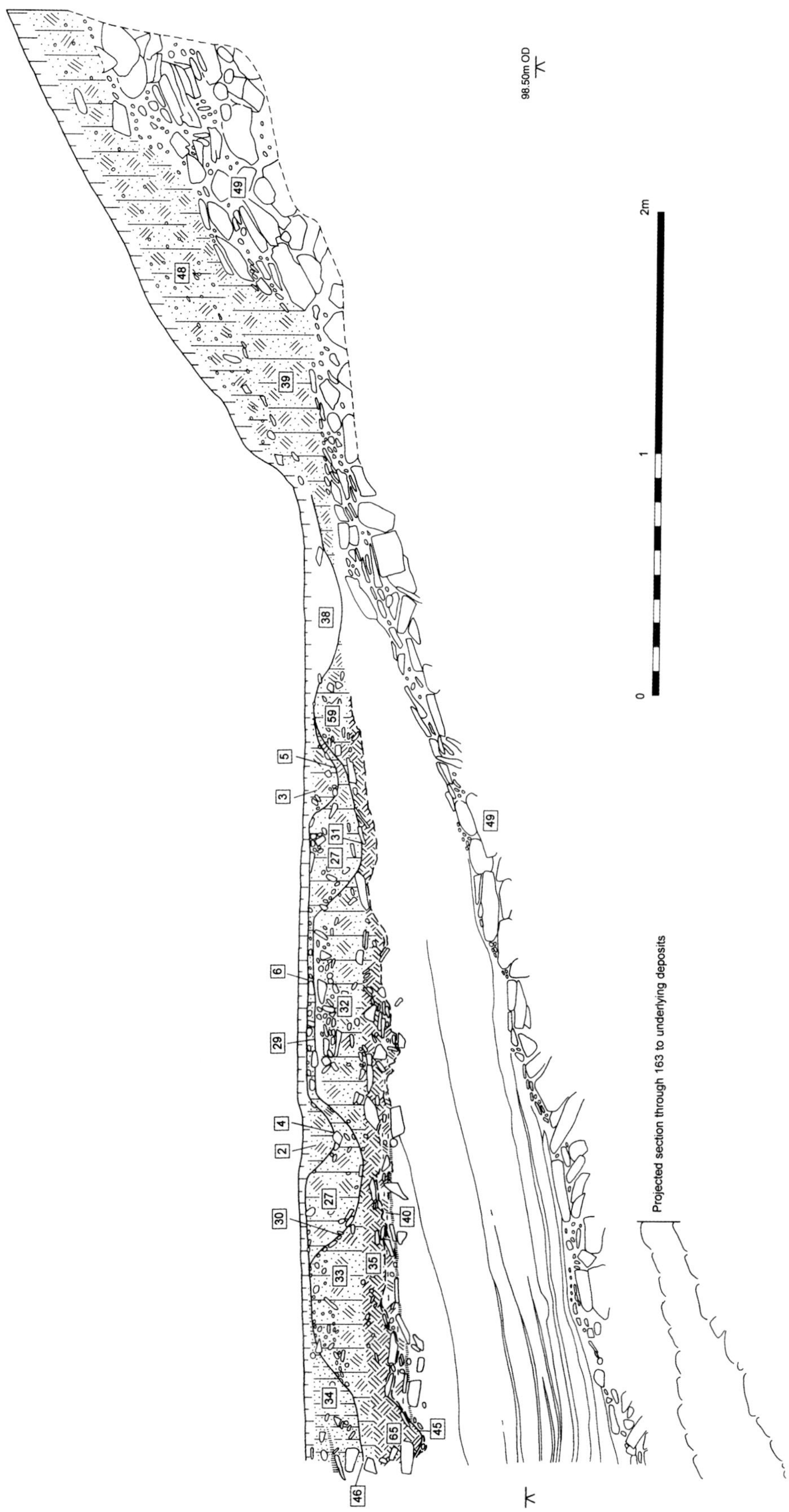

Fig. 4.5B South section: western half; For a detail of the central part including the processing horizons see Fig. 4.11

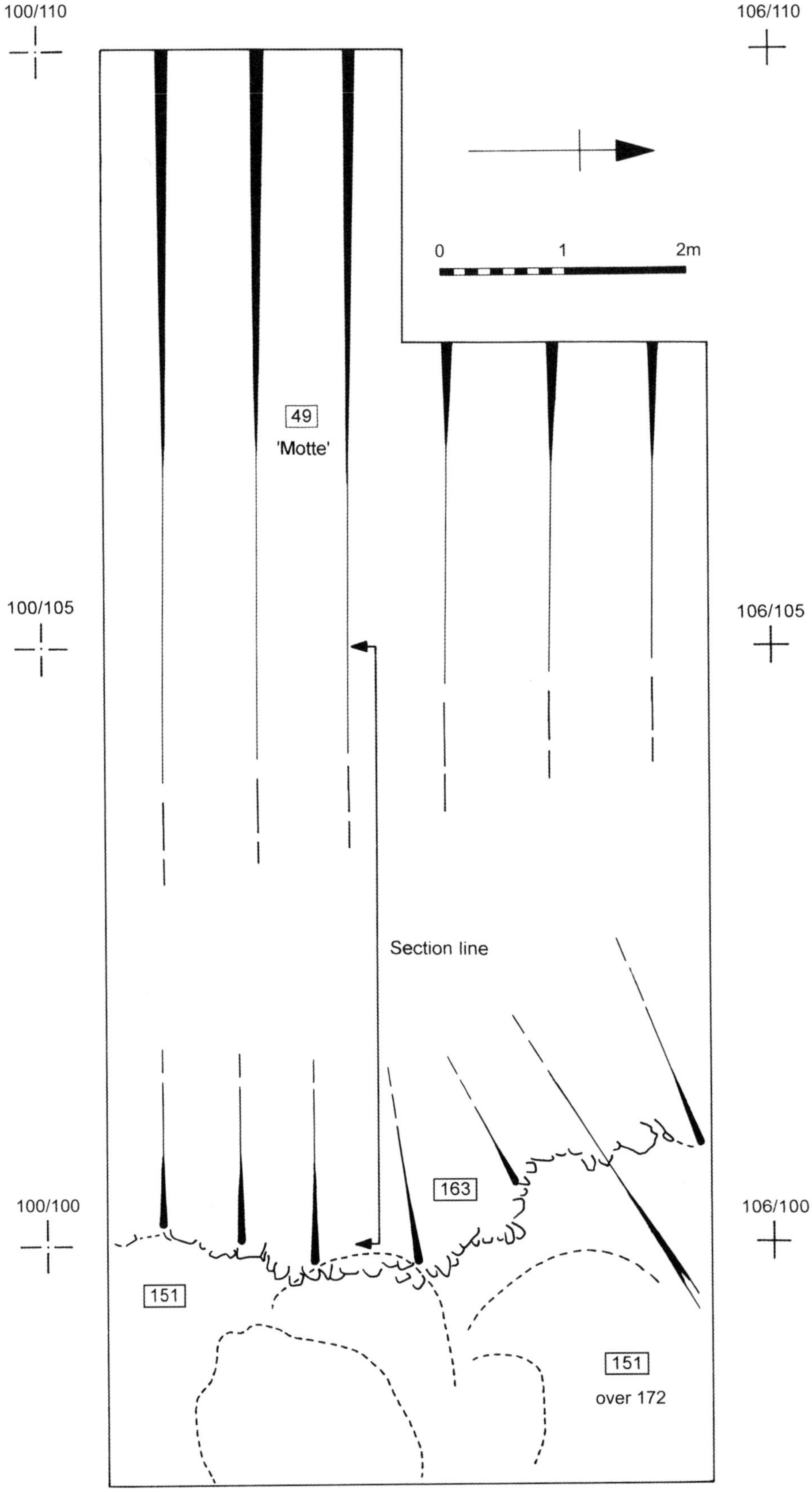

Fig. 4.6 Earliest dumped horizons across trench, including location of east-west section through stone spread [163]

Fig. 4.7 Stone spreads at base of excavations, looking west

maximum depth of 0.70m, before thinning out again upslope, where it was separated from the underlying dump of shale [174] by the thin lens of silty clay [173] already described. Beyond this, both dumps merged with the vacuous shale makeup of the 'motte' [49], without any discernible breaks.

The overlying spreads, which were sandwiched stratigraphically between this stone horizon [163] and the two layers of shale and frost shatter [157 and 158], proved generally unremarkable and can thus be treated summarily here (Fig. 4.8):

162 a thin lens of yellowish-brown loamy coarse sand [10 YR 5/6] with a 90% admixture of what appeared to be finely-crushed quartz and shale and small angular shale chippings, located over an area no more than *c.* 1.00m square at the base of layer [160] in the south-east corner of the trench. It did not show in section.
161 a thin lens of yellowish-brown sandy clay loam [10 YR 5/6] with a 90% admixture of very small to medium-sized angular shale fragments, first located on the slope of the mound associated with the Carreg Pumsaint (Fig. 4.3), where it was sandwiched between the two main stone spreads [157 and 163]. Across the trench to the south, it proved difficult to separate from layer [160].
160 a dump of yellowish-brown sandy clay loam [10 YR 5/6] with an admixture of up to 95% very small to medium-sized shale fragments and some quartz chippings, first located in the south-eastern corner of the trench, where it was at its thickest (Fig. 4.5), and subsequently traced as it thinned out onto the slope above the main stone spread [163], where it effectively merged with layer [161].
159 a dump of small to medium-sized angular shale fragments interspersed with a dark brown sandy clay loam [7.5 YR 3/3], first seen in plan between the exposed edges of the overlying stone spreads [157 and 158]. In the event, only its western edge proved easy to define. Elsewhere, it progressively merged with the makeup of the overlying horizons extending across the eastern side of the hollow.

Across the northern side of the central hollow, the uppermost horizon in this particular sequence proved to be a spread of frost-shattered shale [157], intermixed with medium to large-sized angular shale fragments, which had clearly been dumped into the hollow to a maximum depth of 0.50m. It rapidly thinned out to the east, however, where it rode up onto the mound associated with the Carreg Pumsaint, while to the west it merged imperceptibly with the underlying stone spread [163] and the shale make-up of the 'motte' [49], without any obvious intervening horizons. Stratigraphically, it must also pre-date the construction of the 'built-up' edge [155] at the base of the 'motte', since both clearly underlay its associated trodden surface [153].

Its counterpart [158] across the southern side of

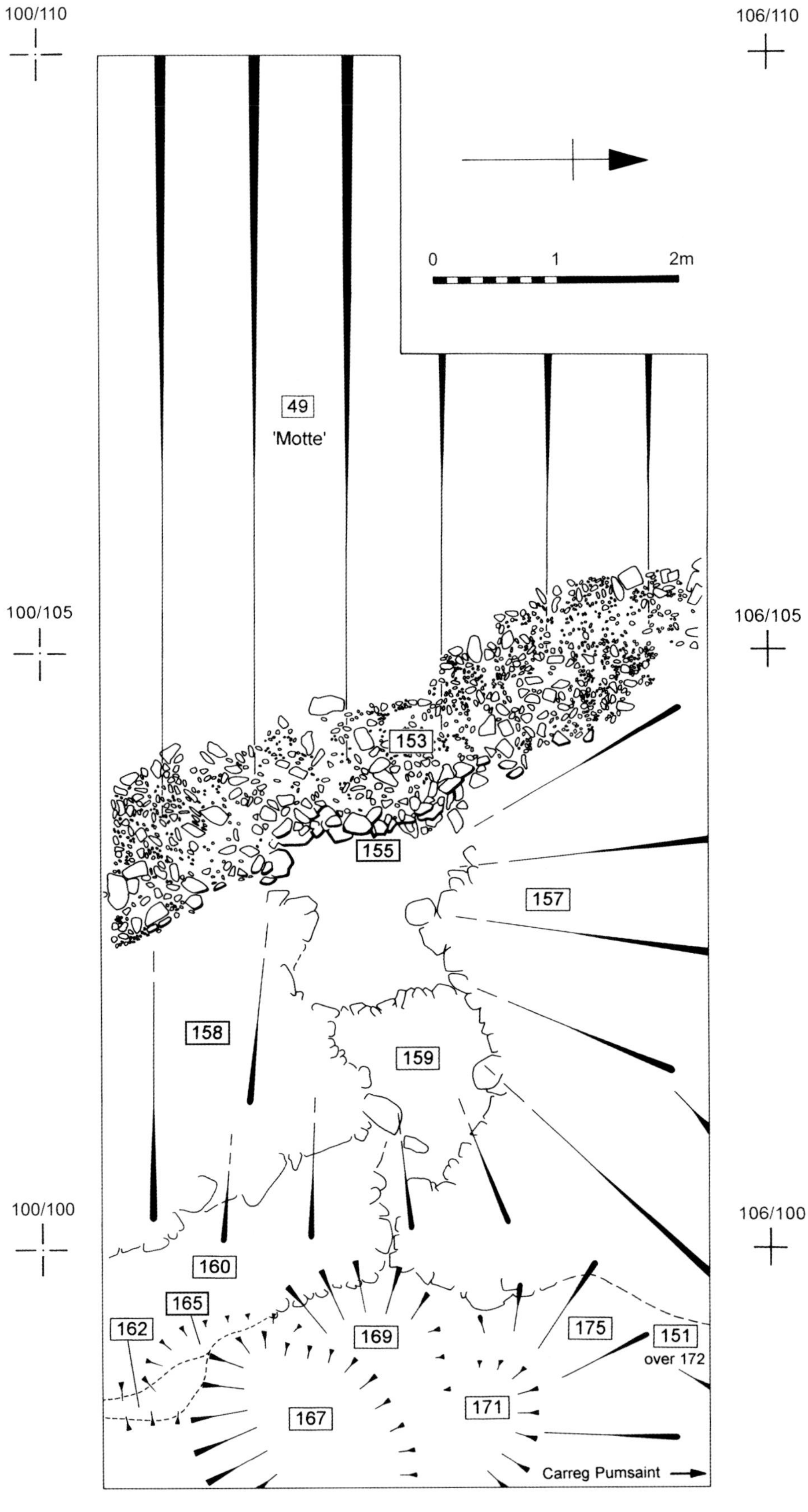

Fig. 4.8 Composite plan of successive dumped horizons across central part of trench, the 'built-up' edge around the base of the 'motte', and the four hollows at the eastern end of the site

the central hollow proved to be a spread of generally brown sandy clay loam [7.5 YR 4/4] with an 80% admixture of small to medium-sized shale fragments. Though generally brown in colour, the matrix actually varied from yellowish-brown [10 YR 5/4] to dark brown [10 YR 3/3], this latter the result of extensive manganese panning. In plan it merged with the underlying stone spread [159], while in the south section it proved difficult to separate from the overlying horizon [156]. Unfortunately, no clear stratigraphic relationship could be established with the 'built-up' edge.

The two remaining deposits associated with this broad phase of dumping both lay east of the 'built-up' edge, against which they had clearly been dumped (Fig. 4.9):

156 a spread of yellowish-brown sandy clay loam [10 YR 5/6] with a 60 to 70% admixture of very small to medium-sized shale fragments, located along the south and south-eastern sides of the central hollow, partly sealed beneath layer [154].

154 an extensive spread of very small to small-sized frost shatter, set in a matrix of dark brown loamy coarse sand [7.5 YR 3/3], apparently derived from a dump of medium to large angular shale fragments with which it is coterminous. It extended across the northern two-thirds of the central hollow, east of the 'built-up' edge. Its eastern and southern edges were clearly defined where they overlay the yellowish-brown matrix of layer [156], but to the north the shale fragments merged with the vacuous stones forming the makeup of layer [157], such that the two proved difficult to disentangle in section or plan.

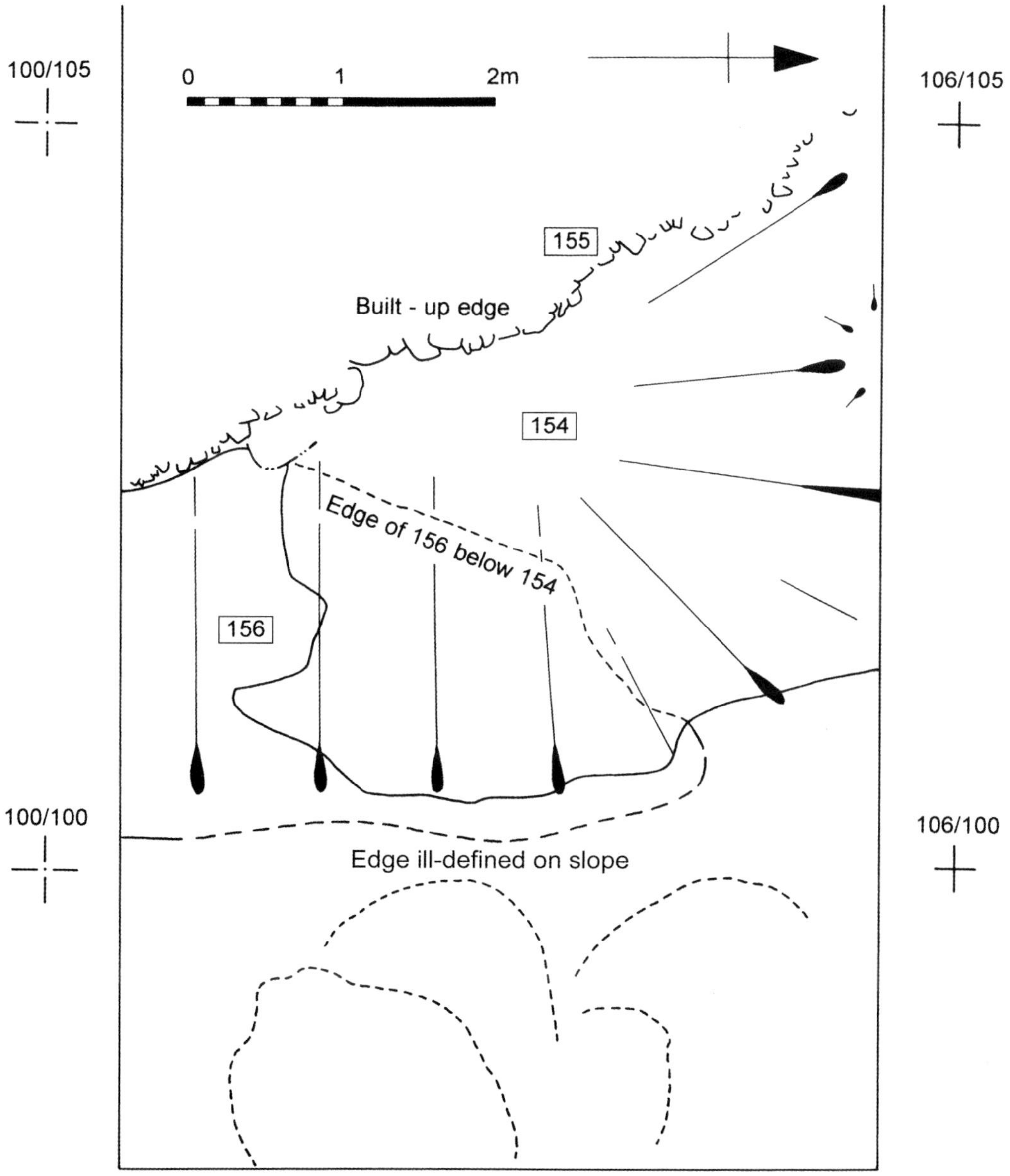

Fig. 4.9 Latest dumped horizons east of 'built-up' edge

Fig. 4.10 'Built-up' edge to 'motte', looking west

(D) The 'built-up' edge at the base of the 'motte'

As has been indicated, quite late in this sequence of dumping operations, the presumed base of the motte seems to have been reinforced with a shallow 'built-up' edge [155], clearly trending from south-south-east to north-north-west across the underlying stone spread [163] onto the makeup associated with layer [157] (Figs 4.8 and 4.10). Its upper surface [153] had clearly been heavily compacted, above a makeup of large angular shale fragments set in a solid matrix composed of frost-shattered shale chippings and a dull yellowish-brown sandy clay loam [10 YR 5/3]. This proved to be very different in character from the structure of the rest of the mound to the west, where the shale and overlying frost shatter were very loose and fragile, and from the corresponding spread of shale and frost shatter [157] to the north. Across the deepest part of the central hollow, as far as the south section (see Fig. 4.5), it seemed as if the makeup had been deliberately banked up to ensure an even gradient on top.

The interpretation of this feature is obviously difficult in so small an area, because so little evidence is available from which to reconstruct its original extent or function. One possibility, however, is that it was created during the dumping process to serve as a convenient rampway onto the mound as it progressively rose upwards, much like the barrow run on a modern-day spoil heap. Some such access up the mound's circumference would certainly have been necessary, given that the feature must have been raised by non-mechanical means.

(E) The 'motte' at the western end of the site

The topography of the western end of the site was necessarily dominated by the mass of angular shale fragments [49] forming the makeup of the presumed motte, the visible profile of which rose progressively steeply above the line of the 'built-up' edge [155], up to and beyond the west section. This proved to have little, if any definable structure, either in plan or in the limited areas where it was cut into along the south section (Figs 4.5 and 4.6). Where it was examined in the narrow trench extending from east to west across the centre of the site, however, it was found to merge imperceptibly with a series of other stone deposits, including the dump of broken shale and frost shatter [157] along the northern end of the site and the extensive stone spread [163] across the centre. This latter, in its turn, also merged with an even lower stone horizon [174].

From the character of the stratigraphy associated with this so-called motte and the absence of any obvious structure in its makeup, it seems highly unlikely that it is anything other than a conical spoil heap, constructed as part of a much more extensive phase of dumping which clearly transformed this part of the site. If so, then this has significant implications for the chronology of this area and for the overlying processing wastes, because it effectively invalidates the TPQ which a genuine motte would have provided (see section 4.2 above). More importantly, to judge from the absence of any turf horizon or other intervening horizons between the successive areas of dumping and the overlying wastes, it would seem likely that the erection of the spoil heap and the accumulation of the wastes were not separated by any significant length of time; it is even possible, but unprovable, that the upper part of the mound continued to be built while the wastes were being put in place. This makes the recovery of several pockets of charcoal from the lowest [146] of the horizons squashed between the shale forming the 'built-up' edge highly significant, because they should provide the critical TPQ in the absence of any other datable finds (see section 4.8.1 below).

(F) The shallow hollows at the eastern end of the site

During the investigation in 1992 of the stone spread [145] towards the base of the large hollow [152] at the eastern end of the site (see section 4.3.4(B) below), several enigmatic contexts were exposed, the precise character of which was not immediately apparent. Excavation initially proceeded on the basis that they represented fills within the hollow, but this became less and less easy to sustain, at which point the various contexts were sketched and their full investigation was reserved for 1993. When work was resumed, it rapidly became obvious that we were dealing with a series of at least four shallow, intercutting hollows [165; 167; 169; 171], each *c.* 1.00 to 2.00m in diameter, with their own distinctive fills (Figs 4.4 and 4.8). To judge from the evidence, only the very bottom of these features had actually survived the cutting of the larger, phase 2 hollow, so that it was not possible to estimate either their original extent or depth.

The uppermost of the four [167], and the best preserved, proved to be an oval hollow, measuring 2.00m by at least 1.70m, part of which lay outside the excavated area. It contained three successive fills [149; 150; 166], the first two of which were inadvertently removed in 1992 during work on the overlying hollow [152]. The lowest of the three [166], examined in 1993, was characterised by a spread of medium to large angular shale fragments, some of the largest of which appeared to have been laid across the base of the hollow, interspersed with a mass of smaller shale chippings. Immediately above this lay a thin lens [150] composed almost entirely of very finely-crushed quartz chippings, no more than 1 to 3mm in size, and a few small shale fragments, which must represent the residue of nearby processing activities. This in its turn was sealed by a layer [149] of very small to medium-sized shale fragments, intermixed with a dull yellowish-brown sandy clay loam [10 YR 5/4], which could either represent a deliberate infill, or be the result of natural weathering, derived perhaps from the surrounding matrix [164] in the south-eastern corner of the trench.

This uppermost feature had truncated the greater part of a second hollow [165], measuring at least 1.40 by 0.80m, which lay at a slightly higher level to the south-west. Traces of what may have been a shallow gully associated with this feature were also noted, extending south into the main section. Both would seem to have been cut into one of the successive episodes of dumping [160] already described across this part of the site (see section 4.3.3.(C) above). Unfortunately, what little of the relevant fill had survived later disturbance, was inadvertently removed in 1992 during the work on the fills of the overlying hollow [152]. From the south section, however, it seems as if the gully had been infilled, in part at least, by the layer [164] of dull yellowish-brown sandy clay loam [10 YR 5/6] and small shale fragments located in the south-eastern corner of the trench; the same may well have been true of the hollow itself.

Both hollows described so far had cut into a third, very similar feature [169], measuring in excess of 1.80 by 0.80m, which was located immediately to the north-west. Its fill [168], which was noted during its removal in 1992, comprised a spread of mainly large angular shale fragments, interspersed with a mass of smaller shale chippings, much like that described at the base of the uppermost hollow [167]. This third feature in its turn had cut into and truncated a fourth such hollow [171], measuring more than 1.10 by 1.00m, which lay immediately to the north-east. It too had a distinctive fill [170] of medium to large angular shale fragments, which extended beyond the edge of the excavation. What might have been a further shallow hollow [175] may have lain on the slope to the north, though this was by no means as clear-cut as the rest. Its fill [148], which comprised a mixture of finely-crushed quartz and very small to medium-sized shale fragments, intermixed with a brown sandy clay loam [7.5 YR 4/4], was also distinctive by comparison with its neighbours.

Stratigraphically, all five hollows are broadly contemporary with the main episodes of dumping across the central part of the site, but any further refinement presents problems because so many of the critical relationships were destroyed by the interface of the larger, phase 2 hollow [152]. Of the links which could be established, the two lowest hollows [171; 175] had certainly been cut into the makeup of the ridge [151] at the eastern end of the site, the penultimate pair [165; 169] had been cut into the dump of shale and sandy clay loam [160] at the south-eastern corner of the trench, while the uppermost [167] had truncated a further layer of shale and sandy clay loam [164] in the same area, which itself sealed the gully leading into the penultimate hollow [165]; this latter also merged with the stony deposit [158] across the southern side of the central hollow, but in such a way that their precise relationship could not be established. Given these difficulties, the discovery of a pocket of charcoal in one of the fills [150] of the uppermost hollow [167] assumes a measure of significance (see section 4.8.4 below).

The exact character and function of these successive hollows remains somewhat uncertain in the absence of any diagnostic finds, though it seems probable that they were connected in some way with nearby processing activities, rather than being straightforward pits. It could, of course, be argued that their distinctive stone fills (certainly in the cases of [167], [169] and [171]) were simply the result of the extensive dumping which so characterises this phase, but this would be to ignore the presence of a similar layer [145] at the bottom of the much larger, phase 2 hollow [152], which incorporated a quern fragment and was overlain by a distinctive horizon [144] of processing wastes (see section 4.3.4(B) below). The repeated occurrence of such a stone layer might suggest that they had all been deliberately laid, perhaps to allow water to soak away from stockpiles of processed material, which could then be recovered for further treatment at a secondary stage. This could well account for the presence of the distinctive lens of finely-crushed quartz [150] in the makeup of the uppermost hollow [167], which must surely represent the residue from nearby processing activities.

4.3.4 Phase 2 (Figs 4.3 to 4.5, 4.11 to 4.41)

(A) The processing wastes across the centre of the site

The horizons discussed so far in connection with phase 1, in particular the ridge at the eastern end of the site, the spoil heap and its 'built-up' edge at the west, and the intervening stone spreads across the centre, together defined a large and convenient, bowl-shaped depression, which clearly extended beyond the main south section. This proved to contain a complicated sequence of interleaved deposits, some no more than a few millimetres thick in places, which had accumulated to a maximum depth of 0.75m. Two distinctive types of material were represented, one composed of finely-crushed clay slimes, the other of more coarsely-crushed quartz and shale (much like brown sugar in consistency), both apparently derived from extensive ore-processing in the immediate vicinity of the site.

These deposits were first explored towards the end of the 1991 season, when considerable problems were encountered in trying to differentiate the successive spreads at those points where they had been laid directly one on top of another. A useful preview was provided, however, by the excavation of an intrusive bath-shaped feature [70], which had been cut down through them from a higher level. In view of these difficulties, considerable effort was expended during the 1992 season, not just in recording and planning the inter-relationships of the different interleaved lenses, but also in tying their stratigraphy securely to the main section across the south end of the site. In the process, it became clear that we were dealing with two very broad sequences, one above and one below a spread of coarsely-crushed material [97] which extended across much of the trench (Fig. 4.11).

(i) The early sequence

While the last of the overlying clay spreads was still being removed from its south-western side in 1992, work on the extensive spread [97] was initially confined to the northern half of the site. This latter sectioned what seemed to be a fairly thick, homogeneous deposit, interspersed with numerous discontinuous, thin sandy lenses, none of which could be pursued as coherent, individual contexts. At the time, it was thought to be the result of extensive dumping, made up of many separate barrow loads of material, all of which overlay an earlier sequence of wastes glimpsed in the sides of the bath-shaped feature [70]. In the event, as excavation progressed and extended to the southern half of the site, it soon became apparent that the picture was more complex and that we had been cutting into several superimposed horizons of variable thickness and extent, which could only really be isolated across the central north-west to south-east axis of the trench where they became interleaved with an equally complex sequence of thin clay-like spreads (see Figs 4.12 and 4.13). This problem was further complicated, however, by the identification of a possible shallow feature [105], which had clearly cut through and isolated various deposits on the sloping profile of the 'motte' to the west and must also have affected some of the successive horizons across the central part of the hollow; the implications of this will be considered later in this section.

Careful definition and recording of the edges of these successive spreads of clay and crushed material, and of their inter-relationships, revealed a coherent sequence, with fine clay spreads predominating on the south and south-western sides of the hollow and coarsely-crushed horizons on the north and north-east. The full stratigraphic details are shown in the accompanying matrix (Fig. 4.14), while the contexts themselves are summarised briefly below (see Figs 4.15 to 4.29):

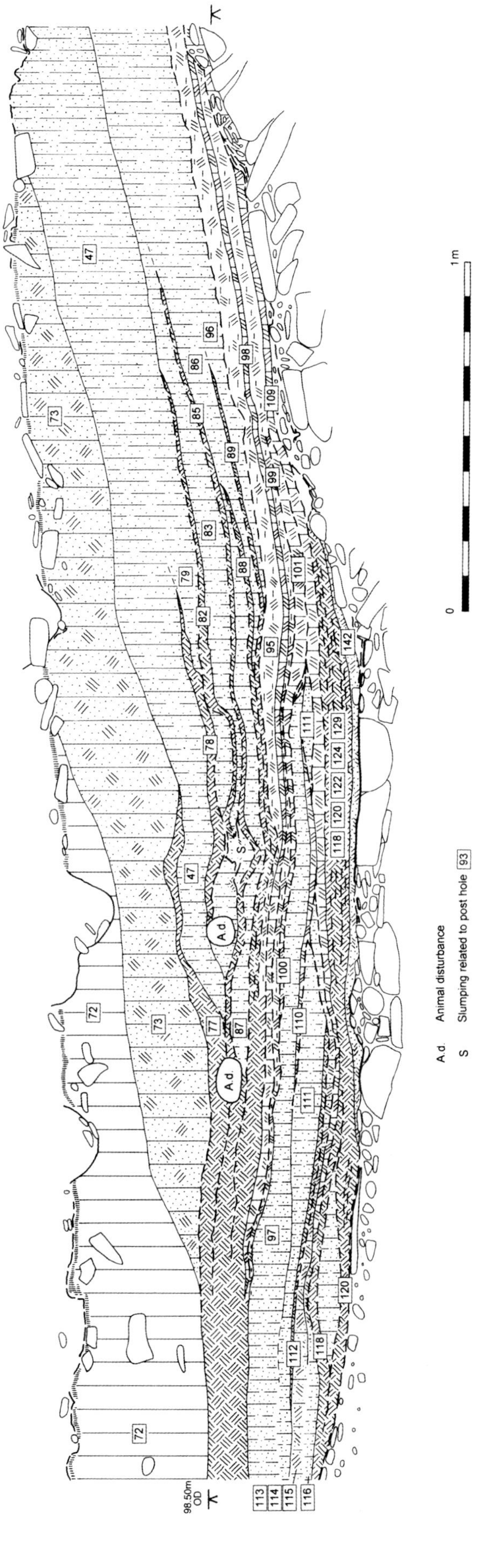

Fig. 4.11 Detail of processing horizons across central part of south section

Fig. 4.12 Successive clay spreads interspersed with lenses of coarsely-crushed material, looking west (at level of Fig. 4.28)

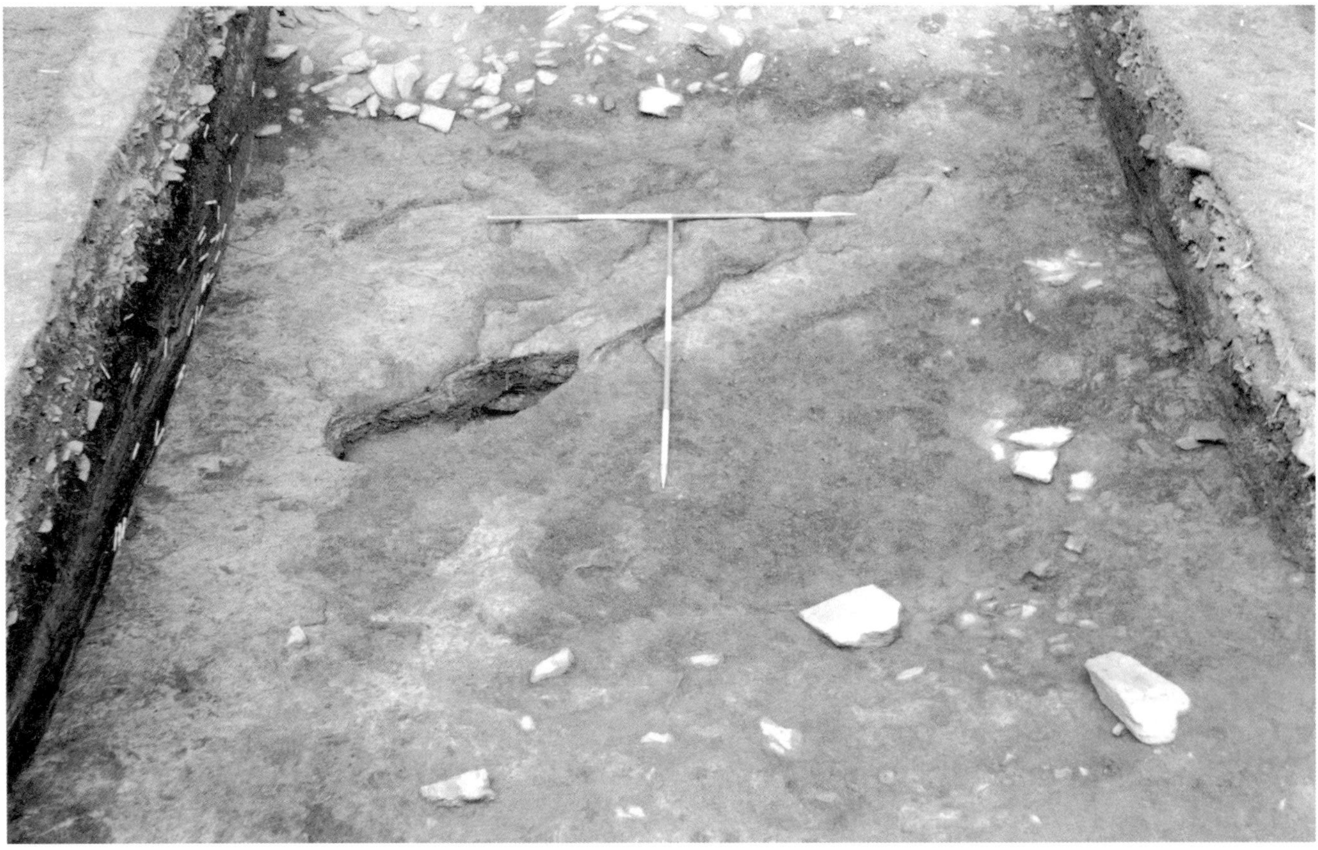

Fig. 4.13 Successive clay spreads interspersed with lenses of coarsely-crushed material, looking west (at level of Figs 4.22 and 4.24)

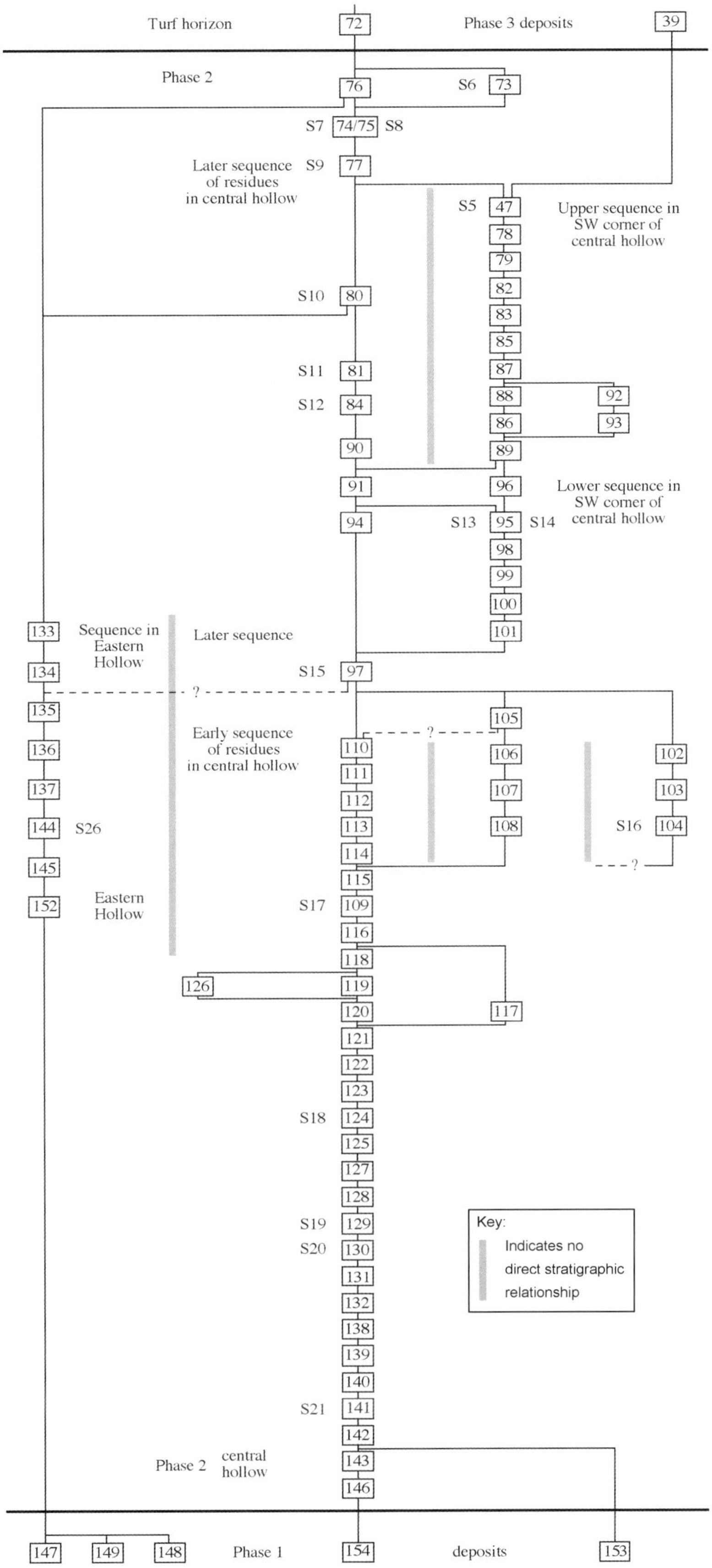

Fig. 4.14 Stratigraphic matrix of processing wastes across the central hollow

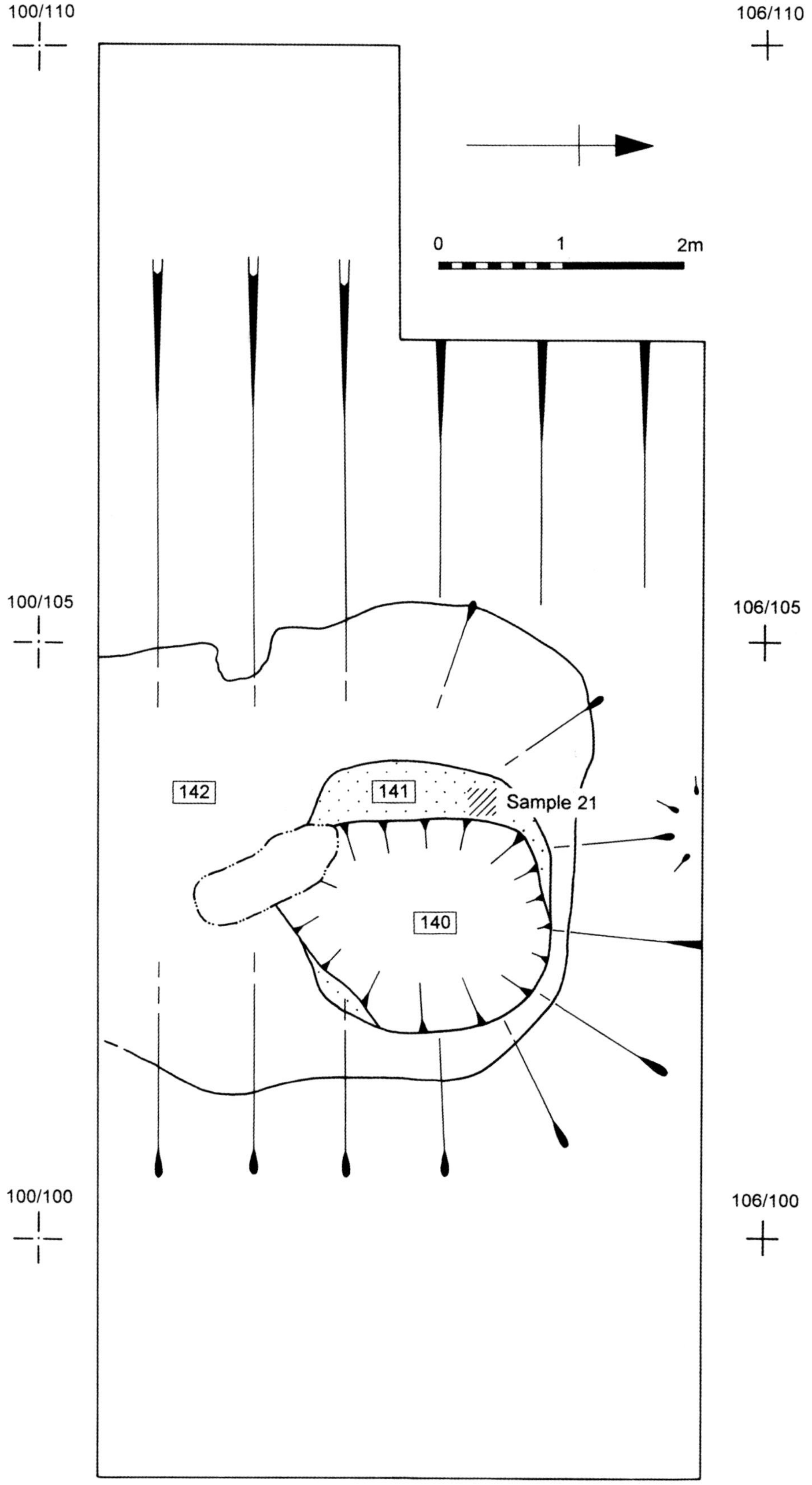

Fig. 4.15 Extent of contexts [140] to [142] across central hollow

146 a variable lens of dull yellowish-brown clay [10 YR 5/4] forming the lowest definable horizon above and in between the stones at the base of the hollow (not illustrated). Did not extend as far as the section. Included various pockets of charcoal where it was squashed into the makeup associated with the built-up edge [155]. May well be a natural silting prior to the deposition of the processing wastes.
143 an irregular spread of dark reddish-brown sandy clay loam [5 YR 3/4] with a 90% admixture of small angular shale fragments and crushed quartz (not illustrated). Extended across the base of the hollow, but not identified as far as the section.
142 an extensive spread of brownish-grey sandy clay [10 YR 5/1] with a 10% admixture of very small angular shale fragments (Fig. 4.15). Extended as a thin spread into the section (Fig. 4.11). Probably represents the lowest of the clay sequences.
141 a distinctive lens of dull reddish-brown loamy fine sand [5 YR 4/3] composed almost entirely of very finely-crushed material. Confined to the edges of a shallow feature [140] by which it had been cut away, suggesting that it was originally more extensive (Fig. 4.15).
140 a shallow, roughly circular hollow, *c.* 1.80 to 2.00m in diameter, partly cut away by the bath-shaped feature (Fig. 4.15). Filled by two successive deposits [139; 138].
139 a bucket-full dump of dark reddish-brown, finely-crushed material [5 YR 3/4], apparently tipped into the shallow hollow from the north-east (not illustrated).
138 a spread of brownish-grey silty clay [10 YR 6/1], infilling and extending beyond the shallow hollow (not illustrated). Its extent beyond this, where it effectively merged with two other clay horizons [129 and 131], could not be determined.
132 a dump of predominantly dark reddish-brown loamy coarse sand [5 YR 3/4] with a 95% admixture of crushed shale and quartz. On excavation, it proved to be much more mixed and interleaved with convoluted lenses of siltiness and clay, probably representing a sequence of dumps into the sinkage above the shallow hollow from the north-east. Its edge across the south-western side of the feature formed quite a pronounced ridge beneath clay spread [131] (Fig. 4.16).
131 a spread of brownish-grey clay [7.5 YR 6/1], otherwise indistinguishable from the underlying [138] and overlying [129] clays, except where it was sandwiched between successive dumps of crushed material [132 and 130] (Figs 4.16 and 4.17).
130 a dump of dull reddish-brown loamy coarse sand [5 YR 4/3] with a 95% admixture of crushed shale and quartz. Its extent corresponded closely with that of its earlier counterpart [132] (Figs 4.17 and 4.18).
129 an extensive spread of brownish-grey clay [7.5 YR 6/1] traced reasonably coherently across the south side of the central hollow (Figs 4.11, 4.19, 4.20 and 4.22), though it was not always easy to separate from the overlying clay spread [127].
128 a dump of dark reddish-brown, crushed shale and

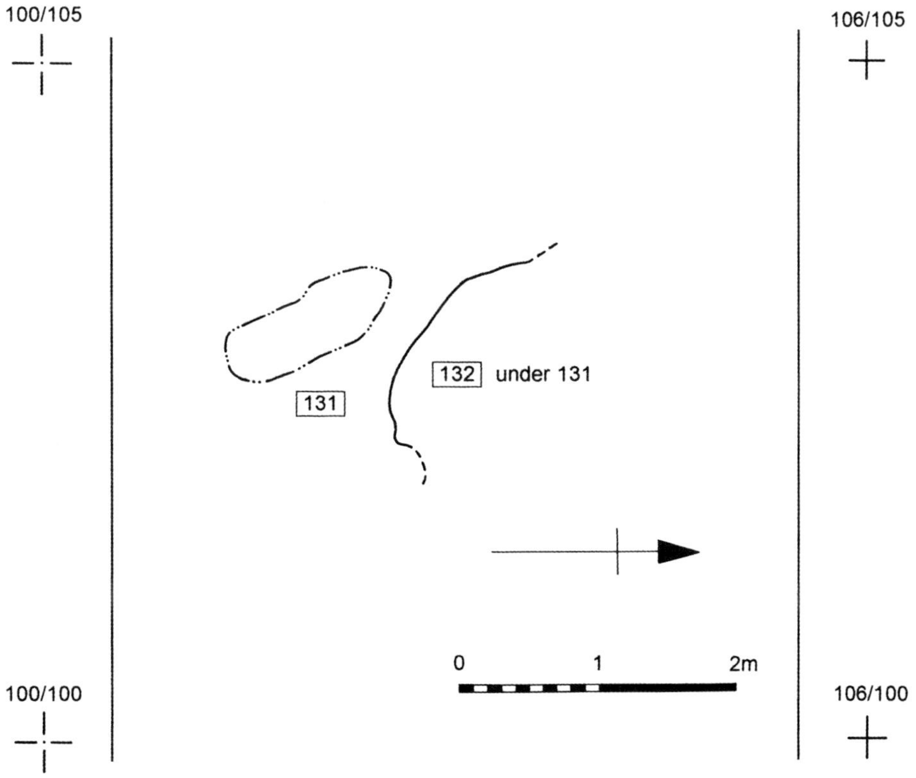

Fig. 4.16 Established edge of clay spread [131] above crushed horizon [132]

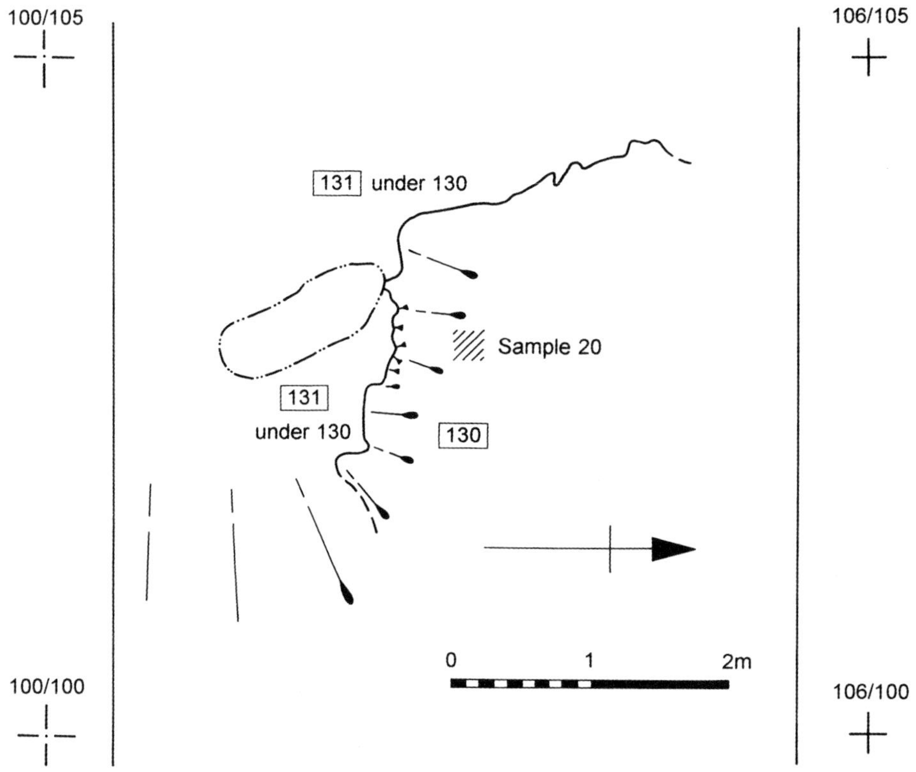

Fig. 4.17 Established edge of crushed horizon [130] above clay spread [131]

Fig. 4.18 Dump of coarsely-crushed material [130], looking west

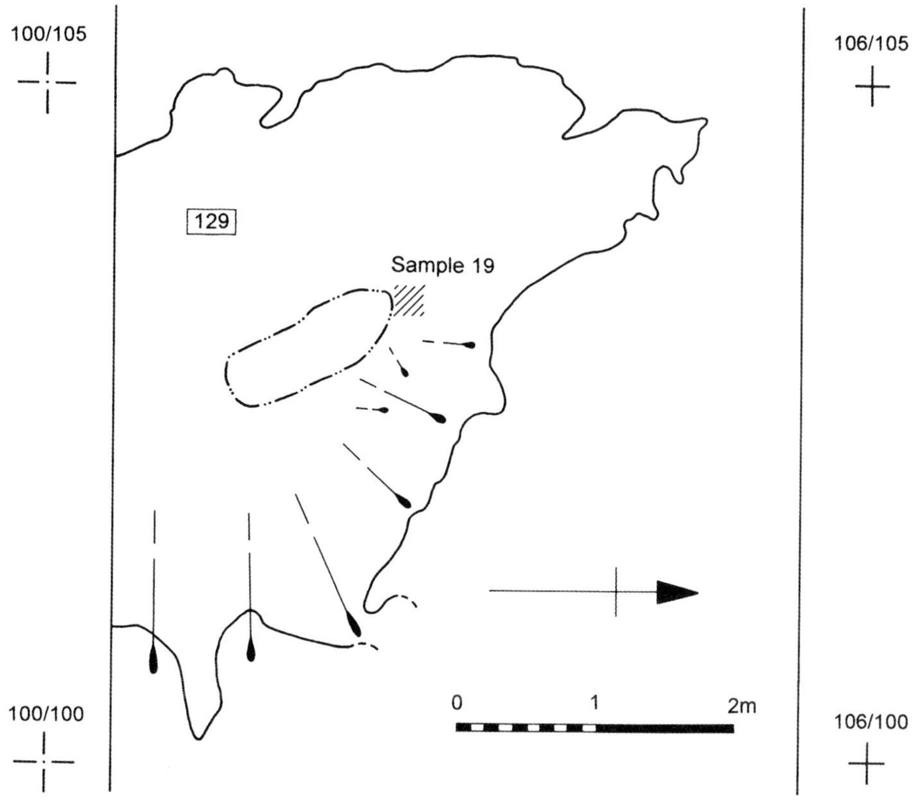

Fig. 4.19 Extent of clay spread [129]

Fig. 4.20 Clay spread [129], looking south

quartz [5 YR 3/4] (not illustrated), only recognised where it intervened between successive clay spreads [127 and 129].

127 a spread of brownish-grey clay [10 YR 5/1], only distinguished from the overlying clay spread [124] during the removal of the intervening dump of crushed material [125] (Figs 4.21 and 4.22). Must originally have extended to the section.

125 a layer of dull reddish-brown loamy coarse sand [5 YR 4/4] with a 95% admixture of crushed shale and quartz, characterised by a heavy iron pan at its base (Fig. 4.21). It separated two successive clay spreads [127 and 124] which were otherwise indistinguishable to the south.

124 a spread of dull yellowish-brown clay [10 YR 5/3], not easily distinguished from the underlying [127] and overlying [122] clay spreads, except where it was sandwiched between successive dumps of crushed material [125 and 123] (Figs 4.11 and 4.22). Must originally have extended to the section.

123 a layer of brown coarse sandy loam [7.5 YR 4/4] with a 90% admixture of crushed shale and quartz (Fig. 4.22). It separated two successive clay spreads [124 and 122] which were otherwise indistinguishable to the south.

122 an extensive spread of dull yellowish-brown clay [10 YR 5/3] traced coherently across the south side of the central hollow (Figs 4.11, 4.22 and 4.23).

121 an extensive layer of dark reddish-brown crushed shale and quartz [5 YR 3/4], characterised by a very heavy iron pan at its base (Fig. 4.24). Its south-western edge petered out on a line running north-west to south-east, where it separated two successive clay spreads [122 and 120].

120 a thin spread of dull brown clay [7.5 YR 5/4] traced coherently across the south side of the central hollow (Figs 4.11, 4.24 and 4.25). Likely to be the same as context [117], a very thin lens of brown sandy clay [10 YR 4/4], from which it was separated by an area of possible disturbance.

126 a thin pocket of dull reddish-brown sandy clay loam [5 YR 4/4] with a 90% admixture of crushed shale and quartz (not illustrated), only recognised near the south section, where it was sandwiched between two successive clay horizons [120 and 118].

119 a very localised patch of dark reddish-brown coarse sandy loam [5 YR 3/4] immediately south of the bath-shaped feature [70] (Fig. 4.24), where it was sandwiched between two successive clay horizons [120 and 118].

118 a thin spread of dull yellowish-brown clay [10 YR 5/4] located close to the south section, south of the bath-shaped feature (Figs 4.11 and 4.24).

116 a spread of brown loamy coarse sand [7.5 YR 4/4] with a 90% admixture of crushed shale and quartz, which petered out along a line running north-west to south-east, to the west of the bath-shaped feature (Figs 4.11 and 4.25). It clearly overlay the edge of the underlying clay [118].

109 an extensive spread of yellowish-brown clay loam [10 YR 5/6] traced right across the south end of the site and, as a narrow band, along the sloping profile of the 'motte' to the west (Figs 4.11, 4.26 and 4.27). In this

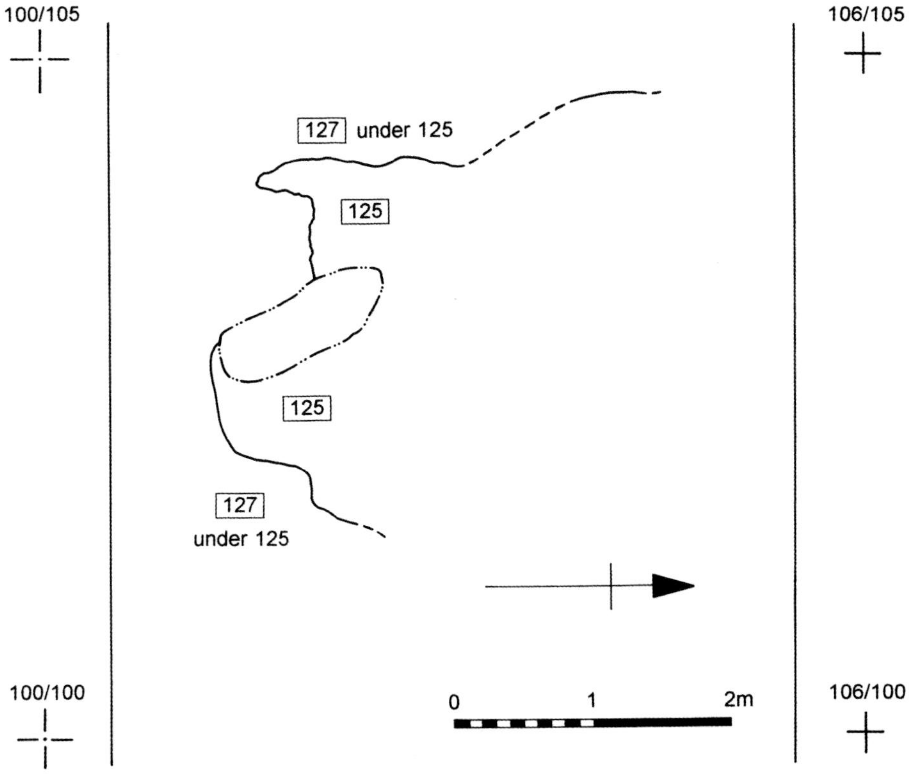

Fig. 4.21 Established edge of crushed horizon [125] above clay spread [127]

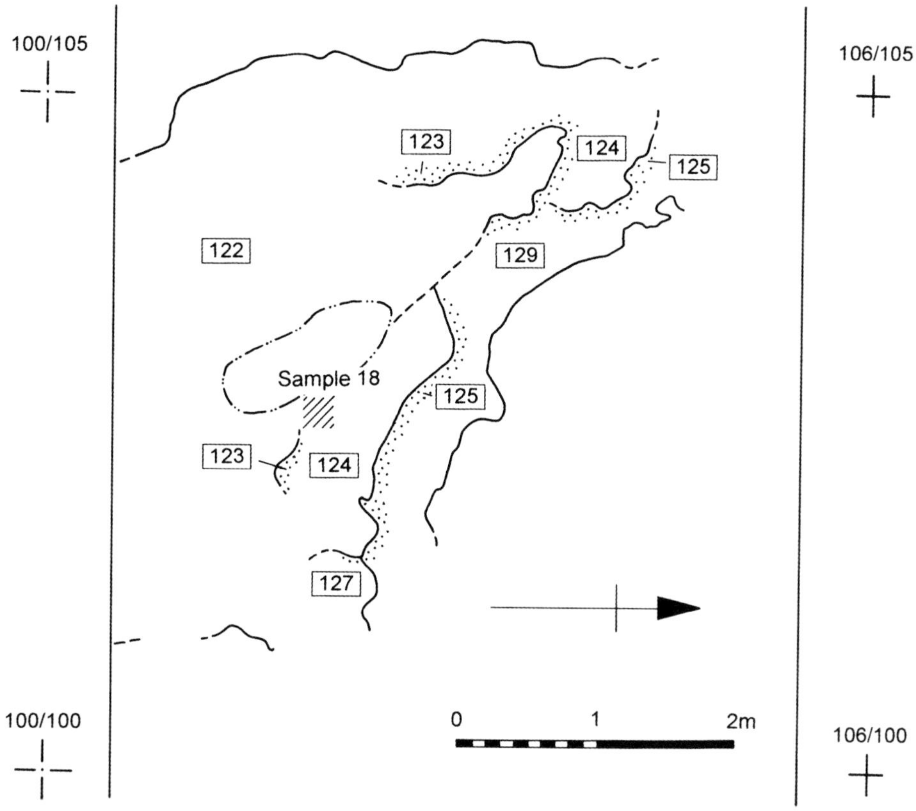

Fig. 4.22 Extent of successive clay spreads [122, 124 and 129] with intervening crushed horizons [123 and 125] indicated

Fig. 4.23 Clay spread [122], looking south

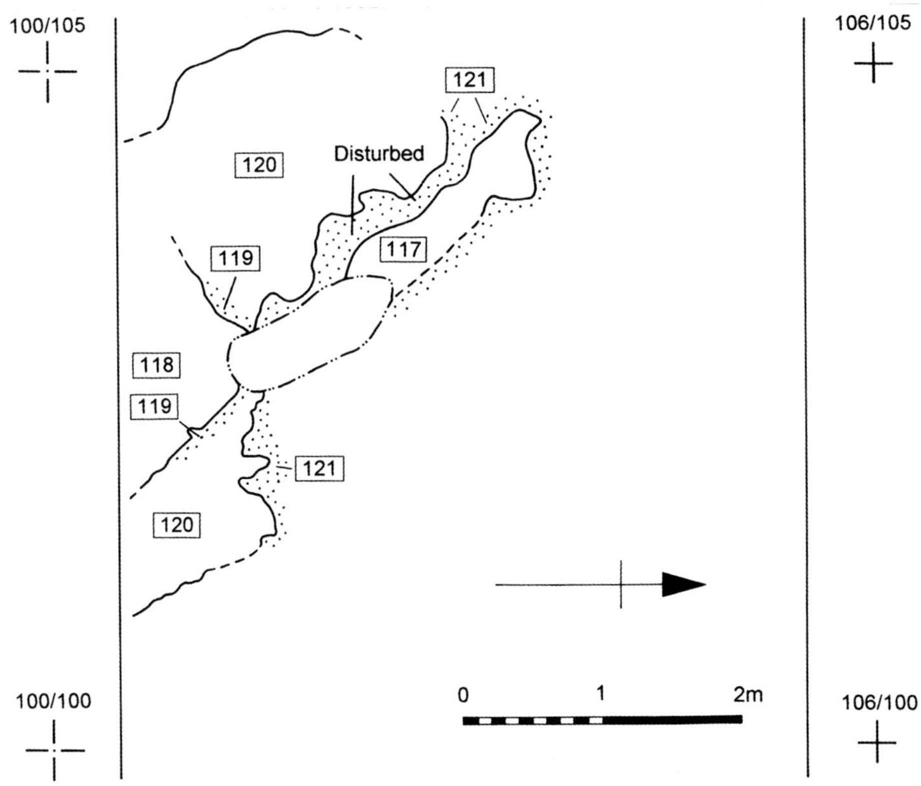

Fig. 4.24 Extent of clay spreads [117, 118 and 120] with intervening crushed horizons [119 and 121] indicated

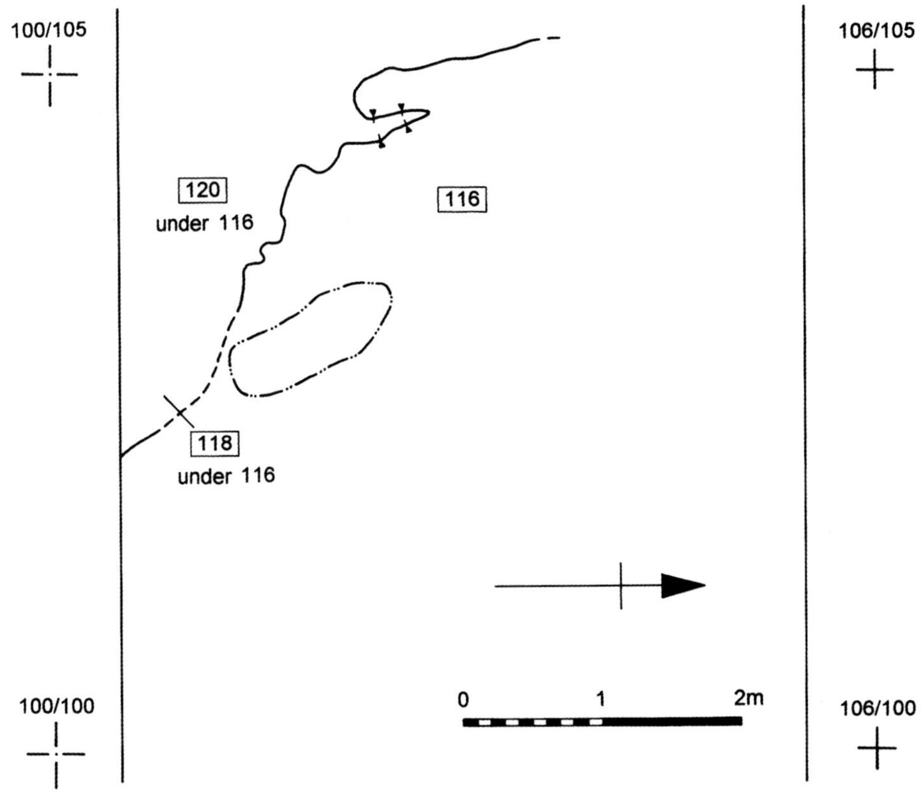

Fig. 4.25 Established edge of crushed horizon [116] above clay spreads [118 and 120]

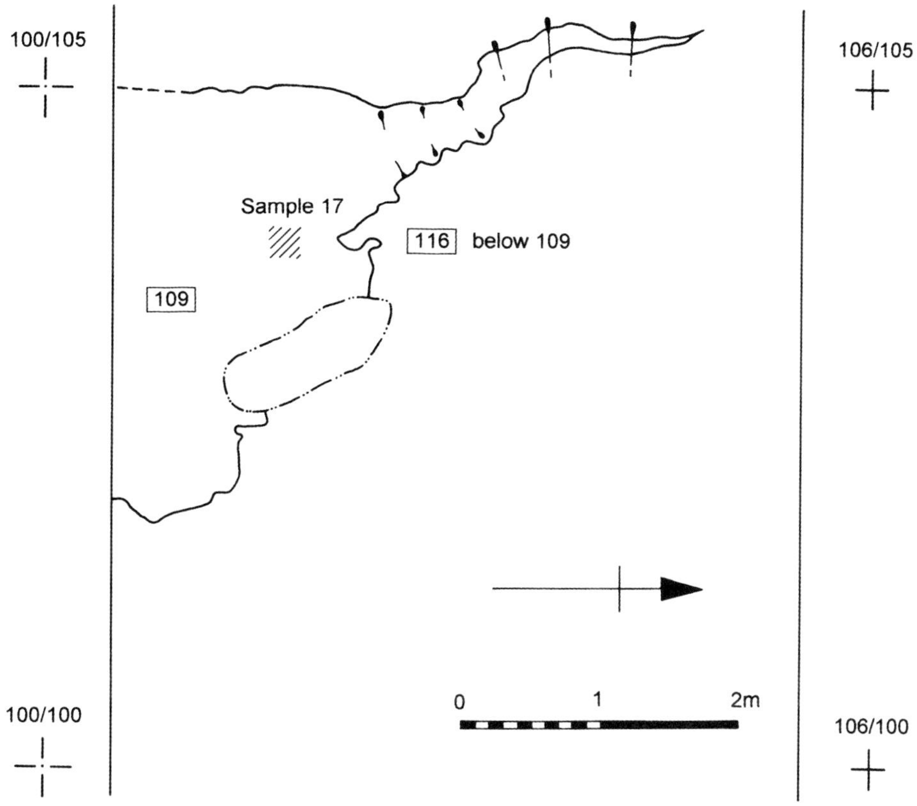

Fig. 4.26 Extent of clay spread [109]

Fig. 4.27 Clay spread [109], looking south

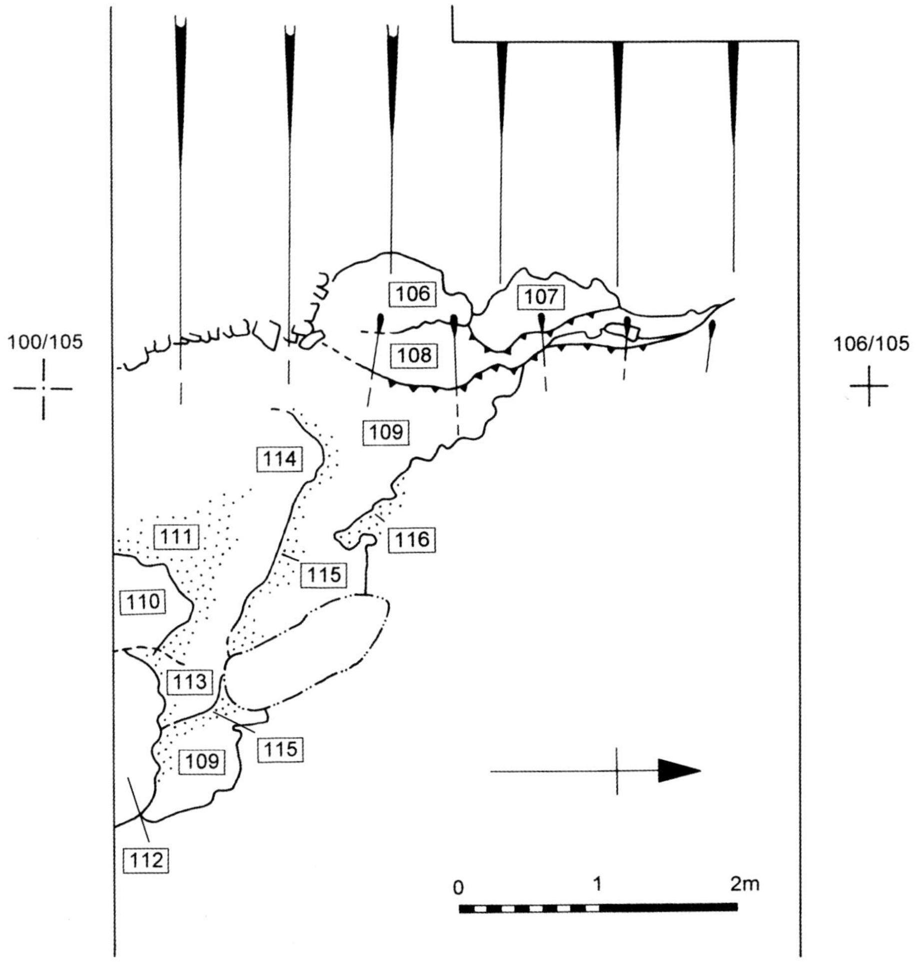

Fig. 4.28 Extent of successive clay spreads [110, 112, 114 and 109] with intervening crushed horizons [111, 113, 115] indicated, together with clay horizons [106–108] apparently truncated by interface [105]

latter area, there was a suggestion that the clay may have been disturbed or truncated from a higher level by the cutting of a shallow feature [105]. Other horizons might have been similarly affected.

115 a spread of dark reddish-brown crushed shale and quartz [5 YR 5/3] only recognised in plan at the south-western end of the site, where it was sandwiched between the underlying clay [109] and the overlying horizon [114] (Fig. 4.28). Indistinguishable in section from either of the overlying spreads [114 and 113] (Fig. 4.11).

114 a very thin skim of dull reddish-brown sandy clay loam [5 YR 5/3] with a 20% admixture of small shale and quartz fragments (Fig. 4.28), seen to be sandwiched between successive horizons of crushed material [115, 113 and 111]. Not otherwise visible in section.

113 a thin spread of dark reddish-brown, crushed shale and quartz [5 YR 3/3] only recognised in plan to the south of the bath-shaped feature (Fig. 4.28), where it separated the underlying sandy clay loam [114] from an overlying clay patch [112]. Indistinguishable in section from either of the underlying spreads [114 and 115].

112 a semi-circular patch of greyish-brown clay loam [7.5 YR 5/2] extending into the section south of the bath-shaped feature (Figs 4.11 and 4.28), where it was sandwiched between two successive spreads of crushed material [113 and 111].

111 a spread of dark brown loamy coarse sand [10 YR 3/4] with a 90% admixture of crushed shale and quartz, only recognised in plan to the south-west of the bath-shaped feature (Fig. 4.28), where it clearly separated the underlying sandy clay loam [114] from an overlying lens of clay [110]. In the section, however, it was clear that it overlay both the spread of crushed material [113] and the patch of clay loam [112] (Fig. 4.11).

110 a small patch of dark brown clay loam [10 YR 3/4] extending into the section south-west of the bath-shaped feature (Figs 4.11 and 4.28). This proved to be the uppermost clay horizon in the early sequence, clearly sealed beneath the main spread of crushed shale and quartz [97] which extended across most of the central hollow.

During the examination of the horizons on the sloping profile of the 'motte' at the western end of the site, it became clear that several deposits had been truncated by a possible shallow feature [105], effectively isolating them in the process from the main sequence across the rest of the central hollow. These included three successive spreads separated by ephemeral lenses of crushed shale and quartz (Fig. 4.28):

108 a layer of dull yellowish-brown silty clay [10 YR 4/3] with a 10% admixture of gravels and very small angular shale and quartz fragments. Seen to overlie a spread of crushed material, perhaps an extension of context [115].
107 a thin spread of greyish clay, removed before sampling.
106 a layer of brown silty clay [7.5 YR 4/3] with a 10 % admixture of gravels and very small angular shale and quartz fragments, clearly sealed beneath the main spread of crushed shale and quartz [**97**].

The interpretation of this feature presents several problems, because its interface was not identified with any confidence anywhere else across the central hollow, though it is self-evident that it must have affected some, at least, of the uppermost horizons. Drawing upon the evidence from the better-known tin industry in Devon and Cornwall, however, it is not unusual to find processing wastes being dug out and reworked over time (G. A. M. Gerrard, *pers. comm.*), an operation which would not be out of place in the current context where recovery rates are likely to have been low. Such a process might well have resulted in the distinctive wave-like effect recorded in plan across the north-west to south-east axis of the site, and for the rather enigmatic extension of clay spread [109] as a narrow band along the sloping profile of the 'motte' to the west, which could only be the result of some sort of truncation from a higher level. An even earlier episode of this type may well be indicated by various other confused deposits noted at the western end of the site, immediately above another of the extensive clay spreads [120].

Any such reworking necessarily created new basins of deposition, which clearly provides the context for the main spread [**97**] of dark brown loamy coarse sand [7.5 YR 3/4] with its 90% admixture of coarsely-crushed shale and quartz, found to extend across much of the central hollow (Figs 4.11 and 4.29) and probably also above the lowest fills of the large hollow [152] at

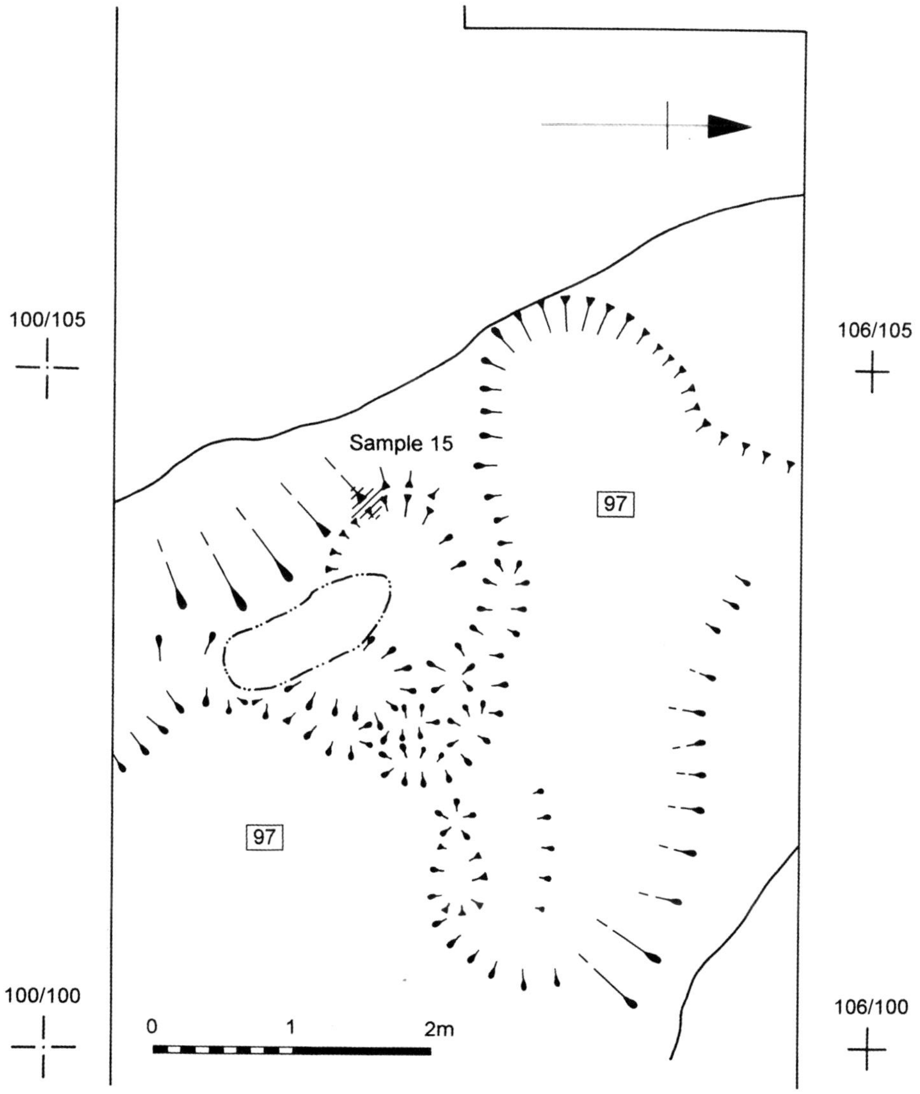

Fig. 4.29 Extent of major crushed horizon [97]

the eastern end of the site. It is also likely to be the context for two small and isolated patches of silty clay loam located within the makeup of [97] in the north-western part of the site (not illustrated):

104 a thin patch of silty clay loam [7.5 YR 4/3] extending into the north section.
103 a thin lens of brown loamy coarse sand [7.5 YR 4/4] with a 90% admixture of crushed shale and quartz, only recognised where it intervened between layers [104] and [102]. Otherwise indistinguishable from the main spread [97].
102 a thin spread of dull reddish-brown silty clay loam [5 YR 4/4], extending into the north section. Clearly truncated to the south by the interface of a shallow feature [90] which had been cut into the main spread [97].

(ii) The later sequence
Above the distinctive makeup of the marker horizon [97] yet more layers of interleaved processing wastes were recorded. They included at least two discrete sequences of coarsely-crushed material and fine clay slimes, clearly separated by the digging of a shallow feature [90] across the central part of the site. All three elements proved difficult to disentangle in places, particularly where the successive clays had been laid directly one on top of another.

The successive deposits associated with the lower of the two main sequences had only survived across the south and south-western sides of the central hollow, where they had not been truncated by the excavation of the shallow feature. Most were difficult to plan as discrete layers because their edges progressively thinned and merged with the makeup of the uppermost clay horizon [95], though the stratigraphic sequence was clear nevertheless in the south section (Figs 4.11 and 4.30).

101 a layer of dull brown clay loam [7.5 YR 5/4] with a small proportion of charcoal flecks, forming the lowest horizon in the sequence.

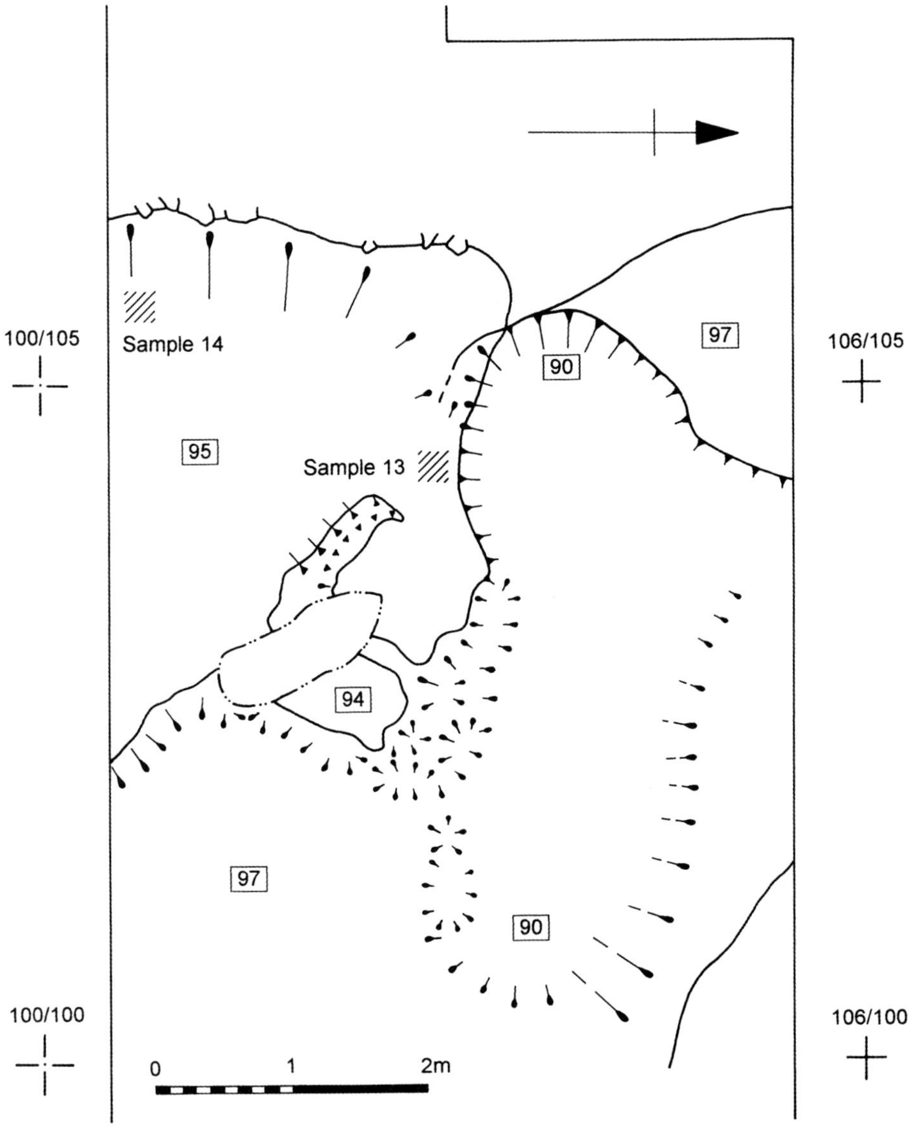

Fig. 4.30 Extent of clay spreads [94 and 95], truncated by oval-shaped hollow [90]

100 a layer of gravel and very small shale fragments within a dark reddish-brown matrix [5 YR 3/4], forming a distinct iron pan between layers [100] and [99].
99 a band of predominantly dark brown silty clay [7.5 YR 3/4] with some specks of charcoal. The overall colour varied from golden to dark reddish-brown depending on the extent of iron-panning.
98 a layer of predominantly brownish-black clay loam [10 YR 2/3], characterised by an increasing proportion of charcoal closer to the section. The colour actually varied from brownish-black to dark reddish-brown depending upon the charcoal content.
95 an extensive spread of brown silty clay [10 YR 4/3], forming the uppermost clay horizon in the sequence. Likely to be the same as a discrete patch [94] of olive-brown silty clay [2.5 YR 4/3] to the east of the bath-shaped intrusion [70], since both were overlain by the same ridge of crushed material [91].

Across its south-western side, this uppermost clay horizon was overlain by a thin spread [96] of brown sandy clay loam [10 YR 4/6] with a 90% admixture of crushed quartz and shale, which effectively separated it from the lowest of the clay spreads [89] associated with the second of the two sequences of processing wastes described in this section.

Greater definition of these two phases was also

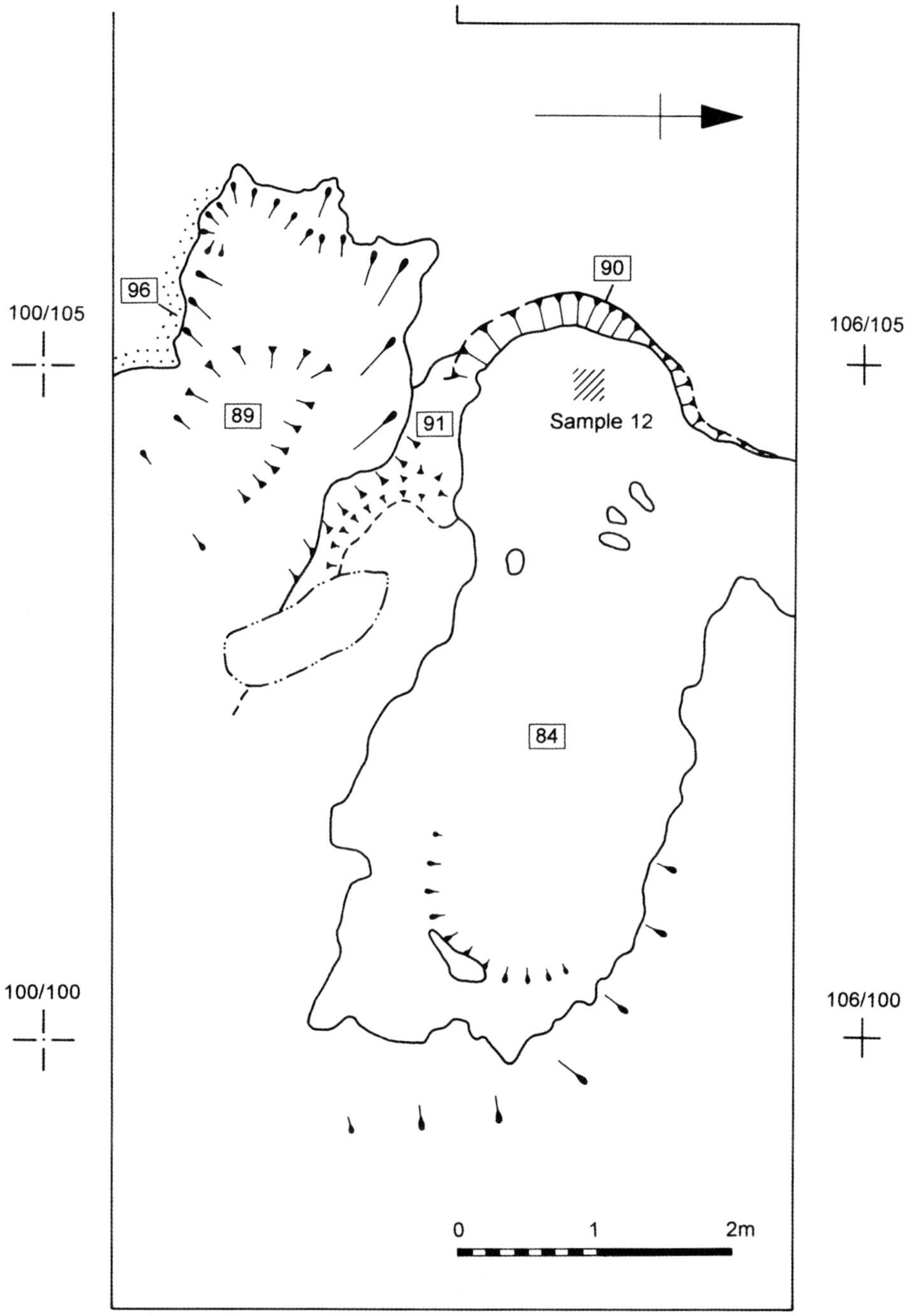

Fig. 4.31 Extent of clay spread [84] infilling oval-shaped hollow [90] and clay spread [89] in south-west corner, separated by intervening ridge of crushed material [91]

provided by a spread [91] of dark reddish-brown sandy clay loam [5 YR 3/3] with a 70% admixture of crushed quartz and charcoal flecks, which formed a slight ridge across the north-eastern side of the uppermost clay spread [95]. Originally this must have been more extensive to the north but, like the underlying horizons, it had clearly been cut away by the digging of a shallow feature [90] extending roughly east-west across the central part of the site (Figs 4.30 and 4.31). Excavation showed that this latter was an hour-glass-shaped hollow with a fairly level base, which had penetrated to a maximum depth of 0.15m into the irregular surface of the main spread of crushed material beneath [97]. Overall, it measured 2.50m from east to west by at least 1.00m from north to south, with a prominent curving edge across the west and a less pronounced profile elsewhere. At one point along the north side, towards its western end, the feature opened outwards to extend beyond the section, though why this should be the case remains uncertain.

There was nothing to suggest the original function of this shallow hollow. One possibility is that it represents a case of processing wastes being dug out for reworking; an alternative might be that it created a deliberately levelled area capable of supporting some form of movable washing table associated with gold extraction. Whatever the truth, it proved to contain two successive fills. The lower one [84] comprised a spread of brown clay loam [7.5 YR 4/3] with some charcoal flecks, which extended right across the base of the feature and its northward projection (Figs 4.31 and 4.37). Its extent was most easily defined to the north and east, where it had been sealed beneath an extensive dump of crushed shale and quartz [80] (Fig. 4.32), but to the south and west, where no such distinctive horizon intervened and where successive clay layers merged with one another, it proved much more difficult to differentiate from the overlying material. The second of the two fills [81] comprised a distinctive brownish-black clay loam [10 YR 2/3] with a pronounced charcoal content in places (not illustrated), sandwiched between the underlying clay loam and the overlying spread of crushed shale and

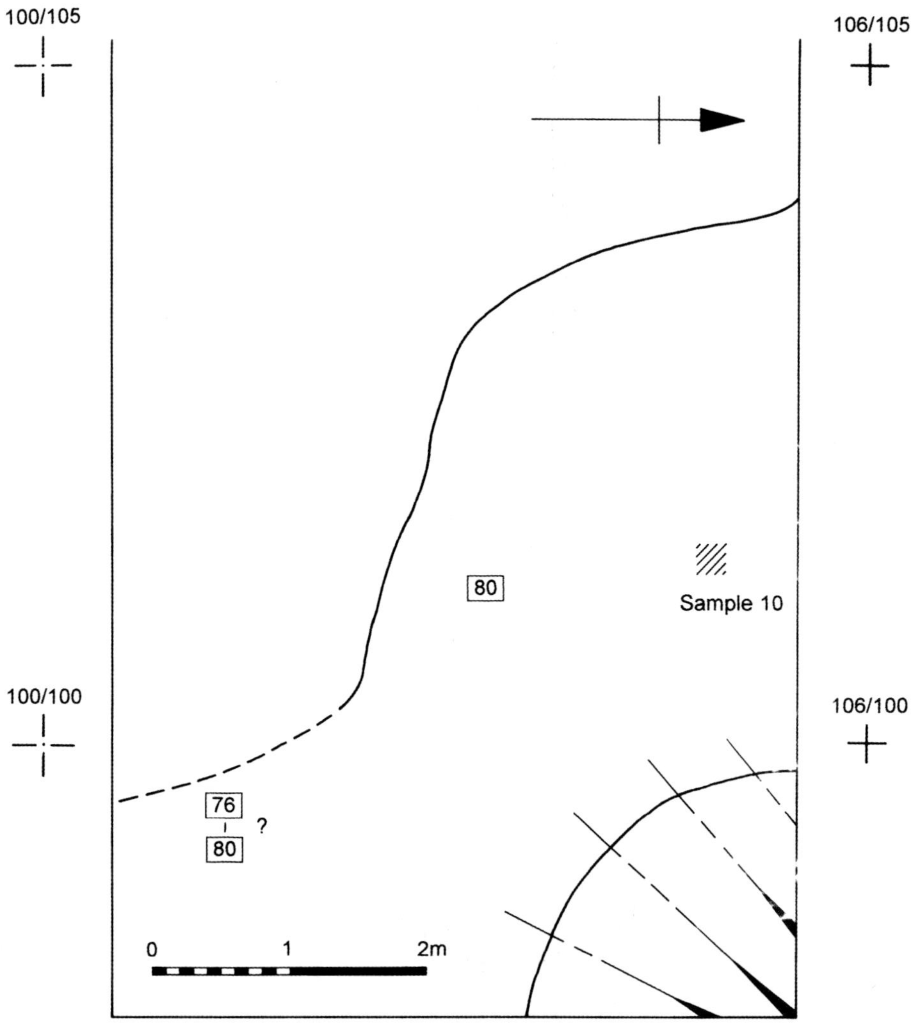

Fig. 4.32 Extent of crushed horizon [80] across eastern end of trench

quartz [80]. At its junction with the latter, there was an extensive iron pan of almost clinker-like consistency, not unlike that which formed later across the extensive clay spread [77] at the top of the sequence.

Much of the eastern end of the trench and the greater part of the shallow hollow and its associated fills was now overlain by an extensive dump [80] of dark brown loamy coarse sand [10 YR 3/4] with a 90% admixture of coarsely-crushed shale and quartz and some small stones (Fig. 4.32). This petered out on its southern and western sides, where it was not always easy to differentiate from the very similar underlying [97] and overlying [76] horizons (see Fig. 4.4), reached a maximum depth of 0.20m at its centre across the hollow, and varied between 0.10 and 0.15m along the northern section. It also petered out onto the mound associated with the Carreg Pumsaint at the north-east corner of the trench. Throughout it was heavily iron-panned, especially on its upper horizon, suggesting that it was a series of separate dumps which had congealed into one over time. It was also especially clinker-like at its base, where it overlay various clay horizons.

Above these various contexts extended the second of the two discrete sequences of coarsely-crushed material and fine clay slimes mentioned at the outset of this section. This proved easiest to define along its western side and close to the south edge of the excavation, despite being disturbed in places by a series of animal runs, but further away, towards the east and north-east, where the lenses of crushed material petered out, it proved extremely difficult to separate the successive clays from one another. While the lowest of the clays may well have been confined to the south-western quadrant, possibly infilling a second, shallow depression which ran into the main section, higher up the sequence it seems likely that they extended right across the central area, effectively levelling up what remained of the original bowl-shaped depression to a uniform height. The broad sequence is recorded in the south section (Fig. 4.11), though it is likely that some of the clay spreads represent more than one horizon.

89 an irregular spread of dull yellowish-brown clay [10 YR 4/3] probably confined to the south-western part of the site, where its central axis seemed to be slumping into a linear depression (Figs 4.31 and 4.33). Separated from the underlying clay horizon by two discrete spreads of crushed material, one to the south-west [96], the other forming a ridge to the north-east [91].

86 a lens of dark brown loamy coarse sand [10 YR 3/3] with a 95% admixture of crushed quartz and shale separating successive clay horizons [89 and 88] along their south-western edges (Fig. 4.34).

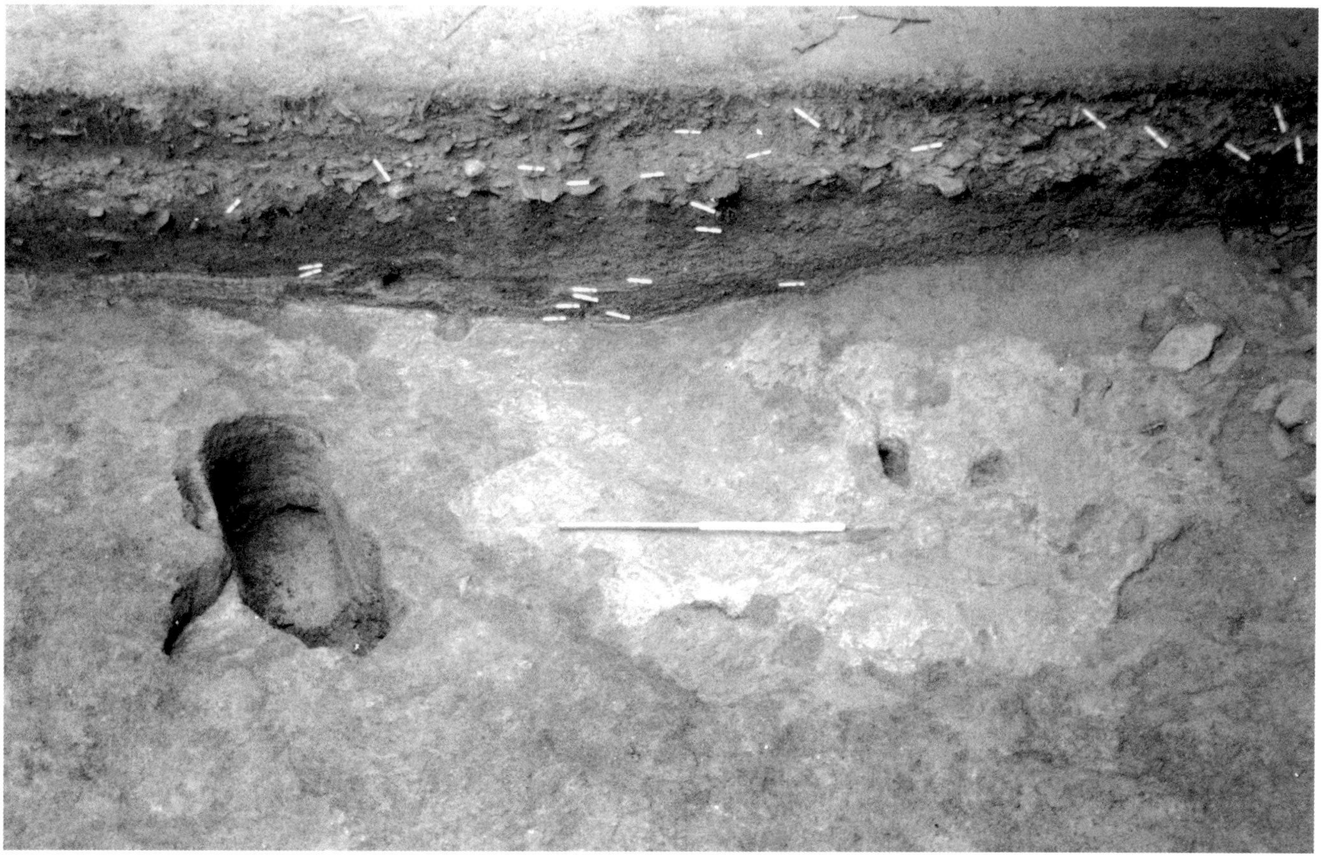

Fig. 4.33 Clay spread [89], looking south

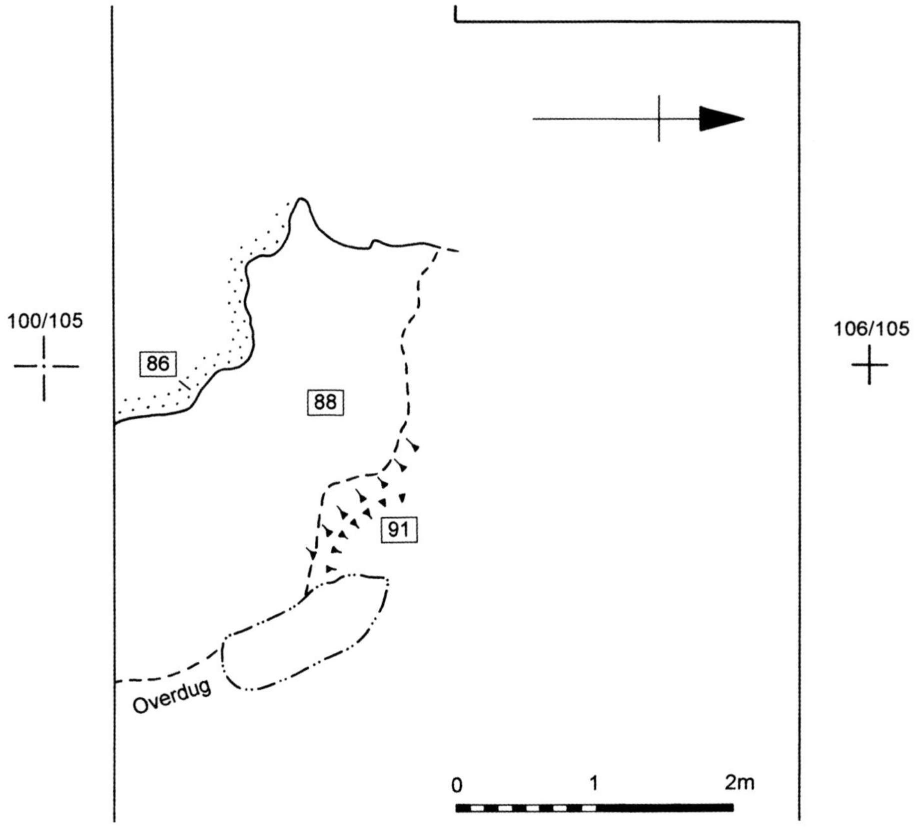

Fig. 4.34 Established edge of clay spread [88] in south-west corner of trench

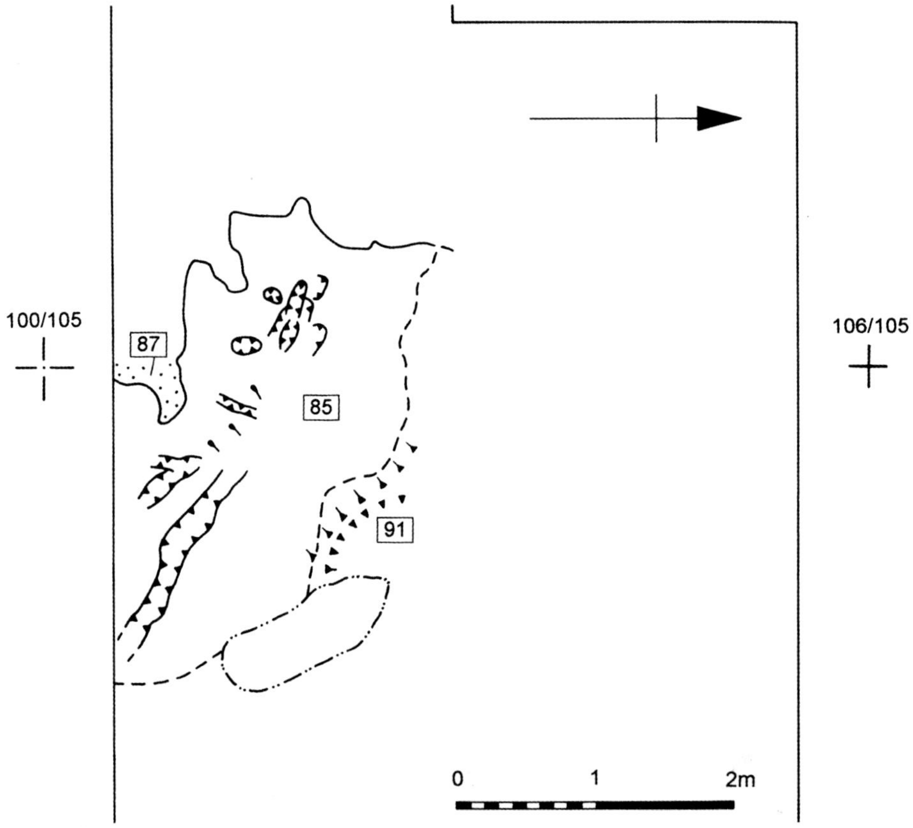

Fig. 4.35 Established edge of clay spread [85] in south-west corner of trench

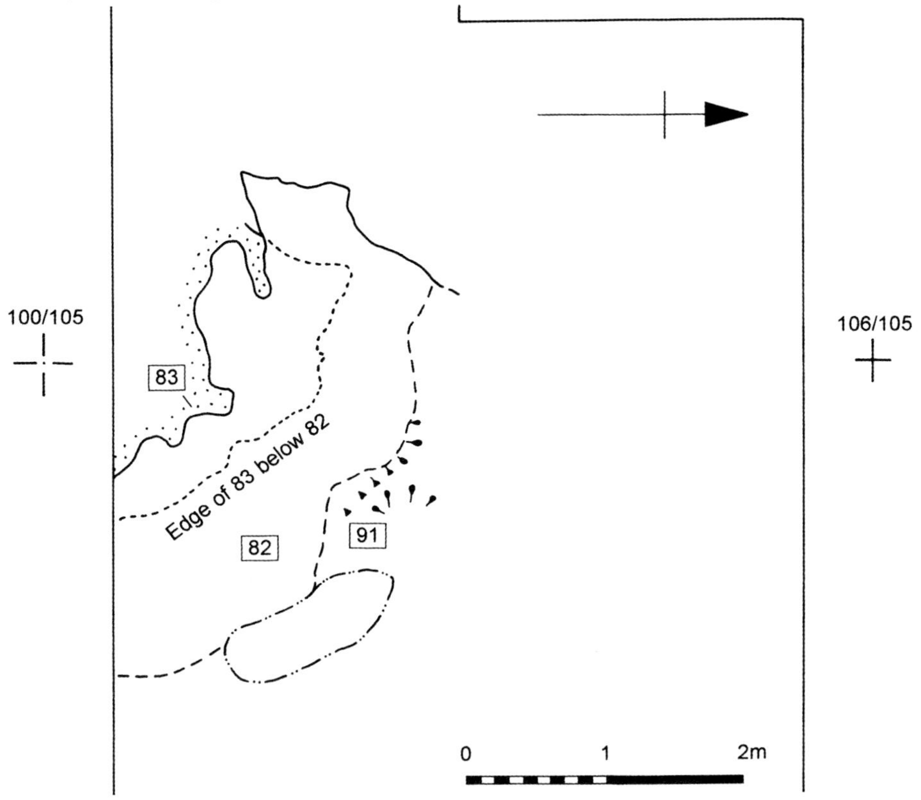

Fig. 4.36 Established edge of clay spread [82] in south-west corner of trench, above crushed horizon [83]

Fig. 4.37 Clay spreads [82 and 84], looking north-east

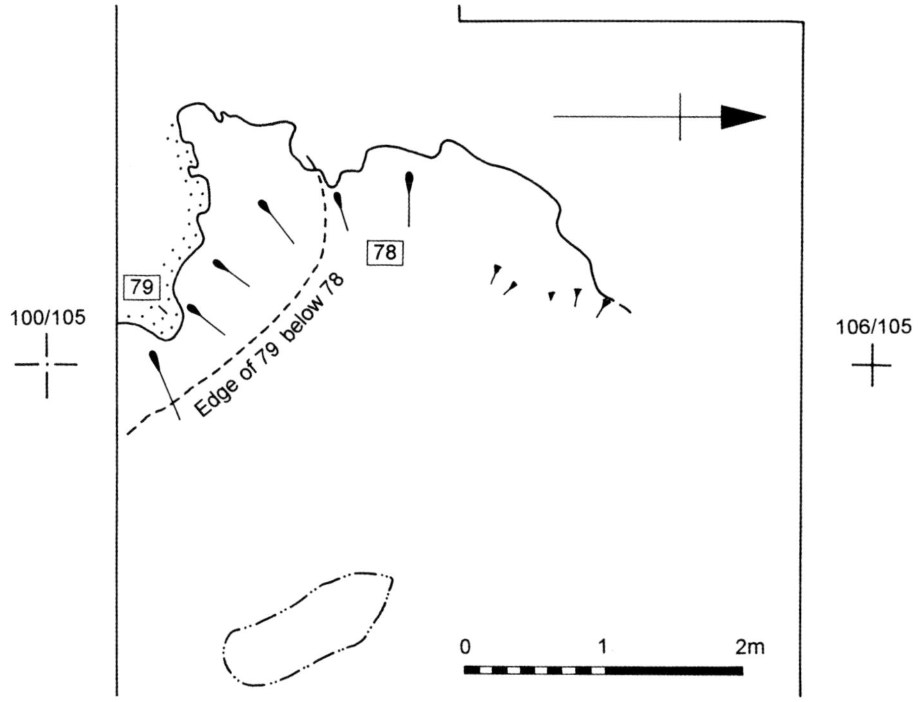

Fig. 4.38 Established edge of clay spread [78]

88 an irregular spread of brown clay [10 YR 4/4] clearly defined along its south-western side (Fig. 4.34). Probably more extensive than planned to the north and east, but impossible to differentiate from the surrounding clays.

87 a thin lens of dark reddish-brown loamy coarse sand [5 YR 3/6] with a 95% admixture of crushed quartz and shale separating successive clay horizons [88 and 85] along their south-western edges.

85 a spread of dull yellowish-brown clay [10 YR 5/4] clearly defined along its south-western side (Fig. 4.35). Must have been more extensive originally than planned to the north and east, but impossible to differentiate from the surrounding clays. Its upper surface was much disturbed by animal runs trending in a north-west to south-east direction.

83 a lens of dark brown loamy coarse sand [7.5 YR 3/3] with a 95% admixture of crushed quartz and shale separating successive clay horizons [85 and 82] along their south-western edges (Fig. 4.36).

82 a thin lens of dull yellowish-brown silty clay [10 YR 4/3] more or less clearly defined along its south and west sides (Figs 4.36 and 4.37). Must have been more extensive originally than planned to the north and east, but impossible to differentiate from the surrounding clays.

79 a dump of brown loamy coarse sand, crushed quartz and shale [10 YR 4/6] separating successive clay horizons [82 and 78] along their south and west edges.

78 a layer of brown clay [10 YR 4/4] clearly defined along its western edge (Fig. 4.38). Like its upper counterpart [77] it must originally have been more extensive to the east, but this proved impossible to differentiate from the surrounding clays.

47 a spread of brown loamy coarse sand, crushed quartz and shale [10 YR 5/4] separating successive clay horizons [78 and 77] along their south-western edge.

77 an extensive spread of brown clay [10 YR 4/6], covered in part at its western end by a patchy clay-flecked horizon **[74/75]**; otherwise it formed the uppermost clay horizon in this sequence of processing wastes (Figs 4.39 and 4.40). Its extent proved easy to define on the south-west, north-east and east, where it overlay distinctive crushed horizons [47 and 80], but elsewhere across the site, where clay had been laid directly on clay, it proved impossible to separate from the underlying horizons which were inadvertently removed in some cases. This problem is particularly evident towards the east end of the main section (Fig. 4.11), where root and worm action had further exacerbated the problem at the interface of the subsoil associated with the overlying turf horizon [72].

73 a final dump of brownish-black sandy clay loam [7.5 YR 3/2] with a 70% admixture of gravels, crushed quartz and shale, charcoal and ash, located across the western part of contexts [74, 75 and 77] beneath the turf horizon [72]. Nowhere was it more than 2.00m wide (for its western edge see Fig. 4.42). Although it petered out to east and west, it reached a maximum thickness of 0.25m as it spilled downslope.

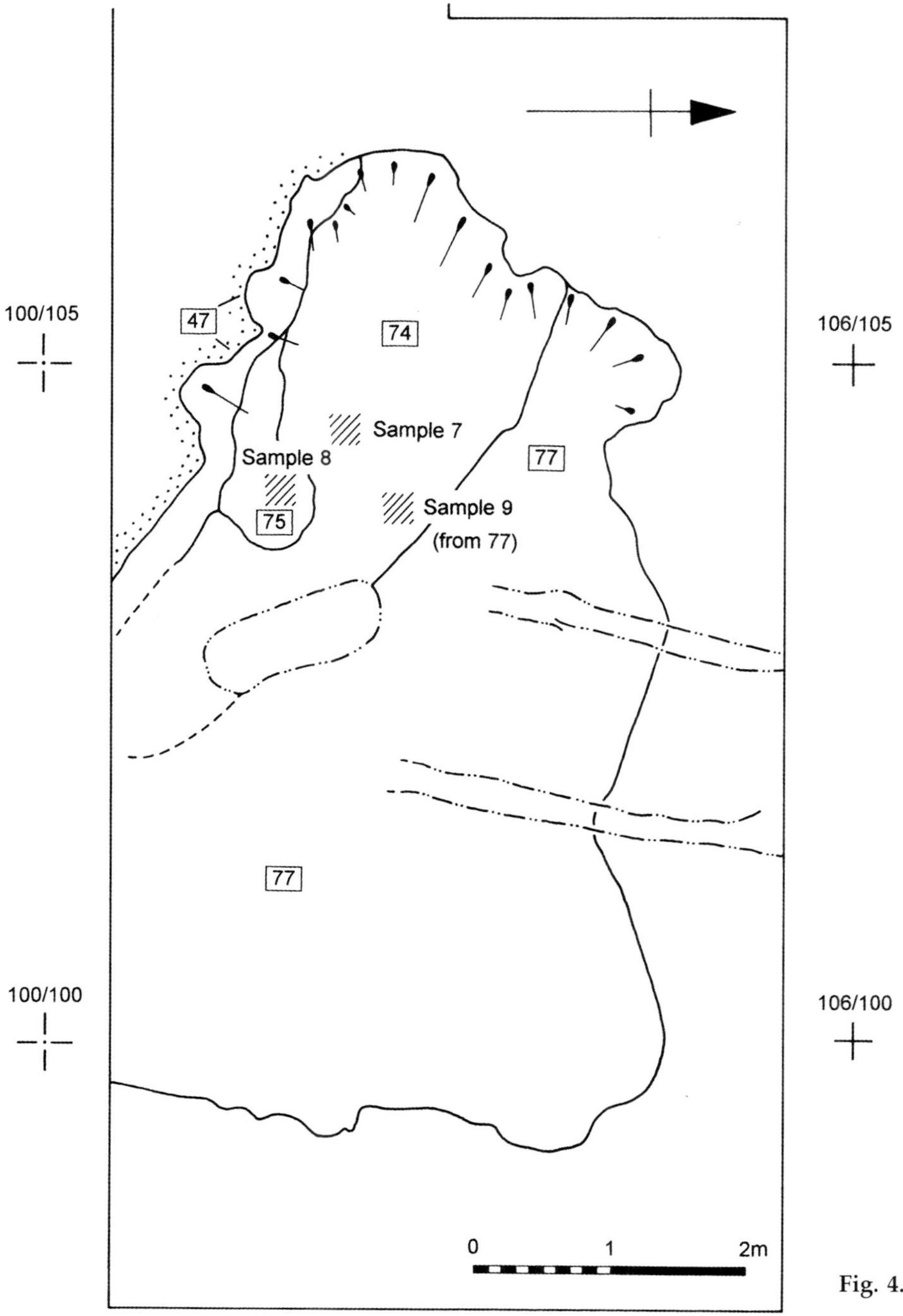

Fig. 4.39 Extent of clay spread [77] and overlying horizons [74–75]

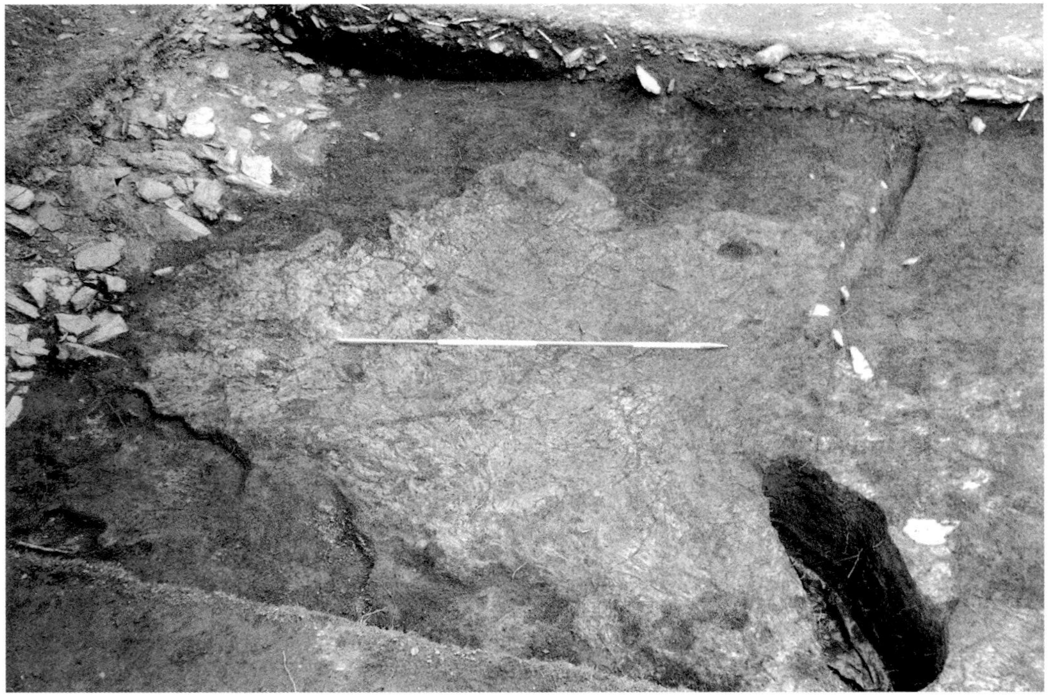

Fig. 4.40 Western part of clay spread [77], looking north

Discussion

The analysis of these interleaved deposits leaves little doubt that the bulk of the material in the central hollow is derived from ore-crushing and related processing activities (for a fuller discussion see sections 4.6 and 4.8.3 below). The coarsely-crushed lenses of quartz and pyritic shale are highly suggestive of material which has been mechanically pounded to a small fraction, prior to being processed and then dumped (or stockpiled?) across the hollow. The fine clay slimes are most likely to be the residue from very finely-milled material which may have been water sorted, prior to being allowed to run down into the hollow or being dumped there in a semi-liquid state and left to settle. The way in which the two distinctive types of deposit are interleaved strongly suggests that different processing activities were being carried out in the immediate vicinity in fairly rapid succession, even if it is impossible on the surviving evidence to define the timescales involved with any degree of precision.

(B) The large hollow at the eastern end of the site

While the various processing wastes were accumulating across the base of the central hollow, the ridge of shale and sandy clay loam [151] at the east end of the site was apparently transformed by the excavation of a large, roughly oval hollow [152], part of which clearly lay outside the excavated area (see Fig. 4.4). Overall it measured *c.* 3.80m from north to south by at least 2.00m from east to west and was up to 0.60m deep. At its base it had cut into the fills associated with a series of earlier, shallow hollows now assigned to phase 1 (see section 4.3.3.(F) above). It proved to contain a succession of primary and secondary infills, only two of which could be directly correlated with the stratigraphic sequence recognised to the west. This makes it difficult to tie in with absolute precision to the wider picture of the site's development in phase 2, though enough links exist to suggest a broad sequence.

Two primary fills were identified, both integral to the feature and unconnected with the wider sequence (Fig. 4.41). The lower one [145] comprised an irregular spread of medium to large angular shale fragments, which had all the appearance of being deliberately laid across the base of the hollow as if to serve a specific function. In this respect, it resembles the very similar stone spreads recorded in the earlier, shallow hollows. Significantly, it also yielded a fragment from the upper stone of a rotary quern (see section 4.4, no. 1, below). Above this, the second of the primary fills [144 a-d] proved to be a succession of predominantly dull yellowish-brown clay or silty clay spreads [10 YR 5/4], interspersed with thin and intermittent bands of coarsely-crushed material,

Fig. 4.41 Stone spread [145] partially revealed during excavation of clay spreads [144a–d], looking south-east

much like the more extensive sequence identified across the central hollow. To judge from the way they lie, they would seem to have been tipped or spread in from the south-east, becoming progressively thinner towards the centre of the hollow before petering out altogether to the north and west. From this horizon came a possible circular disc or counter (see section 4.4, no. 2, below).

The secondary fills within the eastern hollow can be divided into two apparently interleaved sequences, one consisting of alternating tips of vacuous stone and crushed material [133 to 137] only located on its northern side, the other comprising a thick spread of coarsely-crushed material representing an amalgam of contexts [80] and [97], which extended across the rest of the hollow and beyond. Though this latter offers an important opportunity for correlating the stratigraphic sequences across different parts of the site, the situation was complicated by the difficulty of differentiating the various horizons of crushed material where they had been laid directly one on top of the other (see Fig. 4.4). The most likely position is set out below:

137 an extensive tip of small to medium-sized angular shale fragments visible in both the east and north sections, extending across the slope of the mound associated with the Carreg Pumsaint. Clearly the lowest of the secondary fills.

136 a layer of brown coarse sand [10 YR 4/4] with a 90% admixture of finely-crushed quartz and shale, sandwiched between the successive stone tips [137 and 135] and confined to the northern slope. Seemed to be petering out across the base of the hollow, though it was extremely difficult to differentiate from the surrounding contexts [134 and 97] once the stone tips ran out.

135 a narrow tip of small to medium-sized angular shale fragments only recorded in section. To judge from the amount of stone visible, especially at the foot of the tip line, the section probably corresponds closely with its edge, which explains why it was inadvertently removed during the excavation of context [97].

134 a spread of dark brown loamy coarse sand [7.5 YR 3/4] with a 90% admixture of crushed quartz and shale, sandwiched between successive stone spreads [135 and 133]. Virtually identical in character with the more extensive spread [97] and so probably derived from the same process.

133 a tip of small to medium-sized angular shale fragments first noted in 1991 as a loose spread extending a short way out from the section, but not really understood during the removal of context [80] which overlay it.

Such evidence would suggest that the large hollow at the eastern end of the site was originally dug sometime during the accumulation of the processing wastes associated with the early sequence across the central hollow, probably later rather than earlier, and that it had been largely infilled by the extensive horizon of crushed material [97] which preceded the later sequence. Further precision is impossible. There was little in the surviving evidence to indicate the function of this large hollow, though it seems likely that it was connected with the processing activities, as the incorporation of a quern fragment suggests.

(C) The turf horizon

At the end of phase 2 there was a marked break in the character of the activities associated with the site, represented by the development of a pronounced turf horizon [72] which extended right across the site onto the exposed shale matrix forming the upper part of the so-called motte. At its western and eastern ends, this characteristic brown loam [7.5 YR 3/2] overlay the last of the dumps of crushed material [73 and 76 respectively], but in the centre it merged with a deep loamy subsoil horizon much disturbed by root and animal action. Its upper surface was heavily iron-panned. To judge from the evidence, this turf horizon seems to represent a considerable episode of stability post-dating the end of the processing activities and pre-dating all the other levels on the site. Unfortunately nothing was found to date this episode beyond the 19th- and 20th-century finds associated with the phase 3 horizons.

4.3.5 Phase 3 (Figs 4.5, 4.42 to 4.48)

(A) Early features above the turf horizon (Fig. 4.42)

Across the central part of the site, the turf horizon [72] was overlain by a layer [40] of brown silty clay [10 YR 4/4], originally perhaps as much as 7.00m wide by up to 0.15m thick in places. This appeared to respect the remains of a shallow, curving bank [66] along its north-east side, made up of medium to large shale fragments intermixed with a brown coarse sandy loam [7.5 YR 4/4], which also overlay the turf horizon at the point where it skirted the extant base of the mound associated with the Carreg Pumsaint. This would suggest, though the relationship was never established for certain on the ground, that the bank had been put in place first before any of the silty clay. As recorded, the bank proved to be thickest at its base, progressively thinning out upslope where it overlay the truncated tips of stone and crushed material [133 to 137] described in the preceding section.

No clear purpose could be established for this banking, unless it is to be connected in some way with a deliberate reinforcing of the mound contemporary with the decision to relocate the Carreg Pumsaint into its current upright position. This is said to have occurred for certain some time between *c.* 1835 and 1855, on the basis of two separate antiquarian illustrations recorded by Bosanquet (1917, 32): the earlier of the two clearly shows the Carreg lying on its side, but not certainly on the site it now occupies; the other depicts it upright as now, but

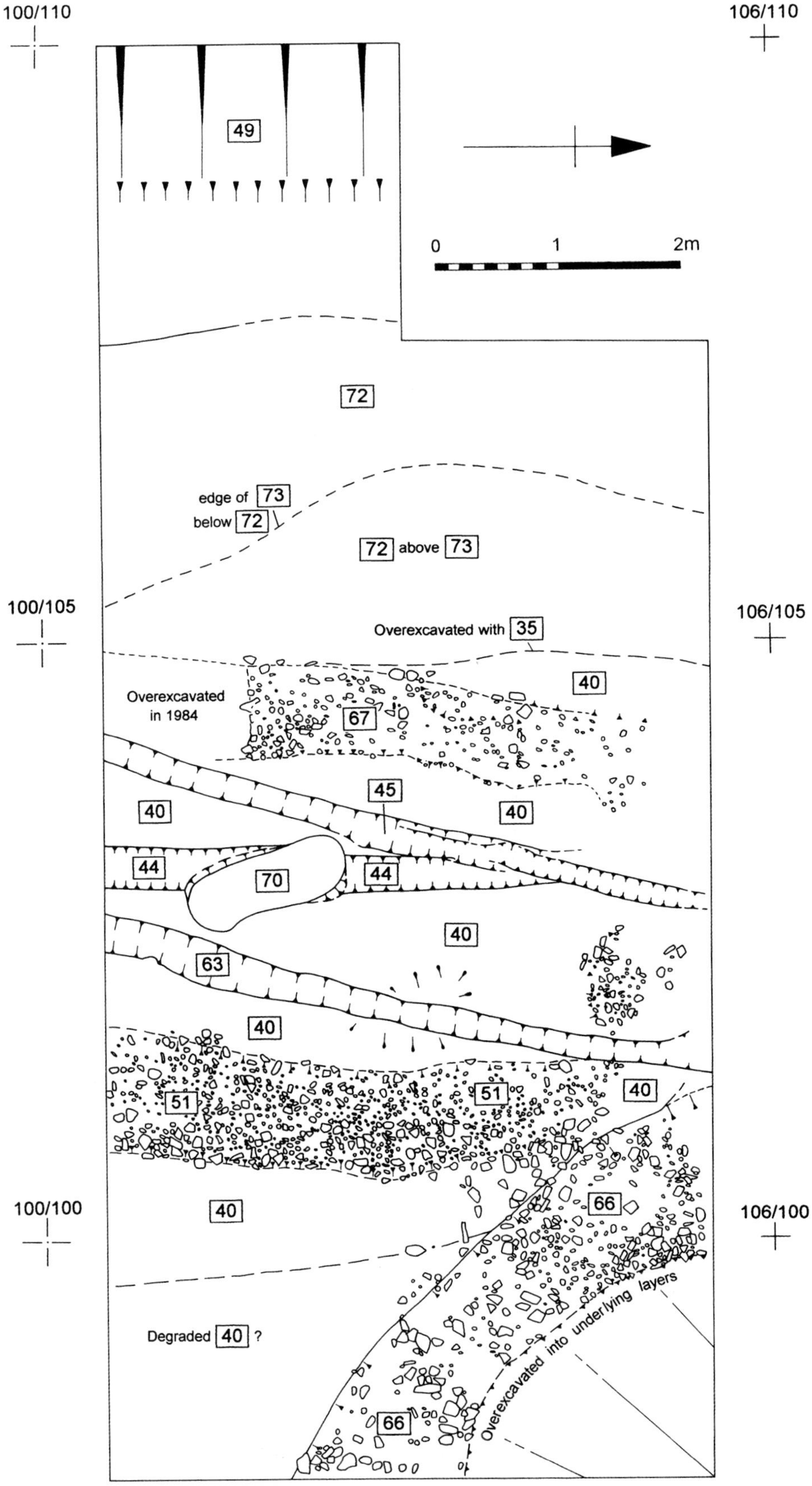

Fig. 4.42 Early features above the turf horizon

without any obvious background context (Knight 1856, fig. facing p. 264). It is also shown upright in Smyth's note on the mine (1846, 483), again without any context. More importantly a pencil sketch by Charlotte Johnes, apparently made between 1844 and 1847, seems to indicate that by then the stone was standing on a low mound with a treeless mound behind (R. Bird, *pers. comm.*, quoting NLW Drawings, vol. 174). Such a date would not be out of line with the known stratigraphic sequence, nor with the evidence from the nearby Ogofau Lodge, which carries a date of 1839.

Above the silty clay horizon [40] a number of interesting features were examined, sealed beneath another extensive clay spread [35, 36 and 43] recorded across much of the trench. Two of the earliest proved to be a pair of shallow stone spreads [51 and 67], trending roughly north-south over the site (Fig. 4.43). That to the east [51] was first noted in 1984, during the partial examination of the overlying clay [43]; when re-examined in 1991, it was shown to comprise a band of medium to large stones up to 1.00m wide, set in a matrix of olive-brown silty clay [2.5 YR 4/4]. Towards its northern end, it appeared to be petering out before reaching the section, though an irregular patch of stones to the west might suggest that it originally swung through a right angle to connect up with a second linear feature [67] first recorded in 1991. This latter also comprised a spread of medium to large stones varying between 0.70 and 1.00m in width, set in a matrix of yellowish-brown silty clay [2.5 Y 5/3], which had been much disturbed from above by later activities. It too appeared to be petering out before reaching the northern section, though an eastward thickening at this point might confirm the suggestion that it originally joined up with its eastern counterpart.

Assuming that what has survived is representative of the original character and extent of this feature, it would seem to have been part of an ephemeral rectangular enclosure of uncertain function, sandwiched between two successive silty clay/clay spreads across the site. A less likely alternative is that it was originally a more extensive surface, perhaps connected with the two cart ruts [45 and 63] cut from a higher level, though there is little to corroborate the sort of truncation that this would have necessitated.

Equally unlikely is any association with a box-shaped gully [44], some 0.36m wide by 0.15 to 0.20m deep, which also trended north-south across the site, between the two stone spreads. It was first located in 1984, south of a bath-shaped feature [70] with which it was thought to have been associated. Detailed examination in 1991, however, showed that it extended north beyond the latter, at least as far as the line of one of the later cart ruts [45] which had cut it

Fig. 4.43 Cart ruts cut through ephemeral rectangular structure, looking south

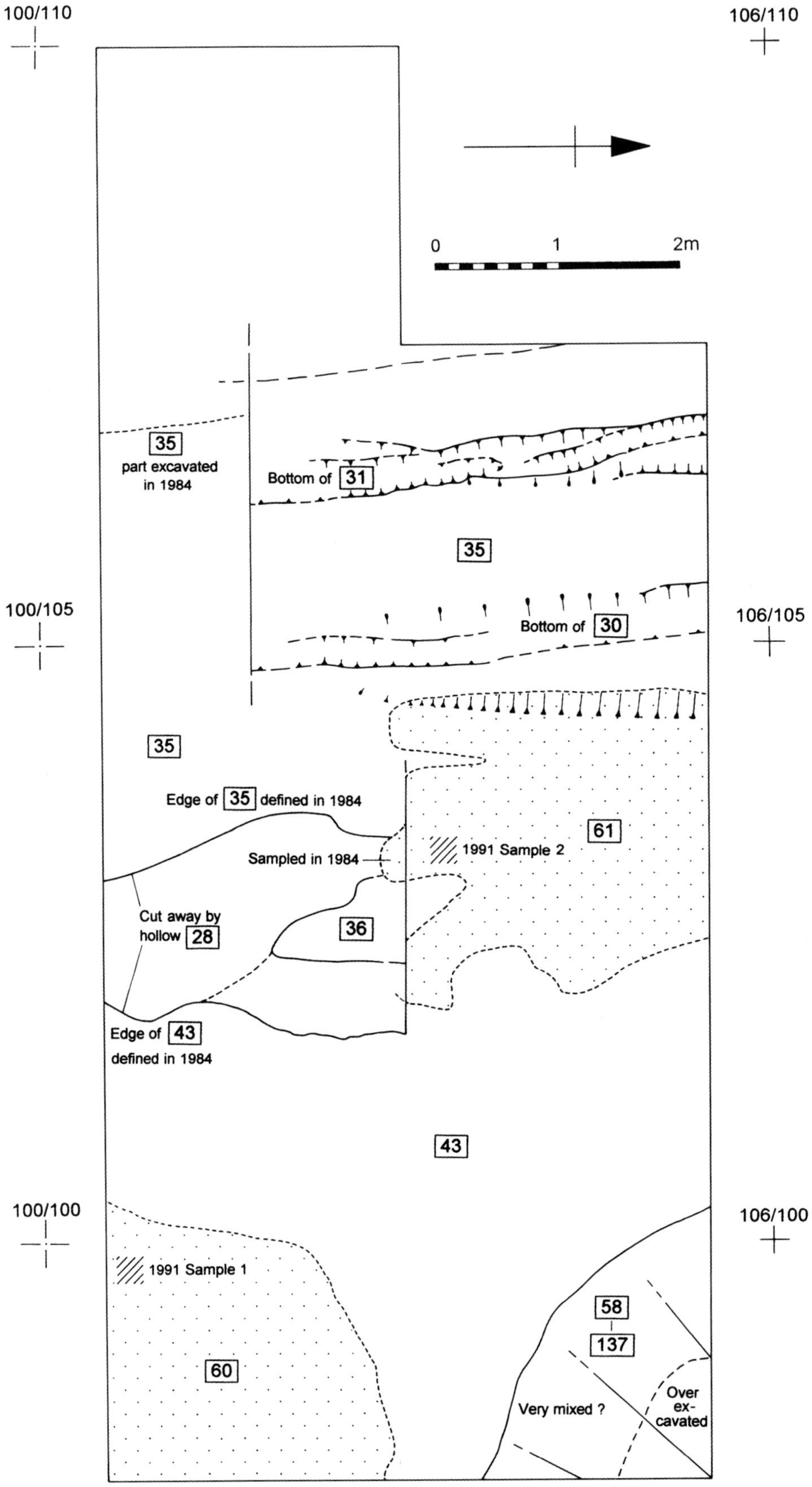

Fig. 4.44 Extent of clay spreads [35], [36] and [43]

away. Further north its line could not be traced for certain, though it seems likely that it originally extended right across the site. Although nothing survived to indicate its function, its box-shaped profile was sufficiently distinctive from that of other gullies to suggest that it may once have been timber-lined, to be connected perhaps with some form of water management. Its infilling varied somewhat, from a brown sandy loam [7.5 YR 4/4] with several medium-sized sub-rounded stones [37] to the south of the bath-shaped feature, to an olive-brown silty clay [2.5 Y 4/4] with very few stones [68] to the north.

The latest features in this section proved to be a pair of parallel, U-shaped gullies [45 and 63], both trending north-north-east to south-south-west across the site (Fig. 4.43). That to the west [45] was first recognised in 1984, but its full extent did not become apparent until 1991, when it was shown to be 0.20m wide and up to 0.14m deep. Its counterpart to the east [63] only became apparent in 1991. It proved to be somewhat more variable in width, from 0.20 to 0.35m, but its depth remained fairly constant at 0.25m. Each had been cut down through one of the shallow stone spreads [67 and 51] into the underlying silty clay [40]. Both had a very mixed fill [65 and 62 respectively] of dull yellowish-brown clay loam [10 YR 5/3] with a 10% admixture of small coarse components, probably representing the churned-up remnants of the underlying horizons. Their interpretation as cart ruts seems fairly certain from their spacing at a gauge of 4 ft 8.5 ins, even if they were not obviously connected with any recognisable road makeup.

These early features above the turf horizon were all sealed beneath an extensive spread of clay [35, 36 and 43], which was recorded across much of the trench (Fig. 4.44). It varied in colour from olive brown on the west [2.5 Y 4/3] to yellowish brown on the east [2.5 Y 5/3], influenced in part by water percolation from the overlying shale horizons which differed markedly from west to east. At two points, one in the south-east corner [60], the other towards the north-west [61], this clay was overlain by a discrete patch of dark greyish-yellow coarse sand [2.5 Y 4/2], composed almost entirely of very finely-crushed quartz and sand granules. Both were no more than 0.05m thick, with a minimum extent of at least 2.50m square. Both the clay and the crushed quartz are best interpreted as processing wastes, though the fineness of the fraction involved is very different from the more coarsely-crushed material associated with phase 2. More than likely these horizons derive from one of known phases of activity at the mines in the mid to late part of the 19th century (see section 4.6 below).

(B) The later roads above the clay horizon

Above the clay horizon [35, 36 and 43] a series of road surfaces was examined, two trending north-south in succession across the western end of the trench, and a third running north-north-west to south-south-east across the east (Figs 4.45 and 4.48). The earliest was located above the clay at the western end of the site [35], skirting the foot of the extant slope of the 'motte' (Fig. 4.46). Its makeup comprised a compact layer [59, 32 and 33] of medium to large-sized shattered shale fragments and small shale chippings, intermixed with a dull yellowish-brown sandy clay loam [10 YR 4/3], which proved to be over 3.00m wide and up to 0.20m thick across the centre, though it rapidly petered out on either side, towards the edge of the camber. Along its eastern side, this had been cut into by a linear gully [46], *c.* 3.60m long at least, which petered out *c.* 1.40m short of the north section. It was shown to be somewhat variable in width, from 0.32 to 0.46m, with a maximum depth of 0.15m.

This makeup formed the foundation for a compacted road surface [29], which had obviously seen considerable usage to judge by the depth, width and repeated infillings of its associated wheel ruts [30 and 31]. These latter and their respective fills proved extremely difficult to excavate in plan, particularly where they had cut through the makeup into the softer, underlying layers, thereby necessitating frequent repairs in the form of successive dumps of broken shale. Traces of this were recognised at various points [40 and 41], intermixed with a sandy clay loam ranging in colour from dark greyish yellow to olive brown [2.5 Y 4/2 to 4/3]. All this suggests that neither was an individual rut *per se*, so much as the cumulative end product of successive rutting and infilling over a considerable period of time. Even so, both ruts had a consistent profile, up to 0.50m wide and over 0.20m deep.

It seems highly likely that this road surface and its associated ruts should be identified with the trackway shown on the 1906 Ordnance Survey as skirting this side of the 'motte' and leading into the Ogofau Pit from the main Cothi valley road in the vicinity of Ogofau Lodge. This would suggest a link with renewed exploitation of the mines during the later 19th and early 20th century, much of it known to have been focused in the main opencast. Though its date of construction remains uncertain, it has already been argued that it saw long usage.

The second of the three different road surfaces was located above the clay [35/43] at the eastern end of the site (Fig. 4.47). Its makeup was first examined in 1984, as a distinct spread [26] of medium to large angular shale fragments and small shale chippings, intermixed with a dull yellowish-brown loamy very fine sand [10 YR 4/3], which incorporated what was

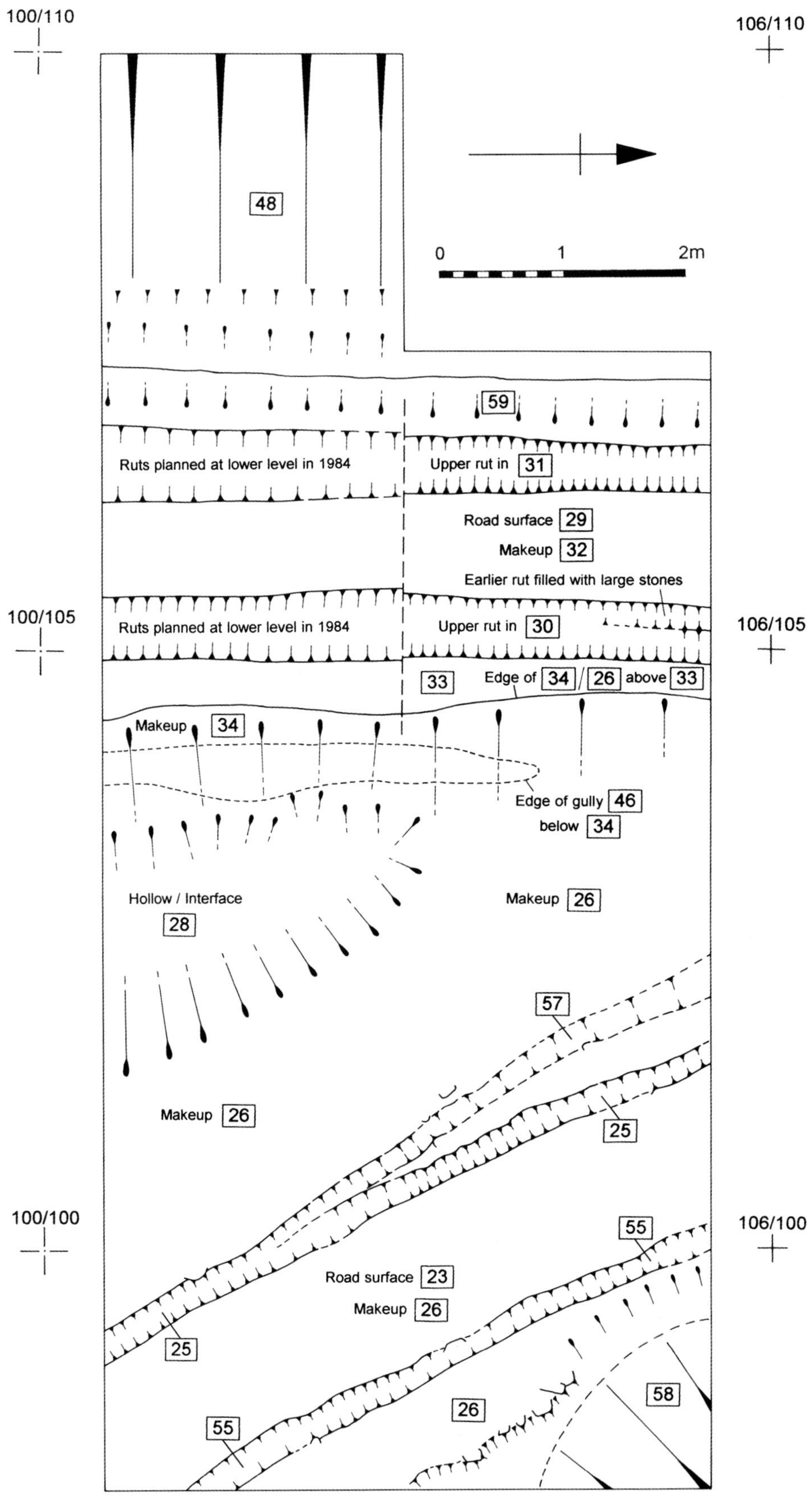

Fig. 4.45 Early road surface into the Ogofau Pit and secondary track into the horseshoe-shaped opencast

Fig. 4.46 Deeply-rutted road skirting base of 'motte', looking north

Fig. 4.47 Road surface and associated cart ruts leading south-east towards small opencast, looking south-east

thought to be a short stretch of wheel rut [25] in the north-east corner of the trench. This makeup appeared to have been truncated along its western side by an elongated depression [28] extending across the centre of the trench. This latter had also truncated an identical shale spread to the west [34], which proved to be the fill of the linear gully [46], cut into the makeup [33] of the first road. Though the nature of these two shale layers was not entirely clear in 1984, they were tentatively interpreted as part of a more extensive horizon laid in connection with the exploitation of a nearby horseshoe-shaped opencast to the south-east of the excavations.

This interpretation was duly confirmed in 1991, when the two distinct spreads were shown to be one and the same across the northern half of the extended trench, as was the fact that it had been constructed later than its counterpart to the west. To the north-east, the shale petered out across the extant slope of the mound associated with the Carreg Pumsaint. This makeup formed the foundation for a compacted road surface [23], which incorporated at least three shallow, U-shaped wheel ruts [25, 55 and 57] trending south-south-east away from a junction with the main road, which must have lain just to the north of the current site. All three were no more than 0.20 to 0.30m wide, with depths ranging from 0.05m [57] to 0.10 to 0.15m [25 and 57]. To judge from their relative shallowness and the stratigraphic relationships of the other road makeups, the exploitation of the opencast was probably relatively shortlived. They proved to have been infilled with an identical greyish-brown sandy clay loam [10 YR 4/2], with a 20% admixture of small shale fragments [53, 54 and 56 respectively].

Following the demise of the second road and the accumulation of two successive clay/clay loam fills [22 and 24] across the bottom of the elongated depression [28], a well-defined layer [21] of dark brown silty clay loam [10 YR 3/3] developed over the eastern half of the site. This turf horizon clearly represents a period of stability across much of the site, possibly coinciding with the closure of the mines in the early part of the 20th century. Somewhat later, the third and final road surface was laid above its predecessor [29] across the western end of the site (Fig. 4.48). Its makeup comprised a compact layer [27] of small to medium-sized stones and angular shale fragments, intermixed with a brownish-black sandy clay loam [2.5 YR 3/2]. This formed the foundation for a compacted road surface [6] *c.* 3.00m wide, associated with two shallow wheel ruts [4 and 5] on a slightly different line from their predecessors in the underlying surface [29]. Both proved to be up to 0.30m wide and 0.10m deep, with an almost identical fill [2 and 3 respectively] of brownish-black sandy clay loam [2.5 Y 3/1 and 3/2] and small to medium stones. This latest road seems to have served a garage to the south of the site within living memory (G. Morgan, *pers. comm.*).

To the east of this latest road, the last significant event was the laying of an extensive spread [19] of medium to large angular shale fragments intermixed with a matrix of dark brown sandy clay loam [7.5 YR 3/3]. This petered out above the sloping edge of the latest road and proved to be at its deepest, up to 0.25m, where it infilled the depression above the earlier hollow [28]. More than likely, this layer represents a relatively recent levelling-up operation by the National Trust. In 1984, several small features [8, 10, 12, 14, 16 and 18] were shown to have been cut into its surface, though none conformed to any coherent plan. No attempt was made to search for comparable features in 1991. Above all this lay the modern turf horizon [1].

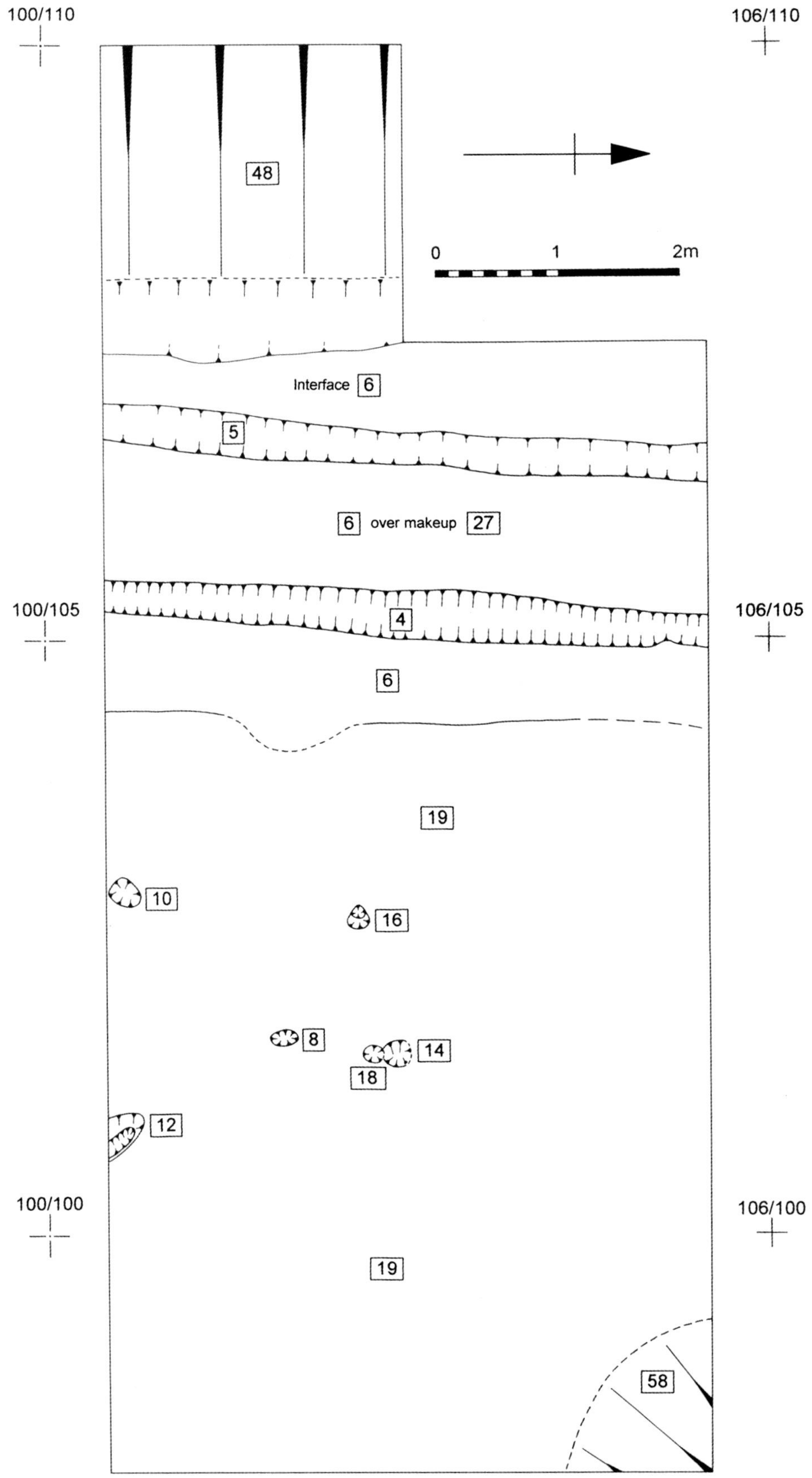

Fig. 4.48 Later road surface and associated features to the east

4.4 THE FINDS

With the exception of the material found in association with the latest phase on the site, most of it of 19th- and 20th-century date, very few finds were made in the earlier levels. Three pieces were of sufficient significance, however, to merit discussion here.

No 1 a fragment from the upper stone of a rotary quern, about one-sixth or one-seventh complete (Fig. 4.49), apparently discarded amidst the stone spread [145] in the large hollow [152] at the eastern end of the site (find no. 8). At its maximum it measures 190mm from edge to eye, 230mm along its surviving length and 60mm thick at the outer edge, but originally it must have been over 500mm in overall diameter with a central hole less than 150mm across. It is made of conglomerate, a locally available, hard stone ideal for grinding mineral-bearing ores. Its upper surface and edge are very rough and worn, whereas its well-smoothed under surface has seen considerable use and still bears characteristic striations running concentrically across it. For similar stones from the vicinity of Ogofau Lodge and elsewhere at Dolaucothi, see the catalogue attached as an appendix in section 4.9 below.

No 2 a roughly circular flat stone disc or counter (Fig. 4.50), 17 to 19mm in diameter, found at the base of the clay horizon [144] above the stone spread [145] in the large hollow [152] at the eastern end of the site (find no. 9). Both its upper and lower surfaces are well polished.

Fig. 4.49 Fragment from upper stone of a rotary quern from [145]; lower surface with characteristic striations

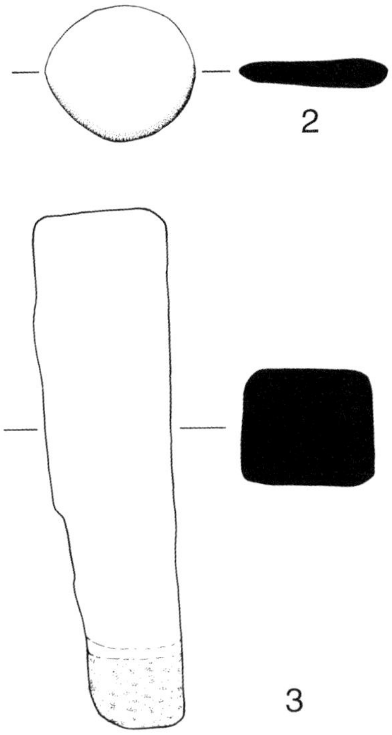

Fig. 4.50 Stone objects. Scale: 1:1 (no. 2); 1:2 (no. 3)

No 3 a roughly rectangular whetstone (Fig. 4.50), found during the excavation of one of the spreads [147] in the south-east corner of the trench (find no. 10). It measures up to 135mm in length and 25mm thick, but its width tapers from 35 to 25mm. Its upper surface is slightly convex and generally well smoothed. For a discussion with parallels see Marvell and Owen John (1997, 418–24).

4.5 RADIOICARBON DETERMINATIONS
(by Quentin Dresser)

Method

The Dolaucothi measurements were made at the Department of Geography, University of Wales, Swansea. All ^{14}C ages are referenced to 1950 (NBS standard) and were corrected for isotopic fractionation. Calibrations and statistical tests were done using the Radiocarbon Calibration Program Rev. 3.0 produced by the University of Washington, Quaternary Isotope Laboratory (Stuiver and Reimer 1993, 215–230). The decadal dataset was used (Stuiver and Becker 1993, 35–65). A laboratory error multiplier of 1.25 and a sample age range of 20 years were used.

Results

The calibrated results are given as a 2 standard deviation range, providing a 95% confidence interval (Table 4.1).

Table 4.1 Radiocarbon calibrations

Designation	Lab. No.	^{14}C Age	Calibrated range (2 σ)	Intercepts
Sample 24 Context [146]	SWAN-37	1890±60 BP	Cal 43 BC – AD 331	130 AD
Sample 28 Context [150]	SWAN-38	2630±70 BP	Cal 925 BC – 427 BC	800 BC

The results for the Dolaucothi samples are obviously dissimilar, and the samples, coming from different parts of the site, probably reflect long term activity at the site.

4.6 THE ANALYSIS OF THE PROCESSING RESIDUES (by Alwyn E. Annels)

Sample description and analysis

A total of 24 samples were submitted for analysis, numbered 1–19 inclusive, 20, 21, 26, 27 and 34 (Table 4.2). They were sub-divided into two main groups, *i.e.* 'crushed material' and 'clays', with a further sub-division into those underlying the turf horizon [72], which are thought to be of Roman origin, and those overlying it.

All samples were dried and weighed before being wet fractioned into seven fractions, *i.e.* >5.6mm, 5.6–2.8mm, 2.8–1mm, 1–0.5mm, 0.5– 0.125mm, 0.125–0.063mm, and <0.063mm. The latter is the fine silt and clay fraction, which was not recovered for analysis. This procedure was undertaken to characterise the grain size distribution in the samples and also to facilitate the mineralogical identification and modal analysis of this 'dirty' material. After drying, the weight of each fraction for each sample was determined; the resulting raw data from this exercise are included in the site archive and synthesised in Tables 4.3 and 4.4 for the crushed material and clays respectively.

The six coarser size fractions for each sample were then examined under the binocular microscope in order to identify their main constituents and to make semi-quantitative visual estimations of the proportions of each.

Table 4.2 Sample description

Sample No.	Context	clay/ crushed material	Located on fig. no.
Above turf horizon [72]			
1	60	crushed material	4.44
2	61	crushed material	4.44
3	43	clay	
4	40	clay	
Below turf horizon [72] in central hollow			
6	73	crushed material	
7	74	clay	4.39
8	75	crushed material	4.39
9	77	clay	4.39
5	47	crushed material	
10	80	crushed material	4.32
11	81	clay	
12	84	clay	4.31
13	95	clay	4.30
14	95	clay	4.30
15	97	crushed material	4.29
16	104	clay	
17	109	clay	4.26
18	124	clay	4.22
19	129	clay	4.19
20	130	crushed material	4.17
21	141	crushed material	4.15
From eastern hollow [152]			
26	144	clay	
From early hollows/deposits at eastern end of site, below [152]			
27	150	crushed material	
34	162	crushed material	

General overview of sample constituents
The following constituents were identified in variable amounts within each sample.

Lithic fragments: these consist of shale and siltstone fragments which show a variable degree of iron-staining and decomposition. Many are, however, subrounded, which is unusual for such material in the natural environment, suggesting that they may have been ground or abraded by some rotary action, as opposed to a vertical crushing action.

'Milky' quartz: this is disaggregated vein material and thus frequently shows some degree of crystal form. The milky colour renders the grain opaque and many have some degree of superficial iron-staining. Cavities or cast-like structures indicate the original presence of carbonate (ankerite) and sulphides (pyrite) which have been completely destroyed by weathering.

Vitric quartz: these quartz grains are characterised by being translucent and by appearing remarkably fresh-looking. They can be angular with conchoidal fracture surfaces or they can be plate- or shard-like in form. The platelets result from fracture along the crystallographic planes and give the samples an almost micaceous appearance. All transitions may be present from milky to vitreous quartz as the intensity of crushing increases.

Charcoal: carbonised wood (charcoal) appears in most samples but in some it is highly abundant (11, 13, 14, 17, 19 and 26).

Ferruginous aggregates: these consist of fine particles of quartz, charcoal, mica and lithic material which have been cemented together by iron-rich precipitates from the ground water. These are thus orange to deep reddish-brown in colour. Other aggregates with a light yellow-brown colour may be decomposed bedrock fragments.

Iron oxide grains and pseudomorphs after pyrite: small granules of iron oxide can often be seen, which sometimes appear to have subhedral to euhedral form reflecting their provenance from grains of pyrite. Some of this material is associated with heavily stained quartz suggesting provenance from vein mineralisation, while other discrete grains may have been dissociated from shale host rocks (such grains are common in the 0.5–0.125mm fractions). Similar examples are seen locked in ferruginous aggregates and/or very iron-stained and decomposed country rock.

Rootlet casts: tubular casts from long-decayed rootlets are common in some samples. These delicate tubes are probably made up of some combination of secondary silica, chelates and iron oxides, which originally accumulated around the roots of plants. All appear in samples designated as clays (3, 4, 7, 9, 11, 13, 16, 17 and 19).

Egg cases: ovoid white egg cases were observed in some samples, some complete with a small hole at one end while others were fragmentary. Complete specimens were generally restricted to the 0.5–0.125mm fraction, indicating their general size. All records are from clay layers at the bottom of the excavated sequence (17, 18, 19 and 26).

Vesicular glass/slag: occasional examples of dark grey to black glassy material were recorded, containing gas bubbles or vesicles, which appear to pass gradationally into partially fused quartz. Many contain inclusions of other material. Such material comes almost exclusively from the clay samples (7, 11, 13, 17 and 19), with the exception of one from the crushed material (10).

Results of fractionation analysis
The crushed material: The data for this material are presented in numerical and graphical form (Table 4.3 and Fig. 4.51). Here and in succeeding tables, samples are arranged in stratigraphic order from top to bottom. Crushed samples 1 and 2 above the turf horizon are clearly different from those below, in that they have a truncated size distribution with very little material of greater than 2.8mm grain size. Another abnormal sample is no. 21, from the very bottom of the sequence in the central hollow, which has a truncation above 1mm, a huge excess of material in the range 0.5 to 0.125mm, and a deficit below 0.063mm. This sample would appear either to have been well sorted (by crushing to a uniform size with winnowing of fines), or to be the result of natural processes at the bottom of the hollow. All other layers show very similar size and distributions.

The clays: The data for this material are presented in numerical and graphical form (Table 4.4 and Fig. 4.52). This once again highlights the difference between the two samples 3 and 4 from above the turf horizon, which have a greater admixture of material over 1mm in size than the older clays beneath it.

Mineralogical changes between the size fractions: The detailed modal analyses of each size fraction for each sample are shown in Table 4.5. These reveal systematic changes occurring from the coarse (>5.8mm) to the fine fractions studied (0.125–0.063mm) which can be summarised as follows:

1. decrease in lithic fragments (shale and siltstones);
2. increase in the amount of crushed quartz (angular, shard like and platelets) at the expense of milky crystalline quartz;
3. increase in the amount of ferruginous aggregates;
4. increase in the content of pyrite pseudomorphs both in the ferruginous aggregates and as free grains.

Table 4.3 Size fractionation of the crushed material

Sample No.	Weight (gms)	>5.6 mm	5.6–2.8 mm	2.8–1 mm	1–0.5 mm	0.5–0.125 mm	0.125–0.063 mm	<0.063 mm
Above turf horizon [72]								
1	1058	9.3g 0.88%	14.7g 1.39%	308g 29.11%	229g 21.64%	289.5g 27.36%	75.2g 7.11%	132.3g 12.5%
2	1002.5	4.7g 0.47%	10.3g 1.03%	383.5g 38.25%	249.5g 24.89%	171.2g 17.08%	37.7g 3.76%	145.6g 14.52%
Below turf horizon [72] in central hollow								
6	627.5	102.2g 16.29%	109.7g 17.48%	118.5g 18.88%	74.1g 11.81%	92.4g 14.73%	30.4g 4.84%	100.2g 15.97%
8	642	76.5g 11.92%	139.9g 21.79%	176g 27.41%	77.2g 12.02%	75.0g 11.68%	23.1g 3.60%	74.3g 11.57%
5	1057	162.6g 15.38%	194.1g 18.36%	232g 21.95%	156.6g 14.82%	150.8g 14.27%	34.7g 3.28%	126.2g 11.94%
10	908	170.6g 18.79%	140.2g 15.44%	176.3g 19.42%	113g 12.44%	126.7g 13.95%	36.1g 3.98%	145.1g 15.98%
15	946	145.8g 15.41%	211.2g 22.33%	204.2g 21.59%	105.5g 11.15%	114.6g 12.11%	35.5g 3.75%	129.2g 13.66%
20	968	190.4g 19.67%	189.2g 19.55%	217.2g 22.44%	119g 12.29%	125.3g 12.94%	34.7g 3.58%	92.2g 9.52%
21	968.5	3.2g 0.33%	16.7g 1.72%	50.1g 5.17%	162.8g 16.81%	615.5g 63.55%	62.1g 6.41%	58.1g 6.00%
From early hollows/deposits at eastern end of site, below [152]								
27	492.5	171.2g 34.76%	41.3g 8.39%	65g 13.20%	82.7g 16.79%	63.2g 12.83%	12.7g 2.58%	56.4g 11.45%
34	1577	326.2g 20.68%	219.5g 13.92%	466.5g 29.58%	191.7g 12.16%	129.4g 8.21%	49g 3.11%	194.7g 12.35%

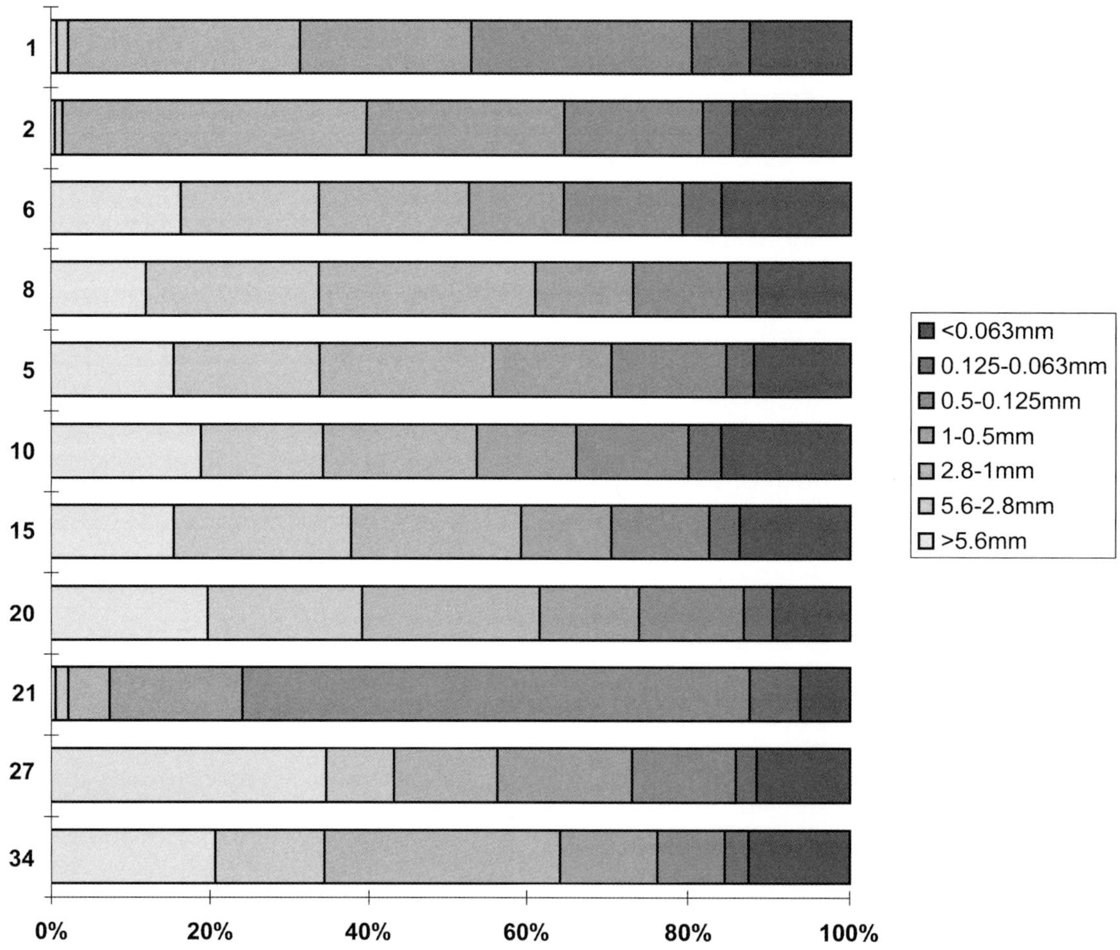

Fig. 4.51 Size fractionation of the crushed material

Table 4.4 Size fractionation of the clays

Sample No.	Weight (gms)	>5.6 mm	5.6–2.8 mm	2.8 –1 mm	1 –0.5 mm	0.5 –0.125 mm	0.125–0.063 mm	<0.063 mm
Above turf horizon [72]								
3	825	67.1g 8.13%	62.1g 7.53%	73.8g 8.95%	64.5g 7.82%	85.4g 10.35%	39.7g 4.81%	432.4g 52.41%
4	704	24.7g 3.51%	52.2g 7.41%	70.8g 10.06%	40.7g 5.78%	69.9g 9.93%	30.6g 4.35%	415.1g 58.96%
Below turf horizon [72] in central hollow								
7	368.5	8.8g 2.39%	16.6g 4.50%	24.1g 6.54%	19.1g 5.18%	66.7g 18.1%	52.1g 14.14%	181.1g 49.15%
9	484	0.4g 0.08%	9.4g 1.94%	17.7g 3.66%	19g 3.93%	20.1g 4.15%	21.4g 4.42%	396g 81.82%
11	473.5	9.9g 2.09%	15.2g 3.21%	21.3g 4.50%	37.9g 8.00%	91.2g 19.26%	67.1g 14.17%	230.9g 48.76%
12	584	1.1g 0.19%	6.5g 1.11%	22.5g 3.85%	30.1g 5.15%	47g 8.05%	33.7g 5.77%	443.1g 75.87%
13	514.5	1.7g 0.33%	10.4g 2.02%	24.6g 4.78%	30.8g 5.99%	68.9g 13.39%	64.8g 12.59%	313.3g 60.89%
14	412	0.5g 0.12%	7.5g 1.82%	17.6g 4.27%	15.2g 3.69%	28.1g 6.82%	38.4g 9.32%	304.7g 73.96%
16	469	15.4g 3.28%	18.4g 3.92%	23.7g 5.05%	16.6g 3.54%	56.1g 11.96%	58.9g 12.56%	279.9g 59.68%
17	574	0.4g 0.07%	6.9g 1.20%	15.4g 2.68%	10.3g 1.79%	23.1g 4.02%	55.3g 9.63%	462.6g 80.59%
18	686.5	12.1g 1.76%	22.8g 3.32%	36.6g 5.33%	21.3g 3.10%	58.6g 8.54%	83.7g 12.19%	451.4g 65.75%
19	649	7.7g 1.19%	11.1g 1.71%	10.2g 1.57%	9.9g 1.53%	53.4g 8.23%	93.8g 14.45%	462.9g 71.33%
From eastern hollow [152]								
26	454.5	2.5g 0.55%	4.4g 0.97%	15.9g 3.50%	14.1g 3.10%	20.9g 4.60%	32.0g 7.04%	364.7g 80.24%

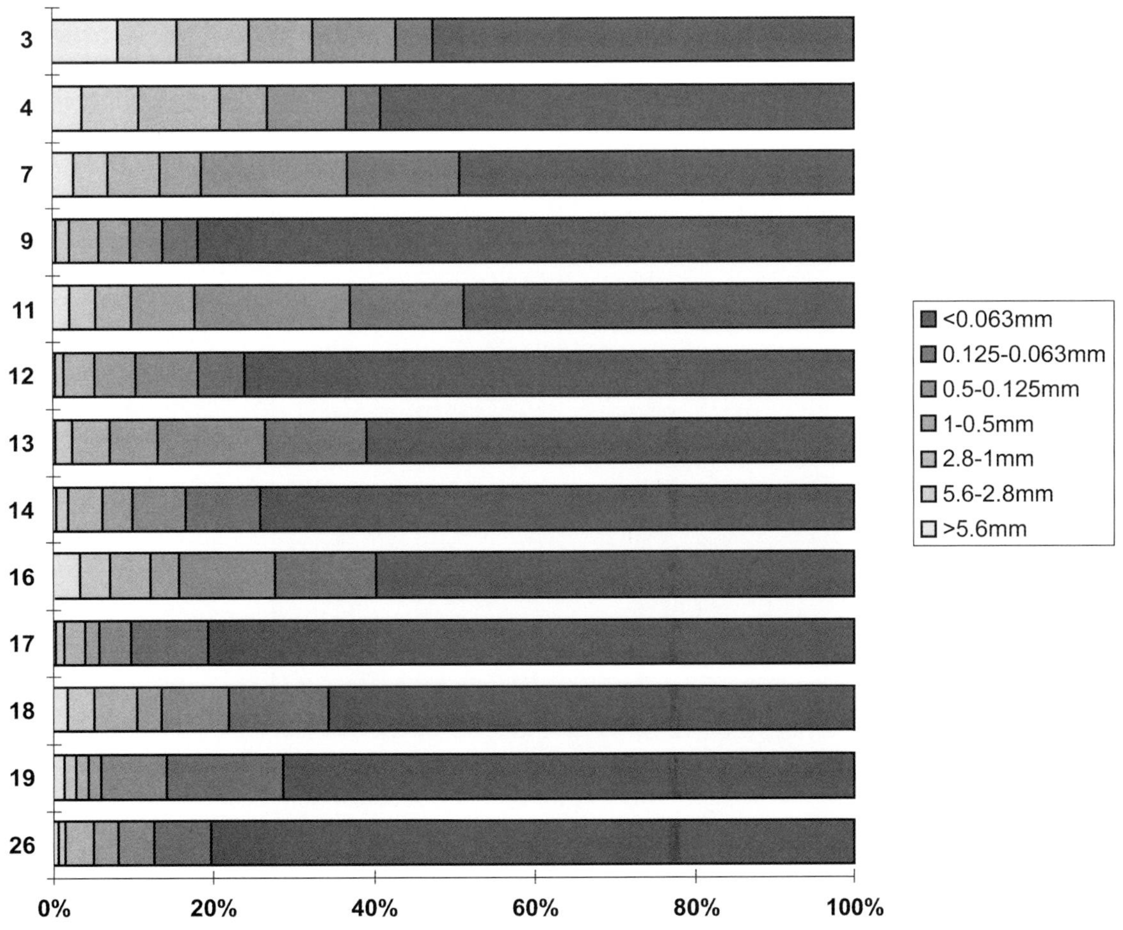

Fig. 4.52 Size fractionation of the clays

Table 4.5 Modal analyses and size fractionation

Sample	Lithic	Carbon	Fe Agg	Fe Oxide/ Pyrite	Milky Quartz	Vitric Quartz	%Sample
Above turf horizon [72]							
1	crushed material						
>5.6					100		0.88
5.6–2.8	45				55		1.39
2.8–1.0	20			1	69	10	29.11
1.0–0.5	35	tr	1	1	3	60	21.64
0.5-0.125	10			5		85	27.36
0.125–0.063	10		1	2		87	7.11
<0.063	-	-	-	-	-	-	12.5
2	crushed material						
>5.6	50				50		0.47
5.6–2.8	45				55		1.03
2.8–1.0	34			1	5	60	38.25
1.0–0.5	40	1		2		57	24.89
0.5–0.125	28		5	2		65	17.08
0.125-0.063	14		1	2		83	3.76
<0.063	-	-	-	-	-	-	14.52
3	clay						
>5.6	85				5		8.13
5.6–2.8	75		10		15		7.53
2.8–1.0	50	1	24		20	5	8.95
1.0–0.5	24	1	35		5	35	7.82
0.5-0.125	5		60			35	10.35
0.125-0.063	5		65			30	4.81
<0.063	-	-	-	-	-	-	52.41
4	clay						
>5.6	60		25		15		3.51
5.6-2.8	35		55		10		7.41
2.8–1.0	35		55		5	5	10.06
1.0–0.5	30	1	34	1	10	25	5.78
0.5–0.125	20		54	1	5	15	9.93
0.125-0.063	7	1	76	1		15	4.35
<0.063	-	-	-	-	-	-	58.96
Below turf horizon [72] in central hollow							
5	crushed material						
>5.6	67		8		25		15.38
5.6–2.8	60		10		30		18.36
2.8-1.0	62	1	5	2	20	10	21.95
1.0–0.5	25	1	15	2	17	40	14.82
0.5-0.125	15	1	13	2	15	55	14.27
0.125–0.063	2	1	17	5	20	55	3.28
<0.063	-	-	-	-	-	-	11.94
6	crushed material						
>5.6	70		10		20		16.29
5.6–2.8	60		15		25		17.48
2.8–1.0	54	1	10		25	10	18.88
1.0–0.5	5	2	51	2	10	30	11.81
0.5-0.125	5	1	50	10	4	30	14.73
0.125–0.063	1	1	76	15	5	2	4.84
<0.063	-	-	-	-	-	-	15.97
7	clay						
>5.6	10	1	84		5		2.39
5.6–2.8	5	5	85		5		4.50
2.8–1.0	5	10	84		1		6.54
1.0–0.5	4	18	67	1	7	3	5.18
0.5–0.125	1	27	60	7	2	3	18.10
0.125–0.063		10	57	5	5	23	14.14
<0.063	-	-	-	-	-	-	49.15

Sample	Lithic	Carbon	Fe Agg	Fe Oxide/ Pyrite	Milky Quartz	Vitric Quartz	%Sample
8	crushed material						
>5.6	40	1	45		15		11.92
5.6–2.8	50	1	34		15		21.79
2.8–1.0	50	2	28		15	5	27.41
1.0–0.5	50	3	12	5	25	5	12.02
0.5–0.125	28	2	10	20	25	15	11.68
0.125–0.063		1	65	20	4	10	3.60
<0.063	-	-	-	-	-	-	11.57
9	clay						
>5.6			100				0.08
5.6–2.8			100				1.94
2.8–1.0			100				3.66
1.0–0.5		1	97		2		3.93
0.5–0.125		1	88	10		1	4.15
0.125–0.063		5	72			23	4.42
<0.063	-	-	-	-	-	-	81.82
10	crushed material						
>5.6	45				55		18.79
5.6–2.8	55		10		35		15.44
2.8–1.0	55		5	5	25	10	19.42
1.0–0.5	20		25	10	20	25	12.44
0.5–0.125	2		48		15	35	13.95
0.125–0.063			74	11	5	10	3.98
<0.063	-	-	-	-	-	-	15.98
11	clay						
>5.6	40		25		35		2.09
5.6–2.8	20	30	34	1	15		3.21
2.8–1.0	15	35	38	1	10	1	4.50
1.0-0.5	2	20	71	5	1	1	8.00
0.5–0.125	1	15	62	15	5	2	19.26
0.125–0.063		15	53	12	5	15	14.17
<0.063	-	-	-	-	-	-	48.76
12	clay						
>5.6	55		45				0.19
5.6–2.8			100				1.11
2.8–1.0			100				3.85
1.0–0.5			95	4		1	5.15
0.5-0.125			84	15		1	8.05
0.125–0.063			85	10		5	5.77
<0.063	-	-	-	-	-	-	75.87
13	clay						
>5.6		45	55				0.33
5.6–2.8	40	25	30		5		2.02
2.8–1.0	10	30	55		3	2	4.78
1.0–0.5	2	20	73			5	5.99
0.5–0.125		20	77		2	1	13.39
0.125–0.063		5	45	8	25	17	12.59
<0.063	-	-	-	-	-	-	60.89
14	clay						
>5.6					100		0.12
5.6–2.8	15	15	45		25		1.82
2.8–1.0	15	15	45		20	5	4.27
1.0–0.5	9	10	60	1	10	10	3.69
0.5–0.125	1	10	70		4	15	6.82
0.125–0.063		5	35		25	35	9.32
<0.063	-	-	-	-	-	-	73.96

Sample	Lithic	Carbon	Fe Agg	Fe Oxide/ Pyrite	Milky Quartz	Vitric Quartz	%Sample
15	crushed material						
>5.6	90		2		8		15.41
5.6–2.8	60		10		30		22.33
2.8–1.0	60	1	9		30		21.59
1.0–0.5	35	2	13		15	35	11.15
0.5–0.125	15	1	20	9	20	35	12.11
0.125–0.063		1	50	5	34	10	3.75
<0.063	-	-	-	-	-	-	13.66
16	clay						
>5.6	35		10		55		3.28
5.6–2.8	15	1	48		35	1	3.92
2.8–1.0	5	1	70		24		5.05
1.0–0.5	1		90		8	1	3.54
0.5–0.125	2	2	60	1	0	35	11.96
0.125–0.063	10		50		0	40	12.56
<0.063	-	-	-	-	-	-	59.68
17	clay						
>5.6			100				0.07
5.6–2.8	2	20	85		13		1.20
2.8–1.0	5	10	75		10		2.68
1.0–0.5	2	18	75			5	1.79
0.5–0.125	5	25	60			10	4.02
0.125–0.063		10	40			50	9.63
<0.063	-	-	-	-	-	-	80.59
18	clay						
>5.6	20		40		40		1.76
5.6–2.8	20	2	40		38		3.32
2.8–1.0	12	5	58		25		5.33
1.0–0.5	20	6	50	1	10	13	3.10
0.5–0.125		4	31	10	15	40	8.54
0.125–0.063		1	10	1	28	60	12.19
<0.063	-	-	-	-	-	-	65.75
19	clay						
>5.6	35		20		45		1.19
5.6-2.8	25	20	30		25		1.71
2.8–1.0	10	30	30			30	1.57
1.0–0.5		35	50			15	1.53
0.5–0.125	5	20	27			48	8.23
0.125–0.063	1	5	1		5	88	14.45
<0.063	-	-	-	-	-	-	71.33
20	crushed material						
>5.6	40				60		19.67
5.6–2.8	55				45		19.55
2.8–1.0	55				45		22.44
1.0–0.5	50				50		12.29
0.5–0.125	5			35	50	10	12.94
0.125–0.063			50	20	20	10	3.58
<0.063	-	-	-	-	-	-	9.52
21	crushed material						
>5.6	20				80		0.33
5.6–2.8	54	1	10		35		1.72
2.8–1.0	50		10		40		5.17
1.0–0.5	20		5	5	60	10	16.81
0.5–0.125	5			2	93		63.55
0.125–0.063			10	5	85		6.41
<0.063	-	-	-	-	-	-	6.00

Sample	Lithic	Carbon	Fe Agg	Fe Oxide/ Pyrite	Milky Quartz	Vitric Quartz	%Sample
From eastern hollow [152]							
26	clay						
>5.6	100						0.55
5.6–2.8	10	8	80		2		0.97
2.8–1.0	2	8	88		2		3.50
1.0–0.5	1	6	92		1		3.10
0.5–0.125	1	8	90			1	4.60
0.125–0.063		5	80			15	7.04
<0.063	-	-	-	-	-	-	80.24
From early hollows/deposits at eastern end of site, below [152]							
27	crushed material						
>5.6	100						34.76
5.6-2.8	90	5			5		8.39
2.8–1.0	75	3	2		20		13.20
1.0–0.5	15	2	18		55	10	16.79
0.5-0.125	10	1	39		5	45	12.83
0.125–0.063			85			10	2.58
<0.063	-	-	-	-	-	-	11.45
34	crushed material						
>5.6	75				25		20.68
5.6–2.8	40		2		58		13.92
2.8–1.0	20		10		65	5	29.58
1.0–0.5	25		15		50	10	12.16
0.5-0.125	15		25			60	8.21
0.125–0.063	2		5	8		60	3.11
<0.063	-	-	-	-	-	-	12.35

Modal analyses

Tables 4.6 and 4.7 and Figs 4.53 and 4.54 show the back-calculated compositions of the samples based on the visual modal analyses of the >0.063mm fractions and from the proportion of fines (<0.063mm) listed in Table 4.5. This involved calculating the weights of each mineral type in each fraction as a percentage of the whole sample before fractionation, after which the weights were summed to give the back-calculated contents of the original sample. These values do not, therefore, reflect the amount of quartz etc that occurs in the fines. The raw data from this exercise are included in the site archive.

The crushed material: Table 4.6 and Fig. 4.53 present the semi-quantitative modal analyses (back-calculated) of this material placed in broad stratigraphic order from top to bottom. This material is characterised by the presence of less than 16% of the <0.063mm material (clay and silt). The top two samples (1 and 2) are characterised by their lack of ferruginous aggregates, reflecting their younger age, their shallow depth with respect to the water table, and/or the lack of iron oxides after pyrite. With the exception of one other sample, they also contain significantly higher quantities of quartz than the older material. The one abnormal sample from below the turf horizon is sample 21, which has very high percentages of quartz. It also appeared abnormal in the grain size distribution, in being very well sorted with a deficit of both fine and coarse material. All samples have less than 1.5% of charcoal (carbon).

The clay material: Table 4.7 and Fig. 4.54 present the semi-quantitative modal analyses of this material placed in broad stratigraphic order from top to bottom. They show that the two layers above the turf horizon (samples 3 and 4) are once again significantly different in that they lack charcoal, have enhanced proportions of lithic material (shale fragments) and lack vesicular/scoriaceous glass or slag. Examples of the latter have, with the exception of one sample, been exclusively found with the clay material and more specifically in those clays which have high concentrations of carbon. This is a significant correlation for the interpretation of the site, for, although both materials are lightweight and thus easily transported by water (hence their occurrence in the silty clay bands), they must be directly related.

Another trend which can be distinguished in the central hollow below the turf horizon is a downward increase in quartz, concomitant with a decrease in ferruginous material. As might be expected, lithic material forms very low concentrations in this fine material. Though the ferruginous material could be an iron pan related to the oxidation zone close to the upper limit of the water table, it also coincided with an interval in which a large amount of pyritic shale

Table 4.6 Modal analysis of crushed material

Sample no	% Fines	% Quartz	% Lithic	% Fe Agg	% Carbon
Above turf horizon [72]					
1	12.5	67.7	17.5	2.3	0.0
2	14.5	54.1	29.0	2.2	0.2
Below turf horizon [72] in central hollow					
6	16.0	24.3	33.5	25.6	0.6
8	11.6	19.3	38.6	29.1	1.5
5	11.9	36.8	40.8	10.0	0.5
10	16.0	35.7	30.4	17.9	0.0
15	13.7	28.3	45.9	11.5	0.6
20	9.5	45.7	37.8	7.0	0.0
21	6.0	79.4	10.1	4.6	0.0
From early hollows/deposits at eastern end of site, below [152]					
27	11.5	20.6	56.0	10.5	1.3
34	12.4	48.0	31.3	7.5	0.0

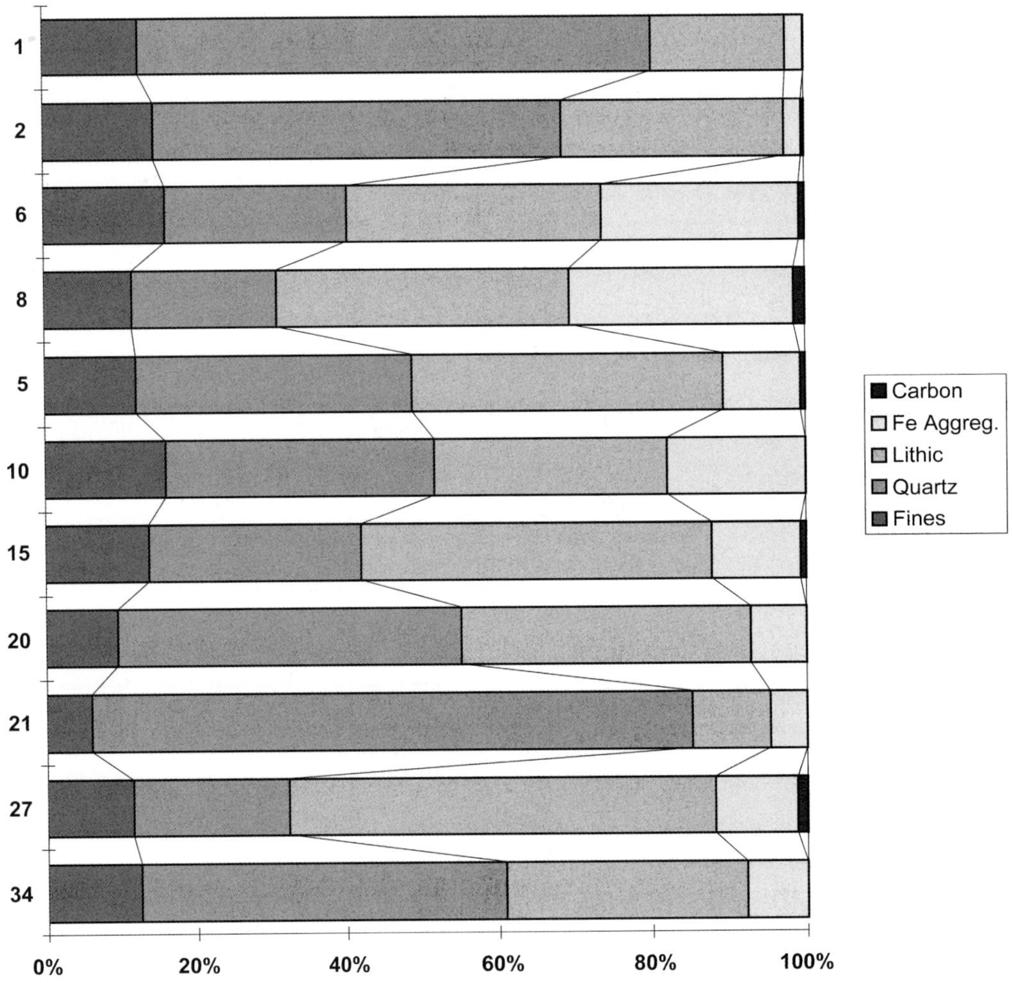

Fig. 4.53 Modal analysis of crushed material

was processed (samples 7, 9, 11, 12 and 13 in Fig. 4.54). This also coincides with a similar increase in pyritic shale and pseudomorphs after pyrite in samples 6, 5, 10, 15 and 20 of the crushed material shown in Fig. 4.53.

Implication of the vertical changes: such evidence indicates that, at an early stage, gold extraction was largely from quartz vein rich ore, but that as time progressed, the proportion of pyritised shale within the ore mined was increasing. The association of the partially fused siliceous material (slag) with charcoal may reflect the use of fire-setting techniques for the fragmentation of the *in-situ* rock. This would suggest

Table 4.7 Modal analysis of clays

Sample no	% Fines	% Quartz	% Lithic	% Fe Agg	% Carbon
Above turf horizon [72]					
3	52.4	12.0	19.7	15.0	0.2
4	59.0	6.9	12.2	21.3	0.1
Below turf horizon [72] in central hollow					
7	49.2	5.8	1.2	35.7	8.1
9	81.8	1.1	0.0	16.7	0.3
11	48.8	6.1	2.5	33.5	9.2
12	75.9	0.4	0.1	23.6	0.0
13	60.9	6.3	1.4	24.8	6.6
14	74.0	9.3	1.3	13.0	2.4
16	59.7	14.0	3.5	22.5	0.3
17	80.6	5.7	0.4	10.7	2.8
18	65.8	19.4	2.3	11.5	1.0
19	71.3	19.1	1.6	4.4	3.7
From eastern hollow [152]					
26	80.3	1.2	0.8	16.5	1.3

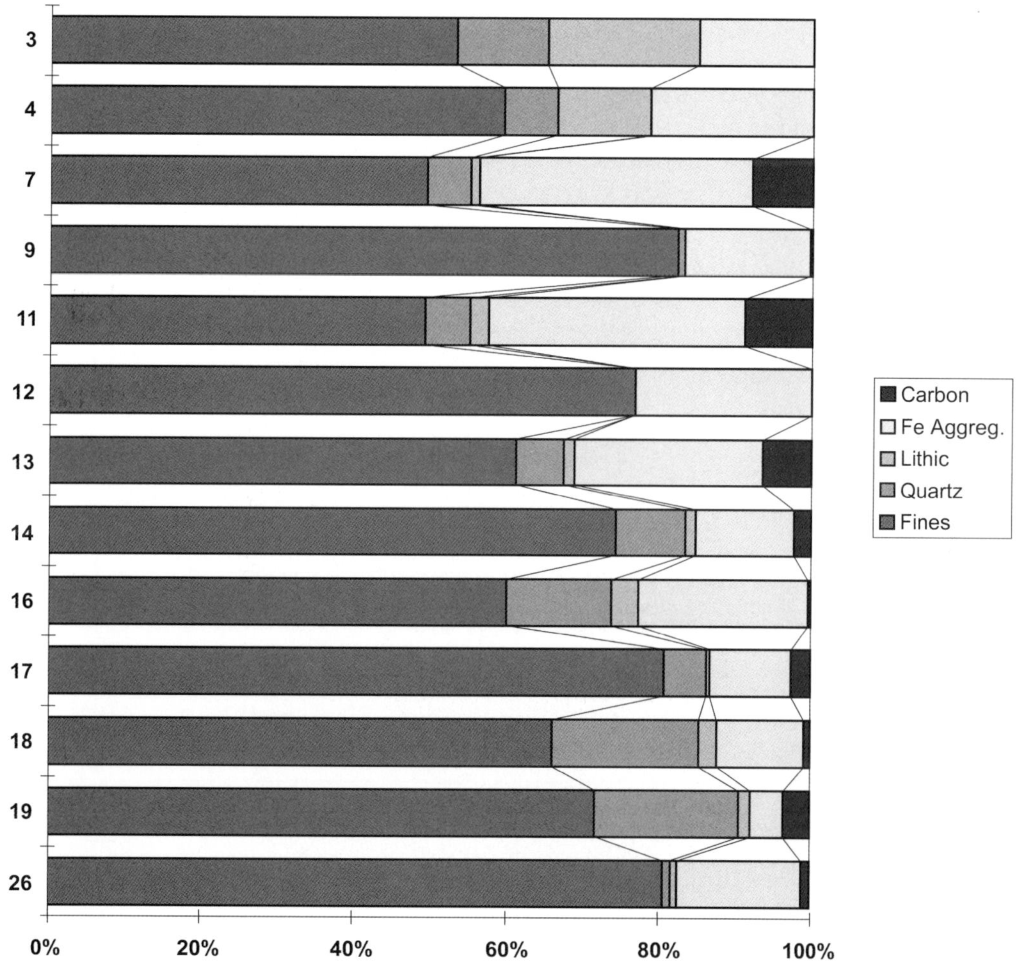

Fig. 4.54 Modal analysis of clays

that locally the temperature at the working face was high enough to cause partial melting of a thin skin directly at the face, thus explaining the apparent intermixture of charcoal, quartz and fused material. An alternative suggestion, that the evidence represents direct smelting or roasting of the auriferous concentrates, cannot be totally disregarded, however.

The two layers above the turf horizon are clearly different: here, the different grain size distributions, the predominance of quartz and shale fragments, and the lack of pyritic shale and charcoal all point to more recent exploitation of the quartz lodes (perhaps in Mitchell Adit) and to the extraction of heavy metals using gravity jigs, probably at the turn of the century.

4.7 THE RESULTS OF GOLD ANALYSES
(by Alwyn E. Annels)

Four samples from the likely Roman horizons (Nos 13, 14, 15 and 19) were subsequently submitted for gold analysis to OMAC Laboratories in Ireland. Of the four, three were 'clays' (13, 14 and 19), while the fourth comprised coarsely-crushed material (15). In each case the different size fractionations were analysed separately, with the exception of the coarsest fractions (>5.6mm) which were mostly composed of barren quartz and shale, and those fractions where the surviving sample proved insufficient (N/A). It should also be remembered that the finest fractions (<0.063mm) were not retained for analysis and that, in some cases, these accounted for up to two-thirds of the original sample. The results are set out in Table 4.8.

Assuming that the gold content of the missing fractions was zero, then it is possible to estimate the grade for the sample as a whole, at least above the 0.063mm level. This gives the following figures: no. 13 = 4.8 g/t; no. 14 = 10.39 g/t; no. 15 = 1.53 g/t; no. 19 = 2.49 g/t. With the exception of no. 14, the gold grade consistently increases into the finer fractions, which makes it highly likely that even higher grades might be expected in the finest fractions (<0.063mm) which were otherwise not available for assay. This would conform with the fine grain size of the gold known from the Dolaucothi ores (generally 10–50µm). For comparison, bulk analyses of the 1930s tailings give grades in the order of 0.5 to 2 g/t. All this suggests that there are high levels of gold still remaining in the samples recovered from the 1991–93 site, no doubt reflecting the very poor recovery rates achieved by the Roman miners. In this respect, it is unlikely that the Romans would have recovered much, if anything, from the finest fractions below 0.063mm.

4.8 DISCUSSION

4.8.1 Dating the processing wastes

Judged against their original objectives (see section 4.3.1 above) the 1991–93 excavations have added considerably to our understanding of this part of the

Table 4.8 Results of gold analyses

Sample no	Fraction size	Sample weight %	Gold Grade g/t	Au Contrib.
13/1	> 5.6mm	0.33	N/A	0.000
13/2	5.6–2.8mm	2.02	N/A	0.000
13/3	2.8–1mm	4.78	3.06	0.146
13/4	1–0.5mm	5.99	4.23	0.253
13/5	0.5–0.125 mm	13.39	4.62	0.619
13/6	0.125–0.063 mm	12.89	6.77	0.873
Total		39.4		1.891
14/1	> 5.6mm	0.12	N/A	0.000
14/2	5.6–2.8mm	1.82	N/A	0.000
14/3	2.8–1mm	4.27	21.26	0.908
14/4	1–0.5mm	3.69	3.51	0.130
14/5	0.5–0.125 mm	6.82	17.33	1.182
14/6	0.125–0.063 mm	9.32	5.23	0.487
Total		26.04		2.707
15/1	> 5.6mm	15.41	N/A	0.000
15/2	5.6–2.8mm	22.33	0.93	0.208
15/3	2.8–1mm	21.59	1.04	0.225
15/4	1–0.5mm	11.15	2.15	0.240
15/5	0.5–0.125 mm	12.11	3.03	0.367
15/6	0.125–0.063 mm	3.75	7.56	0.284
Total		86.34		1.322
19/1	> 5.6mm	1.19	N/A	?0.000
19/2	5.6–2.8mm	1.71	N/A	?0.000
19/3	2.8–1mm	1.57	N/A	?0.000
19/4	1–0.5mm	1.53	N/A	?0.000
19/5	0.5–0.125 mm	8.23	2.56	0.211
19/6	0.125–0.063 mm	14.15	3.5	0.495
Total		28.38		0.706

Dolaucothi complex, by confirming the presence of a sequence of *in-situ* processing wastes in the immediate vicinity of the postulated mill and by rejecting the traditional interpretation of the mound at the western end of the site as a Norman motte. This latter has obvious repercussions, however, for the proposed dating of the mill, because it effectively invalidates the one critical stratigraphic argument for assigning it to the medieval or early modern period, namely that any associated processing wastes must necessarily post-date the construction of the 'motte' and its presumed ditch. Nor was the situation clarified by the recovery of any closely datable finds from the relevant horizons associated with phases 1 and 2, to such an extent that the date of the complex was left almost entirely dependent upon the technological argument that cam-driven, trip-hammer mills were unknown in the Roman period, an argument which is itself now open to debate as will become clear (see section 4.8.2 below). Under these circumstances, the presence of several charcoal samples in key contexts sealed beneath the processing wastes clearly assumed considerable significance, by offering an independent means of establishing a TPQ for the residues themselves, if not also for the mill, since it seemed inherently unlikely that they were unrelated.

Two separate samples were therefore submitted for analysis to the Radiocarbon Dating Laboratory at the University of Wales, Swansea, one (DOL 92, no. 28) from the lens of crushed material [150] infilling one of the shallow hollows [167] at the eastern end of the site, the other (DOL 92, no. 24) from the lowest of the definable horizons [146] located across the central part of the site, where it was squashed into the shale forming the 'built-up' edge [155] of the so-called motte. The resulting uncalibrated determinations of 2630±70 BP (SWAN 38) and 1890±60 BP (SWAN 37) produce a calibrated age range of 925 Cal BC – 427 Cal BC and 43 Cal BC – 331 Cal AD respectively at the 95% confidence level, the latter clearly suggesting a much earlier TPQ than hitherto expected for the overlying processing wastes across the centre of the site, most probably in the early Roman period. The alternative, that a significant length of time had elapsed between the construction of the built-up edge [155], incorporating the charcoal pockets, and the onset of waste dumping into the central depression, seems highly unlikely for reasons which have already been outlined (see section 4.3.3(E) above). The earlier of the two dates from one of the shallow hollows is also sufficiently discrete as to suggest the tantalising possibility of a pre-Roman phase of mining – a suggestion often made but hard to substantiate with any certainty (see section 4.8.4 below).

Such a redating of the processing wastes from the 1991–93 site necessarily raises interesting questions about the dating of the proposed mill complex itself and about its associated technology, since it is now clear from the stratigraphic sequence that there are no obvious medieval or early modern processing horizons with which it could be linked. At the time of the original survey in 1982, the principal objection to a Roman date for the mill stemmed from a widely accepted belief that cam-operated, trip-hammer mills of the type represented by the Carreg Pumsaint were a medieval innovation and, therefore, unavailable in the Roman era (see *e.g.* Davies 1935, 39; Horn 1975). In addition, in his valuable collation of the ancient evidence, Healy (1978, 142–3) could instance no concrete example of the use of any form of water-powered crushing or grinding technology in the sphere of Roman mining, this despite its acknowledged application in grain milling, where the gearing necessary for turning horizontal millstones was well known (Hodges 1974, 190–96; White 1984, 65–6, 196–201). The argument has been repeated recently by Gerrard (2000, 104), who notes that the earliest known references to water-driven trip hammer technologies come from China in AD 290, while the earliest examples in Europe are found in the forges of Schmidmulen in AD 1010 and in ore-crushing by the late 13th century. In view of the potential importance attaching to the new evidence from Dolaucothi, however, the debate needs to be widened to take account of recent thinking on cam-operated devices and the extent to which stones of the Carreg Pumsaint type have been recorded in closely datable Roman contexts elsewhere.

4.8.2 Water-powered technologies and Iberian parallels

This extension of the debate has been considerably facilitated by the publication of a volume specifically devoted to the origins of water power (Lewis 1997), which makes a plausible case for believing that the trip hammer and the cam were both known to the Romans, even if their use might not have been widespread. This is not the place to repeat any of his arguments in detail, merely to note his persuasive conclusion that: (i) both basic types of water mill (horizontal and vertical), mechanical water-lifting in all its forms, and probably the industrial mill (*i.e.* cam-operated), can all be traced back to the mid-3rd century BC, to a phase of significant technological advance focused around Alexandria and other centres; (ii) water-powered fulling stocks very probably existed at Antioch in AD 73/4; (iii) a water-powered grain pounder is probably recorded in c. AD 70 by Pliny; (iv) the cam-operated ore stamp, powered by water, probably existed by at least the early 1st century AD in Spain and Portugal, and almost certainly by the later 1st century at Dolaucothi (Lewis 1997, 123–4). This latter suggestion, of course, was heavily dependent upon the radiocarbon dates from the 1991–93 site and the support they provided for assigning the Carreg Pumsaint and the associated mill complex to the Roman period.

The critical section from our point of view concerns his discussion of ore stamps and, in particular, the recognition in recent years of characteristic blocks of Carreg Pumsaint type in several mining contexts exclusively in Spain and Portugal (*ibid.*, 106–8). The current authors were already aware of possible parallels from the gold mine at Minas dos Mouros (Campo de Jales) in northern Portugal, where innumerable large rectangular blocks of hard porphyritic granite had apparently been recorded, each *c.* 1.00 by 0.50 by 0.50m in size, with four very characteristic cup-shaped depressions across the centre of their long sides; these had been found in association with many broken fragments of circular millstones, apparently from a source not local to the area (Harrison 1931). They were identified at the time as the coarse- and fine-crushing machinery respectively of ancient mine operations. The accompanying illustration of one of the blocks (*ibid.*, fig. 4), though rather schematic, certainly resembles the Carreg Pumsaint and makes a similar origin and function likely, but there was nothing to connect it with water-powered operations and little to justify a Roman rather than a later date.

Far more convincing parallels have now been illustrated from other sites in the immediate vicinity, however, including a granite block with four parallel depressions from the nearby processing site at Forno dos Mouros (Fig. 4.55B; Wahl 1993, 141 and pl. 19), and a similar (if less regular) stone from the mine at Três Minas, also with characteristic rows of indentations on two sides (Sanchez-Palencia 1984–5, 356 and fig. 4). Further afield in north-west Spain, other examples have been illustrated from Bachicón de Fresnedo (Oviedo; Fig. 4.55A; Sanchez-Palencia 1984–5, 356 and fig. 5; 1989, 41 and fig. 4) and Cecos (Oviedo; Sanchez-Palencia 1984–5, 356 and pl. 5), both clearly with four parallel depressions, El Molinillo (Toledo-Ciudad Real; Sanchez-Palencia 1984–5, 356 and pl. 8) and, less certainly, Pozos (León; Domergue 1990, 497, and pl. XXVIIb). The authors concerned also list possible examples at several other sites in north-west Spain, but in the absence of verifiable illustrations, the case must remain unproven. The presence also of numerous circular millstones at many of the sites strongly indicates a close association between the two in the processing technology.

There would seem to be little doubt now that such blocks functioned as mortar stones or anvils in connection with ore-crushing. More importantly, Lewis has made a plausible case for believing that the regularity and spacing of the depressions ought to indicate a mechanically-driven device, most likely involving a set of cam-operated trip hammers much like their medieval and early modern counterparts in the tin industry and elsewhere. In this he is supported by Wahl (1993), who explicitly juxtaposes his illustration of the stone from Forno dos Mouros beside a representation of a dry-stamping mill drawn from Agricola (1556). As yet, however, none of the examples from Spain or Portugal can be convincingly associated with a mill site, though most mines certainly had convenient water sources and associated leat systems, the power of which could easily have been harnessed. Clearly the identification of the remains of such mill structures must remain a matter of high priority in the context of any future archaeological research work.

The occurrence of such mortar stones in closely datable Roman contexts has obvious implications for the Carreg Pumsaint at Dolaucothi and lends considerable weight to the suggestion that it too might be assigned to the Roman period. This would bring it neatly into line with the proposed dating for the processing wastes on the 1991–93 site. As with the Spanish and Portuguese sites, Dolaucothi has also produced an increasing number of the characteristic circular millstones, suggestive of other stages in the processing operations: at least three were published by Bosanquet (1917, fig. 38); further examples including a possible axle stone were illustrated by Boon (1971, figs 13 and 14) and Lewis (1977, fig. 15) from the area of Ogofau Lodge, immediately north of the mill site; one is preserved in the garden wall of Ogofau Lodge itself; a single fragment was recovered from the stone spread at the base of the large hollow [152] at the east end of the 1991–93 site (see section 4.4, no. 1); and one more has recently been recovered from the vicinity of the 'bath-house', south of the river Cothi (for a catalogue of all the known examples from the Dolaucothi area, see section 4.9 below). Although their Roman date cannot be absolutely assumed (see *e.g.* Worth 1967, fig. 96 and pl. 67B, for similar examples of medieval date), the balance is increasingly tipping in that direction in the light of current work.

All this brings us back to the mill complex itself, the discovery of which prompted the current programme of research. Far from corroborating its assumed medieval or early modern date, the evidence presented here opens up the significant possibility that it too might be assigned to the Roman period, as an integral part of a much more extensive operation embracing the Carreg Pumsaint, the processing wastes on the 1991–93 site, and the extensive spreads of crushed quartz known from the fields to the north (Lewis and Jones 1969, 263). This would obviously make it unique, at present, in the sphere of ancient mining. Some element of caution is perhaps necessary, however, in view of previously misplaced(?) pronouncements as to its date, until such time as the site has been properly investigated by controlled archaeological excavation. Even then, should the link with the proposed mill prove un-

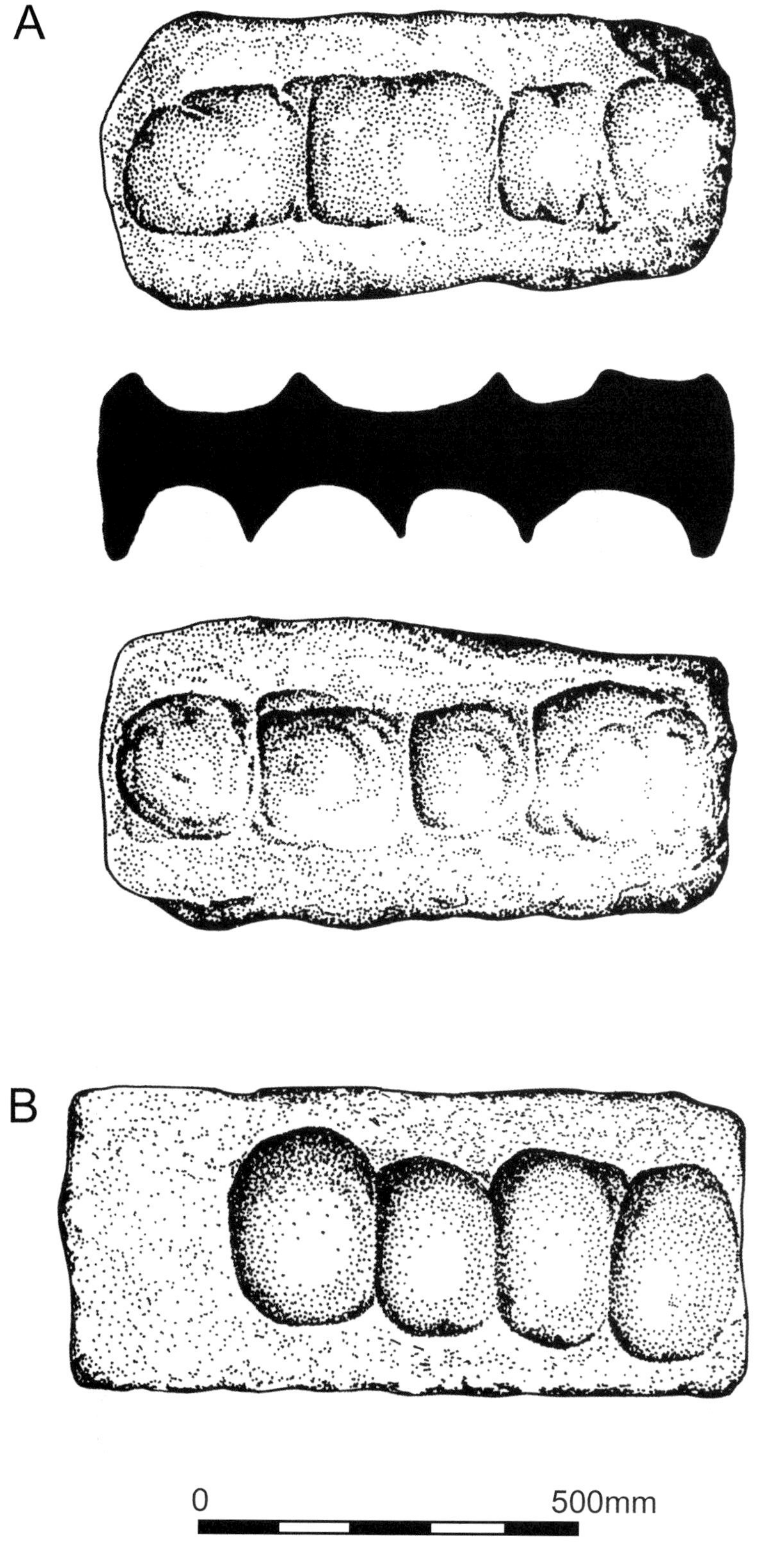

Fig. 4.55 Mortar stones of Carreg-Pumsaint type from the Iberian peninsula. A: Bachicón de Fresnedo (after Sanchez-Palencia 1984–5, pl. 5). B: Forno dos Mouros (after Wahl 1993, pl. 19)

founded, the discovery of an *in-situ* sequence of processing wastes of undoubted Roman date, the first so far recorded from the site, would still represent a very significant addition to our understanding of the site in its own right.

4.8.3 The nature of the processing residues

The analysis of the two very different types of residue by Alwyn Annels (see section 4.6 above) has already made it clear that we are dealing with material derived from processing activities. All the evidence would seem to suggest that such activities were being undertaken at, or relatively near to, the site in question, with the resulting residues being dumped into a convenient hollow formed by the mining debris. Some consideration must now be given to assessing the nature of the different residues, the stages of processing from which they were derived, and the level of gold recovery which was achieved.

Once mined, gold ores have to be processed to produce a usable product, a process which involves separating the gold from the associated waste materials. For much of its history gold has been processed by relatively crude methods, all of which have ultimately depended upon the metal's high density with respect to the other rocks and minerals with which it is most commonly found. This difference is most easily exploited where the gold is well liberated from the associated minerals (*i.e.* where it is in the form of free gold), essentially because this facilitates recovery when the crushed ores are mixed with water and then panned or washed. Such a situation is known to prevail at Dolaucothi, where the bulk of the ancient workings were confined to those areas of the ore bodies where surface weathering had been sufficiently extensive to decompose the sulphides and so release the fine grains of gold trapped within. Though the extent and depth of this clearly varied, drilling has indicated that typical weathering and oxidation occurs to depths of 20 to 30m while locally, in heavily fractured ground, greater depths are achieved (Alwyn Annels, *pers. comm.*). Beyond this the gold grains are known to become increasingly sulphide-locked (*i.e.* they are actually contained within the crystal structure of the associated minerals), a condition which would have posed increasing problems when it came to separating the gold from the associated waste materials.

Where the gold was sufficiently well liberated from the associated minerals, the ancient sources make it clear that there were four basic aspects to processing, which were used as necessary depending on the nature of the respective ore bodies: (i) initial crushing; (ii) grinding to a flour; (iii) washing; and (iv) smelting. This much is confirmed by Diodorus' account of the concentration and washing process both in Egypt (III, 13–14; in Ptolemaic times) and in Gaul (V, 27, 2), while a similar sequence seems to be described by Pliny (*Natural History* 33, 68), despite difficulties in the ordering of the various stages in the original text (Wahl 1993, 139).

Corroboration of the four broad stages is also provided by an extensive range of archaeological evidence, much of it conveniently synthesised in general accounts of processing and refining (*e.g.* Healy 1978, ch. 6). These show that, after an initial sort to remove the bulk of the barren rock, the remaining material was further broken up to facilitate the separation of the metal-bearing parts of the ore from the remaining wastes. Traditionally such crushing was done by hand, using stone mortars and iron pestles, with the resulting fraction being passed through sieves to ensure a broad consistency of size, though obviously a much more efficient reduction would have been possible with the advent of water-powered technologies of the type well known from the pages of Agricola (1556). Once crushed, the ore-bearing element might then be washed or panned, to enable the recovery of such free gold as was already visible within the coarse fraction. This often involved placing the crushed material on some sort of inclined washing table (natural or artificial), across which water was allowed to flow at a constant rate; as the material was agitated, the lighter, barren wastes were carried away downslope and into the tailings, while the heavier gold particles settled out along the slope or towards the sides. After a suitable period the washing was halted and the gold remaining on the slope was collected and transported off for smelting.

Alternatively, once crushed, the ore-bearing element could be taken directly to the milling stage, whereby it would be ground to the consistency of a fine flour or powder, either by using simple hand mills or the more developed hour-glass mills of Pompeian type. The resulting flour could then be washed or panned. Where the gold particles were particularly fine it was often necessary to wash the powder over some sort of animal hide (such as a sheep fleece) or a layer of heather/gorse, so as to increase the efficiency of recovery. Even when the crushed material had already been washed, the resulting tailings could also be milled and rewashed to increase recovery rates. In the case of suitably weathered ore deposits which could be ground-sluiced, as has been argued for much of Dolaucothi by Bird (2001), the relative importance of crushing and milling would have been correspondingly reduced before the washing stage.

Following these various concentration and washing stages the liberated gold was then sent on for further treatment, which might have included amalgamation and/or smelting as necessary. In the case of sulphide ores, however, the remaining concentrates were sometimes subjected to a preliminary roasting before smelting, though the process was far from universal (Davies 1935, 41–2). This was designed to expel the sulphur and other impurities, which were not otherwise separated by the traditional mechanical processes, and thereby to produce the more easily reducible oxide.

It is within this general framework that we have to contextualise the specific evidence from the 1991–93 excavation and its immediate environs, *viz.* (i) the Carreg Pumsaint; (ii) the various millstones from the area of the Ogofau Lodge; and (iii) the sequence of processing wastes. To these we might also add the stepped washing tables which have been tentatively identified in the vicinity of Tank C and elsewhere (Lewis and Jones 1969, 255, fig. 7), and the spread of crushed quartz found on the flood-plain beyond the Ogofau Lodge (Jones and Little 1973, 14, pl. VIb). Arguments have long been advanced for connecting the Carreg Pumsaint with the primary stage of hand-crushing the ore (Lewis 1977, 12), though a case has been made here for its association with a water-powered mill. Either way, such activities would easily account for the nature of the coarsely-crushed material found on the 1991–93 site, both in terms of its overall size fraction and the changing composition of the successive samples; this latter, in particular, must be a direct consequence of the process by which the metal-bearing parts of the ore were selectively separated from the wastes at the crushing stage. What remains unclear from the analysis, however, is the extent to which the resulting material was still awaiting milling, or had already been washed, prior to being dumped or stockpiled for secondary milling, once any ore-rich concentrate had been removed.

The discovery of so many millstones in the vicinity of the Ogofau Lodge, and the recovery of a fragment during the 1992 excavations, makes it likely that they played an integral part in the processing activities taking place in the immediate vicinity. Their presence must surely account for the character of the clay slimes located on the 1991–93 site, since these latter are broadly consistent with the residues which would result from the milling of crushed ores to a fine flour. That said, however, their stratigraphic layering within the central hollow can only be explained if they had also been water washed in the immediate area, with the resulting slimes effectively being allowed to accumulate as thin lenses, one on top of another. An alternative suggestion by Bird (2001, 269), that they represent phases of ground sluicing higher up the site, seems unlikely in view of the consistently interleaved nature of the two different types of residue, and the very localised nature and complexity of the resulting stratigraphy. The situation is further complicated, however, by Alwyn Annels' analysis of the wastes, which reveals the presence of carbonised wood in many of the samples; it is especially abundant in some of the clay horizons, where there are also occasional examples of partially fused siliceous material or slag which might not have been expected if the residues had simply been water processed. Annels tends towards fire-setting as a possible explanation, but notes that the question of roasting should not be excluded at Dolaucothi, where the gold became increasingly pyrite-locked with depth.

Two further points are worth stressing from the analyses. The first concerns the vertical variations in the samples below the turf horizon, which reveal a downward increase in quartz concomitant with a decrease in ferruginous material. This indicates very clearly that, as far as this sequence is concerned, gold extraction was initially from quartz vein rich ore, but that over time the proportion of pyritic shale increased significantly. Taken alongside the presence of slag, this might argue for an increasingly pyrite-locked element in the ore body, hence the potential significance of roasting. The second point is the low recovery rates achieved by the ancient miners, with the gold grade in the clays consistently increasing into the fine fractions; all this conforms with the fine grain size of much of the gold known from the later operations at Dolaucothi and indicates that the Romans failed to recover much if anything from the finest fractions, which were otherwise lost to the tailings.

4.8.4 The pre-processing landscape

The complete removal of the processing residues in 1993 afforded an important opportunity for examining the underlying horizons and for reconstructing something of the pre-existing landscape in this part of the site. It has already been shown that the wastes occupied a convenient hollow at the centre of the site, which had been created by a number of distinctive deposits: (i) the core of the mound in the north-east corner, on which, ultimately, the Carreg Pumsaint was located; (ii) a ridge of shale and sandy clay loam extending south of Carreg along the eastern side, into the top of which had been cut a series of four or five shallow hollows; (iii) several successive shale spreads which merged together across the centre and, ultimately, formed the basal core and lower slopes of the 'motte' flanking the western side. Though these deposits were sampled, essentially by means of several narrow trenches, at no point was the 'natural' encountered, this despite the work reaching a depth of *c.* 2.50m below current ground level. It also proved extremely difficult to separate the different stone spreads, which effectively merged one into another, strongly suggesting that the build-up of the underlying horizons was a continuous process, even if the precise timescales remain uncertain.

While none of this allows us to reconstruct the character of the underlying topography, let alone estimate the depth of the overburden, it is the first indication that this part of the site had been significantly remodelled by extensive dumping operations, virtually all of it barren mining debris suggestive of nearby opencast working. This represents an important extension to our knowledge of the early exploitation, at a date which clearly precedes the onset of the processing activities described in the previous section. Since these latter have already been plausibly assigned to the early Roman era on radiocarbon grounds (see section 4.8.1 above), interesting questions necessarily arise as to date of the earlier activity. In the absence of diagnostic finds from the related horizons, however, we are forced to rely upon a single charcoal sample (DOL 92, no. 28), derived from the fill of one of the four shallow hollows [167] which were cut into the ridge of shale and sandy clay loam across the eastern end of the site. All four have already been shown to be relatively early on purely stratigraphic grounds, so the resulting calibrated date of 925 Cal BC – 427 Cal BC at the 95% confidence level is not as surprising as it might have appeared in isolation. In his discussion of the two dates from the excavation Quentin Dresser notes that the two are sufficiently discrete as to indicate long-term activity at the site, which must surely raise the very real possibility that we are dealing with a pre-Roman mining landscape, with significant implications for the history of exploitation at Dolaucothi.

It is obviously dangerous to be over-reliant on a single date in advancing such a far-reaching conclusion, even if a measure of support is provided by the stratigraphy and by the site's proximity to the main focus of the workings in and around the Ogofau Pit. We should not be surprised, however, at the prospect of pre-Roman mining, given that recent work elsewhere in Wales has identified significant traces of bronze-age activity, not least at Cwmystwyth (Timberlake 2003) and Great Orme (Dutton and Fasham 1994), where extensive underground galleries have been excavated. Important research has also been carried out in the Limousin area of Central France, revealing in the process evidence for extensive mining operations in the later iron age, from the 5th to 1st centuries BC, alongside significant data on the associated mining and ore-processing techniques (Cauuet 1999). All this should remind us of Cauuet's recent assertion that much of the Dolaucothi landscape of opencasts and pits could have had a pre-Roman origin (Ancel *et al.* 2000), to which the current radiocarbon date might add some measure of support (see section 6.2.4(C) below). Clearly any opportunity for further exploring this pre-existing landscape must be accorded a high priority.

4.9 APPENDIX: EARLY MILLSTONES RECOVERED FROM THE DOLAUCOTHI AREA

As early as 1855 antiquarian sources were recording the presence of many large fragments of millstone, including both upper and lower stones, from the Dolaucothi area (Knight 1856, 206). Three examples were published by the RCAHMW (Bosanquet 1917, fig. 38), while Davies (1936, 51–2, note 7) noted their close affinities with parallels elsewhere in the Roman world. Further examples were located during the digging of deep foundations at Ogofau Lodge, in a sealed context alongside probable processing wastes (Lewis and Jones 1969, 263, pl. 50c; Boon 1971, fig. 14; Lewis 1977, fig. 14), though all were subsequently lost. Their fortunate 'rediscovery' in 1993, courtesy of Messers C. Page and T. Cleminson, was prompted by the recovery of another fragment from a sealed context during the excavations near the Carreg Pumsaint in the previous year (no. 13 below). In view of the importance of these and other discoveries, this volume provides a timely opportunity to catalogue all the known fragments from the Dolaucothi area, together with a photographic record where possible. The authors are grateful to Sue Watts for access to her unpublished catalogue of millstones in Carmarthen Museum (1997), from which details have been drawn for Nos 1 and 3–6.

CATALOGUE (Fig. 4.56)

No 1 Lower stone in fine grey sandstone. 460mm diameter, 120mm thick at edge, rising to c. 145mm in middle. Concavo-convex grinding surface with concentric scoring. ?Post base chopped out in middle, 140mm diameter and 30mm maximum depth. Underside unworked, intended to sit in ground? Edges very rough, possibly damaged or knocked off. First published by RCAHMW (1917, fig. 38, lefthand stone). Now in Carmarthen Museum, no. 1977.0379. Catalogued by Watts.

No 2 Upper stone, complete. Roughly circular, though tending towards triangular. 520mm maximum diameter, 70mm thick at eye, tapering across convex surface, steepest towards edge, virtually to a point. Eye 140–150mm diameter. Upper surface roughly worked, lower surface slightly concave with concentric scoring over about two-thirds of area, smooth elsewhere. Two single upright handle holes and two double holes, each with a deeper and a shallower component. First published by RCAHMW (1917, fig. 38, central stone). Now in Carmarthen Museum, no. 1977.0582 (normally on display).

No 3 Lower stone, reused, with scratch dial on grinding surface. Convex grinding face *c.* 330mm diameter, originally pecked. Perforated spindle hole 25mm diameter. Stone *c.* 80mm thick at centre. Edge probably trimmed down for later re-use as sundial. Underside smooth and otherwise unworked. First published by RCAHMW (1917, fig. 38, righthand stone). Now in Carmarthen Museum, no. 1977.0626. Catalogued by Watts.

No 4 Upper stone, damaged, with some similarities to no. 2. Maximum diameter *c.* 530mm but has rounded edge; diameter of grinding face 480–490mm. 70mm thick at eye and *c.* 90mm at outer edge, which is thickest part of skirt. Eye slightly oval, *c.* 130–140mm diameter, and slightly off-centre as stone survives. Flat to slightly convex grinding surface with concentric wear marks. Very rough top. Two upright handle holes; may have been another where top of stone has broken away (museum records say 3). Now in Carmarthen Museum, no. 1977.0380, attributed to 'Dolaucothi'. Catalogued by Watts.

No 5 Upper stone in medium-fine sandstone. Diameter 490mm, 65mm thick at eye with slight collar. 70–90mm thick at edge, which is quite rounded. Eye not entirely circular, 90mm maximum diameter. Concave grinding face with concentric wear marks, very faint. No handle hole or rynd chases, but upper surface clearly somewhat dressed. Now in Carmarthen Museum, no. 1977.0581, attributed to 'probably Dolaucothi'. Catalogued by Watts.

No 6 Lower stone, at least 460mm diameter (using radius; maximum surviving 410mm) but now cut down to a square on three sides; the fourth roughly rounded but possibly chipped away. 30–50mm thick at edge, *c.* 50mm at centre with small, perforated spindle hole 15mm diameter. Concavo-convex grinding surface, furrow dressing, probably in harps, appears virtually unworn. Probably reused as building stone, hence its discovery during the excavations at Pumsaint; said to come from Jones and Little's Trench 4, from the trench side. Now in Carmarthen Museum, no. 1975.1749. Catalogued by Watts.

No 7 Lower stone of local conglomerate. Maximum diameter 600mm (from radius), maximum surviving 560mm. Roughly conical shape (to base), giving considerable maximum thickness of 330mm, with conical indentation at centre 45mm diameter at surface and 40mm deep. Flattish working surface with slight rise towards centre. Underside very roughly worked, probably intended to be set in ground. Stone has been squared off slightly for re-use and is now set into NW corner of garden wall alongside road at Ogofau Lodge. Not previously recorded.

No 8 (Not illustrated). Fragment of upper stone (?) shown in Boon (1971, fig. 14 top right). Current location unknown.

No 9 Fragment of upper stone, about one-fifth or one-sixth of the stone. Curve of the edge somewhat uneven, reminiscent of no. 2; surviving radius from eye to edge 200mm. Diameter of eye *c.* 150mm (curve matches no. 10), giving overall diameter *c.* 550mm. Thickness at eye 65mm, at edge 50–90mm. Surface worn smooth, no obvious scoring. Single handle hole on upper surface. Illustrated in Boon (1971, fig. 14 lower right), but is not part of no. 10 as picture seems to suggest. Recovered from Ogofau Lodge in 1993, courtesy of Mr T. Cleminson.

No 10 Upper stone in two fragments, incomplete (about 60° missing). Local conglomerate. 570mm diameter, 50mm thick at eye, tapering across convex surface, steepest towards edge, almost to a point in places, elsewhere rounded. Eye slightly oval, 140–150mm diameter. Working surface worn flat and probably slightly domed, reflecting lower stone raised in centre. Two double upright handle holes on upper surface, each with a shallower and a deeper component, almost but not quite opposed across the diameter. Found in a sealed context during digging at Ogofau Lodge (Lewis and Jones 1969, pl. 50c; Boon 1971, fig. 14 left and centre; Lewis 1977 fig. 15 lower left). Recovered from Ogofau Lodge in 1993, courtesy of Mr T. Cleminson.

No 11 ?Axle stone; just over half of a quadrilateral, broken irregularly across two not-quite-opposed central socket holes or post bases. Maximum 580mm between corners of surviving long side, *c.* 280mm from break to centre of this side. Thickness 150–160mm all over. One side (with deep socket) fairly flat, other side (with shallow socket) has slight depression or area of wear along the surviving long side. The unbroken edges are generally worn and rounded. The deep circular socket is slightly oval, 150–160mm diameter, with a depth in the slight groove around the edge of its generally flat base of *c.* 80mm, dropping to 90mm around part of the circumference. The shallow socket, which is similar but not precisely opposed, is 140mm diameter and 25–30mm deep; at one point on the edge of its base there is a slightly deeper indentation measuring 30mm around the socket's circumference by 15mm and 5mm deep. Around the edge of this socket, small pieces of finely-crushed quartz are embedded in iron-staining. Found in a sealed context during digging at Ogofau Lodge (Lewis and Jones 1969, pl. 50c right; Boon 1971, fig. 13; Lewis 1977, fig. 15 right). Recovered from adjoining property in 1993, courtesy of Mr C. Page.

No 12 (Not illustrated). Lower stone, half complete. Shown in Lewis and Jones (1969, pl. 50c, top); Lewis (1977, fig. 15 centre). Current location unknown.

No 13 Fragment of upper stone of local conglomerate, about one-sixth or one-seventh of the stone. Surviving radius from edge to eye 190mm maximum. Diameter of eye a little less than 150mm, by analogy with curvature of no. 10. Surviving thickness 60mm maximum at outer edge. Both upper surface and outer edge very rough, probably damaged; no evidence of handle holes or rynd chases. Working surface slightly convex and well worn with concentric scoring clearly visible, especially towards outer edge. Found during excavations in 1992 (see section 4.4, no. 1, above).

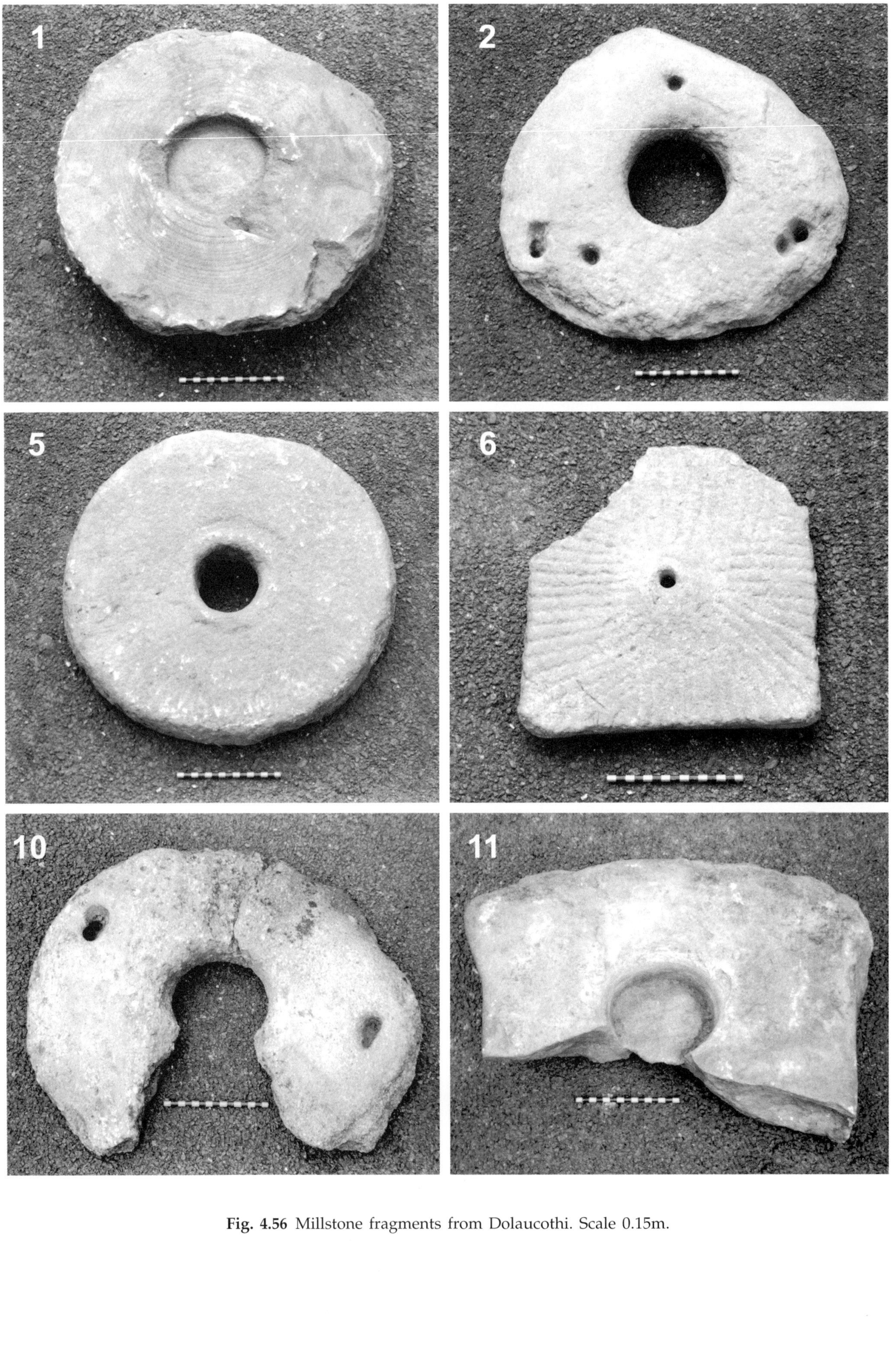

Fig. 4.56 Millstone fragments from Dolaucothi. Scale 0.15m.

3

4

7

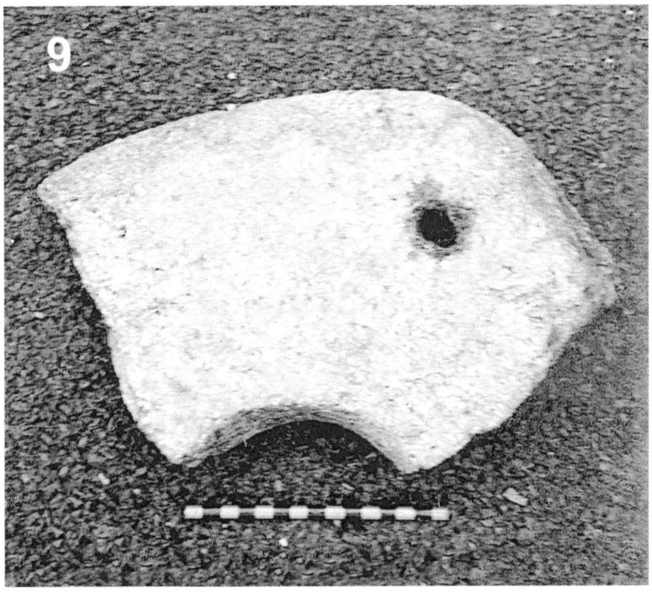
9

13

14

No 14 Lower stone, roughly half complete, broken across centre through spindle hollow and generally much damaged around edge. Maximum surviving diameter 590mm. Maximum thickness 120mm in centre, tapering to 50–70mm at edges. Upper surface largely smooth with traces of concentric scoring on inner half, slightly concave with slight rise towards centre. Small conical hollow for spindle *c.* 40mm diameter at surface and *c.* 50mm deep. Slightly domed under-surface, apparently unworked. Found on surface near Roman road in field to W of Ynysau Uchaf in 1997.

5 TRIAL EXCAVATION AND GEOPHYSICAL SURVEY IN THE VICINITY OF THE BATH-HOUSE, 1997

5.1 INTRODUCTION

Despite the extensive investigations which have been carried out in the Dolaucothi-Pumsaint area, surprisingly little was known until recently about two important aspects of the wider complex, namely the location of any civilian settlement(s) associated with the mines, and the course of the Roman road from Llandovery to Llanio in the immediate environs of the site. Both have now been considerably clarified, however, by the coincidence of several unrelated strands of evidence, culminating in the trial excavations undertaken in the summer of 1997.

Hitherto, the nature and location of any civilian settlement(s) had largely been a matter of speculation, even if two sites had logically commended themselves in the literature. One lay in the immediate vicinity of the fort at Pumsaint, north of the river Cothi, where a *vicus* might have been expected to develop along lines better known at other military sites (Sommer 1984; Davies 1991b). That this was the case south of the fort had long been suggested by Roman finds recovered from the eroding river section east of the modern bridge (Lewis and Jones 1969, 260; Burnham and Burnham 1987a), but little or nothing was known about the area east of the fort prior to some limited geophysical work undertaken during 1997 and 1998; this hinted at the presence of an extensive area of settlement, the details of which were confirmed by more extensive surveys in 1999 (see section 2.6 above).

The other possible focus lay to the south of the Cothi, somewhat closer to the mine workings, in an area where the remains of a stone building were dug up in the 1830s (Bosanquet 1917, 30-1). Though its precise location was not recorded at the time, it seems likely that it lay on the very edge of a fossil river terrace, *c.* 200m west of Ynysau-Uchaf Farm (at *c.* SN 6564 4026), where there is still a scatter of brick and tile across an area of disturbed ground. Contemporary drawings suggest that it may have been a bath-house, an hypothesis broadly in line with the recent recovery of several fragments of box-flue tile, which raises interesting questions about its wider links with the fort, the mines and the communications network. Such issues are further highlighted by what little is known of the associated finds, which apparently indicate occupation extending beyond the known life of the fort down to the end of the 2nd century at least.

The significance of this area was further emphasised during the ploughing of Field 5532 in 1993, when a scatter of brick and tile was discovered on a headland formed by the same fossil river terrace, west of the site of the postulated bath-house. This discovery prompted the National Trust to commission a geophysical survey in an effort to define the extent of the buried archaeology and the layout of any buildings associated with the baths. This was duly carried out by Geophysical Surveys of Bradford in March 1994 (Report No. 94/19), across an area straddling Fields 5532, 6015 and 7020 (Fig. 5.1).

The resulting resistivity survey identified several interesting features (Fig. 5.2A), including a band of relatively high resistance readings (F) across the western part of the site, measuring *c.* 50 by 8m, with an adjoining block of high responses (G) to the east. This was interpreted as the remains of a range of buildings, possibly associated with a villa of familiar corridor type. Also of interest were two further areas of high resistance readings (H and I), at the eastern end of the site. These were tentatively associated with former buildings, though it was noted that they might represent a response to natural variations as was the case with other areas of high resistance on the raised ground (D and E). A parallel gradiometer survey across the site also recorded several areas of strong magnetic anomaly (Fig. 5.2B), including at least three (K, L and M) which were thought to represent either small furnaces or fires, or small-scale industrial working, though the postulated villa showed only as a poorly defined band of magnetic noise (N). While some of the other anomalies noted in the survey area are likely to represent pits and short lengths of ditch, it remains far from certain that all of them are archaeologically significant in view of the presence of numerous magnetic rocks in the topsoil.

By a happy coincidence, these surveys have been complemented by recent fieldwork and aerial reconnaissance along the line of the Roman road from

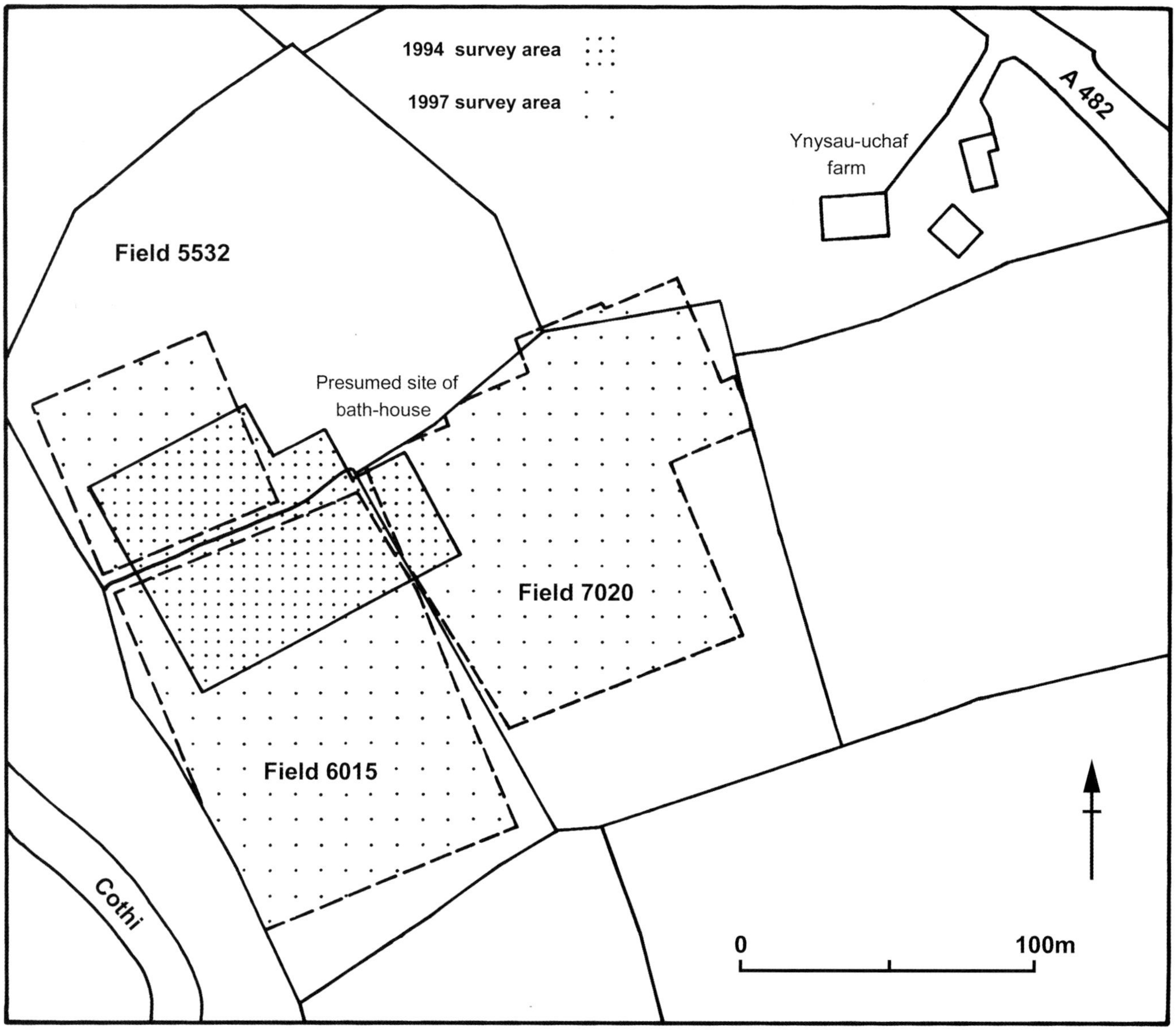

Fig. 5.1 Location plan of 1994 and 1997 geophysical surveys

Llandovery, which have done much to clarify its course in the immediate environs of the site, en route for the fort at Pumsaint; the full details have been conveniently summarised by Heather James in an appendix to this chapter (see section 5.6 below). The new photographs clearly show the road diverging away from the A482 at SN 656 391, to take a more northerly line than its modern counterpart (see Figs 5.15 and 5.19), whence it can be traced across the fields heading directly for a point on the headland west of the postulated bath-house. At this point, a stretch could be seen on the photographs (Fig. 5.3), coinciding so closely with the linear anomalies already noted on the resistivity and gradiometer surveys, that any identification as a possible villa of corridor type had to be in serious doubt. From here, the road must have aimed for a point near the west gate of the fort, in the area now lost to river erosion.

Such discoveries obviously contributed to the debate about the character of this part of the Dolaucothi complex, without resolving the issues one way or the other. They also raised a concern about the extent to which continued ploughing might pose a threat to any archaeological horizons, given their potential significance. As a response to this, the National Trust commissioned the authors, in conjunction with the Dyfed Archaeological Trust, to undertake small-scale trial excavations in Field 5532, on the headland west of the bath-house, with the declared aim of evaluating the nature and depth of the archaeology and of determining an appropriate management regime for the foreseeable future. The work was duly carried out over two weeks in June 1997. Independent of the commission, the chance was also taken to undertake a more extensive geophysical survey across Fields 5532, 6015 and 7020, in an

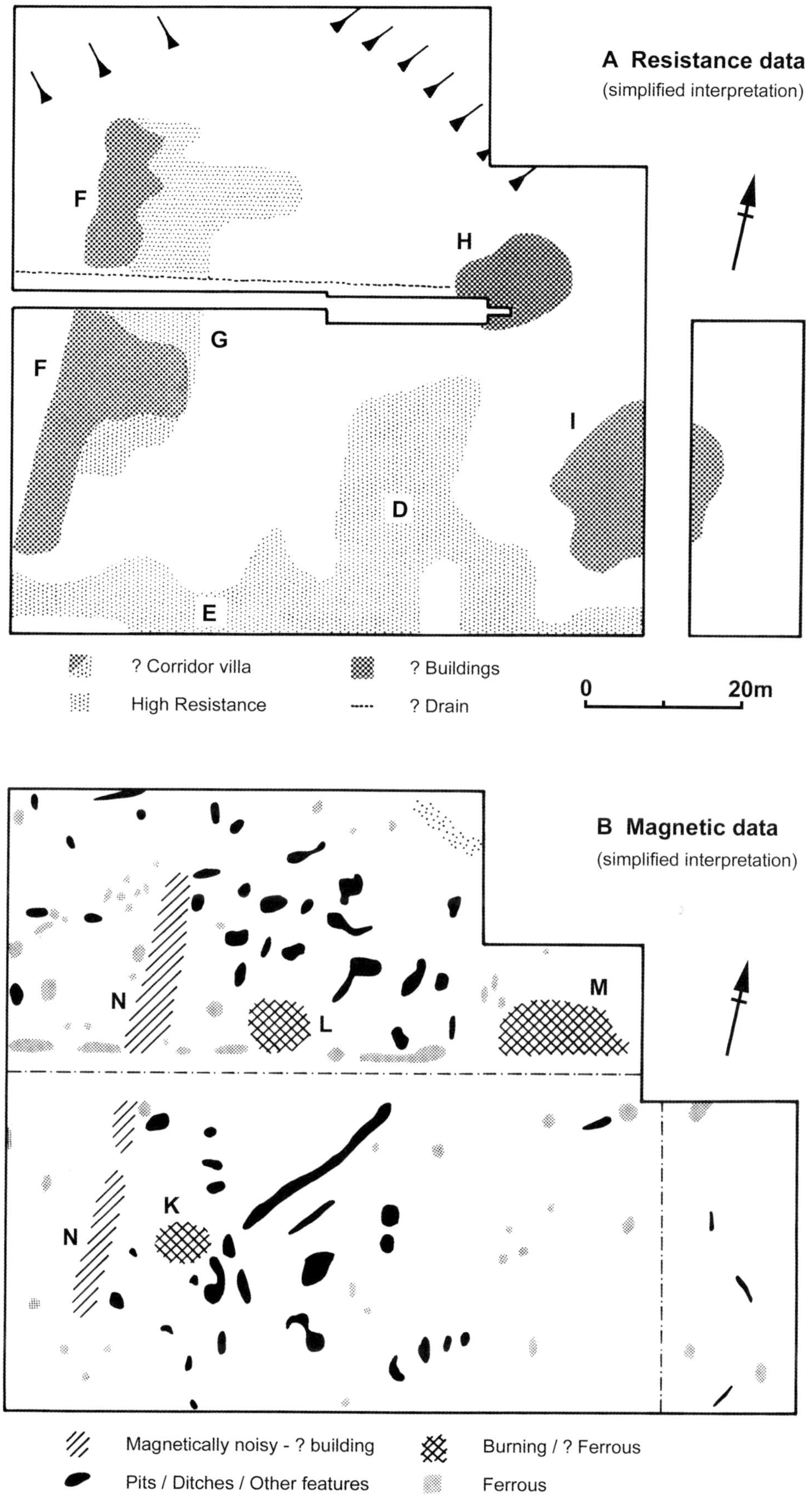

Fig. 5.2 Details of 1994 geophysical survey (after Geophysical Surveys of Bradford Report No. 94/19)

Fig. 5.3 Aerial view showing faint traces of Roman Road extending onto headland south of the flood plain of the river Cothi; the site of the bath-house is said to lie in the clump of trees at centre (APDAT 95–91.25). Reproduced by permission of Cambria Archaeology

attempt to establish the nature and extent of the associated archaeology. Most of this work was carried out in August 1997.

5.2 TRIAL EXCAVATION, 1997

The evaluation took the form of a single trench, 22.00m long by 2.00m wide, excavated parallel with, and 6.00m north of, the field boundary separating fields 5532 and 6015 (Fig. 5.4). This was designed to sample both the pronounced linear feature (F) across the western part of the site and the roughly rectangular anomaly (L) at its eastern end. It had also been hoped to investigate the equally interesting anomaly (M) at the extreme eastern end of the headland, closest to the site of the presumed bath-house, but this proved impossible in view of the prevailing weather conditions and time constraints. Once the overall nature and extent of the surviving remains was clarified, no attempt was made to penetrate the intact archaeological horizons, thereby minimising disturbance and ensuring the integrity of the site for future investigation.

Beneath the modern turf horizon [1] and the mixed stony subsoil [2 and 3], excavation encountered the tell-tale signs of the 1993 ploughing (Fig. 5.5). This took the form of a decayed and overturned layer of turf [4], up to 0.12m thick, which was found to extend right across the excavated area. In its turn, this overlay a series of distinctive plough furrows, trending obliquely across much of the trench, which together denote the lower limit of modern ploughing [5] at a depth of 0.15m. Across much of the area, this plough horizon had cut into a distinctive layer of sandy clay (variously contexted [6], [7] and [8]), which proved to contain a high proportion of Roman material alongside only a handful of later finds. This seems to represent a much-disturbed and older plough soil, itself sandwiched between the majority of the Roman features and the overlying layers. The

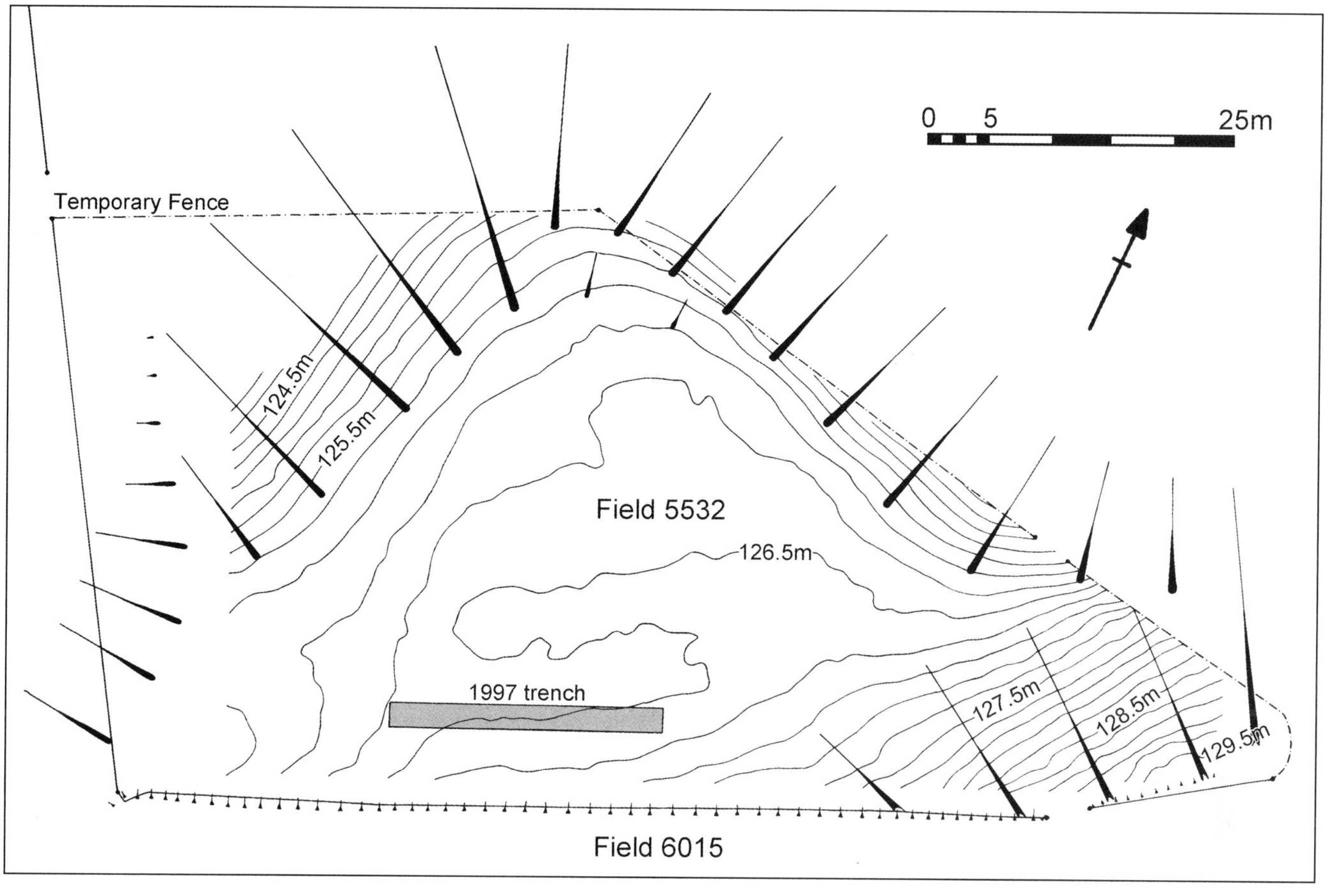

Fig. 5.4 Location of 1997 evaluation trench in south-west corner of Field 5532

only exception to this was at the west end of the trench, where [7] and [8] respectively petered out across the western and eastern sides of a compacted gravel and clay surface [9 and 11], which proved to be the cambered remains of the Roman road. At this point, it was clear that the modern plough horizon [5] coincided with the intact upper limits of Roman archaeology, though fortunately the amount of recent disturbance seems to have been minimal.

Few features were found in association with the older plough soil, except for three probable post-holes [23; 29; 33] towards its eastern end. They were not immediately obvious in the upper surface of [6], but only became visible during its removal and cleaning onto the clay floor [10]. There can be no doubt, however, that they represent late (probably post-Roman) features, more than likely associated with some of the otherwise undated field boundaries identified by the geophysical surveys in the fields to the south (see section 5.4 below).

23 a roughly oval feature, measuring 0.28 by 0.22m, with a fill [22] composed of a brown sandy clay (10 YR 4/4) and a single large stone. It was not excavated beyond the point where it cut into the fill [24] of an underlying Roman feature.

29 an elongated oval feature, measuring 0.40 by 0.20m, with a fill [28] composed of a yellowish-brown sandy clay (10 YR 5/6) and numerous pitched packing stones. It was not excavated beyond the point where it cut into the clay spread [10].

33 a roughly circular feature, measuring 0.42 by 0.40m, with a fill [32] of brown sandy clay (10 YR 4/4) and numerous pitched packing stones. It was not excavated beyond the point where it cut into the fill [34] of an underlying Roman feature.

Beneath the older plough soil, excavation revealed a relatively intact archaeological horizon, with evidence for numerous features which were clearly not all contemporary with one another (Figs 5.5 to 5.7). These were not always easy to interpret within the narrow confines of an evaluation trench, though a reasonably coherent picture did emerge. For convenience the details will be described from east to west.

The most prominent feature across the eastern part of the trench was a layer [10] of yellow-orange clay (10 YR 7/8) with traces of charcoal flecking and burnt patches, which coincides closely with the high magnetometer anomaly (L) identified in the Bradford survey. It extended at least 4.50m into the trench, to the point where it had been cut away by an irregular gully [35]. It may originally have continued beyond

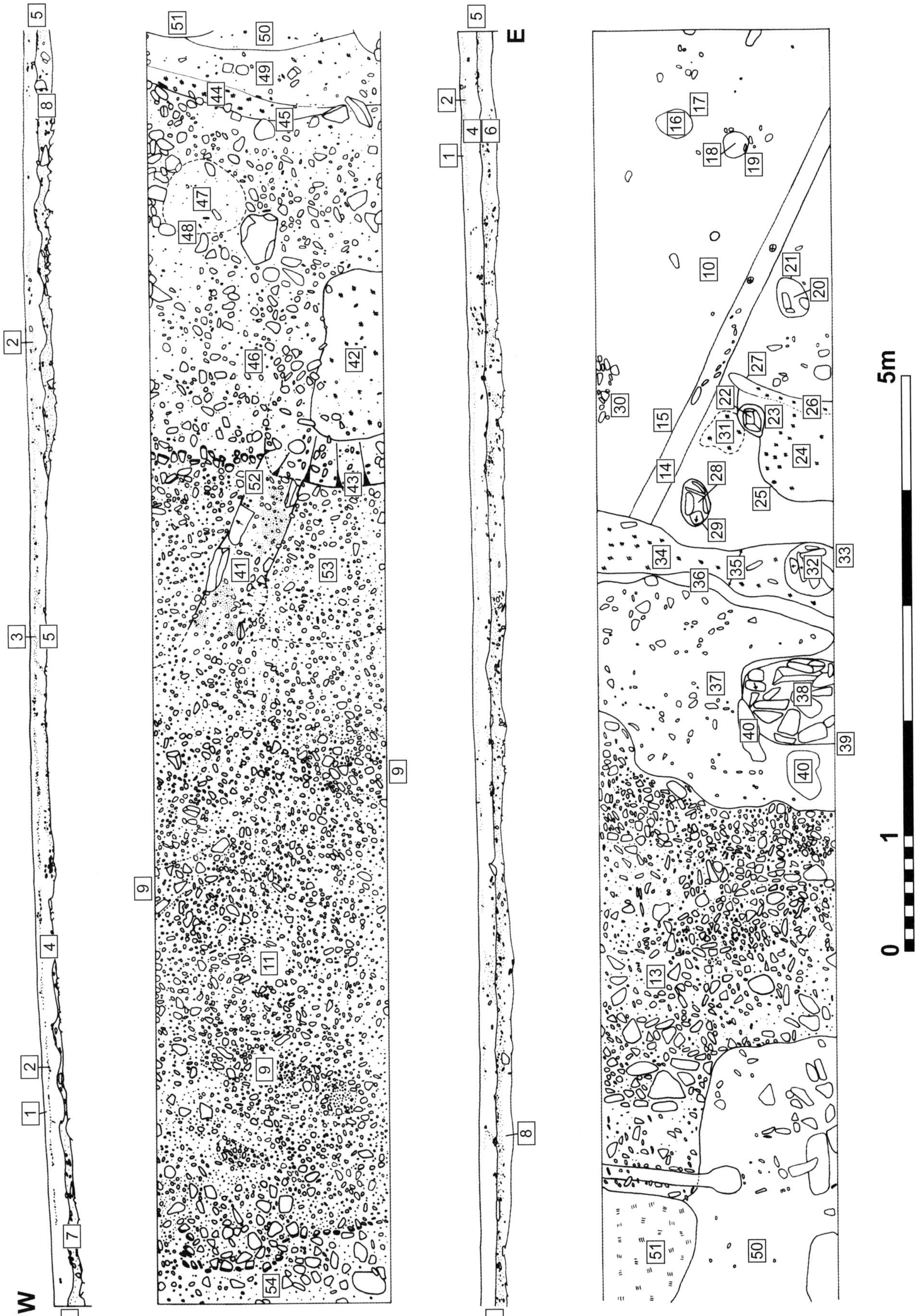

Fig. 5.5 Plan and section of 1997 trench

Fig. 5.6 Roman road surfaces at west end of trench, looking west

Fig. 5.7 Roman features east of the road, looking east

this, however, in the form of a pronounced band [36] of yellowish-brown sand (10 YR 5/6), containing patches of the same distinctive clay, which clearly extended westwards beneath the overlying stone horizon [37]. Both layers may well have formed part of the same 'floor' surface, connected with a structure across the eastern end of the trench, set back from, but probably aligned upon, the Roman road.

Other features may have been associated with this proposed structure, including the large post-hole [39] and the stone spread [37]. The post-hole proved to be almost circular, 0.74 by 0.70m in size, with a fill [38] of brown sandy clay (10 YR 4/4) intermixed with numerous medium to large, pitched limestone fragments, derived from the packing around the original post. It seemed to have been cut from the top of the sand 'floor' [36], suggesting that it was a secondary feature. Its central location, however, relative to the Bradford anomaly, might indicate that it was one of the principal structural members, perhaps associated with an entrance at this point. This might help to explain the function of the spread [37] of small to medium rounded pebbles interspersed with a brown sandy clay (10 YR 4/4), which had been laid across the top of the sand 'floor' and appeared to respect the line of the post-hole. This is best interpreted as a secondary makeup layer, infilling and repairing what may have been a well-trodden area astride a main entrance.

Across the clay floor [10] as a whole, several secondary features were identified, either overlying or cut into its surface. These were very difficult to interpret within the confines of the trench.

15 an apparently shallow linear gully, *c.* 0.30m wide, which extends at an oblique angle from north-west to south east across the trench. It contained a fill [14] of brown sandy clay (10 YR 4/4) with traces of charcoal and burnt clay throughout, as well as two nails. At its north-western end, it seemed to be cut away by another, wider gully [35].

17 a possible post-hole, 0.32 by 0.28m in size, filled with a roughly circular patch [16] of light brown clay (10 YR 3/4) and some limited post-packing.

19 a possible post-hole with rather ill-defined edges, *c.* 0.26 by 0.23m in size, filled by a roughly circular patch [18] of brown clay, mixed with a quantity of charcoal and burnt clay.

21 an oval-shaped post-hole, 0.45 by 0.35m in size, with a fill [20] of dark brown sandy clay and some limited post-packing.

25/27 the relationship between these two features and

their respective fills [24/26] remains very uncertain. For recording purposes they have been treated as separate entities, but they could equally be part of a single feature with successive fills. The former [25] seemed to be a roughly rectangular feature, *c.* 0.77 by 0.75m in size, extending south beyond the section, with a fill [24] of dull yellowish-brown sandy clay (5 YR 5/5) and some heavy charcoal flecking, especially in a band running east-west. The second element [27], which extended along the eastern side of [25], appeared to be a narrow linear feature, at least 0.83m long by 0.08 to 0.14m wide, with a mixed fill [26] of heavy charcoal flecks and brown sandy clay (5 YR 4/4). Nothing survived to indicate their respective functions.

30 a small patch of rounded pebbles intermixed with a small quantity of brown sandy clay (10 YR 4/4), charcoal and burnt clay, which continues beyond the section. It might represent a discrete area of patching above the clay floor.

31 a shallow patch of brown clay (10 YR 4/4) with traces of charcoal and burning, which may represent an area of trampling above the clay floor.

35 an irregular linear gully extending roughly north-south across the trench, immediately to the east of the stone spread [37], where it had truncated the critical junction between the separate elements of the clay and sand floor [10 and 36]. It contained a fill [34] of brownish-grey clay (10 YR 5/1), sandy patches and extensive charcoal flecking, not unlike that encountered in [24].

At its western end, the stone spread [37] associated with the possible building merged with a second layer of compacted small stones [13], in such a way that it was impossible to demonstrate their stratigraphic relationship with any certainty. It would seem to have formed a possible 'cobbled' surface or external yard, extending at least 2.00m beyond the entrance to the building. To the west, however, it ran under two other layers [50 and 51] so that its full extent could not be determined. At one point where its upper surface had formed a slight hollow, this yard was overlain by a thin band [12] of dark brown silty sandy clay (10 YR 4/4), up to 0.05m thick. This was clearly traced over an area *c.* 1.10 by 0.42m, but may originally have been somewhat more extensive across the surface as a whole.

The archaeology of the central part of the trench, sandwiched between the possible building to the east and the contexts associated with the road to the west, proved far less easy to disentangle and interpret. On its western side, the 'cobbled' yard [13] was overlain by two successive horizons. The lower of the pair [50] proved to be a roughly rectangular layer of light orangey-brown sandy clay loam, 2.40m by 1.50m at least, which extended south beyond the section. At two points it peeled off a pair of flat shale slabs, which appeared to have been laid as if to form part of a structure. The later of the two horizons [51], which extended north beyond the section, was found to be a squarish patch of very similar texture and consistency, 1.00 by 0.80m at least, but was clearly differentiated by its frequent inclusion of burnt clay. This was very reminiscent of building debris, suggesting that the different horizons may have been derived from a clay-walled or daub structure in the immediate vicinity.

At their western end, both horizons overlay a spread [49] of mid greyish-brown silty clay loam, with occasional charcoal flecks and stone inclusions, which extended roughly north-south across the trench. This was very similar in its character and consistency to a narrow band [44] of dark greyish-brown silty clay loam, immediately to the west, suggesting a common origin. Both would seem to represent (successive?) fills in a linear gully or ditch [45], of uncertain width, which had been cut through the extensive layer [46] bordering the east side of the Roman road. This latter layer proved to be a stony horizon, intermixed with a matrix of dark brown sandy silt loam, which had been cut into by at least two features, one a large pit [43], the other a possible smaller hollow [48] to the north-east. From the nature of the sandy silt loam, the whole might represent a natural filling in a hollow or ditch extending along the east side of the road, traces of which were indicated on the geophysical surveys.

The larger of the two features [43] proved to be a large rectangular pit, 1.90 by 1.00m at least, which extended south beyond the section. It had been cut into the cambered makeup [53] along the east side of the road and probably also through the fill [41] of what may have been an original drainage culvert. Only its uppermost fill [42] was encountered for certain, consisting of a brownish-black sandy silt loam with some charcoal flecking and small stones. Around the edge of the feature this was interspersed with numerous stones which appeared to be pitched inwards, as if conforming to a tip line. The second of the two features [48] seems to have been a roughly circular pit, defined principally by the extent of its fill [47] of dark greyish-brown silty clay loam, occasional stone inclusions and charcoal flecking. Caution is perhaps necessary about the character and full extent of this feature, which appeared far less distinct in the photographs taken after a period of differential drying.

The western part of the site was dominated by the surviving remains of the Roman road, in the area of relatively high resistance readings (F) identified in the Bradford survey. What appeared to be its uppermost surface [9] survived at various points across the crest of the road and on the camber at its western side, in the form of several small patches of flat, well-bedded limestone pebbles. It had clearly been laid on a compacted layer [11] of mid greyish-brown clay and small to large stone inclusions, which petered out on the gentle camber to either side. Ploughing in the past has certainly disturbed this upper surface, which no doubt accounts for the high concentration of small stones both in the older plough soil [7 and 8] and in the modern subsoil [2], in those areas immediately above the road. An earlier possible surface may be represented by a compacted layer [53] across the eastern side of the road, which was found to comprise a mixture of small to large stones set in a mid orangey-brown sandy clay loam. No comparable surfaces were identified across the western side of the road, where what appeared to be the road's substructure of medium to large stones [54] was encountered.

The earlier of the two possible surfaces [53] had been cut into by a linear feature, 0.40m wide, trending obliquely across the trench from north-west to south-east. Its sides had originally been lined by upright shale slabs, perhaps with some form of capping, suggesting that it may have served as a drainage culvert. This had subsequently been infilled with a layer [41] of mid orangey-brown sandy clay loam, with a marked sandy content in places. To the south-east, this had been truncated by the later pit [43]. To the north-west, no clear relationship could be established with the makeup layer [11] for the latest road, though it seems likely that the culvert had ceased to function and been infilled prior to the latter being laid.

Summary

Despite the constraints imposed by the nature of a narrow assessment trench, a coherent picture emerged from the 1997 excavations, to amplify the story derived from the Bradford survey and recent work on the aerial photographs. The clearest evidence emerged at the western end of the site, where the trench was crossed by the substantial remains of the Roman road from Llandovery to Pumsaint, en route for its crossing over the river Cothi. To the east, this was shown to have been flanked by a complex succession of features and occupation horizons, including at least one clay-floored building, obviously representing several different phases of activity. Such evidence is typical of other roadside settlements known throughout the province, but here represents the first detailed proof of the existence of a comparable site at Dolaucothi. Further details were provided by the geophysical survey in the fields to the south and south-east.

5.3 THE FINDS

Analysis of the finds indicates that the bulk of the associated material is of Roman date. A handful of later finds was confined exclusively to the disturbed plough soils across the site. Material recovered from the archaeological horizons below this proved to be Roman. Most of the Roman material was found in the older plough soil [6, 7 and 8], however, suggesting that there had been some disturbance of the uppermost horizons. Finds included sherds of samian, amphora and mortarium, Black-burnished and Severn Valley wares, various fragments of vessel and bottle glass, and a quantity of brick and tile fragments. The date range for this material would seem to extend from the later 1st through to the mid 2nd century, but with none of the later material said to have been associated with the bath-house (Bosanquet 1917, 31). Interestingly, the somewhat higher percentage of 2nd-century material here provides a contrast with the early material from the fort, suggesting that activity in the settlement south of the river may have outlasted the latter by up to two decades; this raises interesting questions about the intensity of early activity at the mines and about subsequent operations (see section 6.2.3 below).

5.3.1 Pottery

(A) The samian (by Peter Webster)

CATALOGUE

S 1 (Not illustrated). Form 37, Central Gaulish. A small portion of ovolo frieze survives above a wavy line border and foliate decoration. The ovolo may be Rogers 1974, B74 ascribed to Quintilianus. The abraded fragment of foliage would suit the vine scroll, Rogers 1974, M5 (*cf.* Stansfield and Simpson, pl. 69, 14) but the piece is too abraded and fragmentary for certain identification. Hadrianic-Antonine. Context **[8]**.

S 2 (Not illustrated). Knorr form 78, South Gaulish. 12 fragments are all very abraded with some missing both slip and decoration. The better preserved fragments consist of:

a) A panel just below the grooved upper border contains the rear portion of a ?lion (or possibly a dog) running to the right.

b) Panel decoration. The upper left panel contains a ?bird apparently decapitated by the grooved upper border. No decoration survives in the panel below. To the right is a panel which probably ran from top to bottom of the decoration. It includes double opposed buds on drooping stems (similar to a rounded letter W inverted on top of itself). This is similar to, but not necessarily identical to that seen in work stamped by Maccarus (*cf.* Knorr 1952, taf. 41, D and 52, F). The motif is surmounted by a lanceolate bud or small leaf.

c) What is apparently the lower end of a similar panel to b) above contains the opposed double bud overstamping a vertical wavy line with rosette terminal. The lowest buds of the motif are overstamped themselves with lanceolate leaves similar to that used at the top of b) above.

d) A narrow panel edged by wavy lines with rosette terminals contains no decoration. To the right is a baluster 'leg' possibly based on the Germanus small column, Knorr 1919, taf. 35, 78, but with overstamped elements. To the right of this is a narrow frame, possibly part of the 'table' seen on Mees 1995, taf. 77, 5 (Germanus III).

Knorr 78 is a Flavian form which is not particularly common, even at the manufacturing site. Few bear makers' stamps, but there is no reason to suppose that the form was in the repertoire of many makers. Identification in our case is hindered both by the poor preservation of the vessel and by the habit of overstamping employed by the mould maker. Of the known producers of the form, Germanus is the most likely maker for this vessel and a mid-late Flavian date may be suggested. From the interface of context **[8]** and the cobbled surface **[37]**.

S 3 (Not illustrated). Form 37, Central Gaulish. The fabric appears bright orange but is slightly darker in the break with few visible inclusions. Manufacture at Les Martres-de-Veyre seems probable. The ovolo is abraded and was smudged in manufacture. Below is panel decoration defined by wavy lines with the small trifid leaf ornament, Stansfield and Simpson, fig. 15, 3, at the one surviving junction (*cf.* Geminus, Stansfield and Simpson, pl. 66, 30). A panel contains the vine branch, Rogers 1974, M13 also used by Geminus (Stansfield and Simpson, pl. 65, 3). The damaged ovolo may be Ovolo 2 of Geminus (Stansfield and Simpson, fig. 15, Rogers 1974, B28). *c.* AD 120-140. From the interface of context **[8]** and the cobbled surface **[37]**.

(B) The coarse pottery (by Dee Brennan and Peter Webster)

All the pottery from the excavation has been examined and listed for archive purposes (DB). The archive list will be deposited with the pottery. The list formed the basis for the selection of pottery for drawing and cataloguing (PVW) and for all the general comments made below.

CATALOGUE (Fig. 5.8)

CP 1 (Not illustrated). Mortarium fragments in slightly granular creamy-buff fabric with white and colourless quartz trituration grits. Probably a Verulamium product and Flavian to Flavian-Trajanic in date. Context **[8]**.

CP 2 (Not illustrated). Bowl in orange Caerleon ware with faint orange-red slip. The complete bowl will have resembled Webster and Webster 1998, no. 65. Caerleon ware was in production *c.* AD 110-160/70. This piece is most likely to be early to mid Antonine. From the interface of context **[8]** and the cobbled surface **[37]**.

CP 3 Base of a roughcast beaker in light brown Caerleon ware. For a discussion of Caerleon ware with associated dating, see Manning 1993, 255–7; also 262–3 for roughcast beakers. The most likely date for this piece is Hadrianic to mid Antonine *c.* AD 120-60. Context **[8]**.

CP 4 Jar in Black-burnished ware, *cf.* Gillam 1970, no. 120. Early to mid 2nd century, or possibly earlier in south Wales. From context **[8]**, above the cobbled surface **[37]**.

CP 5 Flat-rimmed dish in Black-burnished ware. 2nd century. From context **[8]**, above the cobbled surface **[37]**.

CP 6 Dish in Black-burnished ware broadly similar to Gillam 1976, no. 59. Context **[8]**.

CP 7 Dish in Black-burnished ware, *cf.* Gillam 1976, no. 75. Faint signs of burnished lattice decoration. Early to mid 2nd century. Context **[8]**.

CP 8 Dressel 20 amphora base and other fragments. A small stamp has been impressed alongside the basal 'button' externally. Stamps on the body of Dressel 20 amphorae are extremely rare. Funari lists only two other examples from Britain (1996, nos 174 and 194j). It is, therefore, regrettable that the stamp is all but illegible. A possible reading of ATPF. is tentatively offered but without a drawing of the stamp as this would impose an undue degree of certainty upon the suggestion. Context **[8]**.

CP 9 Wide-mouthed jar in orange Severn Valley ware with a grey core. A similar type was being produced at Malvern North End Farm kiln, Evans *et al.* 2000, fig. 23, type JWM6. The complete vessel probably resembled Webster 1976, nos 23–4, in which case a mid 2nd-century or later date is possible. Context **[8]**.

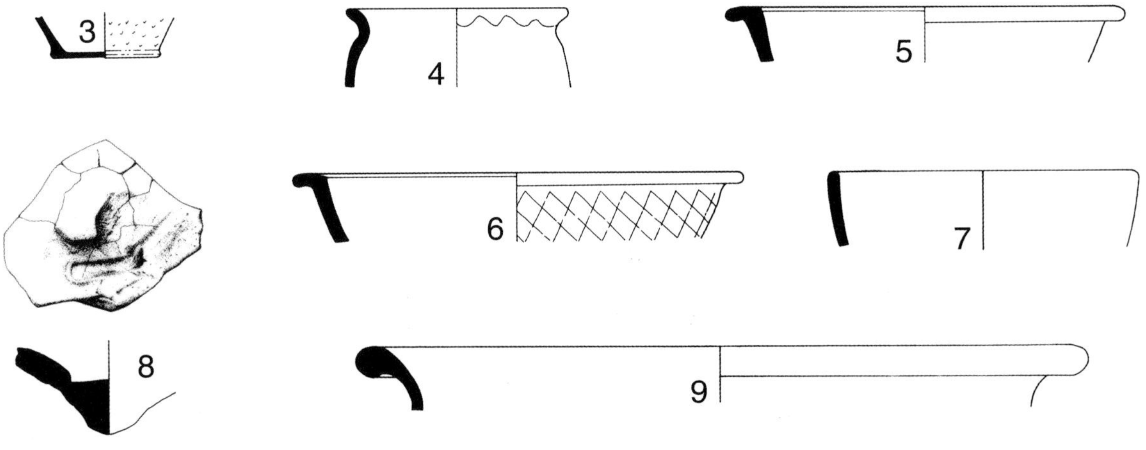

Fig. 5.8 Coarse pottery. Scale 1:4

5.3.2 Glass (by Dee Brennan)

The 1997 excavation yielded a total of 12 fragments of Roman vessel glass. This small assemblage is dated by association with 1st- and 2nd-century pottery found on the site though the types of glass vessel present are those which were made and used beyond the first two centuries.

All fragments are of natural-coloured green and bluish-green glass. Seven are body fragments from bottles of Isings (1957) type 50 (not catalogued). The remaining five are from thin-walled table vessels. They comprise two bowls (Gl 1 and 2) and three featureless body fragments of indeterminate form (not catalogued).

CATALOGUE (Fig. 5.9)
Vessel glass; tablewares

Gl 1 Fragment from rim and body of small bowl; natural-coloured pale green. Tubular rim. Rim edge bent out and down. Blown. Rim diameter 120–130mm; wall thickness 1.5mm. Context **[8]** above courtyard.

Gl 2 Fragment from rim of bowl; natural-coloured pale green. Tubular rim as Gl 1, but of slightly larger diameter. Blown. Rim diameter 130–140mm; wall thickness 1.5mm. Context **[8]**.

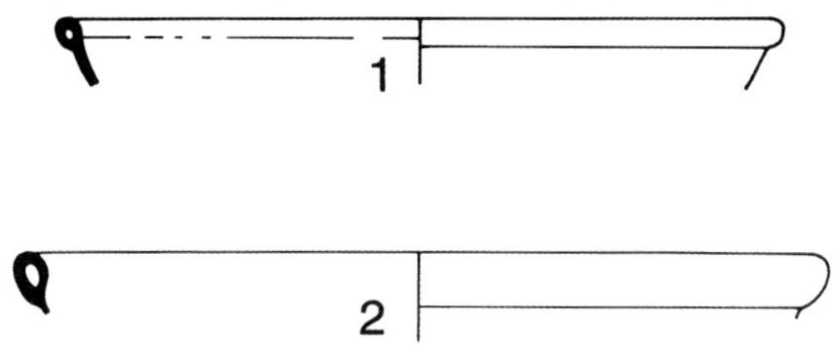

Fig. 5.9 Glass. Scale 1:2

Gl 1 and 2 are both from bowls of a type common throughout the Roman period but most frequently found in late 1st- and early 2nd-century contexts.

5.4 GEOPHYSICAL SURVEY, 1997
(by David Jordan)

Magnetic and electrical resistivity survey was carried out under the direction of David Jordan of Terra Nova Ltd. across the two pasture fields to the south and south-east of the excavated area (OS parcels 6015 and 7020), and in the vicinity of the excavation itself (see Fig. 5.1). The area covered by the resistivity survey extended slightly further to the south than that covered by the magnetic survey, although the latter extended further to the north across the excavated area. The magnetic survey was carried out by David Jordan, and the resistivity by Helen Burnham and a team of Lampeter students.

It was hoped that this survey would clarify the line of the Roman road, on the west of the area, and would show whether there was any substantial settlement alongside it. It was also hoped to locate more precisely the position of the presumed bath-house excavated in the 1830s, material from which, possibly excavation spoil, was identified among the trees on the north-facing slope to the east of the excavation. This material included box-flue and roof tiles, suggesting that the original structure might be expected to show up as a magnetic anomaly.

The site lies on the dissected terraces of the Cothi valley on mixed gravels and sands derived from glacial outwash. The soils are typical brown earths, moderately stony and relatively well drained, but tending to wetness in a few small depressions.

The magnetic survey was carried out using a fluxgate gradiometer, the Geoscan FM18. This instrument has a precision of 0.1 nanoTesla (nT), and was able to detect magnetic anomalies as small as 0.5 nT, against a background variation of the order of 1 nT. There was very weak natural magnetic susceptibility at Dolaucothi. Readings were taken at every 0.25m along survey lines 1.00m apart. The survey was designed to detect structures producing magnetic anomalies buried within the upper 1.50m of the soil and to reject deeper effects which might confuse the picture. The information was assembled and processed by computer. A linear histogram stretch was used to increase contrast between high and low values, and the images were smoothed using an adaptive convolution filter to bring out trends in the data and suppress poor values.

The resistivity survey was carried out with a lightweight, high frequency resistivity meter, the Geoscan RM15, using the Twin Electrode system with a mobile electrode spacing of 0.50m. The electrode arrangement is sensitive to structures, particularly walls and ditches, up to about 1.50m in depth, and can resolve resistivity anomalies of less than 1% of the background of this site. The range of resistance values was unusually large, with values of more than 600 Ohms being recorded. It was, however, possible to detect archaeological features, because they produced correspondingly large anomalies. Readings were taken over a grid at 1.00m intervals. The results were then assembled by computer, and treated using a variety of mathematical routines, including a histogram stretch and adaptive mean and medial smoothing kernal filters.

Results

The results of the resistivity and magnetometer surveys are reproduced as Figs 5.10 and 5.11. Though the magnetometer results contained a great deal of information, the majority of the interpretation was carried out from the computer screen, which is

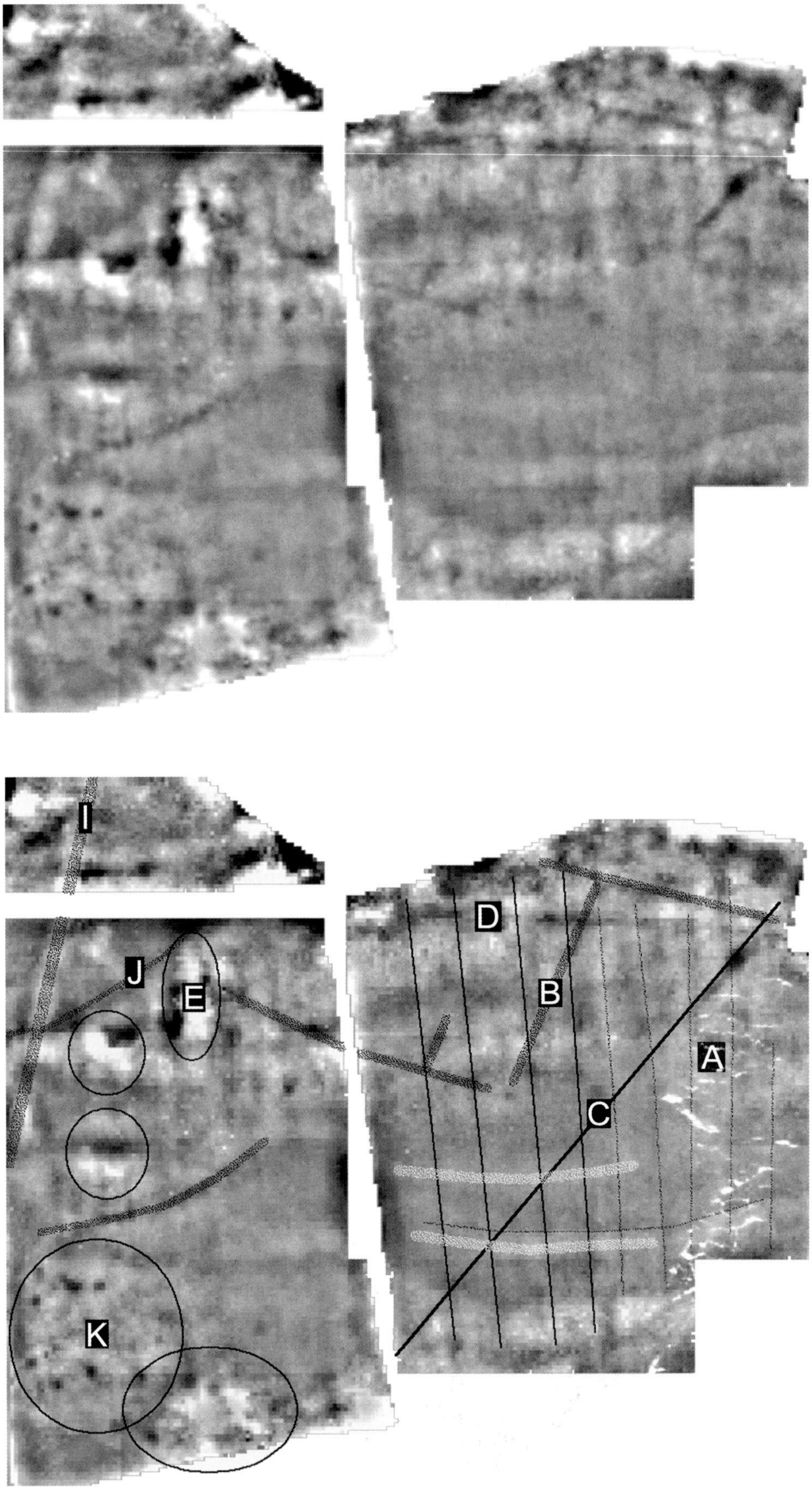

Fig. 5.10 Electrical resistance Survey. Top: processed resisitivity results. Bottom: results with interpretative overlay

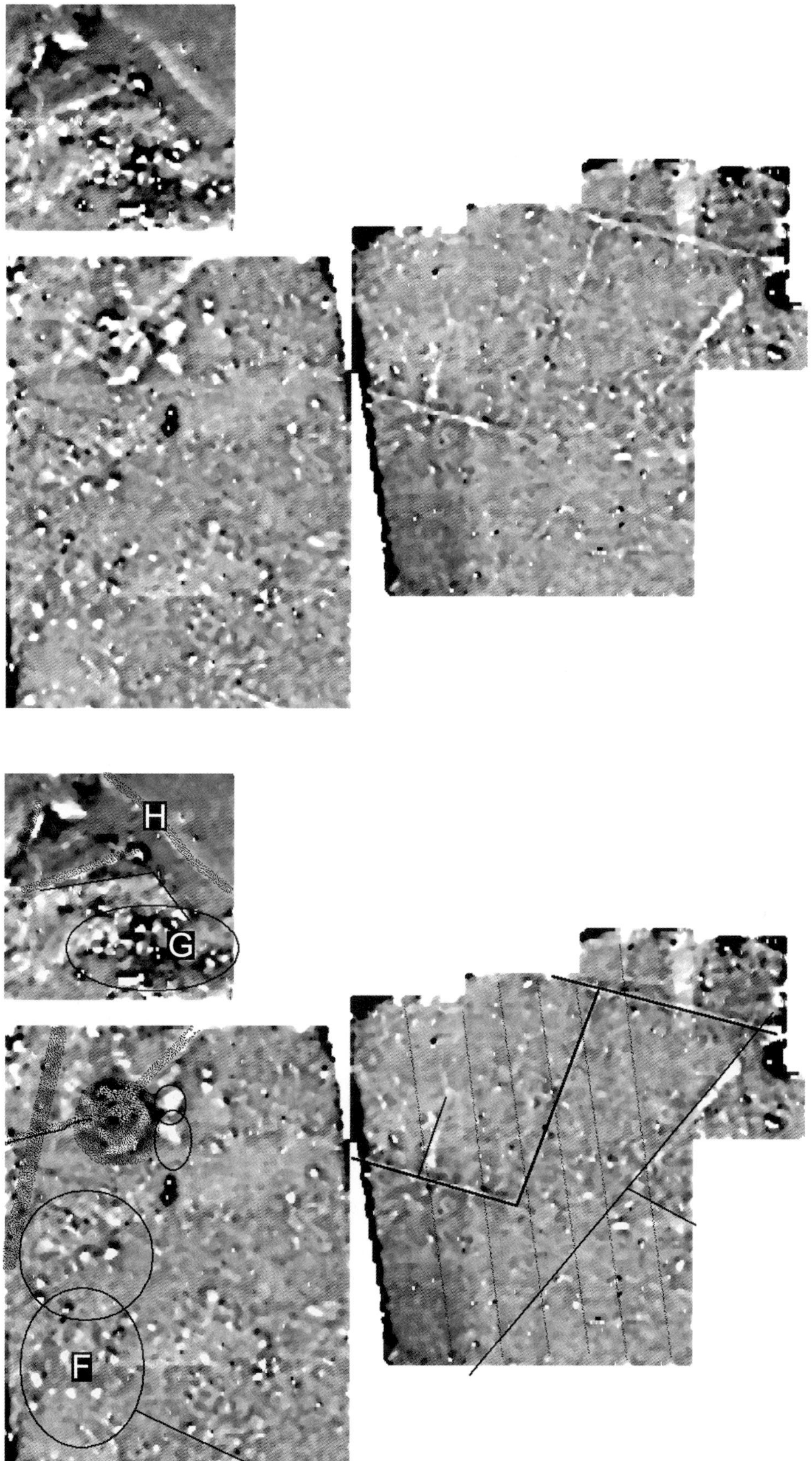

Fig. 5.11 Magnetometer Survey. Top: processed magnetometer results. Bottom: results with interpretative overlay

capable of much higher resolution than the printed image. The resistivity also showed many of the same anomalies, as well as others relating to different aspects of the same archaeological structures or the underlying geology or soil conditions. Among the numerous anomalies located across the survey area, the following were interpretable:

- A Traces of possible cultivation which may represent rig and furrow, although their alignment with the present boundary might suggest more recent drainage works.
- B Probable field boundary ditches.
- C A modern pipe, observed on the ground to lead to a trough.
- D A larger boundary ditch.
- E A concentration of large anomalies which may represent Roman building remains.
- F An area of isolated, positive anomalies which may represent buried pits.
- G A number of strong, sharply defined anomalies due to buried ferrous metal.
- H A weak, linear, positive anomaly which coincides with the edge of a terrace and which may either be due to the geology exposed at this point or, less plausibly, to a retaining structure.
- I The Roman road, a positive anomaly accompanied by a weakly negative anomaly on its west, perhaps suggesting a ditch.
- J A well-defined boundary between resistivity zones which might normally be interpreted as a natural change in the underlying gravels but which here accompanies a possible ditch apparent in the magnetic images.
- K A group of low resistance anomalies which relate closely to the positive magnetic anomalies at (F).

Summary

The magnetic and resistivity surveys complement one another to give a useful picture of the buried features. The underlying trends in the resistivity results reveal a pattern of broad gravel ridges relating to the fluvioglacial history of the valley. The depressions between these ridges have become filled with finer soils which retain moisture to a greater depth, and thus have lower electrical resistance. This pattern has been smoothed by natural erosion and by cultivation.

There are a number of distinct areas of archaeological remains. To the north-west is a cluster of anomalies around the Roman road, indicating the remains of buildings. The most likely candidate for the bath-house is anomaly (E) but the magnetic image does not contain the kind of strong positive anomaly which would be expected from a bath-house furnace. Excavation and cultivation may, however, have eroded the remains. A number of closely defined anomalies to the south-west, showing on both surveys, may be due to pits. This suggests that occupation alongside the road continued to the south of the area of the probable buildings around (E), although no remains of structures could be detected. Finally, to the north-east are the ditches of a rectangular field system. This has no obvious association with the other remains and is orientated differently.

5.5 DISCUSSION

The work undertaken in 1997 clearly made a significant contribution to our wider understanding of the complex at Dolaucothi-Pumsaint in several respects, whilst at the same time raising important management issues for the National Trust if the remains are to be protected and preserved *in situ* for the future.

The trial excavations in field 5532 revealed the presence of an intact horizon of Roman date, at a depth which varies between 0.15 and 0.30m; this includes the remains of a substantial road running roughly north-south, flanked by an area of settlement along its eastern side. So far modern disturbance has been minimal, except where the modern plough horizon [5] coincides closely with the uppermost levels of the road, parts of which have been damaged and turned in the process; further ploughing will inevitably exacerbate this problem. Elsewhere, beyond the line of the road, the older plough soil [6, 7 and 8] probably affords some measure of protection, though even this may not extend right across the headland, leaving the archaeology at risk. Given the importance of the remains for our understanding of the lifespan of the surrounding settlement, a strong case has been made for restricting future ploughing on the headland, so as to maximise protection.

In field 6015 to the south, the geophysical surveys have shown that the Roman settlement was originally more extensive, apparently continuing south-southwest along the line of the road at least as far as the modern field boundary. This clearly includes the remains of several other buildings which were revealed as distinctive geophysical anomalies along its eastern side. More than likely similar features would also have existed along its western side in the area now lost to erosion by the river Cothi. Such ribbon developments, with buildings alongside and beyond the main frontages, are a well-attested feature of the province (Smith 1987), though here their association with both a fort complex north of the river and a mining area to the south may be more reminiscent of the mining complex at Charterhouse (Burnham and Wacher 1990, 208–11). Not enough is yet known about the extent and depth of the associated remains to make a definitive case for restricting the farming regime, though more than likely the archaeology will continue to be threatened.

Given the importance of the settlement, however, two courses of action should form the focus of future work: (i) if and when the field is next ploughed, the chance must be taken to carry out a full-scale fieldwalking exercise, in an effort to record scatters of building debris and to recover any associated dating evidence; advance warning will be essential to ensure the best results; (ii) independent of this, a case can be made for further test excavations, focused primarily on the cluster of probable buildings away from the road frontage; such work would need to be sufficiently extensive, not just to deal with the management issues, but also to establish sufficient details about their function, extent and overall chronology.

To the east of this, in field 7020, the density of settlement seems to drop off, though even here there are traces of characteristic field boundaries of probable post-Roman date, indicating agricultural activity on a different alignment from that of the modern field system. Though these are probably not threatened by modern ploughing, the chance should be taken to sample some of the boundaries in an effort to establish their character and to recover any associated dating evidence. This is all relevant to our wider understanding of the evolving landscape of the Dolaucothi area, particularly as traces of such post-Roman, pre-modern ploughing regimes have now been located not just here, but also in work north of the river and east of the modern village of Pumsaint (see section 2.6 above).

5.6 APPENDIX: The Roman road from Llandovery to Pumsaint, Carmarthenshire (by Heather James)

The Roman forts of Pumsaint and Llandovery lie *c.* 13 km apart as the crow flies. A pre-turnpike road, mapped by Ogilby in the late 17th century, takes a direct course between the two via Caio. This was the Roman road line proposed by Margary (1973, 335, RR62c). Recent fieldwork by Hugh Toller, corroborated and extended by air photography, has identified a new, less direct route that runs north of Llandovery before fording the Tywi to take a broadly west-north-west course to Pumsaint utilising valley sides and encountering only two major gradients.

5.6.1 Historiography of the road

Finds of Roman material at Llandovery were recorded by the antiquary Edward Lhwyd in the late 17th century and published in his additions to Camden's *Britannia* of 1695. He also speaks of 'a notable Roman way of gravel and small pebbles from that Church (*i.e.* Llanfair-ar-y-Bryn within the Roman fort) to Llanbran' (Camden 1984, 51). This is the Roman road heading north-east from the fort towards Beulah, mainly below the present A483. Sir Richard Colt Hoare published a Map of Roman roads and 'stations' in Wales (Colt Hoare 1806). Although bedevilled by errors from the forged Itinerary of 'Richard of Cirencester' (really Charles Bertram), he shows five roads radiating from Llandovery, one of which went to the 'gold mines of Ogofau'. Throughout the 19th century there was continued interest in the Roman road network fostered by the Cambrian Archaeological Association. William Rees, scholar and publisher, owner of the celebrated Tonn Press at Llandovery, was quite certain that the line of the Roman road from Llandovery to 'the gold mines of Caio' was that of Ogilby's pre-turnpike road (Rees 1873) (Fig. 5.12). Yet he also seems to have been aware of other lines of road, straight lengths of banks, identified by Richard Fenton, Colt Hoare's colleague, near Porthyrhyd, described thus: 'called on a Mr. Lewis of Gilvach, who accompanied me to shew me a bit of the Roman Road, the very portion alluded to in Ed. Lhwyd's MS., near a place called Pentre Cwn, a little to the North of the present Turnpike Road from Llandovery to Caio. Another portion has been observed in the low grounds nearer Caio, under a farm called Porth y Rhyd. The portion I saw was a raised Ridge from 15 to 20 feet wide, passing through a moory piece of ground, yet retaining a very solid and firm appearance' (Fenton 1917, 63–64). Wedded to the direct pre-turnpike line, Rees thought that Fenton's lengths of road 'might be a branch from the main Sarn Helen' which 'although not noticed by the Ordnance Surveyors, is most clearly defined' (Rees 1873, 136). There was also a tradition of Roman working at Rhandirmwyn silver-lead mines 10 km north of Llandovery on the eastern side of the Tywi valley. 'Roman levels' were marked on Rolley's late 18th-century plan (Hall 1993).

Although valuable work was carried out by J. F. Jones, Curator of the Carmarthen Museum in the 1950s and 1960s, modern work on the Roman road network and Roman mines began with excavations and aerial reconnaissance by the late Professor Barri Jones in the late 1960s and 1970s. In 1971, he photographed what was locally known as a Roman road, visible from the air as the parch mark of a substantial agger at Dolauhirion Farm just over a kilometre north-north-west of the fort site close to the Tywi, where the valley narrows. Subsequent air photography, discussed in detail below, has established more of the course of the road between Dolauhirion and the fort. At the time, Jones was convinced that its 'alignment makes it highly unlikely that this represents part of the road linking Llandovery to Dolaucothi; more probably an entirely new route is involved running a mile or so up the Towy valley to service some as yet unidentified Roman working in the Rhandirmwyn area' (Jones 1971, 15).

Hugh Toller's fieldwork, written up in a privately circulated report (Toller 1993), has pulled all these strands together and convincingly established a new line of Roman road between Llandovery and Pum-

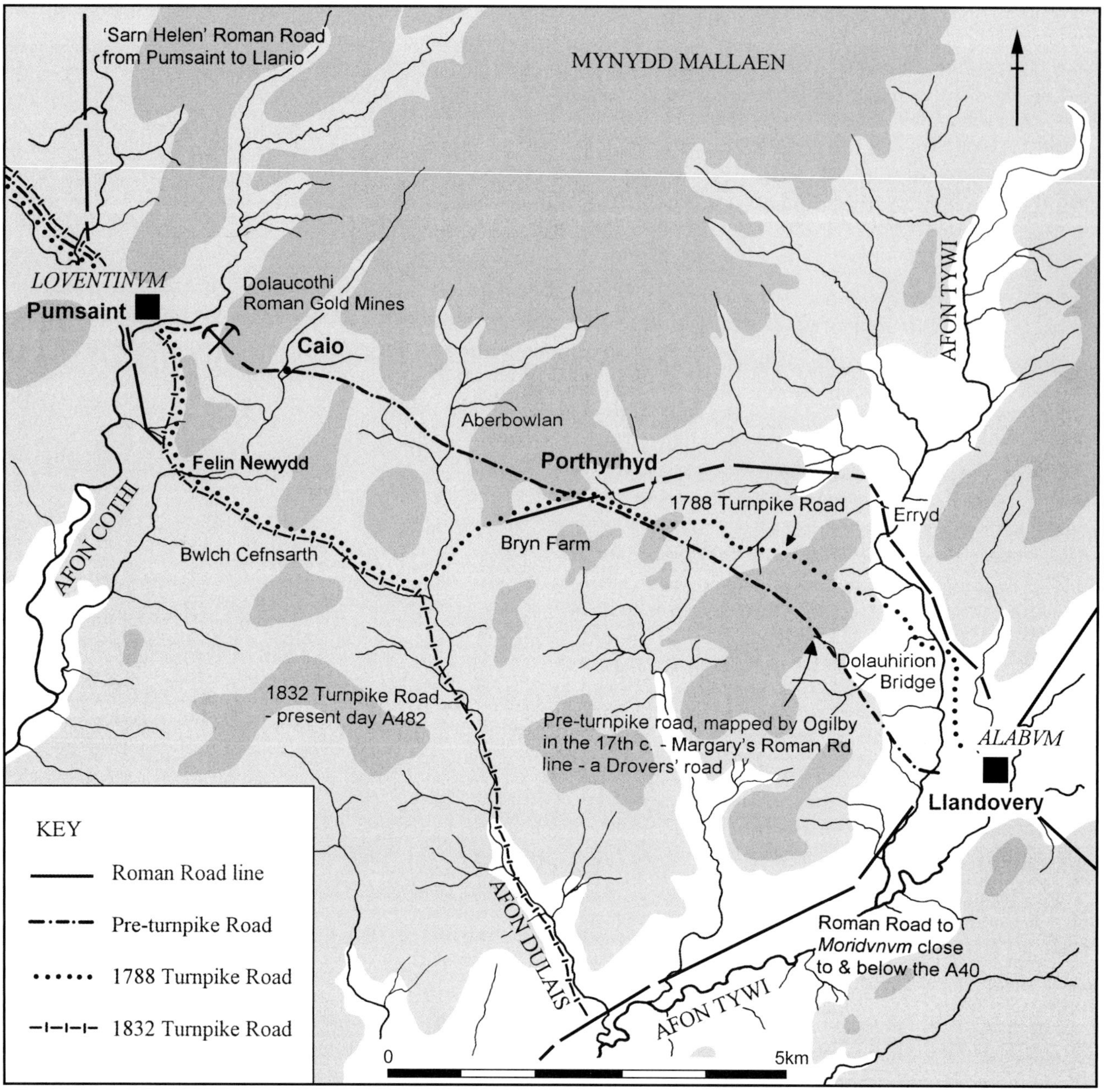

Fig. 5.12 Map showing the line of the Roman Road as established by 1999 from Llandovery to Pumsaint, with pre-turnpike and turnpike roads between the two places

saint which includes the Dolauhirion stretch and those pronounced aggers noted by Fenton. However, he is also careful to retain the possibility that 'Ogilby's road' might have been used for foot and packhorse traffic in the Roman period as it was later and also that there may be other lines of road advancing up the Tywi valley. I am most grateful to him for use of his report from which quotations are made below. Air photography by myself and T. A. James in 1995, and later that year by C. R. Musson, then Investigator, Air Photography, for the Royal Commission on Ancient and Historic Monuments Wales, substantiated and extended Toller's evidence. These, and all other oblique air photographs from flying in the 1980s and 1990s relating to Roman roads in west Wales were mapped by K. Murphy of Dyfed Archaeological Trust, part of a grant-aided project funded through the Royal Commission's Air Photography Mapping programme. Some small additional detail has been added from examination of 1946 RAF high level vertical air photographs. The process of mapping itself creates insights and observations of the route planning and constructional details of the road and these were registered as individually numbered 'site' records within the Trust's Sites and Monuments Records (PRN or Primary Record Number) with descriptive text attached by K. Murphy. What follows is thus very much a collaborative effort. An overall map of the Roman road line (Fig. 5.12) and three more detailed strip maps (Figs 5.13 to 5.15) have been drawn for this volume, but anyone interested in the route details should also consult the National Monuments Record (RCAHMW) and the Sites and Monuments Record (Dyfed Archaeological Trust). As dynamic databases and archives they will contain any new work.

5.6.2 The establishment of the route line

The Roman fort of Llandovery, probably the *Alabum* of the Ravenna Cosmography (Rivet and Smith 1979, 242) is sited at Llanfair-ar-y-Bryn, a short distance to the north of the medieval (and modern) town of Llandovery. The fort lies on a fairly level raised area at the southern end of a long ridge of high ground, Fforest, which separates the valley of the upper river Tywi to the west and the river Brân to the east. Small-scale excavations by Jarrett in 1961 and 1962 (Jarrett 1969, 95–6) and recording of a watermain trench in 1983 (James *et al.* 1983) suggest, though do not prove, that there was an early, pre-Flavian fort on the site, established as a bridgehead in forward campaigning from the Usk valley. Even if that is not the case, it is likely that a fort was established fairly early in AD 75 at Llandovery and that a further advance from there to the known gold-bearing area of Dolaucothi was an early military objective. Certainly the siting of the fort suggests that the road west down the Tywi valley might have been constructed later than routes to the north-east and north-west.

On the basis of his field experience of much of the Roman road network of Wales, Hugh Toller notes the 'substantial scale' of most of the roads from Llandovery compared to a slighter construction elsewhere, particularly in hill country. He suggests that 'this may well be peculiar to a local Roman command and not of especial significance, but it may have some connection with the mines at Dolaucothi'.

Llandovery is sited close to the confluence of a number of rivers and streams, the most important being the river Tywi. The difficulties of fording and bridging streams and rivers and the evident skill with which this was achieved by Roman military engineers are the main factors in the choice of site for the fort at Llanfair-ar-y-Bryn and the course of the road to Pumsaint. The northerly line initially taken on the eastern bank of the Tywi was necessary to find a safe, narrow crossing point before heading west. The Tywi in particular is liable to flood and has been subject to major course changes, which documentary sources from the late middle ages allow us to map (James 1991). Its course in Roman times is unknown but it is possible that in the early Roman period at least it was flowing on the eastern side of its flood-plain, the opposite extreme to today's course.

5.6.3 Description of the area between Llandovery and Pumsaint

The area between Llandovery and Pumsaint is bounded to the north by ridges of high ground extending south-westwards from Mynyddd Mallaen, and, further north, the Cambrian Mountains. The river Tywi, one of the major rivers of Wales, forms an eastern and a southern boundary, due to the change in its course from a south to a south-westward flow at Llandovery. To the west lies the river Cothi, Pumsaint being sited on the north side of the confluence of the rivers Cothi and Twrch. The ground is hilly and broken between the two main rivers, with many small to medium-sized steep-sided valleys, rounded hills, and, reflecting the underlying geology, south-west tending ridges. In the early middle ages it was part of Cantref Mawr, remote, wooded, difficult of access and long a stronghold of native Welsh power. Certainly the distribution of hillforts and defended settlements of assumed iron-age date is sparse. There has been no excavation of such sites nor any systematic fieldwork or palaeo-environmental sampling, apart from pollen sampling of sediments in Talley Lakes (Butler 1984). Little therefore is known of the late iron-age environment of the area. It is reasonable to suppose that the Roman military surveyors and road engineers encountered well-wooded and marshy conditions in addition to ranges of hills and ridges to be crossed.

5.6.4 Detailed description of the Roman road line (Figs 5.13 to 5.15)

The low-lying area north-west of the Roman fort at Llanfair has produced a reasonable amount of Roman material, hinting at a possible *vicus* settlement there. But its topography, and probably drainage, have been altered by the construction of the Central Wales railway in 1868, which is carried past the west side of the fort on a massive embankment. A footpath shown on the first edition (1880) of the OS 1:2500 map led from the then Vicarage (whose construction destroyed the Roman Bath-house site – PRN 4071) to Gilfach and thence north along the steep-sided Allt-y-Gilfach. It is likely that this reflects the Roman line. The route is now one of Carmarthenshire County Council's 'County Walks' and is thus easily accessible. Several possible lengths of causeway and terrace are visible. Just below the path along the foot of Allt-y-Gilfach is the Nant Bawddwr. This stream rises *c.* 4 km to the north in the heights of Fforest. At SN 7659 3542, where there is a footbridge, the Bawddwr takes a sharp change of course curving to the south-east and south. This may mark a former point of confluence with the Tywi so that the stream now flows in an old meander loop. Alternatively it may mark a Roman diversion to take water to the bath-house site,

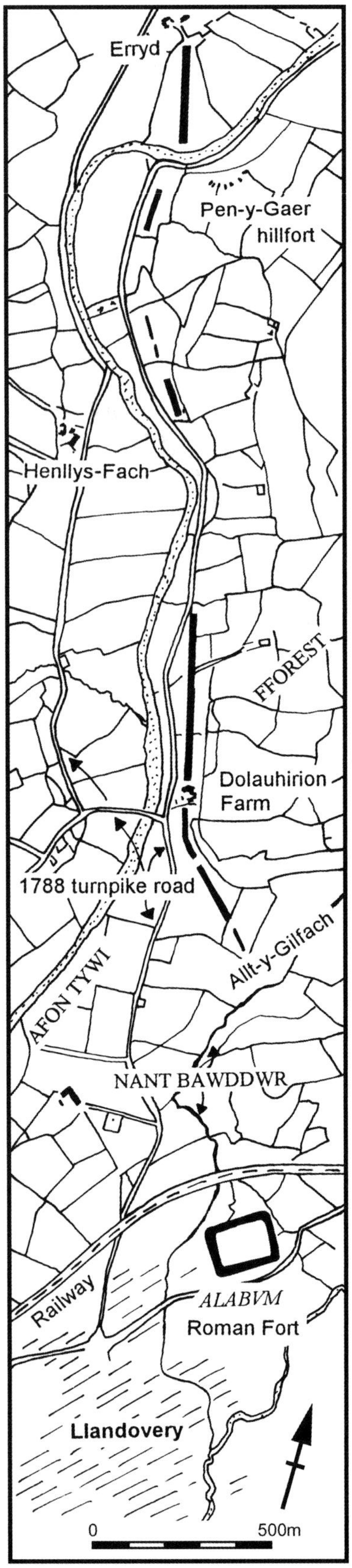

Fig. 5.13 Detailed strip map of the line of the Roman Road between Llandovery and Erryd

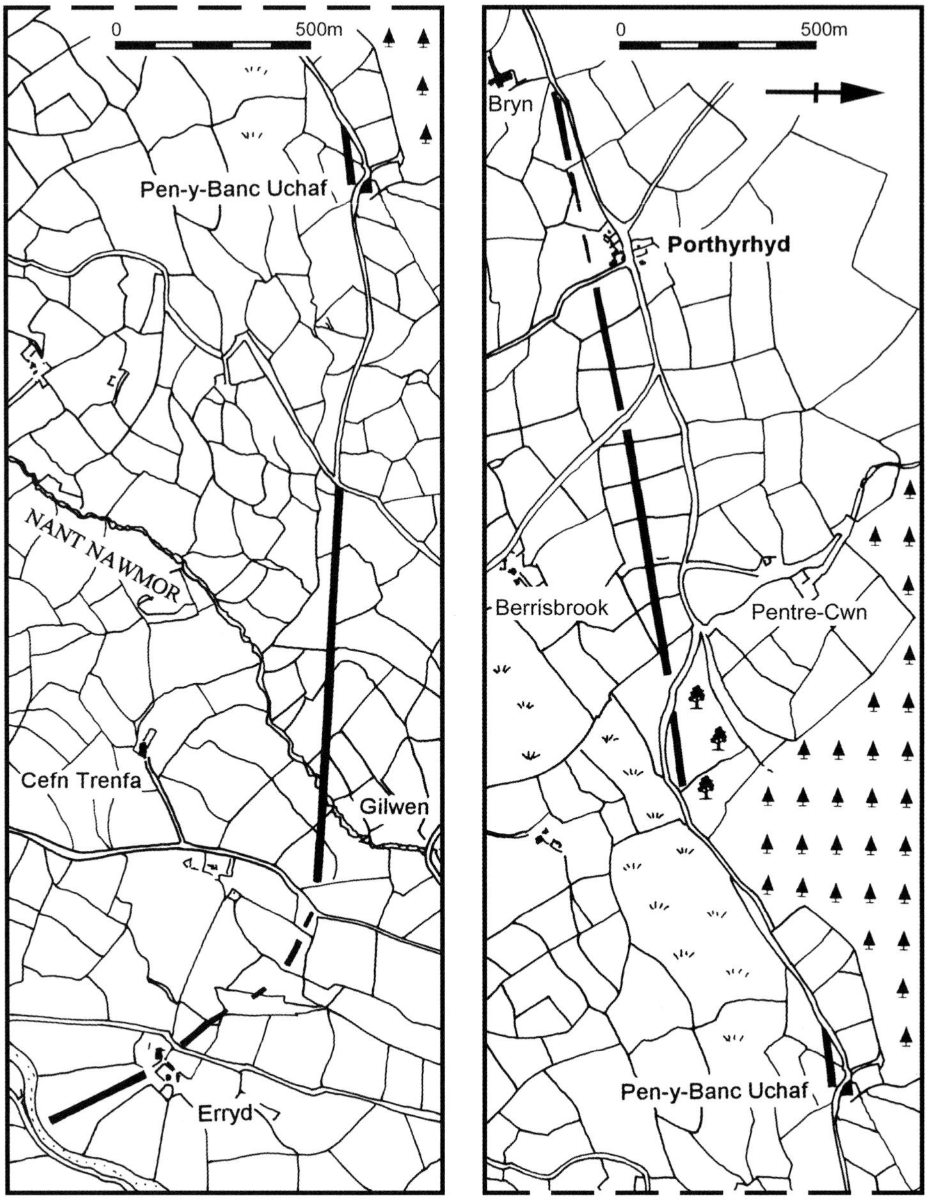

Fig. 5.14 Detailed strip map showing the line of the Roman Road between Erryd and Porthyrhyd

all trace of which has now gone. It is quite likely that the river Tywi was flowing much further east in the Roman period, closer to Allt-y-Gilfach. Parch-marks of former meanders are evident in the air photograph of the Roman road close to Dolauhirion Farm (Fig. 5.16). The Roman road line would thus have been confined to the foot of the steep slopes of Allt-y-Gilfach.

Around SN 765 358, presumably having circumvented a meander loop, the Roman road struck out north-westwards and from SN 7659 3556 took a very slightly curving course before once again making a sharp angled change to head north-north-westwards. This new length to the already well-known line (still visible as a wide, low, agger either side of Dolauhirion Farm) was discovered from the air in the dry summer

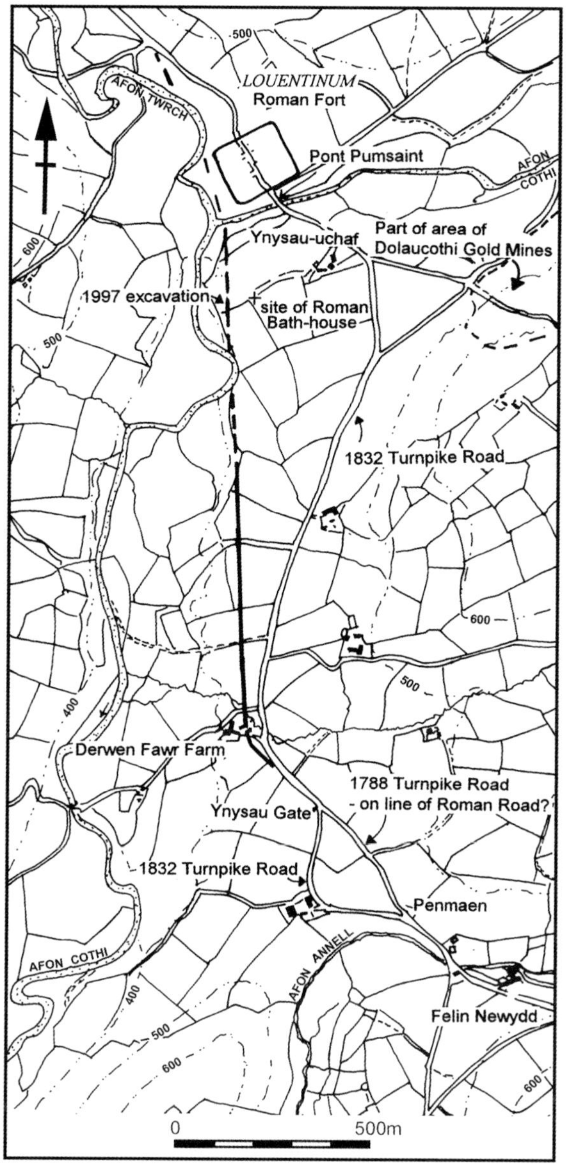

Fig. 5.15 Detailed strip map showing the line of the Roman Road on its final approach to Pumsaint

Fig. 5.16 Aerial view south-eastwards towards Llandovery from Dolauhirion Farm (SN 763 361) bottom right, showing the line of the Roman Road as a parchmark with a change of course south-westwards and parch and cropmarks of former river courses of the Afon Tywi (APDAT 84–15.20). Reproduced by permission of Cambria Archaeology

of 1984 by T. A. James (Fig. 5.16). From Dolauhirion Farm, the road takes a straight north-north-west course across four fields and is then overlain by the modern road line. Opposite Henllys-Fach (on the west side of the Tywi) the modern and Roman road line changes course again to the north-west, mirroring the course of the Tywi in this narrow part of the valley. Although faint, some short lengths of parchmarks have been identified from RCAHMW oblique colour air photographs which show that the Roman line took a straighter course than the more curving modern line and thus survives across fields to the east of the modern road, both however heading for a fording point of the Tywi at Glan-Rhyd-Erryd.

As the place-name element *rhyd* (ford) suggests (an element probably present in Erryd farm), this was an ancient crossing point and hollow-ways down to and up from the opposite bank substantiate this. The crossing is also overlooked by Pen-y-Gaer hillfort (PRN 4074) a few hundred metres to the north-east on raised ground. Although almost 3 km north-west of the fort, this is the first possible fording area of the river above both its wide flood-plain and a deep rock-cut gorge near Henllys-Fach where river levels can rise and fall alarmingly quickly. The Roman crossing point may have been a little further to the north-east than that below Glan-Rhyd-Erryd where the line of a track (shown 1817) leading up to the farm is preserved by a hedgeline, because, across the middle of the field between Erryd and the river, is a clear straight parch mark visible on C. R. Musson's colour slide (RCAHMW 95–CS-2608–9).

From Erryd, a track ascends straight up Cefn Trenfa hill to the farm of that name; but that does not perpetuate the course taken by the Roman road because of the location of the next long, clear length identified between Cefn Trenfa and Gilwen. To approach this definite line, the Roman road must have worked its way around the hill in a series of short lengths. Despite careful fieldwork, Toller could not identify any hillside terraces or cuttings and concluded that modern woodland and hillslope ploughing have obliterated the Roman traces. Fortunately, colour air photography (RCAHMW 95–CS-2934) shows parchmarks of the road running north-westwards in the field across the road west of Erryd Farm and then making an angled change of direction more to the north-north-west at a point now marked by a hedgeline (SN 7525 3818), before being obscured by woodland. At the northern end of the rectangular hillside block of woodland, there must have been a major course change once the hill was rounded and a new, westwards sight line obtained. Again visible from air photographs, but also detectable in the field surface, is a length of westwards-aligned hollow-way running up to the modern north-south line of the lane to Cefn Trenfa.

From here, Toller has successfully established a line from south of Gilwen farm for *c.* 4.5 km as far as Bryn farm west of Porthyrhyd. It runs, he notes, in a remarkably straight line between these two points with a few curves and kinks of course around spurs and hollows. Descending the hill across an area that even today is marshy, barely drained by the little Nant Nawmor, Toller recorded a substantial agger on the west side of the stream and a less well-preserved, but still clearly visible bank on its eastern side as well. From west of the stream to a point of junction (SN 7388 3842) with the minor road from Pen-y-Groes to Pen-y-Banc the causeway of the Roman road has been utilised as both field boundaries and access track to the now defunct holding of Ty-Isaf (Fig. 5.17). Toller noted that the agger had been reduced and spread a little by recent ploughing and comparison between the air photograph and the most recent 1:10,000 Ordnance Survey map shows removal of hedgebanks. Yet the Roman road still remains a substantial earthwork and a prominent feature visible from the air. Murphy identified some possible quarry pits for the road's construction from the air photographs on the edges of the field north of the site of Ty-Isaf.

From the road junction a modern minor road up to Pen-y-Banc overlies the course of the Roman road. The farm lies at the eastern end of a ridge of high ground, one of only two highpoints to be crossed by the Roman road across a route evidently carefully engineered to avoid too many steep ascents and descents. Toller calculates the incline to be 100 in 700 metres. This high point must have been an important sighting position during the laying out of the road line by the Roman military surveyors, looking back eastwards to Cefn Trenfa and west-south-westwards down the valley to Porthyrhyd and beyond towards the next major north-south aligned valley of the river Dulais. From the farm, the modern route diverges northwards from the Roman line, thus preserving a large, albeit ploughed, length of agger across two fields south-west of Pen-y-Banc. Here too, visible from the air as darker coloured areas either side of the road, are possible quarry pits for road metalling. The Roman road then lies along the modern line from which it diverges at a fine angle into woodland (SN 7275 3820) where Toller identified its line as a cutting. He considers that these two lengths south-west of Pen-y-Banc give an excellent demonstration of the different constructional methods of agger and cutting used across this variable terrain. Further down the slope there are also good examples of the kind of processes that can obscure or destroy the Roman

Fig. 5.17 Aerial view eastwards from the road junction (SN 7390 3845) leading to Pen-y-banc-uchaf showing the line of the Roman Road as a parchmark diagonally across the photograph approaching the crossing of the Nant Nawmor (APDAT 95–71.28). Reproduced by permission of Cambria Archaeology

road. Emerging from the woodland at SN 7248 3811, Toller identified a terrace cut across the north-facing slope and a ploughed agger representing its continuation west-south-west. Water erosion had completely removed parts of the terrace and at another point soil slippage masked the point where the terrace rounded a spur. Changes in field boundaries and ploughing made the line of the agger difficult to map from air photographs.

Once the foot of the slope was reached another long straight length of agger was constructed to carry the road west-south-westwards. The Roman line lies to the south of the present road from Pen-y-Banc to Porthyrhyd and can easily be seen from the road traversing seven fields (Fig. 5.18); it is marked on the Ordnance Survey 1:25,000 maps. At the eastern end of this long length, it is crossed by a footpath south from Pentrecwn to Berrisbrook. This must be the stretch noted by Fenton. West of Porthyrhyd near Bryn, the course of the Roman and modern roads coincide once again. It was evident during mapping of the parchmark of the road between Porthyrhyd and Bryn that there is a very slight curve to the

Fig. 5.18 Aerial view east-north-eastwards from SN 7157 3782 east of Porthyrhyd showing a pronounced *agger* crossing three fields and the area of a slight course change in the Roman Road partly overlain by a modern minor road leading to Pen-y-Banc Uchaf (APDAT 95–70.24). Reproduced by permission of Cambria Archaeology

Roman line. It has been suggested that from Porthyrhyd the Roman road may have then run north-west towards Caio on the Afon Annell, *c.* 2 km distant from Pumsaint, but, as Toller convincingly argues, the clear trace as far as Bryn takes the Roman line past the junction north-west to Caio, leaving the Roman course 'committed to the valley route' (see Fig. 5.12). However, 1 km west of Porthyrhyd, the modern (and Roman) road line crosses the river Dulais and there are two route lines going north to Aberbowlan on the old Drovers' road between Caio and Porthyrhyd. Hugh Toller has examined this area but with negative results from the point of view of a Roman line, and thus a course west, crossing the Dulais at Pont-yr-Efail, seems much more likely.

Having crossed the Dulais the Porthyrhyd road joins the A482 at Pont-yr-Efail, heading north-west to Pumsaint. This was the junction point of the 1832 road up from the Tywi valley at Llanwrda with the 1788 turnpike. It is thought that the turnpike line, constructed in 1788, closely followed the Roman course, but the 1788 course was not exactly followed by the 1832 road. The map shows points of difference which can be located by comparison between the 1812 Ordnance Surveyors' drawings and the first edition One Inch Ordnance Survey map of 1834. Terraces and cuttings are visible at many of these points of divergence in the 1946 vertical RAF air photographs (consulted at the National Monuments Record) but it is impossible to distinguish between Roman and late 18th-century work. From Pont-yr-Efail the road line continues westwards for a short, straight length but soon makes a major course change from west to north-west. This kept the turnpike, and presumably the Roman line, close to the foot of the steep slopes of Banc Maestwynog before crossing a minor tributary of the Dulais to reach Bwlchcefnsarth.

From here the second high point of the route had to be traversed. Bwlch carries the meaning of 'breach, gap or notch' and by extension, topographically, a pass or defile, and with this specific meaning is a common place-name element in Wales. Toller has noted traces of an earlier road as a terrace below the modern road approaching the pass, below Bwlchcefnsarth. The road rounds the side of the hill to approach the pass which it traverses on a straight westerly alignment. Once the high ground was crossed the turnpike road returns to a north-westerly alignment to cross the river Annell at Felin Newydd. Between here and Ynysau Gate there is a gap in the long ridge of high ground between the valleys of the Annell and the much larger river Cothi. The turnpike, and almost certainly the Roman road, used this gap to reach the wide expanse of the Cothi valley.

Between Penmaen and Ynysau the 1788 line, and perhaps the Roman, is preserved as a straight length of track with the 1832 road curving to the west. At Derwen Fawr Farm both modern roads took the same

Fig. 5.19 Aerial view showing line of Roman Road extending across fields south of the 1997 site (APDAT 95–90.30). Reproduced by permission of Cambria Archaeology

course north-eastwards towards Pumsaint, bridging the Cothi at Pont Pumsaint. It is clear from air photographs taken in 1995 that the Roman road diverged here from the modern, and approached the fort at Pumsaint west of the modern line, across the valley floor (Fig. 5.19). This was the last major course change before Pumsaint. Air photographs provide direct evidence here for course changes in the river Cothi actually cutting away the Roman line at one extremity of a meander loop. The final approach to the fort is dealt with in more detail elsewhere in this chapter. From Pumsaint the road continues to take a valley side route before ascending the high ground to reach Llanio on the river Teifi. As 'Sarn Helen' this line has long been known and recognised as Roman and a review of its course has recently been published (Davies 1994, 309–11).

5.6.5 Conclusion

Difficulties in finding a safe, all-season fording point of the Tywi undoubtedly lie behind the initial northward course of the Llandovery to Pumsaint route rather than, necessarily, demonstrating a Roman route to Rhandirmwyn. The seemingly indirect course between the two Roman forts was also carefully selected to reduce the number of hills encountered, with only three areas of high ground to be crossed or skirted: Cefn Trenfa, Pen-y-Banc and Bwlchcefnsarth. Each seems to have served as a sighting point in the laying-out stage of a Roman road that followed on from general reconnaissance over the route area. Such vantage points may have been all the more necessary in wooded and marshy country. The Roman road must then have been capable of taking wheeled traffic and being used all year round, yet the line fell out of use. Its successor, the direct 'Ogilby' route shows all the characteristics of pre-turnpike roads in its steep ascents and descents designed for foot travel and pack-horse transport. It may also have come into use in the medieval period because of course changes in the Tywi allowing a crossing point to 'Heol Rhos' close to the modern bridge carrying the A40 across the river Tywi, nearer to the core of the medieval Castle-borough of Llandovery.

6 OVERVIEW OF 1987–99 SURVEY AND EXCAVATIONS

6.1 DATING EVIDENCE FROM DOLAUCOTHI-PUMSAINT, 1972–1997
(by Peter Webster)

It is clear from a glance at the dating evidence offered here and in other publications of recent work (Burnham and Burnham 1987a, Lewis 1977) that not all sites associated with the fort, fortlet or mines are producing dating evidence which precisely matches one with the other. The reason is presumably that not all areas were occupied at the same time. Some summary of the significant available evidence may be in order and that is offered briefly here. The information on the 1972, 1989 and 1997 sites is derived from material published in this volume. Samian evidence from 1973–74 is derived from a report kindly made available by Mrs Felicity Wild on the samian from the millpond, Melin-y-Milwyr, immediately adjacent to the main opencast. Roman coarse pottery from Melin-y-Milwyr is summarised in section 6.1.2 below. The pond material includes pottery of post-Roman date and does not necessarily date the pond itself. It does, however, presumably reflect Roman activity at or near the main mine site (Lewis 1977).

6.1.1 Samian

The samian evidence currently available is summarised by site, source and form in Table 6.1.

Differences between the 1989 and 1972 assemblages have already been discussed in section 2.2.4(Ciii) above. The small quantity of material from Les Martres-de-Veyre and elsewhere in Central Gaul is apparent from the table. Within this small sample, the greater quantity of later material from 1972 is also clear (6% of the total 1972 assemblage as opposed to 1% of the 1989). The discrepancy cannot be adequately explained if one assumes that both areas are within the later fortlet but it must be noted that the overall sample size in both cases is fairly small. The Central Gaulish material from 1972 carries the date of the fortlet into the 120s AD but seems too small to suggest a prolonged occupation, even in the Hadrianic period. Abandonment of the fortlet *c.* AD 125 is, therefore, tentatively suggested.

It is difficult to compare the 1997 material with any of the other assemblages due to its limited quantity. The pattern of the samian looks closer to that of 1972 than to that of 1989. The coarse pottery, however, contains a few pieces which might be mid rather than early 2nd century and, overall, a higher percentage of 2nd-century material. If this truly reflects the occupation of the baths area (and only further excavation can demonstrate this), then such occupation might outlast that of the fort and fortlet by perhaps up to two decades. However, like the 1972 and 1989 material, that from 1997 stands in complete contrast to the assemblage from Melin-y-Milwyr close to the main opencast.

It is natural to assume that the occupation of the fort and fortlet reflects the main period of activity at the mines. The Melin-y-Milwyr material suggests that this is not necessarily the case. The samian shows a scatter of material contemporary with that from the 1972 and 1989 fort/fortlet sites, as well as a comparative wealth of later samian, including forms 31R, 45 and 79, all of which belong to the later half of the 2nd century. The sequence does, in fact, go beyond the samian-exporting period as the coarse pottery section below demonstrates. The exact interpretation of this material is difficult. It is unstratified and the deposit includes later medieval and post-medieval pottery. The Roman pottery does not, therefore, date the millpond from which it came. Nevertheless, it is unlikely to have travelled far and should indicate Roman interest in the mines area well after the associated military bases were abandoned. Is it our sole evidence for the leasing of the mines to a civilian from the mid 2nd century?

6.1.2 Coarse Pottery

The coarse pottery from the 1989 and 1997 sites is published above in sections 2.2.4(Cii) and 5.3.1(B) respectively. That from 1972 is summarised in section 2.2.4(Ciii), as part of the 1989 report. An archive list will be deposited with the pottery itself pending a fuller structural report. In the main, these

Fig. 6.1 Summary of samian evidence from all sites

	1989 Fort/Fortlet		1972 Fort/Fortlet		1997 (Baths area)		1973-4 (Mine area)	
Form/Source	**No.**	**%**	**No.**	**%**	**No.**	**%**	**No.**	**%**
South Gaul								
15/17	-	-	2	2	-	-	-	-
15/17 or 18	-	-	16	15	-	-	-	-
18/18R	14	20	17	15	1	7	3	6
18/31	7	10	5	5	2	14	-	-
18 or 18/31	3	4	-	-	-	-	1	2
27	10	14	28	25	-	-	1	2
29	6	9	5*	5	1	7	-	-
33	-	-	1	1	-	-	-	-
35	3	4	3	3	-	-	-	-
36	5	7	3	3	1	7	-	-
37	12	17	17	15	-	-	4	8
46	-	-	1	1	-	-	-	-
67	-	-	1	1	-	-	-	-
78	-	-	-	-	1	7	-	-
C11	8	11	2	2	-	-	-	-
C15	-	-	1	1	-	-	-	-
Bowl	1	1	-	-	-	-	-	-
Les Martres								
18/31	-	-	1	1	2	14	1	2
33	1	1	-	-	-	-	-	-
37	-	-	1	1	1	7	-	-
C11	-	-	-	-	-	-	1	2
Central Gaul								
18/31	-	-	1	1	-	-	4	8
18/31R	-	-	-	-	-	-	3	6
18/31 or 31	-	-	-	-	-	-	4	8
31	-	-	-	-	-	-	1	2
31R	-	-	-	-	-	-	3	6
33	-	-	2	2	-	-	6	12
36	-	-	-	-	-	-	1	2
37	-	-	2	2	2	14	11	22
38	-	-	-	-	-	-	1	2
42	-	-	1	1	-	-	-	-
45	-	-	-	-	-	-	1	2
79	-	-	-	-	-	-	1	2
East Gaul								
18/31R or 31R	-	-	-	-	-	-	1	2
37	-	-	-	-	-	-	1	2
Totals	70	98	110	102	13	99	47	98

* 3 added from notes made in 1972

assemblages support but do not greatly extend the chronological conclusions which can be deduced from the samian evidence. All are predominantly Flavian to Flavian-Trajanic, but contain a few certainly 2nd-century pieces (such as the Caerleon ware from the 1972 and 1997 sites). The Black-burnished ware, particularly that from the 1972 excavations, includes examples of the bowl or dish with flange at the rim (the 'flat-rimmed' variety of Gillam's 1976 corpus) which seem likely to be of the developed and probably 2nd-century class. As far as can be ascertained the Black-burnished ware from the 1972 excavations, like that from 1989, is predominantly from later levels. Black-burnished ware does not, however, appear on either the 1972, 1989 or 1997 sites in the quantities which one associates with prolonged Hadrianic or Antonine occupation and in this, the coarse pottery supports the date of *c.* AD 125 suggested for abandonment of the fortlet.

The contrast between the fort/fortlet and baths sites and Melin-y-Milwyr is as clear in the coarse pottery as in the samian. Here, it seems appropriate to illustrate only a few pieces to show the chronological spread (Fig. 6.1). Among the pieces chosen, Black-burnished ware predominates, not only because this provides the best chronological indicator but also because it demonstrates most clearly what is missing from the BB1 collections from the other sites.

Amphorae

No 1 Dressel 20 amphora rim; Martin-Kilcher type 17 (Peacock and Williams 1986, 137). Mid-late 1st to early 2nd century. MM74.

No 2 (Not illustrated). Base of an amphora of the Gauloise 4 (Pélichet 47) type; a South Gaulish wine amphora. A later 1st- or 2nd-century date is most probable. MM73.

Late first or early second century reduced wares

No 3 Everted rim jar in soft light grey fabric. MM74.

No 4 Flanged bowl in fairly soft light grey fabric. The form is typical of the Flavian and Flavian-Trajanic period. MM73 and MM74.

Black-burnished ware

No 5 Jar in Black-burnished ware; *cf.* Gillam 1976, 2–3. The rim lacks the wavy line of the Gillam examples but is, nevertheless, most likely to be 2nd century in date. MM74.

No 6 Jar in Black-burnished ware; *cf.* Gillam 1976, 4 (late 2nd century). MM73.

No 7 Jar in Black-burnished ware; *cf.* Gillam 1976, 7–8 (early to mid 3rd century). MM73.

No 8 Jar in Black-burnished ware; *cf.* Gillam 1976, 10 (late 3rd century). One of two very similar rims. MM73.

No 9 Jar in Black-burnished ware; *cf.* Gillam 1976, 10-12 (late 3rd to early 4th century). With a fragment of the same or a very similar vessel (from MM74). MM73.

No 10 Flanged bowl or dish in Black-burnished ware; for the general type see Gillam 1976, 54 (early to mid 2nd century). MM73.

No 11 Flanged dish in Black-burnished ware. The intersecting chevrons suggest a date in the second half of the 2nd century. MM73.

No 12 Flanged and grooved bowl in Black-burnished ware; *cf.* Gillam 1976, 42 (late 2nd to early 3rd century). MM74.

No 13 Flanged and grooved bowl in Black-burnished ware; *cf.* Gillam 1976, 43 (early to mid 3rd century). MM73.

No 14 Flanged and grooved bowl in Black-burnished ware; *cf.* Gillam 1976, 44 (mid to late 3rd century). MM73.

No 15 Dish in Black-burnished ware; *cf.* Gillam 1976, 68–70 (2nd century). MM73.

No 16 Dish in Black-burnished ware; *cf.* Gillam 1976, 79 (early 3rd century). MM73.

Oxfordshire ware

No 17 Dish in light red fabric with traces of a red colour coat; *cf.* Young 1977, C45 (*c.* AD 270-400). MM73.

No 18 Mortarium in hard off-white fabric with red grits; *cf.* Young 1977, M12 (late 2nd to mid 3rd century). MM74.

Other fabrics

No 19 Jar in light orange fabric with a greyish core. This is probably Severn Valley Ware. The ware is not well represented in south-west Wales, but of the pieces present, jars seem to predominate. This jar is likely to be 3rd or 4th century (*cf.* Webster 1976, 9–11). MM74.

No 20 Mortarium in pink to light red fabric with red and grey grits. Mrs K. F. Hartley examined the piece soon after its recovery and suggested manufacture in the west of England. 3rd to 4th century. MM73.

The contrast with the other sites is clear enough. Here there is material which carries Roman occupation in the area through into the later third century at least. What is most interesting about this material and its accompanying samian summarised in Table 6.1, however, is the way in which it seems to complement the other sites. While it is obviously unwise to place too much emphasis on unstratified material, the Melin-y-Milwyr collection seems the best evidence currently available for a movement of occupation away from the fort area (Pumsaint village) and towards the mines. It does not seem too fanciful to suggest that the pottery points to a military occupation on the fort and fortlet site up to the reign of Hadrian and occupation of another kind and in another place thereafter up to the late 3rd century at least.

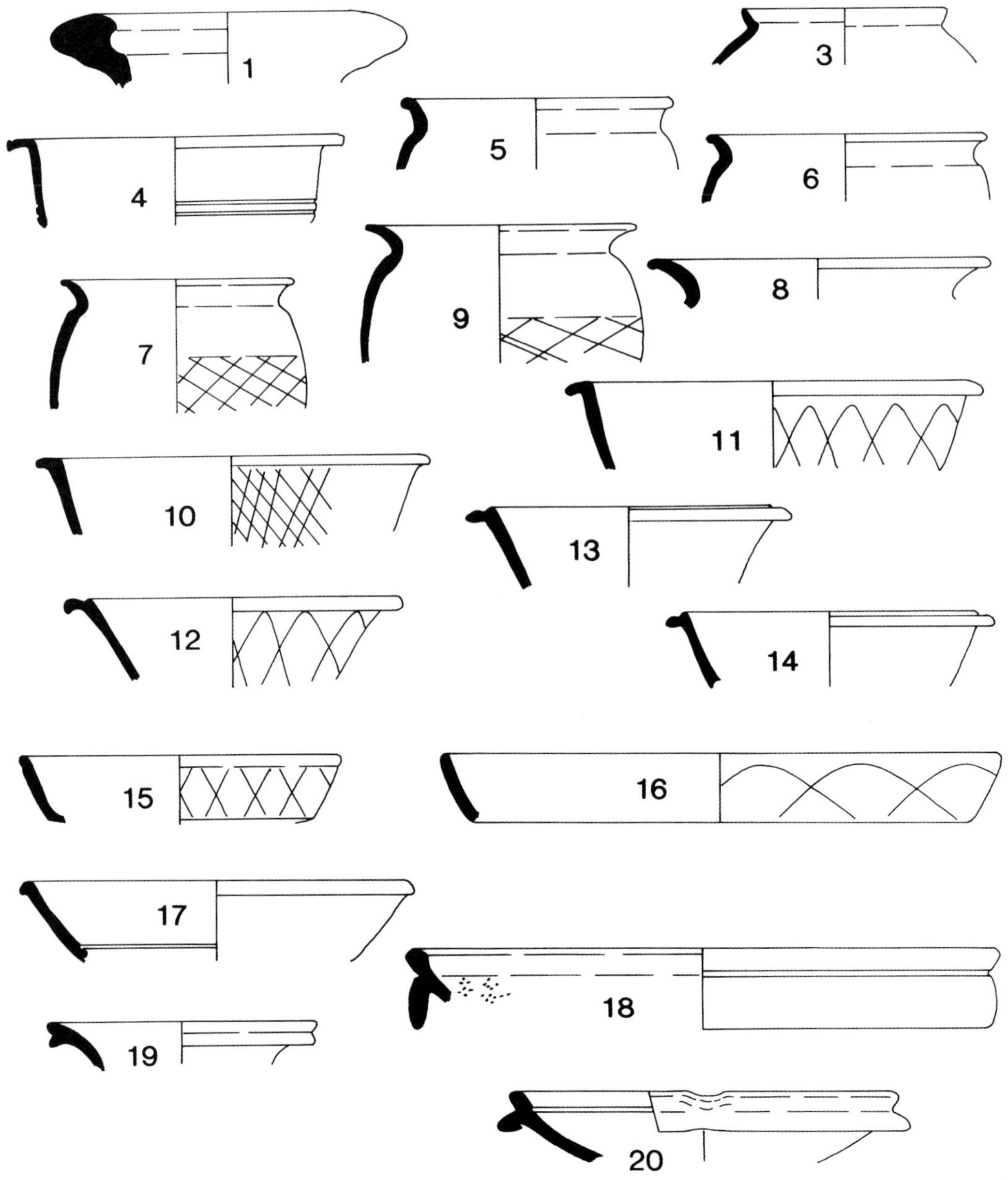

Fig. 6.1 Pottery from Melin-y-Milwyr. Scale 1:4

6.2 DOLAUCOTHI-PUMSAINT, 1987–99: RETROSPECT AND PROSPECT

6.2.1 Introduction – the Lewis and Jones model

The starting point for any discussion of the Dolaucothi-Pumsaint complex must be the pioneering work undertaken by Lewis and Jones, which established a simple developmental model for the successive stages of Roman exploitation (see Fig. 1.3), based on a logical progression from prospecting, through opencast development, to underground stoping, all in close association with a complex system of leats supplying water to the workings (see section 1.3 above). Though neat, this framework was ultimately dependent on several assumptions, including (i) a belief that the basic layout of the mines as we see them today had originated in the Roman period, (ii) a presumed relationship between the individual opencasts and the relative sequence supplied by the leat systems and their associated features, and (iii) an overall chronology largely derived from material peripheral to the workings themselves. Such views have been reiterated so often in the literature (*cf.* Jones and Maude 1991), however, that they have increasingly been taken as gospel, despite growing concerns over their general validity.

The question of dating the physical remains is obviously central to any reconstruction of the early exploitation at Dolaucothi. The Lewis and Jones model relied in part on the use of analogy to Roman workings elsewhere, especially Las Medulas in Spain, in part on a reference in Pliny (*Natural History* 33, 70-74) to the importance of water power, also in Spain, and in part on several sources of archaeological evidence from different parts of the complex. Though these latter were never set down in full in a single publication, the principal strands can be conveniently summarised as follows:

(a) the fragment from the rim of an oak drainage wheel, found in the stopes beneath the Ogofau Pit in the 1930s, which has produced a radiocarbon date of 2040±40 bp, centred on 90±70 bc (HAR-2709);
(b) the fragment of glass from a Roman square bottle, of 1st- to 3rd-century date, said to have been found 'stratified' in a primary context in the Mitchell Pit in 1970, but never published;
(c) the two radiocarbon dates derived from a peat horizon infilling a section across the lower end of the Gwenlais leat (said to be the upper reaches of the Annell when first published by Jones and Maude 1991); the peat itself produced a date of 1175±15 bp, centred on ad 775±15 (Grn-16553), while an alkali extract yielded a reading of 1085±20 bp, centred on ad 865±20 (Grn-16720), indicating that by *c.* AD 800 the leat channel was already three-quarters infilled;
(d) the Roman pottery and other material found in association with the fort/fortlet at Pumsaint and in the vicinity of the bath-house south of the river Cothi; inclusive of the 1989 and 1997 material, this indicates a Flavian-Trajanic date for the former, with the bath-house area perhaps outlasting the fortlet by up to two decades;
(e) the Roman pottery from Melin-y-Milwyr, discovered in 1973–74, which includes some 96 sherds of 1st- to probable 4th-century date; the samian shows a clear peak in the latter half of the 2nd century, while the coarse wares extend down to the later 3rd century at least;
(f) the well-known hoard of gold jewellery, first discovered in 1796 or 1797 and subsequently added to in 1819, which is usually dated to the 2nd to 3rd centuries (Bosanquet 1917, 25–7);
(g) certainly one, and possibly two, late coin hoards, both essentially closed in the later 3rd century (Boon 1966).

Such evidence was clearly seen as demonstrating both a significant Roman presence in the vicinity of the mines and an extensive Roman involvement in their exploitation. The material from the workings, in particular, was said to make it 'impossible to treat the overall layout of the site as anything other than Roman', with the implication that 'the basic form of the mines had already taken shape in the Roman period, probably before 150' (Jones and Maude 1991). The fragment from the drainage wheel was especially critical to this line of argument, because it clearly had major implications for the date of the underground stope in which it was found and, thereby, for the extent of earlier mining above this, in an already worked-out opencast. Likewise, the fragment of glass bottle was deemed equally critical for the date of the Mitchell Pit, and for the associated Cothi leat immediately above its rock-cut rim, though some caution is necessary perhaps, because so little has been published about its findspot and the related stratigraphy. Jones also attached considerable importance to the radiocarbon dates from what was thought to be the upper reaches of the Annell leat, because they appeared to confirm a Roman date for a feature which had already been shown to be early in the overall sequence, in association with the exploitation of the Allt Cwmhenog Pit.

What is surprising in all this is just how limited the primary *in-situ* material from the mines actually is, even if it seems to confirm the relative sequence derived from a study of the various opencasts and their presumed relationship with the associated leat systems. This problem is compounded by the doubts surrounding the context of the glass fragment, and by the uncertainties which remain over the exact relationship between the Gwenlais leat (the true location of the radiocarbon dates) and the lower reaches of the so-called Annell. Effectively only the wheel fragment stands up to detailed scrutiny, but technically it can only date the end of the sequence, not necessarily its beginning, a point to which we will return in section 6.2.4 below. Equally surprising is the extent to which the Lewis and Jones model is ultimately dependent upon the large assemblages of finds derived from the fort/fortlet, the settlement south of the river, and Melin-y-Milwyr. These latter were obviously crucial to fleshing out the broader chronology of Roman activity, even though strictly speaking they came from potentially peripheral non-mining contexts, which might conform to other regional or provincial sequences.

Such evidence is clearly suggestive of a significant Roman interest in the mines, but whether it is sufficient to substantiate the view that most, if not all, the surviving surface features took shape in the Roman period must remain open to debate, in part because of the lack of any large-scale work within the main opencasts and trenches, and in part because of the inaccessibility of the critical underground stopes. The appearance of this volume, however, detailing the results of the most recent investigations at the site, provides a welcome opportunity for reviewing some of these issues, in the hopes of clarifying our wider understanding of the mine's development, the technologies involved, and the nature and extent of the associated settlements.

6.2.2 Scope of the 1987–99 investigations

Given the particular gaps in our knowledge of the Dolaucothi-Pumsaint complex, it is a matter of regret that so little of the 1987–99 work was directly concerned with the mine workings *per se*. Part of the reason for this lay in the fact that, until very recently, no full-scale survey of the mines existed for archaeological purposes (or for any other!), which made it problematical to undertake any detailed assessment of the inter-relationship between key areas of exploitation or surface features, let alone envisage any excavation at the critical points; moreover, with so much of the mines now extensively scheduled as an ancient monument, the prospect of large-scale research intervention (primarily excavation) was neither feasible nor desirable without the appropriate survey framework. Another part of the reason for the focus of the work stemmed from the fact that opportunities had to be taken where and when they were afforded to investigate the wider context of the mines as a whole, driven as much by the underlying research questions, as prompted by the rescue and evaluation requirements of the National Trust and, on occasion, Cadw: Welsh Historic Monuments. Only the 1991–93 excavations near the Carreg Pumsaint, themselves building on the 1982 mill survey, and the 1999 geophysical work on the east side of the fort were driven by outright research questions; the 1987/89 excavations on the fort were conducted as a rescue operation for the National Trust in advance of a proposed housing development; the 1990 work on the Annell-Gwenlais was prompted by Cadw, prior to a decision being taken on scheduling; and the 1997 work south of the river arose initially as an evaluation for the National Trust, though it was extended by geophysical survey over a more extensive area to input a wider research dimension.

As a consequence, the primary focus of the 1987–99 work has been concerned with four aspects: (1) the fort and its associated settlement north of the river; (2) the road system and the associated settlement south of the river; (3) the various leat systems serving the mines; and (4) the related processing areas and their associated waste residues. This inevitably limits some of the conclusions which can legitimately be drawn about aspects of the exploitation, since the results still present a rather lopsided picture of those parts of the site at the periphery of the workings, even if we accept that the processing aspects impinge more directly than most. This apparent imbalance can be offset to some extent, however, by the fact that much recent work has been undertaken on Roman (and iron-age) mining sites across Spain (Domergue 1990) and France (Cauuet 1999), while equally important research has produced valuable insights into prehistoric (bronze-age) mining elsewhere in Wales, most notably at Cwmystwyth (Timberlake 2003) and Great Orme (Dutton and Fasham 1994). Even more significantly, building upon the National Trust's decision to commission a full-scale survey of the workings during 1999, with an appropriate overlay of archaeological interpretation (Landmark Surveys and Cambria Archaeology 1999, supplemented by Cambria Archaeology 2002), a French team led by Béatrice Cauuet has carried out a short season of survey on the main areas of surface and underground workings, the preliminary results of which raise some interesting questions of relevance to the wider framework (Ancel *et al.* 2000). Considered all together, we now possess a significant body of data with which to undertake a reassessment of the Lewis and Jones model for Dolaucothi and to explore alternative hypotheses with respect to the nature, extent and chronology of the Roman presence. What follows is necessarily, however, no more than a working model for the development of the complex, no doubt providing as many questions as it does answers; the first part will concern itself with wider settlement and landscape issues, the second with aspects of exploitation and processing.

6.2.3 Settlement, landscape and environment: the Roman presence

By comparison with the situation in the early 1980s, our knowledge of the Roman landscape in the vicinity of Dolaucothi, and its chronological development, has been significantly enhanced by the 1987–99 work. We are now in a position, for instance, to appreciate not just the character of the military presence at Pumsaint, north of the river Cothi, but also the extent of the civilian settlement stretching along the road towards Llandovery, south of the river and adjacent to the main mine workings. If this latter is plausibly associated with the exploitation of the gold-bearing ores, then we can further speculate on the nature of the associated industrial landscapes and their impact on the wider cultural and environmental setting.

(A) The fort/fortlet sequence and the associated *vicus*

It is now abundantly clear, both from the earlier excavations and from the 1989 assemblage, that the fort at Pumsaint was founded in the mid 70s AD, on a level gravel terrace above the confluence of the rivers Cothi and Twrch, which today extend along the south and west sides respectively. In this respect, it conforms with the main Flavian thrust into south-west and west Wales, under the governor Julius Frontinus (Davies 1980), contemporary with the known forts at Carmarthen to the south-west (James 2003, 13), Loughor and Neath to the south (Marvell and Owen John 1997, 206), and Llanio and Trawscoed on the main road to the north (Davies 1984, 275; 1994, 282). Nothing in its assemblage can be taken to indicate activity in the immediate vicinity before the mid 70s, despite the presence of a probable pre-Flavian fort at Llandovery to the south-east, which might be interpreted as evidence for an early interest in the potential of the gold-mines (Jarrett 1969, 95–6; James *et al.* 1983); such an early interest might also be indicated by the existence of two successive marching camps at Y Pigwn, on the route east from Llandovery, though this could equally reflect the importance of such sites in a wider strategy controlling the Tywi-Usk river corridor.

Enough is now known of the primary defences of the Pumsaint fort from excavation and trial trenching to establish their line with reasonable certainty on the north, west and south, where they have been shown to comprise a clay, gravel and turf bank, laid in places on a corduroy of log timbers. The situation on the east side is still not fully resolved, however, despite some suggestive evidence from the 1999 survey that they lay along the predicted line; for them to have lain any further to the west, one would either have to explain the absence of any appropriate features in the geophysical results or to call into question the very nature and extent of the fort as currently reconstructed (see Fig. 2.115). If all four sides are as predicted, this would confirm Jones and Little's estimate of its size as *c.* 1.9ha (4.75 acres), somewhat smaller than contemporary examples at Llandovery (2.38ha), Loughor (2.15ha) and Neath (2.22ha), but larger than Llanio (1.55ha). In the absence of a detailed internal plan, this would seem to be appropriate for a *cohors peditata quingenaria*, to follow the statistical survey of auxiliary forts and their garrisons by Bennet (1986); such a garrison would have required only six barracks, well within the space available in the proposed *praetentura* and *retentura* of the fort at Pumsaint. Any reduction in the postulated size of the fort would probably mean that it could only have accommodated a part unit.

Within the interior of the fort/fortlet, the excavations of 1972 and 1989 revealed a complex sequence of up to five phases of building, while those of 1973 in the central range recorded only two. The buildings so recovered included a mixture of barracks, workshops and structures of uncertain function, some of which on the 1989 site encroached onto and across the *via sagularis* in phases 4 and 5. If we accept the shorter chronology proposed by Peter Webster for the 1989 site, with occupation apparently not extending beyond the first decade of the 2nd century, this would indicate an average lifespan of roughly seven years per phase; on the longer chronology for the 1972 site, this would rise to about ten years, all of this at a time which saw the reduction in the size of the fort to fortlet status, accompanied by the refurbishment of the defences with a stone wall constructed probably *c.* AD 100. Unless we assume significant levels of wear and tear being responsible, such rapid changes in the internal arrangements of the site much surely correlate with parallel changes either in the nature of the associated garrison, or in the function of the site itself over a period of 30 to 50 years. Parallel complexity has also been noted and discussed for Loughor and Neath (Marvell and Owen John 1997, 226–7) and for Carmarthen (James 2003, 31), though a close analysis of the different sites does not reveal any consistent or recognisable pattern, except at the most generalised of levels.

It is known, however, that the Welsh garrison underwent a process of gradual reduction during the later Flavian to Trajanic era, consequent in part upon the aftermath of the Agricolan advance, in part upon the reduction in the size of the provincial garrison as imperial interests were focused elsewhere, and in part upon a reassessment of needs locally within Wales (Davies 1980, 264–5). In the process several forts in the south-west were reduced in size, including Carmarthen (James 1992, 8–9; 2000, 28; 2003, 31–2), Llanio (Davies 1994, 304) and, most probably, Llandovery (Jarrett 1969, 96), which clearly provides a partial context for the recognisable changes at Pumsaint. Problems remain, however, over the extent to which the whole site was intensively used during the last 10 to 15 years of its existence, given the apparent differences in the pottery assemblages on the 1972 and 1989 sites.

In their reconstruction of the fort Jones and Little (1973, fig. 1) argued that it faced east, onto the level ground now occupied by the village's cricket field. It was also assumed that the main Roman road, approaching from the south from Llandovery, reached the fort somewhere in the vicinity of, or upstream from, the modern bridge, before presumably by-passing the site on the east. Not unsurprisingly, traces of timber buildings and Roman finds had been exposed in this area by the eroding action of the river (Burnham and Burnham 1987a), which raised the distinct possibility that there was an extra-mural *vicus* on this side at least. There the matter rested, however, until the trial work on the Roman road south of the river in 1997 and the geophysical survey across the area east of the fort in 1999, both of which have radically transformed our understanding of the site and its immediate environs. The 1997 work clearly confirmed the line of the road on an alignment well to the west of the modern A482, running roughly north-south as it approached the point where it had been truncated by a recent meander of the river (see Figs 5.3 and 5.15). As it seems unlikely that this road changed direction beyond here, it is clear that it had originally been aligned on a point outside the west gate of the fort, such that it would effectively by-pass the site on this side before heading north up the valley of the Twrch. This must surely mean that the fort originally faced west, no doubt with a short spur road connecting to the gate; this would have provided a convenient focus for a *vicus*, though any evidence for this is now lost to river erosion on this side. Parallels for this type of arrangement are known at other sites, however, among them Llanio (Davies 1994, 303; Arnold and Davies 2000, 60, fig. 2.7).

Significant new details were also forthcoming from the 1999 geophysical survey on the east side of the fort/fortlet, not least in terms of the identification of an extensive area of civilian settlement potentially post-dating the reduction in the size of the fort (see Fig. 2.115). The key element in this was an east-west road or track extending across the cricket field, which seems to have been the focus for a whole series of timber structures set within individual plots defined by ditches. There is also evidence for at least one large, stone-built structure, with a linear arrangement of rooms and possible side chambers, reminiscent of a row-type bath-house (*cf.* Jarrett 1969, 166–72, fig. 89). Parallel activity, clearly post-dating the demolition arrangements of *c.* AD 100, was recognised in the eroding river section south of the fort in 1986, suggesting that the reorganisation was more extensive here as well (Burnham and Burnham 1987a). The presence of a mixture of both official-looking structures and possible civilian enclosures is fairly typical of other *vici*, both within Wales and further afield in northern England and along Hadrian's Wall, as recent large-scale geophysical surveys have increasingly emphasised at sites like Cefn Caer, Pennal (Hopewell 2002, 99–101 and fig.), Maryport (Burnham *et al.* 2001, 337–9, fig. 13) and Birdoswald (Biggins *et al.* 1999, fig. 51). At Pumsaint, however, it adds a significant new element to the military landscape north of the river, which in its turn has to be set in context alongside the equally extensive area of civilian settlement now known to the south of the river, adjacent to the mine workings themselves.

From an analysis of the finds associated with the 1970s excavations, Jones and Little dated the abandonment of the fortlet towards the middle of the 2nd century, a date accepted by Davies (1980, 269), even though it effectively retained the site long after its neighbours had been lost; no doubt, in part, this was a reflection of the site's assumed importance and continued role in the exploitation of the associated mines. From a recent reassessment of all the material now available from Pumsaint, Peter Webster has concluded that the fortlet continued at best into the 120s, but that the small amount of Central Gaulish samian in the assemblage makes it unlikely that occupation lasted much beyond the appearance of Lezoux wares *c.* AD 120/125. If this is so, this would bring Pumsaint into line with those other sites which were abandoned early in the reign of the emperor Hadrian, among them the immediate neighbours at Carmarthen (James 2000, 29; 2003, 45–6), Llanio (Davies 1994, 304) and Trawscoed (Davies 1984, 276), and, most probably, Llandovery (Jarrett 1969, 96), no doubt in connection with the reorganisation of the Welsh garrisons consequent upon the decision to construct Hadrian's Wall and the troop redeployment that this necessarily entailed (Davies 1980, 269, fig. 17.9); as a consequence south-west Wales was effectively devoid of troops by the mid 120s AD. Though the associated *vicus* still remains poorly known, there is little to suggest that it would have survived for long beyond the demise of the fortlet, given our existing knowledge of other sites in Wales (Davies 1991b). Whether or not the military enclosure was demolished or maintained intact in official hands must remain uncertain on current evidence. From the story so far, however, there is nothing remarkable about the foregoing military sequence at Pumsaint, by comparison with the general Welsh sequence for the period from AD 75 through to the 120s, and certainly there is little as such to indicate its significance (or otherwise) vis-à-vis the development and exploitation of the mines themselves.

(B) The road system and settlement south of the river Cothi

By contrast with the area north of the river and the mine workings themselves, the area along the south side of the Cothi has been rather neglected, even though Lewis and Jones (1969, fig. 2) marked the location of a possible settlement here to the south and west of the site of the postulated bath-house excavated in the 1830s. Although no specific evidence was advanced in support of this, beyond the presence of the structure itself and the location close to the actual workings, its apparent importance was further reinforced in 1994 when a geophysical survey by Bradford raised intriguing possibilities by (mis)identifying a possible corridor building west of the presumed site of the baths. More critical, however, for our understanding of the Roman landscape on this side of the river has been the detailed investigation of the course of the Roman road from Llandovery to Pumsaint, the results of which have been well rehearsed elsewhere in this volume by Heather James (see section 5.6 above). This latter convincingly demonstrated that, in its approach to the mines, the road diverged from the line of the modern A482 at Derwen Fawr Farm, a mile south of Pumsaint, whence it proceeded on a more northerly course heading directly for the river crossing and the western side of the fort. Interestingly, its immediate approach to the flood-plain of the Cothi is recorded on an aerial photograph, extending across the fields to a point west of the site of the baths (see Fig. 5.19), where it had obviously been misidentified by the Bradford survey. In 1997, this site was duly confirmed by trial excavation just south of the point where it had been truncated by a former meander of the river.

The establishment of the road's alignment in the immediate vicinity of the mines, close to the site of the presumed baths, obviously added a significant new piece to the jigsaw, as did the fact that the 1997 trial work also encountered traces of associated occupation debris and structural remains extending away from its eastern frontage. This picture was subsequently reinforced by the extensive geophysical surveys undertaken across the two fields to the south and south-east, which clearly recorded the presence of a more extensive area of settlement apparently stretching for some distance southwards along the eastern side of the road; this settlement clearly included the remains of several structures, represented by distinctive geophysical anomalies, none of which could, however, be associated with any confidence with the site of the baths. There would seem to be little doubt that a similar sort of pattern must originally have existed along the west side of the road, where so much has now been lost to river action by the Cothi. Further work will obviously be necessary to clarify the extent of this site to the south-south-west, though one possible clue might be afforded by Lewis and Jones' reference to 'burials' in the area west of the A482 near the Cwrt-y-Cilion workings (1969, fig. 2); such a suggestion seems to be dependent, however, upon the discovery of a single cinerary urn of Black-burnished ware in 1909, apparently in a prominent alluvial terrace east of the river (Evans, G. Eyre 1909–10, 75 and fig.; RCAHMW 1917, 36–7, no. 124), so some caution is perhaps necessary.

Despite the potential significance of this new area of settlement, much still remains uncertain about its date and character. For the former, in particular, we are primarily dependent upon the small quantity of material recovered during the 1997 excavations, which would seem to indicate a date range extending from the later 1st through to the mid 2nd century. In his comparison with the assemblage from the fort site, however, Peter Webster has noted a somewhat higher percentage of 2nd-century finds overall, which, if truly representative of the whole area south of the river, could potentially suggest that activity here effectively outlasted that of the fort/fortlet by up to two decades or so. Even then it would do no more than bring the site into line with the evidence from other Welsh *vici*, few of which show any sign of occupation after *c.* AD 140 (Arnold and Davies 2000, 62), though on current evidence we cannot obviously rule out the possibility that occupation may have contracted after that date, perhaps northwards towards the original river crossing. That the postulated baths could have survived somewhat later might be suggested by a single fragment of samian attributed to Advocisus (Bosanquet 1917, 31), whose date range is currently assigned to the period AD 160-190. All this makes further excavation essential if the dating issues are to be resolved with any clarity.

The restricted nature of the investigations to date also makes it difficult to speculate about the wider implications of the site, not least its relationship (or otherwise) with the known Roman exploitation of the mines. Part of the reason for this lies in the fact that there is nothing immediately distinctive in the excavated evidence to establish an outright link with the mining operations, with the possible exception of a stray fragment of a lower quernstone recovered from the surface of the modern field prior to the 1997 excavations (see above section 4.9, no. 14), which itself appears to be somewhat of a coincidence. At first glance, indeed, the known remains are reminiscent of a characteristic class of roadside settlements which have been recorded both outside known forts (Sommer 1984; Davies 1991b) and in the civilian parts of the province (Smith 1987), sometimes with buildings and facilities associated with the official *cursus publicus* (Black 1995). This latter could explain the presence of a possible bath suite, apparently occupying conveniently available land immediately behind the buildings lining the main frontages, as was the case at Caves Inn (Lucas 1981, 27–31) and Wall (Round 1992, 13–19), both of them ribbon villages on Watling Street. Such villages were also characterised by their range of everyday structures, built either in timber or stone, many of them of stripbuilding type with evidence of associated small-scale specialisation (Burnham 1989, 38–40), none of which would conflict with the details recorded in the 1997 trial trench.

It would be somewhat surprising, however, if the mines had not encouraged the growth of an associated mining settlement, in part to house some of the workforce and in part to provide a range of related services, over and above the military presence north of the river. Dependent mining villages are certainly known elsewhere in the Roman world, both from archaeological and epigraphic evidence, even though they remain poorly researched. Such sites provide a useful working model for what might have been the case at Dolaucothi. In Spain, for instance, at the Rio Tinto mine, the associated settlement occupied the entire hill of Cerro Salomon, *c.* 1 km in diameter, the workers being housed in multi-roomed structures with paved flooring and an integral hearth (Blanco *et al.* 1970, 10–11 and 20, quoted by Curchin 1991, 125). Closer to home, at Charterhouse, the well-known mining site extended over *c.* 30 acres, complete with its network of internal streets, a possible amphitheatre, and well-made drains, the latter suggestive of a bath-house, all in association with an extensive area of lead-mining in which the military had an early interest (Burnham and Wacher 1990, 208–11). The presence of a possible bath suite should also remind us of the evidence contained in the *Lex metalli Vipascensis*, which is recorded on two bronze plaques found in the slag heaps near modern Aljustrel in Portugal; while one of them is obviously concerned with the conditions of mine tenure and operation, the other covers some of the regulations governing wider economic activities in the mining district, including the concessions for operating the bath-house in the associated village (Lewis and Reinhold 1966, 188–94).

Without appropriate epigraphic evidence from

Dolaucothi we can only speculate on the extent to which parallel arrangements existed for the management of the settlement south of the river, though the broad contemporaneity of its recorded occupation with the Hadrianic dating assigned to the Aljustrel tablets enhances rather then detracts from the possibility. It also emphasises the enhanced significance now attaching to this newly discovered area of settlement within the wider landscape associated with the mines. If indeed it does represent a dependent mining village, then its proposed date range, extending as it does from the later first through to the mid 2nd century, must surely raise important questions both for the chronology and intensity of the associated mining operations and for what happened from the later 2nd century onwards, questions to which we must return later.

(C) Later Roman activity

In the absence of any clearcut evidence for occupation extending beyond the mid to late 2nd century in the area south of the river, much remains uncertain about the nature and extent of any later activity in the vicinity of the mines. Such activity has long been indicated, however, by several strands of evidence, all of it somewhat scattered in the landscape. One of the first indicators was the discovery of a goldwork hoard, said to have been dug up at intervals on the Dolaucothi estate at the turn of the later 18th and earlier 19th centuries (Bosanquet 1917, 25–7). It comprised several pieces, among them two necklaces or chains, two wheels, an armlet with a serpent's head, and a rod in the form of a serpent (Nash-Williams 1952, 81–2). Parallels elsewhere have generally indicated a 2nd- to 3rd-century context for this assemblage, though Manning (2001, 107) has recently raised the intriguing idea that it could have belonged to the wife of one of the military commanders of the area in the later 1st century. If a later date applies, the purpose of such a hoard remains uncertain. It has often been taken to indicate the presence of specialist gold workers at Dolaucothi, but this seems unlikely on current evidence; alternatively, it is possible that it points to the existence of a cult focus (Arnold and Davies 2000, 130) or else represents a ritual deposit, given the well-attested solar symbolism of the spoked wheel and its association with Jupiter in numerous contexts (*cf.* Green 1986, ch. 2).

Equally interesting is the evidence of a late coin hoard, apparently discovered *c.* 190m north-north-east of Erw-Hên near Pumsaint, which proved to contain at least 684 base Antoniniani belonging to the period AD 260–291 (Boon 1966). This clearly closed with issues of Carausius, whereas a second reported hoard of *c.* 3,000 coins from an unknown location in the parish of Cynwyl Gaeo closed with those of Allectus (*ibid.*, 157 and note 2). Such hoards closing either with issues of the Tetrici (AD 270-3) or with those of Carausius/Allectus (AD 286–96) are not uncommon in Wales (Arnold and Davies 2000, 116 and fig. 9.8B), and so may carry no particular significance for mining in the immediate vicinity. Their presence, however, is indicative of some form of settlement in the area, with access to coins which were hoarded at times of insecurity and subsequently never recovered for whatever reason.

Also indicative of settlement is the pottery recovered from the millpond of Melin-y-Milwyr (Lewis 1977, fig. 17). Peter Webster, in his assessment of this material (see section 6.1 above), makes it clear that both the samian and the coarsewares provide a significant contrast with the assemblages recorded from the fort and the settlement south of the river. The samian includes a wealth of material of the second half of the 2nd century, while the various coarsewares certainly extend activity down to the later 3rd century at least, if not potentially beyond. None of this material is demonstrably *in situ*, as the recovery of some later 16th- and 17th-century material from the same location attests, but it must surely have been derived from a settlement site somewhere in the vicinity. Webster sees this as evidence of a progressive migration away from the fort area and the settlement south of the river towards the mines, though there is nothing here to indicate any link with ongoing mining activity. Its wider significance lies in the fact that it represents continuing activity at or near the mines, activity which was sufficiently engaged in extended socio-economic transactions to bring a variety of pottery types to the site.

Such scattered evidence makes the 1989 discovery of the burnt timber building within the confines of the earlier fortlet all the more interesting, even if its precise date remains difficult to establish. The radiocarbon dates would certainly seem to indicate activity somewhere in the period spanning the later 2nd through to the early 5th century, but there is nothing in the associated archaeology to help us narrow this further. What is striking, however, is its location immediately adjacent to one of the fortlet's north-south streets, with which it is aligned, not far from the presumed north gate. This might suggest that either the fortlet's stone rampart was retained intact in army hands following the demise of the occupation in the earlier 2nd century, or that elements of its defences were recommissioned at a later stage. Whichever view one takes, it is hard to see this large structure as anything other than military or official in character, even if there is nothing to connect it directly with the mines.

Finding the context for such an enigmatic building, and for a renewed phase of military or official interest in the site, remains problematical. Much obviously depends upon the extent to which the site conforms to wider trends, or is a response to some specific set of local circumstances. While the latter is almost impossible to assess, the former is susceptible to analysis in the light of recent accounts of the changing military deployment within Wales (Davies 1991a; Arnold and Davies 2000, 24–34). Davies makes it clear that very few auxiliary forts were occupied after *c.* AD 140, with the apparent exception of a number of garrison posts in east-central Wales (Forden Gaer, Caersws, Castell Collen and Brecon) and of Caernarfon in the north-west (Davies 1991a, fig. 10.1). This broad pattern was apparently retained throughout the Severan era, down to the 230s AD, after which sites like Forden Gaer and Caersws were progressively abandoned. Davies has argued that this diminution of the Welsh garrison in the mid to later 3rd century was a direct consequence of increasing pressures on other European frontiers, with the remaining forces being more directly concerned with the collection of money dues and military supplies at appropriate points along the road network (Arnold and Davies 2000, 28).

There are various signs of change in these arrangements in the last quarter of the 3rd century (Davies 1991a, figs 10.2 and 10.3). Principal among them was a need to deploy troops against a perceived threat from across the Irish Sea; this is most clearly seen at Cardiff (Webster 1981), but is also likely to account for renewed activity at Loughor (Marvell and Owen-John 1997, 216 and 227–8) and Neath (Heywood and Marvell 1992, 291–2), both of them located on navigable estuaries. Several other forts (Bryn y Gefeiliau, Caersws, Castell Collen, Llandovery and Gelligaer) have produced material suggestive of a short-lived re-occupation in parts of Wales in the later 3rd and early 4th centuries, to add to the evidence of continuing activity at Brecon (Arnold and Davies 2000, 29), while an unfinished bath-house at Caernarfon suggests some level of activity also in the north (Casey and Davies 1993, 15, 62–5). Although none of these are certainly contemporary, it has been suggested that the frequency of Carausian hoards in Wales may point to at least one specific set of local circumstances (amongst others) which would have necessitated a short-term military response. After this apparent burst of activity the numismatic evidence would seem to suggest that few of the forts with a 3rd-century sequence were occupied beyond the 330s AD at the latest, with the emphasis henceforth resting on key strategic sites designed to meet specific localised problems (Davies 1991a, figs 10.5 and 10.6).

Unless the burnt timber building at Pumsaint had very specific local purposes, a date after *c.* AD 330 seems unlikely on several grounds, in part because of the restricted character of later troop deployments and in part because of the absence of all but a handful of possible 4th-century finds in association with the Melin-y-Milwyr material. Such a building could obviously have acted as a military collection post at any stage from the later 2nd century through to the early 4th, but perhaps the most plausible context would seem to lie in the later 3rd to 4th century, when there would seem to be a growing body of evidence to indicate activity, however short-lived, at a range of sites across Wales. The combined evidence of the two coin hoards, closing with Carausius and Allectus respectively, and of the pottery extending down to the later 3rd century at least from Melin-y-Milwyr, could be said to point in a similar direction on the balance of probabilities. Nothing, however, points one way or the other to a connection with the mines, so the jury must remain out on the continuity of mining into this later period. Though there is nothing positive to indicate the function of the building, that of storage remains the most likely possibility (see section 2.2.5(D) above).

None of the evidence discussed in this section indicates more than small-scale interest in the Dolaucothi area, pending the discovery of any related settlement sites. This is in marked contrast to the landscape of the later 1st and 2nd centuries, which seems to have been dominated by larger-scale settlements, in part military and in part civilian. Arguably, this contrast ought to be mirrored in the chronology and intensity of the associated mining operations.

(D) The cultural and environmental context

The recognition of a major Roman presence in the Dolaucothi area during the later 1st and 2nd centuries clearly poses some intriguing questions about its impact on the wider cultural and environmental landscape, and about the interrelationship between incomer and native in the immediate vicinity of the mines. Such questions are not easy to answer, however, for a variety of reasons, not least the absence of any systematic fieldwork or palaeo-environmental sampling along the course of the Cothi, our continuing uncertainty over the extent to which the mines were worked in the immediately pre-Roman period (see section 6.2.4 below), and the difficulty of calculating the likely population levels of the fort and its satellite settlements. What follows, therefore, is as much an assessment of our relative ignorance for the Cothi valley as a synthesis of existing knowledge. For ease of convenience the issues will be approached from three directions: from a study of the evidence for later prehistoric and Roman settlement activity in the Cothi valley; from a review of the limited palaeo-environmental sequences in the area; and from the perspective of the impact of the fort and its satellite settlements on the immediate locale.

Any assessment of the cultural context of the Dolaucothi complex is ultimately dependent upon what is known about the local settlement tradition, set against the wider regional picture now available from across south-west Wales (Fig. 6.2; *cf.* Cunliffe 1991; Davies and Hogg 1994; James and Williams 1982; Williams 1988), itself much enhanced by recent survey and excavation (*e.g.* Austin *et al.* 1984–89; Mytum 1999; Williams 2002; Williams and Mytum 1998). This latter has been used to identify discrete 'economic zones' within the territory of the *Demetae*, primarily based on perceived variations in the dominant settlement morphology. Thus the inland and upland areas are said to be characterised by a hierarchy of sites, including hillforts; the inland south-west, at lower altitudes, saw a proliferation of small defended enclosures of the type well known from Walesland Rath; and in coastal locations there were a number of larger and more elaborately defended sites suggestive of greater wealth. From a Dolaucothi perspective the upper reaches of the Cothi valley ought to lie in the first of these zones, paralleling the concentrations of defended enclosures and hillforts known to extend along the flanks of its counterparts, the Teifi and Tywi, to the north and south respectively; something of the potential complexity of these different sites has been recognised from recent work in the Teifi valley, east of Lampeter, at both Cellan (Williams 2002) and Caer Cadwgan (Austin *et al.* 1984–89), though much still remains

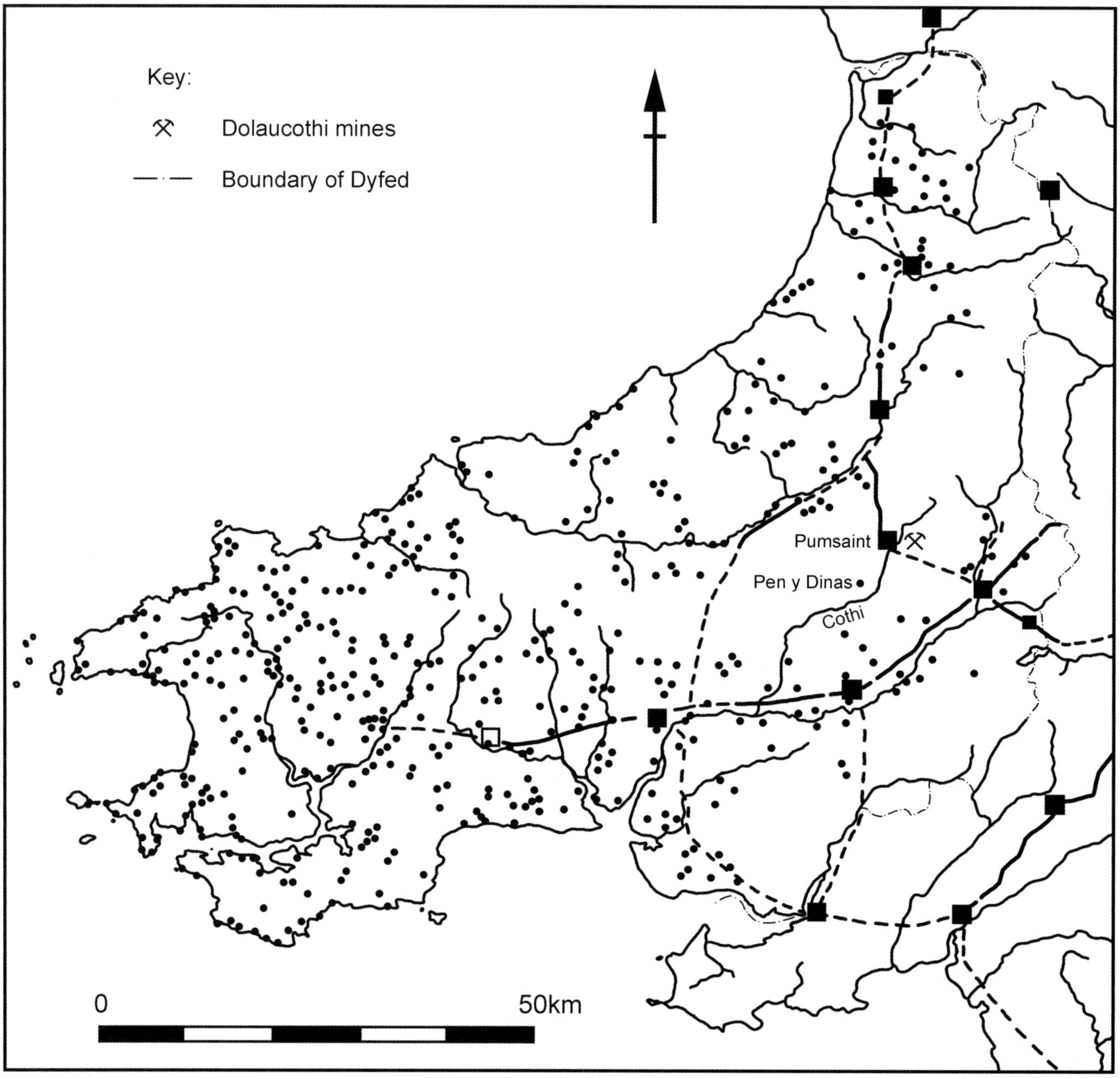

Fig. 6.2 Iron-age defended enclosures in south-west Wales (adapted from Williams and Mytum 1998, fig. 2, with Roman forts and roads superimposed)

uncertain. So few of the sites have been excavated, however, let alone dated, that it is difficult to be at all confident about the precise distribution of settlement at any given point during the later prehistoric or Roman periods.

As interesting as this regional framework is, however, any extension of the analysis to the Dolaucothi area is complicated by our relative ignorance of the settlement tradition in the Cothi valley and by the absence of any systematic programme of fieldwork and aerial reconnaissance. Indeed, a close examination of both the regional SMR, maintained by Cambria Archaeology, and of CARN, a parallel index maintained by the RCAHMW (www.rcahmw.org.uk/data/carn.html), drew a significant blank for the upper reaches of the valley, beyond the obvious sites associated with the fort, the mines and the attendant leats. The single exception is the site of Pen y Dinas, *c.* 5 km downstream from the mines, at SN 629 358 (PRN 1872), where the name is highly suggestive of a former hillfort and where Fenton noted 'some sort of ancient encampment'; sadly, little else is known about the site, and nothing now survives on the ground as most of the hilltop was quarried away during the later 19th and early 20th centuries (RCAHMW 1917, 190-1). It clearly occupied a dominant location, however, and as such it could well have controlled this part of the Cothi valley, though any connection with the putative early exploitation of the mines would be pure speculation. Beyond this the only other potential area of prehistoric settlement would seem to lie on the southern slopes of Allt Ogofau, where several circular huts have been claimed as evidence for pre-Roman activity (Jones 1969).

Such gaps in our knowledge of the cultural landscape make it all but impossible to address any of the key questions about native settlement in the Cothi valley, let alone the interrelationship between incomer and native in the immediate vicinity of the mines. The absence of settlement evidence in the valley also makes it difficult to provide any sort of framework in support of the putative case for pre-Roman mining advanced by Cauuet *et al.* (see section 6.2.4 below). That more remains to be found, however, is emphasised by the recent discovery of a previously unrecorded defended enclosure, close to the summit of Allt Aber-Mangoed (SN 6774 4345), during aerial reconnaissance in December 2002 (Driver 2003). It is reported as being polygonal, *c.* 65m east-west by 80m north-south, with a well-defined single rampart on all but its northern side, beyond which lay a ditch and substantial counterscarp bank (Fig. 6.3). The main gate lies midway along the west side, with terminals which appear to overlap slightly for presumed defensive reasons. No definite internal features are visible, but several hollows in the hillside along the southern side of the enclosed area may represent house platforms. A contemporary field boundary or outer enclosure may

Fig. 6.3 Aerial view of Allt Aber-Mangoed defended enclosure, looking west (AP 2002/5091–58). Crown Copyright: RCAHMW

be represented by a low field bank, connected to the north-west angle of the enclosure.

We are only slightly better off when we come to consider the wider environmental context. Here too there has been no systematic programme of research in the immediate vicinity, so we are left to rely on the palaeo-environmental evidence derived fortuitously from the excavations themselves, both from the fort rampart (see section 2.2.4(R) above) and from the sections across the Gwenlais leat (see 3.3.3 above). These can be cross-matched, however, with several longer sequences from further afield, most notably at Tregaron (Turner 1964; 1965), Cefn Gwernffrwd (Chambers 1982a) and Talley (Butler 1984), though the details necessarily remain generalised.

The material derived from the fort rampart in 1989 has proved to be especially important to our understanding of the local environment. In her discussion of the sequence Astrid Caseldine draws attention to the early dominance of hazel scrub, with some alder and oak woodland, perhaps indicating a landscape with alder growing on the wetter flood-plains of the Cothi and Twrch and oak on the valley sides. While there is also evidence for hazel clearance, the absence of cereal pollen points to pastoral rather than arable farming. Further clearance followed, with the first evidence of cereal type pollen being taken to indicate a phase of arable cultivation on the terrace prior to the arrival of the military. In a final phase, post-dating the construction of the turf rampart, the continuing high levels of Poaceae accompanied by an absence of cereal type pollen suggest a reversion to grassland in the immediate vicinity. Overall Caseldine concludes by noting that the evidence for hazel clearance is not inconsistent with that from other sites in mid and south Wales. It would be dangerous, however, to press the material too far, since the sequence provides only a limited window on the environment during the actual occupation of the fort, with little or nothing on what followed its reduction in size.

A further window has been provided by the material from the Gwenlais leat in 1990, though in this case it relates to the upland beyond the immediate confines of the mines. According to Walker's analysis the data from the buried soil beneath the upcast bank suggests an open heathland landscape, interspersed with grassland and elements of wood and scrub, the latter probably confined to the more sheltered valley slopes bordering the uplands. By contrast the lowermost assemblage from the leat profiles indicates a somewhat different landscape in the period after it went out of use, characterised by a reduction in heathland and more extensive grassland, interspersed with an increase in woody taxa suggesting an element of wood and scrub regeneration. Walker concludes that this might date to the early medieval period, but there is at least a possibility that it is a feature of the Roman era.

As informative as such sequences are, they still offer only very localised insights into the environmental context of the site, to the extent that we remain surprisingly ignorant about how the development of the complex affected the landscape in the immediate vicinity of the mines. A key requirement is the identification of suitable sites for the recovery of extended pollen sequences. This is especially critical for several reasons: first for our understanding of changing agricultural regimes, perhaps in response to an increased imperial interest in this part of the valley; second for the question of local woodland management, given the widespread demand for timber for construction purposes, for use in the bath-house(s), and for activities related to underground mining and the reduction of sulphidic ores; and third for the insights it might offer on the problem of any pre-Roman exploitation of the ore bodies. Techniques other than pollen analysis will also be needed if we are to explore the hydrological consequences for the Cothi, both locally and downstream from the mines, of any extensive exploitation of alluvial deposits, let alone the use of water in ore-processing.

That the impact on the immediate locale was extensive can hardly be in doubt, however, given the fact that the fort and its satellite settlements represent a major island of intrusive population. While it is impossible to establish precise figures, the fort alone would have accounted for up to 500 soldiers, with at least as many people again in the associated settlements if we assume a complement of women and children, and a range of resident specialists and traders; this figure would have risen even further if the settlement south of the river also housed any personnel or an element of the workforce associated with the mines. Moving beyond such 'estimates' to hard facts, however, is necessarily speculative, but the suggestion that a typical 4-acre fort might have required up to 30 acres of forest provides an order of magnitude, at least for the scale of local timber demands (Hanson 1978, 298), let alone the question of other raw materials, provisions and wider supply lines (on which see Casey *et al.* 1993, 75–7, with reference to Caernarfon). Some general points can be established from the Dolaucothi evidence, however, even if much remains uncertain.

In terms of construction most of the necessary raw materials would probably have been readily available in the immediate locale. Turf for the rampart was certainly stripped from the site, as Caseldine's analysis of the pollen has shown; clay, both for the rampart and for flooring and walls, is known to be locally available on the flood-plain of the Twrch; while gravel for the main approach road and the fort's internal streets could easily be quarried from the terraces bordering the river. When the stone wall was added, the foundations made use of water-worn cobbles from the river bed, while the facing and the core were made up of shale and quartz, the presence of which must surely indicate that the mines were in operation and that waste material, including white (gold-free) quartz, was easily accessible. Timber would also have been required in large quantities, both within the settlement areas and within the mines, though little in the evidence adds to the ongoing debate about the organisation of timber supply (*cf.* Hanson 1978; Hobley 1982). More than likely, however, local resources would have been exploited to the full, hence the importance of Caseldine's demonstration that oak was readily available on the valley sides during the early life of the fort; equally important is the exclusive use of oak in the late timber building, which must surely indicate that such resources continued to be available.

Sadly, little can be said about the provisioning of the fort from the excavated evidence, though cattle bones were certainly represented in the burnt debris at the bottom of the late cellar. Bloomery iron-smithing in a military context would seem to be indicated by Peter Crew's identification of several distinctive slags and various fragments of furnace or hearth lining (see sections 2.2.4(I) and (K) above), though none of this material was actually *in situ* and nothing is known about the source of the raw materials. While insufficient pottery has been recovered to facilitate a statistical analysis of the different sources of supply over time (*cf.* Webster 1991), a general assessment of the qualitative range of all the finds is enough to suggest that the Dolaucothi-Pumsaint complex drew on much the same networks as other, better-known sites; indeed, for the 1989 material, Peter Webster even suggests that the site may have been drawing on a wider range of sources than usual, perhaps reflecting a higher than normal throughput of personnel. One final area of interest concerns how the construction of the different leat systems and the appropriation of water resources might have affected local perceptions of landscape and traditional patterns of movement, but in the current vacuum any answers would be purely speculative.

This is perhaps as far as we can legitimately go for the present in our assessment of the site's wider cultural and environmental context. It is worth concluding, however, with the thought that current research, like that which preceded it, continues to privilege the intrusive 'Roman' element in the landscape and, thereby, its dominant place in the resulting narratives of the valley, at the obvious expense of the native framework. It is to be hoped that future research strategies will go some way to redressing the balance.

6.2.4 Exploitation and Processing: the mining landscape

Despite some significant advances in our understanding of the wider Roman presence in the landscape, the 1987–99 work has had a somewhat lesser impact on our appreciation of the mining landscape and related aspects of exploitation and processing. This reflects the fact that the work on the associated leats and the processing areas has been largely peripheral to the mines *per se*, though each has added detail to the wider picture. Much more significant has been the work of a French team under Béatrice Cauuet, which has done much to question the Lewis and Jones model (Ancel *et al.* 2000). Before discussing its wider implications, it is worth assessing the 1987–99 work first.

(A) The leat systems

Work on the different leat systems proposed by Lewis and Jones has been insufficiently extensive to add much to the debate over their extent and function. The Cothi has been incidentally re-examined by one of the authors (HBB) in the course of re-scheduling for Cadw: Welsh Historic Monuments, though without significantly affecting Jones' original conclusions (Jones *et al.* 1960). Similar work in relation to the Nant Dâr, together with a review of some of the available air photographic coverage, has helped to clarify aspects of its line and suggests strongly that it ought to be traceable through the forestry as far as the rim of the Allt Cwmhenog Pit (see section 3.6 above). This might help to explain some of the complexities of the earthworks around its upper edge, perhaps indicating that both the Nant Dâr and the water supply associated with Tank G (the so-called Annell) were involved in the initial exploitation of the pit using hushing as an exploratory technique.

The nature of the so-called Annell leat and its relation to Bick's 'Gwenlais leat' (1989) still remain to be resolved. The 1993 work does appear to indicate a possible water source feeding into Tank G, however ephemeral its remains, while geophysical work by Cardiff's (Mining) Field Centre certainly seems to have traced a possible line around the spur of the hill onto the Annell side (Alwyn Annels, *pers. comm.*). The relationship of these elements to the triple system of leats at Gwarnoethle, and the explanation of what happens upstream along the flanks of Allt Cwmhenog and Allt Cwmgerwyn, is now an urgent priority, not least if we are to understand how (if at all) the supply systems link up with the upper Annell/Gwenlais system. The realisation that Jones' Upper Annell and Bick's Gwenlais leats are one and the same represents one of the most significant advances of recent years, though we remain ignorant of its ultimate destination and function. The publication of the two radiocarbon dates from the upper Annell (Jones and Maude 1991), with their implied Roman date for the associated system, can only help to emphasise the urgency of further work.

Little of what has been done either by Bick or by the 1987–99 work has helped to clarify the function of the various leat systems, though the identification of the possible water-driven mill, if proven, would clearly emphasise the importance of water-driven technologies and the use of water in processing. The processing wastes on the 1991–93 site point in a similar direction, though further work in the vicinity of the flood-plain would seem desirable. The use of the water in hushing still seems assured by the remains associated with the Allt Cwmhenog Pit, but elsewhere the jury must remain out, in part because of Bird's recent suggestion that large-scale ground-sluicing would seem to be a much more effective method of exploiting weathered auriferous deposits (echoing Hall 1993, xii-xiii), and in part because of long-standing doubts about the efficacy of hushing in hard-rock contexts (Boon and Williams 1966, 122, note 6; Manning 1968, 301), now reinforced by Cauuet and her colleagues (see section 6.2.4 below).

(B) The processing aspects

The recovery of a sequence of *in-situ*, interleaved processing wastes from the 1991–93 excavations near the Carreg Pumsaint represents an important addition to our knowledge of the Dolaucothi complex, not least because previous discoveries had suggested their likely presence. Equally significant is the radiocarbon date of 1890±60BP from a sample recovered from the lowermost horizon in the central hollow, sealed directly beneath the wastes, which must surely indicate a Roman date for the subsequent build-up. Though a single date needs to be used with considerable caution, the absence of evidence to the contrary, and the ready availability of timber supplies in the vicinity in the Roman era, add confidence to the conclusion. Irrespective of the discussion about the possible mill site in the vicinity, this is the first such sequence of processing wastes to be excavated and examined in detail at Dolaucothi, which makes the results of Alwyn Annels' analysis all the more significant (see sections 4.6 and 4.7 above). Interest necessarily attaches to three particular aspects: (i) the increasing proportion of pyritised shale within the samples and, thus, within the nature of the ore being mined, vis-à-vis ores rich in quartz; (ii) the presence in several of the clay samples of slag or clinker (partially-fused siliceous material) with charcoal; and (iii) the poor recovery rates achieved by the Roman miners. While it is difficult to interpret these trends in detail, there is some suggestion that, over time, the Romans were attempting to exploit both quartz and pyritic shale, and that as the latter progressively increased, they had to employ roasting techniques perhaps to unlock the gold from the sulphides. This is much in line with what has been confirmed by recent episodes of exploitation, where the percentage of sulphide-locked gold increases with depth, a trend which would equate well with the postulated sequence of ancient mining.

The presence of *in-situ* wastes, together with an associated fragment of quernstone (itself one of several now known from the immediate vicinity), also raises interesting questions about the spatial aspects of processing. The probability that several different activities were going on in the immediate vicinity can now be postulated. There seems good reason to believe that 'dead material' was being dumped not just inside the mines, but also on the nearby conical spoil heap ('motte'), close to where ore-bearing material was being crushed, washed and possibly roasted, perhaps with the aid of mechanically-driven technologies. Material at various stages of processing, including possible roasting, has been recovered, while the presence of material in the outwash area of the flood-plain indicates just how extensive the whole operation must have been. Though the excavations were limited, there is enough evidence to indicate an organised system, managing and limiting the processing aspects.

The case for a water-driven mill in the immediate vicinity, associated with a mortar stone represented by the Carreg Pumsaint (Burnham 1997; see section 4.8.2 above), remains controversial and unproven, despite recent advances in our understanding of the use of water in the Classical era (Lewis 1997; Sim and Ridge 2002, 132–36). This is unsurprising for a number of reasons: (i) the prevailing theories on the invention of cam-operated or trip-hammer devices, which generally do not favour a Hellenistic-Roman origin; (ii) the absence, so far, of clearly-defined parallels in the rest of the Roman world, not least in mining contexts, and the absence of any documentary record in key sources like Pliny who seems otherwise to have been well-versed in the Spanish operations; and (iii) the absence of any actual excavation or sampling of the mill site itself. There also remains the possibility that the Carreg Pumsaint has been moved from a site further away – perhaps from higher up the hillside in the Ogofau Pit – which would remove the argument of association while still leaving open the possibility of later mining elsewhere. It is enough to conclude that the argument for a Roman date is based on a series of interlinked probabilities, any of which could unravel and seriously jeopardise the conclusions drawn here; if accepted, however, the result would be a significant addition to our understanding not just of the site itself but of Roman mining in general, not least because of its association with other well-preserved aspects of processing in the immediate vicinity. Even allowing for a more pessimistic view, the wastes in themselves represent a significant addition to our knowledge of Dolaucothi. Any opportunity to assess further sequences should be sought in the future.

Even more significant, perhaps, but somewhat incidental to the original aims of the 1991–93 research, was the recovery of a pre-existing landscape beneath the processing wastes, which everywhere overlay extensive dumps of material, including the deposits at the base of the so-called motte, all of which must have been in place before the Roman processing activity took place. It is even arguable that the motte spoil heap itself was all pre-processing, though the case is not absolute. In the past, the case for any early, pre-Roman (?) mining has depended upon scant evidence, not least the postulated huts and an associated quartz exposure on the south side of Allt Ogofau (Jones 1969). The recovery of charcoal from a well-stratified early context in the 1991–93 sequence, with a radiocarbon date of 2630±70BP, well back in the 1st millennium BC in calibrated terms, and sufficiently discrete from the other date beneath the wastes, has obvious added significance, therefore, by raising the tantalising possibility of extensive pre-Roman activity in this area close to the main focus of the later workings on Allt Ogofau. In view of recent work over the last few decades at both Cwmystwyth (Timberlake 2003) and Great Orme (Dutton and Fasham 1994), extensive prehistoric working should come as no surprise, but the possibilities at Dolaucothi assume an even greater significance now in view of the recent suggestions emerging from the work of Cauuet and her French counterparts to which we will return (Ancel *et al.* 2000). The fact that such landscapes have not been recognised before must caution us against overly simplistic interpretations of the visible surface remains.

(C) Models of the mining landscape

In the absence of explicit work on the mine workings *per se*, the 1987–99 surveys can do little more than provide an enhanced contextual framework for the complex. At first glance, none of what has been said so far about the associated fort and settlement areas necessarily contradicts the idea that there was a major phase of Roman activity at the site, initially under the military from *c.* AD 75 to 110+, succeeded by a continuing (if declining?) civilian presence south of the river down to the later 2nd century at least; after that activity is less easy to pin down precisely in terms of settlement foci, despite the importance of the Melin-y-Milwyr material. Nor does it preclude elements of the model of exploitation postulated by Lewis and Jones, involving the rapid exploitation of new provincial resources, with a subsequent decline in profitability, related in part to the increasingly sulphidic nature of the ore body and in part to the readily available, and more attractive, new resources opened up by Trajan's conquest of Dacia, hence the possible shift from military to civilian oversight. The presence of the processing wastes in the vicinity of the Ogofau Pit adds weight to the notion of a technologically developed mine, while the importance of water remains well attested by the known remains of leat systems and associated tanks.

This much has been re-emphasised by David Bird, in a recent reassessment of Dolaucothi (2001), written as a tribute to the life and work of the late Professor Barri Jones; in this he seeks to restate the argument that 'the visible remains are largely the result of Roman work' and that 'the aqueducts were intended for use in hydraulic forms of mining'. This is not the place to repeat his arguments in detail, merely the context to note some of the key points: (i) the importance of water supply systems in the gold-mining area of the Asturias, and the use of water in a continuous process of ground sluicing to exploit the oxidised zones of what were massive sulphide ore-bodies; (ii) the significant similarities between the Asturian mines and Dolaucothi, not least in terms of high level aqueducts and water tanks, which must have been used from the outset to attack the ore-bodies through extensive ground sluicing; (iii) the subsequent importance of water power for crushing and washing as the workings increased in depth and progressed underground, with the known fragment of waterwheel representing the latest phase of such operations. Such conclusions necessarily extend and modify some of the original Lewis and Jones ideas, not least with reference to hushing, but essentially the developmental framework remains much the same. If the 1987–99 work does anything at all to this model, it is to suggest that activity may have been intensive and short-lived, with the emphasis on a 60-year period at best from the 70s AD to the 140s, after which activity may have declined sharply.

There the matter might have rested but for the intervention of a French team led by Béatrice Cauuet in 2000, in the wake of Dolaucothi's inclusion in the EU MINET network. Over a two-week period, at Easter, work was carried out both on the surface and on the underground workings. Their conclusions have so far been published only in interim form (Ancel *et al.* 2000), but even so they present a significant challenge to our understanding of the mining landscape as envisaged both by Lewis and Jones and by Bird. Cauuet's primary contention is that all previous attempts to explain the surface morphology of the site have been based on a misunderstanding of the two types of gold-bearing sources at Dolaucothi and of the different techniques of exploitation necessary. She identifies these as: (i) the primary deposit of hard rock which was effectively exploited by tool-working or fire-setting, and (ii) the secondary deposits of alluvial nature which were worked by hydraulic systems. She further stresses that the hydraulic systems, long known from the site, belong not with the primary hard-rock deposits, but rather with the secondary alluvial deposits, often at the periphery of the hard-rock workings. Drawing on her experience with the hard-rock mines of the Limousin, dated to the 5th to 1st centuries BC, Cauuet (1999) contends that a large part of the workings at Dolaucothi could have been extensively exploited in the pre-Roman period, with the Romans arriving in the 70s AD to inherit a site already extensively worked by opencast methods (if not also by underground technology), which they subsequently sought to rework, in part by extending further underground (as the waterwheel attests) and in part by the exploitation of the secondary deposits involving the use of large-scale hydraulic systems. None of this is contradicted by the 1987–99 work on the associated settlement and processing areas, and none of it would be impossible given recent advances in our understanding of prehistoric mining elsewhere in Wales, though at present there is very little in the way of a later prehistoric context in the Cothi valley within which to fit such large-scale activity, a point emphasised in earlier discussions (see section 6.2.3(D) above).

Such opposed models of the mining landscape at Dolaucothi – one emphasising the primary exploitation of the site with large-scale hydraulic techniques (drawn from experience in the Asturias), followed by opencast and underground mining using traditional techniques, all exclusively within the Roman period, the other emphasising the primary exploitation of hard-rock deposits in later prehistory, followed by a large-scale hydraulic episode under the Romans – are incapable of resolution on present evidence. What they emphasise, however, is the urgent need for a detailed investigation both of the opencast and underground workings, and of the alluvial deposits and outwash areas on the flood-plain of the Cothi, for herein must lie potential answers to these unresolved questions. They also emphasise the extent to which caution is ever necessary in privileging the settlement narrative from the fort at Pumsaint and the settlement area south of the river as a proxy for the interpretation and dating of the mining landscape and the hydraulic systems; the former may inform, but can never be taken as absolute. For the present it is enough to conclude that Dolaucothi still continues to spring some profound surprises despite the scope and intensity of effort which has gone into its investigation for more than 30 years. Nor should we forget that the potential exists for later phases of mining which have otherwise not figured in this debate.

6.2.5 Future research directions

As with so much archaeological research, this volume is a staging post in a journey with more questions than answers, not least because of the multi-disciplinary nature of the problem. More than ever before, new avenues of research are now essential, not just to adjudicate between the competing models of the mining landscape, but also to link that landscape with the surrounding contextual, settlement landscape, both prehistoric, Roman and later. The potential for the emergence of new narratives should be obvious both in terms of pure research and in terms of interpreting and presenting the site to the local community and the wider public at large. The following areas emerge as significant for future research, where possible with inter-disciplinary teams combining different areas of expertise:

(i) Concerted mapping of the surface and underground geology, building on the considerable body of evidence collated both by Cardiff University's (Mining) Field Centre and by the French team. This will need to be extended to mapping the extent of the alluvial deposits, since a clearer understanding of both hard-rock and alluvial contexts will be essential to elucidating aspects of the mining landscape over time.

(ii) Exploration of the mine workings themselves, building on the readily available surface surveys produced for the National Trust (Landmark Surveys and Cambria Archaeology 1999, supplemented by Cambria Archaeology 2002). Surface and underground survey in combination should seek to identify plausible locations for excavation, critical to the recovery of (a) diagnostic finds, (b) datable timbers, and (c) charcoal samples for radiocarbon dating. Such methods have been employed effectively at other prehistoric sites in Wales, and by the French in Limousin (Cauuet 1999). Potential areas already identified include the vicinity of the Carreg Pumsaint (in view of the 1991–93 excavations), Mitchell Pit (in part to clarify Jones' unpublished section there), Roman Pit (where a link to underground workings in Middle Adit has been postulated), and the Pen-lan-wen trenches. Without such work it will be difficult to establish the claims of Cauuet *et al.* with respect to the extent of prehistoric mining and the primacy of non-hydraulic opencast working.

(iii) Close examination of the inter-relationship of the different opencasts and the various hydraulic systems, not least the Cothi and its related tanks. This is particularly critical to establishing the primacy of hydraulic mining in the early development of Dolaucothi, in accordance with the Lewis/Jones and Bird model, not least because it has always excited debate over the extent to which the leat systems were actually grafted onto existing (recent) workings at a very high level (G. Boon, *pers. comm.*, echoing Boon 1979, 165). Particularly critical is the relationship of the Cothi leat to the Allt Cwmhenog pit which both schools interpret in different ways – Bird that it contours around an early Roman opencast, itself served by the so-called Annell, Cauuet that it overlies an opencast of much greater antiquity. The situation is not so clear-cut on the ground, however, where a second leat line immediately beneath the Cothi appears to have been cut by the opencast, hence the need for a careful reappraisal. Other presumed tanks around the edges of the opencasts continue to arouse suspicion and either need tying in to the framework or testing by excavation.

(iv) Clarification of the water supply systems and their publication to the standards of the Cothi leat (Jones *et al.* 1960). The Nant Dâr provides the easiest target and has particular relevance in view of its potential link with the Allt Cwmhenog pit. By contrast, the so-called Annell leat provides a major challenge, not just in terms of tracing and tying together the disparate elements along Allt Cwmhenog and Allt Cwmgerwyn, but also in terms of its relationship (or otherwise) to the Upper Annell/Gwenlais sectors. Close examination of the available aerial photographic cover, coupled with the cost-effective use of geophysical survey and small-scale trenching would seem to be the obvious way forward. Nor should the probability of hitherto unrecognised systems be ignored, in view of Bick's recognition of the Gwenlais as late as 1989.

(v) Exploration of the outwash areas below the mine workings, together with any alluvial deposits extending onto the flood-plain of the river Cothi. This is particularly critical in view of the importance which both schools attach to alluvial exploitation and in view of the recognition at various times that this area includes evidence of 'sedimented beds of ground material from large-scale washing activities' (Lewis and Jones 1969, 263). The 1991–93 work points in much the same direction. Every chance should be taken to record pipeline trenches across the area, together with a concerted series of excavated sections to map the stratigraphy and take appropriate samples for gold analysis. Inter-disciplinary work might also profitably explore the hydrological consequences of such alluvial washing and the use of extensive quantities of water for the river Cothi both here and downstream.

(vi) Continued exploration of the settlement area south of the river to clarify its extent and function. This might involve more extensive geophysical survey and fieldwalking in the fields to the south and south-west, along the line of the approach road from Llandovery, close examination of the exposed section of an old river meander, and test-pitting to recover a range of dating evidence with which to chart the chronology and extent of the site. More extensive excavation of key geophysical anomalies might also help to identify other buildings or features associated with mining and processing. The potential for recovering burials should also be borne in mind.

(vii) Further exploration of the settlement area or *vicus* to the east of the fort, not least because doubts remain about the precise location of the eastern defences and about the character and function of the features and buildings identified by the geophysical survey. The potential for extending the geophysics is limited, but a combination of test pitting to recover dating evidence and of carefully located trenches to examine potential defence lines would seem to be desirable. The chance to investigate a probable bath-house is also too good to miss!

(viii) Systematic archaeological and palaeo-environmental survey of the wider landscape context of the site and the mining complex in order to provide a contextual framework for the later prehistoric and Roman period. Deficiencies in this area have already been noted and need to be remedied, if we are to engage with wider narratives of interaction between natives and incomers, agricultural change, environmental impacts, and the intriguing question of how a postulated early medieval 'multiple estate' focused on Caeo might potentially have come into existence (Jones, G. R. 1972, 308–20, fig. 44). Systematic aerial reconnaissance and fieldwork is essential.

Such research themes necessarily focus on the later prehistoric and Roman contexts, as befits a volume which is almost exclusively concerned with the Roman presence, but the potential for addressing other narratives must not be forgotten. Mention might here be made of the ploughing and field systems noted in the 1997 and 1999 geophysical work, the evidence for post-Roman activity located in the village in 1989, and the wider question of the origins of the village of Pumsaint itself, all of which deserve attention; so too does the evidence of the 17th-century pottery in Melin-y-Milwyr, and the evidence for several pillow mounds scattered around the workings (exclusively) on Allt Ogofau (Gerrard 1988).

It is to be hoped that, at a time when the National Trust is actively working on a Conservation Strategy and a Research Design to take the Dolaucothi Estate forward, some at least of the objectives outlined in the preceding section will be progressed. There is a window of opportunity now to locate the site firmly within the wider European context where it so rightly belongs.

BIBLIOGRAPHY

Agricola, G. 1556 *De Re Metallica* (1950 English Edition, translated by H. C. and L. H. Hoover), New York

Allan, J. P. 1984 *Medieval and Post-Medieval Finds from Exeter 1971–1980*, Exeter Archaeological Reports 3, Exeter

Allen, B. D. 1996 *Foul Sewer Replacement at Pumpsaint, Dyfed*, Dyfed Archaeological Trust Report, PRN 32472

Ancel, B., Cauuet, B. and Cowburn, I. 2000 *The Dolaucothi Gold Mines: Intermediate Report, August 2000*, National Trust, unpublished report

Andersen, S. T. 1986 Palaeoecological studies of terrestrial soils, in Berglund, B. E. (ed.) *Handbook of Holocene Palaeoecology and Palaeohydrology*, Chichester & New York, 165–177

Anderson, A. C. 1981 Some continental beakers of the first and second centuries AD, in Anderson, A. C. and Anderson, A. S. (eds) *Roman Pottery Research in Britain and North-West Europe*, BAR International Series 123, Oxford

Annels, A. E. and Burnham, B. C. 1995 *The Dolaucothi Gold Mines: Geology and Mining History* (3rd edition), Cardiff

Arnold, C. J. and Davies, J. L. 2000 *Roman and Early Medieval Wales*, Stroud

Atkinson, D. R. 1975 *Tobacco Pipes of Broseley, Shropshire*, Saffron Walden

Atterbury, P. (ed.) 1980 *English Pottery and Porcelain*, London

Austin, D., Bell, M., Burnham, B. C. and Young, R. 1984–89 *The Caer Cadwgan Project* (Interim Reports), Lampeter

Austin, D. and Burnham, B. C. 1984 A new milling and processing complex at Dolaucothi: some recent fieldwork results, *Bulletin of the Board of Celtic Studies* 31, 304–313

Austin, D. and Burnham, B. C. 1985 Dolaucothi 1984, *Archaeology in Wales* 24 (1984), 81–82

Austin, D., Gerrard, G. A. M. and Greeves, T. A. P. 1989 Tin and agriculture in the middle ages and beyond: landscape archaeology in St Neot Parish, Cornwall, *Cornish Archaeology* 28, 5–251

Ball, D. F. 1964 Loss-on-ignition as an estimate of organic matter and organic carbon in non-calcareous soils, *Journal of Soil Science* 15, 84–92

Banks, Sir J. 1767 Journal of a Tour in Carmarthenshire, National Library of Wales (352 Add. MSS. 147C)

Barton, R. N. E. 1992 *Hengistbury Head Vol. II. The Late Upper Palaeolithic and Early Mesolithic*, Oxford

Bell, M. and Walker, M. J. C. 1992 *Late Quaternary Environmental Change: Physical and Human Perspectives*, London

Bénard, J. and Mangin, M. 1994 Alise-Sainte-Reine, Alésia, in Bénard *et al.* 1994, 29–60

Bénard, J., Mangin, M, Goguey, R. and Roussel, L. (eds) 1994 *Les Agglomérations Antiques de Côte-d'Or*, Annales Litteraires de l'Université de Besançon, 522, Série archéologique No. 39

Bennet, J. 1986 Fort sizes as a guide to garrison types: a preliminary study of selected forts in the European provinces, in Planck, D. (ed.) *Studien zu den Militärgrenzen Roms III: 13 Internationalen Limeskongress (Aalen, 1983)*

Bennett, K. D. 1994 *Annotated Catalogue of Pollen and Pteridophyte Spore Types of the British Isles*, Cambridge

Bennett, K. D., Whittington, G. and Edwards, K. J. 1994 Recent plant nomenclatural changes and pollen morphology in the British Isles, *Quaternary Newsletter* 73, 1–6

Bick, D. 1989 An ancient leat near Dolaucothi gold mine, *Archaeology in Wales* 28 (1988), 20–21

Bick, D. 1991 Further evidence of hushing in mid Wales, *Archaeology in Wales* 30 (1990), 33–34

Bick, D. 1996 Dolaucothi, Gwenlais leat (SN 7055 4255), *Archaeology in Wales* 35 (1995), 50–52

Bick, D. and Boon, G. 1993 Dolaucothi again, *Britannia* 24, 247–249

Bidwell, P. (ed.) 1999 *Hadrian's Wall 1989–1999: a Summary of Recent Excavations and Research*, Kendal

Biggins, J. A., Robinson, J. and Taylor, D. J. A. 1999 Birdoswald: Geophysical survey, in Bidwell 1999, 157–160

Bird, D. G. 1972 The Roman gold mines of north-west Spain, *Bonner Jahrbücher* 172, 36–64

Bird, D. G. 1984 Pliny and the gold-mines of north-west Iberia, in Blagg, T. F. C., Jones, R. F. J. and Keay, S. J. (eds) *Papers in Iberian Archaeology*, BAR International Series 193, Oxford, 341–363

Bird, D. G. 2001 Aspects of Roman gold-mining: Dolaucothi, Asturias and Pliny, in Higham, N. J. (ed.) *Archaeology of the Roman Empire: a Tribute to the Life and Works of Professor Barri Jones*, BAR International Series 940, Oxford, 265–275

Birley, A. 2002 *Garrison Life at Vindolanda: a Band of Brothers*, Stroud

Black, E. W. 1995 *Cursus Publicus: The Infrastructure of Government in Roman Britain*, BAR British Series 241, Oxford

Blanco, A., Luzón, J. M. and Ruiz, D. 1970 *Excavaciones Arqueológicas en el Cerro Salomon*, Universidad de Sevilla, Seville

Boon, G. C. 1966 The Erw-Hên treasure trove of Roman antoniniani, *Numismatic Chronicle* 7th series, 6, 157–163

Boon, G. C. 1968 The Glass, in Hogg, A. H. A. Pen Llystyn,

a Roman fort and other remains, *Archaeological Journal* 125, 178–181
Boon, G. C. 1971 Aperçu sur la production des métaux non ferreux dans la Bretagne Romaine, *Apulum* 9, 453–503
Boon, G. C. 1979 Review of Roman Britain 1:625,000 Fourth Edition OS Map, *Archaeologia Cambrensis* 128 (1979), 165–166
Boon, G. C. 1981 *Cardiganshire Silver and the Aberystwyth Mint in Peace and War*, Cardiff
Boon, G. C. 1982 The Coins, in Boon, G. C. and Hassall, M. W. C. *Report on the Excavations at Usk, 1965–1976: The Coins, Inscriptions and Graffiti*, Cardiff
Boon, G. C. and Williams, C. 1966 The Dolaucothi drainage wheel, *Journal of Roman Studies* 56, 122–127
Bosanquet, R. C. 1917 [Remains at Dolaucothi], in RCAHMW 1917, 25–37
Brunskill, R. W. 1982 *Traditional Farm Buildings of Britain*, London
Burnham, B. C. 1989 A Survey of Building Types in Romano-British Small Towns, *Journal of the British Archaeological Association* 141 (1988), 35–59
Burnham, B. C. 1995 Dolaucothi revisited, in Ford, T. D. and Willies, L. (eds) *Mining Before Powder*, Peak District Mines Historical Society Bulletin 12 no. 3, 41–47
Burnham, B. C. 1997 Roman mining at Dolaucothi: the implications of the 1991–93 excavations near the Carreg Pumsaint, *Britannia* 28, 325–336, pls XXIX and XXX
Burnham, B. C. and Burnham, H. B. 1987a Recent Survey on the Fort and Vicus at Pumsaint, *Carmarthenshire Antiquary* 22 (1986), 3–13
Burnham, B. C. and Burnham, H. B. 1987b *Excavations at Pumsaint 1987*, Lampeter, 14pp
Burnham, B. C. and Burnham, H. B. 1988 Clearance Work at Dolaucothi in 1986, *Carmarthenshire Antiquary* 23 (1987), 57–62
Burnham, B. C. and Burnham, H. B. 1989 *Excavations at Pumsaint, 1989, Interim Report*, Lampeter, 16pp
Burnham, B. C. and Burnham, H. B. 1991a Pumsaint: the excavations of 1989, in Burnham and Davies 1991, 49–56
Burnham, B. C. and Burnham, H. B. 1991b A burnt timber building from within the fort at Pumsaint, *Britannia* 22, 203–207
Burnham, B. C. and Burnham, H. B. 1991c *Excavations at Dolaucothi 1991; Interim Report*, Lampeter, 10pp
Burnham, B. C. and Burnham, H. B. 1992 *Excavations at Dolaucothi 1992; Interim Report*, Lampeter, 12pp
Burnham, B. C. and Burnham, H. B. 1993 *Excavations at Dolaucothi 1993; Interim Report*, Lampeter, 16pp
Burnham, B. C. and Burnham, H. B. 1994 Survey and excavation on the line of the Annell leat at Dolaucothi, *Archaeology in Wales* 33 (1993), 16–19
Burnham, B. C. and H. B. and Walker, M. J. C. 1993 Excavations across the Annell and Gwenlais leats, near Dolaucothi, in 1990, *Archaeology in Wales* 32 (1992), 2–8
Burnham, B. C. and Davies, J. L. (eds) 1991 *Conquest, Co-existence and Change: Recent Work in Roman Wales*, Trivium 25 (1990), Lampeter
Burnham, B. C. and Wacher, J. S. 1990 *The "Small Towns" of Roman Britain*, London
Burnham, B. C., Keppie, L. J. F. and Fitzpatrick, A. P. 2001 Roman Britain in 2000, *Britannia* 32, 312–385
Butler, S. 1984 Preliminary Investigation of the Pollen Record from Talley Lakes, *Carmarthenshire Antiquary* 20, 3–14
Cambria Archaeology 2002 *Dolaucothi Gold Mines, Carmarthenshire, 2002 Survey*, unpublished report by Cambria Archaeology for The National Trust, March-April 2002, Report no. 2002/40
Camden, W. 1984 *Camden's Wales: Welsh chapters of the 1722 edition of the Britannia*, Carmarthen
Casal Garcia, R. 1981 La Iconografia de Nemesis en la Gliptica Romana, in Arce, J. *La Religion Romana en Hispania*, (Symposio organizado por el Instituto de Arquelogia 'Rodrigo Caro' del C. S. I. C. del 17 al 19 de diciembre de 1979), 113–120
Caseldine, A. 1990 *Environmental Archaeology in Wales*, Lampeter
Caseldine, A. 1991 Environmental archaeology and the Roman period in Wales: recent work, in Burnham and Davies 1991, 172–184
Caseldine, C. J. 1983 Pollen analysis and rates of pollen incorporation into a radiocarbon-dated palaeopodzolic soil at Haugarbreen, southern Norway, *Boreas* 12, 233–246
Casey, P. J. and Davies, J. L. 1993 *Excavations at Segontium (Caernarfon) Roman Fort, 1975–79*, CBA Research Report 90, London
Cauuet, B. 1999 L'Exploitation de l'or en Gaule à l'Age du Fer, in Cauuet, B. *L'Or dans L'Antiquité, de la Mine à l'Objet*, Aquitania, supplement IX, 31–70
Chambers, F. M. 1982a Environmental history of Cefn Gwernffrwd, near Rhandirmwyn, mid-Wales, *New Phytologist* 92, 607–615
Chambers, F. M. 1982b Two radiocarbon-dated pollen diagrams from high-altitude blanket peats in South Wales, *Journal of Ecology* 70, 445–459
Chambers, F. M. 1983a Three radiocarbon-dated pollen diagrams from upland peats, north-west of Merthyr Tydfil, South Wales, *Journal of Ecology* 71, 475–487
Chambers, F. M. 1983b The palaeoecological setting of Cefn Gwernffrwd – a prehistoric complex in Mid-Wales, *Proceedings of the Prehistoric Society* 49, 303–316
Chambers, F. M., Dresser, P. Q. and Smith, A. G. 1979 Radiocarbon dating evidence on the impact of atmospheric pollution on upland peats, *Nature* 282, 829–831
Charlesworth, D. 1972 The Glass, in Frere 1972, 196–215
Charlesworth, D. 1979 Glass, in Bidwell, P. T. *The Legionary Bath-house, Basilica and Forum at Exeter*, Exeter, 222–231
Clarke, G. 1979 *The Roman Cemetery at Lankhills, Winchester*, Winchester Studies 3 Pre-Roman and Roman Winchester Part II, Oxford
Colt Hoare, R. 1806 intro and trans *The Itinerary of Archbishop Baldwin through Wales AD MCLXXXVIII by Giraldus de Barri*
Cool, H. E. M. and Price, J. 1995 *Roman Vessel Glass from the Excavations in Colchester, 1971–85*, Colchester Archaeological Report 8, Colchester
Crampton, C. B. and Webley, D. P. 1964 Preliminary studies of the historic succession of plants and soils in selected archaeological sites in south Wales, *Bulletin of the Board of Celtic Studies* 20, 440–449
Crew, P. and S. (eds) 1990 *Early Mining in the British Isles*, Plas Tan y Bwlch Occasional Paper no. 1

Crowther, J. 1987 Soil phosphorus, in Burnham and Burnham 1987a, 13

Crowther, J. 1997 Soil phosphate surveys: critical approaches to sampling, analysis and interpretation, *Archaeological Prospection* 4, 93–102

Crowther, J. and Barker, P. 1995 Magnetic susceptibility: distinguishing anthropogenic effects from the natural, *Archaeological Prospection* 2, 207–215

Crummy, P. 1992 Royal graves, *Colchester Archaeologist* 5 (1991–2), 1–5

Crummy, P. 1997 *City of Victory: the Story of Britain's First Town*, Colchester Archaeological Trust

Cunliffe, B. W. 1991 *Iron Age Communities in Britain* (3rd edition), London and New York

Curchin, L. A. 1991 *Roman Spain. Conquest and Assimilation*, London

Darlington, G. and Noake, P. 1989 *Vernacular Buildings Survey: Dolaucothi Arms and Outbuildings*, National Trust, unpublished report

Davies, J. L. 1980 Roman Military Deployment in Wales and the Marches from Claudius to the Antonines, in Hanson, W. S. and Keppie, L. J. F. (eds) *Roman Frontier Studies 1979*, BAR International Series 71, Oxford, 255–278

Davies, J. L. 1984 Excavations at Trawscoed Roman fort, Dyfed, *Bulletin Board of Celtic Studies* 31, 258–292

Davies, J. L. 1991a The Roman military deployment in Wales and the Marches from Pius to Theodosius, in Maxfield, V. A. and Dobson, M. J. (eds) *Roman Frontier Studies 1989: Proceedings of the XVth International Congress of Roman Frontier Studies*, Exeter, 52–57

Davies, J. L. 1991b Military *Vici*: Recent research and its significance, in Burnham and Davies 1991, 65–74

Davies, J. L. 1994 The Roman Period, in Davies and Kirby 1994, 275–317.

Davies, J. L. and Hogg, A. H. A. 1994 The Iron Age, in Davies and Kirby 1994, 219–274

Davies, J. L. and Kirby, D. (eds) 1994 *The Cardiganshire County History Vol. 1: from the Earliest Times to the Coming of the Normans*, Cardiff

Davies, O. 1935 *Roman Mines in Europe*, Oxford

Davies, O. 1936 Finds at Dolaucothi, *Archaeologia Cambrensis* 91, 51–57

Davison, D. P. 1989 *The Barracks of the Roman Army from the 1st to 3rd Centuries AD*, BAR International Series 472, Oxford

Dick, W. A. and Tabatabai, M. A. 1977 An alkaline oxidation method for the determination of total phosphorus in soils, *Journal of Soil Science Society of America* 41, 511–514

Dillon, J. 1989 A Roman timber building from Southwark, *Britannia* 20, 229–231

Dimbleby, G. W. 1968 Pollen report, in Hogg, A. H. A. Pen Llystyn: a Roman fort and other remains, *Archaeological Journal* 125, 188–190

Domergue, C. 1990 *Les Mines de la Péninsule Ibérique dans l'Antiquité Romaine*, Collection de l'Ecole Française de Rome

Driver, T. 2003 Allt Aber-Mangoed, Cynwyl Gaeo, (nr. Pumsaint), SN 6774 4345, *Archaeology in Wales* 42 (2002), 96–7

Dutton, L. A. and Fasham, P. J. 1994 Prehistoric copper mining on the Great Orme, Llandudno, Gwynedd, *Proceedings of the Prehistoric Society* 60, 245–286

Elliot, J. W. and Henig, M. 1982 Engraved gemstones from the Roman frontier post at Newstead, Roxburghshire, *Proceedings of the Scottish Antiquarian Society* 112, 295–299

Evans, C. J., Jones, L. and Ellis, P. 2000 *Severn Valley Ware Production at Newland Hopfields: Excavation of a Romano-British Kiln Site at North End Farm, Great Malvern, Worcestershire in 1992 and 1994*, BAR British Series 313, Oxford

Evans, D. H. 1979 Gravel-tempered ware: a survey of forms, *Medieval and Later Pottery in Wales* 2, 18–29

Evans, G. 1996 The Clay Pipes, in Brennan, D., Evans, G., James, H. and Dale-Jones, E. 1996 Excavations in Carmarthen, Dyfed 1976–1990: Finds from the Seventeenth to the Nineteenth Centuries: Pottery, Glass, Clay Pipes and Bone, *Medieval and Later Pottery in Wales* 14, 63–104

Evans, G. Eyre 1909–10 Llystroyddyn Urn [Dolaucothi], *Transactions of the Carmarthenshire Antiquarian Society* 5 (1909–10), 75 and plate

Evans, G. Eyre 1921–22 Journal of a Tour in Carmarthenshire, *Transactions of the Carmarthenshire Antiquarian Society* 15 (1921–22), 14–18

Faegri, K. and Iversen, J. 1989 *Textbook of Pollen Analysis*, IV Edition K. Faegri, P. E. Kaland and K. Krzywinski, Chichester

Fenton, R. 1917 *Tours in Wales 1804–1813* (edited from the MS Journals in the Cardiff Free Library by J. Fisher), Cambrian Archaeological Association

Frere, S. S. 1972 *Verulamium Excavations, Volume I*, Society of Antiquaries Research Report 28, London

Frere, S. S. 1983 *Verulamium Excavations, Volume II*, Society of Antiquaries Research Report 41, London

Frere, S. S. and Wilkes, J. J. 1989 *Strageath: Excavations within the Roman Fort 1973–86*, Britannia Monograph 9, London

Funari, P. P. A. 1996 *Dressel 20 Inscriptions from Britain and the Consumption of Spanish Olive Oil*, BAR British Series 250, Oxford

Gage, J. 1834 The Bartlow Hills, Ashdon, Essex, with an account of Roman sepulchral remains recently discovered in the lesser barrows, *Archaeologia* 25, 1–23

Gerrard, G. A. M. 1988 Pillow mounds in the Dolaucothi mining complex, in Austin, D. Excavation and survey at Bryn Cysegrfan, Llanfair Clydogau, Dyfed, *Medieval Archaeology* 32, 161–165

Gerrard, G. A. M. 2000 *The Early British Tin Industry*, Charleston

Gillam, J. P. 1970 *Types of Roman Coarse Pottery Vessels in Northern Britain* (3rd edition), Newcastle upon Tyne

Gillam, J. P. 1976 Coarse fumed ware in north Britain and beyond, *Glasgow Archaeological Journal* 4, 57–80

Gilles, K-J. 1994 Rhénanie, in Petit *et al.* 1994, 268–86

Glasbergen, W. 1972 *De Romeinse Castella te Valkenburg Z. H.*, Groningen

Graham, I. D. G. and Scollar, I. 1976 Limitations on magnetic prospection in archaeology imposed by soil properties, *Archaeo-Physika* 6, 1–124

Grant, A. 1983 *North Devon Pottery: the Seventeenth Century*, University of Exeter

Green, H. J. M. 1957 Romano-British hoard from Godmanchester, *Proceedings of the Cambridge Antiquarian Society* 50, 23–29

Green, M. 1986 *The Gods of the Celts*, Gloucester

Greene, K. T. 1979 *Report on the Excavations at Usk, 1965–76. The Pre-Flavian Fine Wares*, Cardiff

Grove J. M. 1988 *The Little Ice Age*, London

Guido, M. 1978 *The Glass Beads of the Prehistoric and Roman Periods in Britain and Ireland*, Society of Antiquaries Research Report 35, London

Hall, G. 1988 A suggested solution to some problems of the Dolaucothi Gold Mines site, South Wales, *Archaeology in Wales* 27 (1987), 67

Hall, G. W. 1993 *Metal Mines of Southern Wales* (2nd edition), Kington

Hamond, F. W. 1983 Phosphate analysis of archaeological sediments, in Reeves-Smyth, T. and Hamond, F. W. (eds) *Landscape and Archaeology in Ireland*, BAR British Series 116, Oxford, 47–80

Hanson, W. S. 1978 The Roman military timber-supply, *Britannia* 9, 293–306

Hanson, W. S. 1987 Elginhaugh, *Current Archaeology* 104, 268–272

Harden, D. B. 1967 The Glass Jug, in Biddle, M. Two Flavian Burials from Grange Road, Winchester, *Antiquaries Journal* 47, 238–240

Harden, D. B. and Price, J. 1971 The Glass, in Cunliffe, B. W. *Excavations at Fishbourne 1961–1969, vol 2, The Finds*, Society of Antiquaries Research Report 27, London, 317–368

Harden, D. B. 1987 *Glass of the Caesars*, Milan

Harrison, F. A. 1931 Ancient mining activities in Portugal, *The Mining Magazine* 45, 137–145

Hartley, K. F. and Webster, P. V. 1973 The Romano-British pottery kilns near Wilderspool, *Archaeological Journal* 130, 77–103

Harvey, N. 1970 *History of Farm Buildings in England and Wales*, Newton Abbot

Haverfield, F. J. 1909–10 Roman Dolaucothy, *Transactions of the Carmarthenshire Antiquarian Society* 5 (1909–10), 14–15

Healy, J. F. 1978 *Mining and Metallurgy in the Greek and Roman World*, London

Henig, M. 1978 *A Corpus of Roman Engraved Gemstones from British Sites*, BAR British Series 8 (2nd edition), Oxford

Heywood, B, and Marvell, A. 1992 Excavations at Neath, *Bulletin of the Board of Celtic Studies* 39, 171–298

Hillam, J. 1998 *Guidelines for Dendrochronology*, English Heritage, London

Hillam, J., Morgan R. A. and Tyers I. 1987 Sapwood estimates and the dating of short ring sequences, in Ward, R. G. W. (ed.) *Applications of tree-ring studies: current research in dendrochronology and related areas*, BAR International Series 333, Oxford, 165–185

Hobley, B. 1982 Roman military structures at 'The Lunt' Roman fort: experimental simulations 1966–1977, in Drury, P. J. (ed.) *Structural Reconstruction: Approaches to the Interpretation of the Excavated Remains of Buildings*, BAR British Series 110, Oxford, 227–274

Hodges, H. 1974 *Technology in the Ancient World*, London

Hodgson, N. 1999 Wallsend – *Segedunum*, in Bidwell 1999, 83–94

Hogg, A. H. A. and King, D. J. C. 1963 Early castles in Wales and the Marches: a preliminary list, *Archaeologia Cambrensis* 112, 77–124

Hook, D. 1971 Excavations at Caio, Carms., *Carmarthenshire Antiquary* 6 (1970), 71–84

Hopewell, D. 2002 The Roman fort environs project, *Archaeology in Wales* 41 (2001), 99–101

Horn, W. 1975 Water power and the plan of St Gall, *Journal of Medieval History* 1, 219–257

Hurst, J. D. (ed.) (forthcoming) *Roman Droitwich: Dodderhill Fort, Bays Meadow Villa, and a Roadside Settlement*, English Heritage

Isings, C. 1957 *Roman Glass from Dated Finds*, Groningen

Jacobs, J. 1912 Sigillatafunde aus einen römische Keller zu Bregenz, *Jahrbuch für Altertumskunde* 6 (1912), 172–184 and taf. 1–5

James, H. 1991 The Roman Roads of Carmarthenshire, in James, H. (ed.) *Sir Gâr: Studies in Carmarthenshire History*, 53–77

James, H. 1992 Excavations in Roman Carmarthen 1978–1990, *Carmarthenshire Antiquary* 28, 5–36

James, H. 2000 Roman Carmarthenshire, *Carmarthenshire Antiquary* 36, 23–46

James, H. 2003 *Roman Carmarthen: Excavations 1978–1993*, Britannia Monograph 20, London

James, H., Murphy, K. and Davies, J. L. 1983 Llandovery Roman Fort, *Archaeology in Wales* 23, no. 73, 34

James, H. and Williams, G. 1982 Rural settlement in Roman Dyfed, in Miles, D. (ed.) *The Romano-British Countryside: Studies in Rural Settlement and Economy*, BAR British Series 103, Oxford

Jarrett, M. (ed.) 1969 *The Roman Frontier in Wales* (2nd Edn), Cardiff

Johnson, A. 1983 *Roman Forts*, London

Jones, G. D. B. 1971 Fieldwork and Air Photography in Carmarthenshire, *Carmarthenshire Antiquary* 7, 3–40

Jones, G. D. B. 1980 The Roman mines at Riotinto, *Journal of Roman Studies* 70, 146–165

Jones, G. D. B, Blakey, I. J. and Macpherson, E. C. F. 1960 Dolaucothi: the Roman aqueduct, *Bulletin of the Board of Celtic Studies* 19, 71–84

Jones, G. D. B. and Lewis, P. R. 1971a The Dolaucothi Gold-mines, *Bönner Jahrbücher* 171, 288–300

Jones, G. D. B. and Lewis, P. R. 1971b *The Roman Gold Mines at Dolaucothi*, Carmarthen

Jones, G. D. B. and Little, J. H. 1973 Excavations on the Roman fort at Pumsaint, Carmarthenshire: Interim Report 1972, *Carmarthenshire Antiquary* 9, 3–27

Jones, G. D. B. and Little, J. H. 1974 Excavations at Pumsaint 1973: Interim Report, *Carmarthenshire Antiquary* 10, 3–16

Jones, G. D. B. and Maude, K. 1991 Dolaucothi: the dating problem, in Burnham and Davies 1991, 169–171

Jones, G. R. J. 1972 Post-Roman Wales, in Finberg, H. P. R. (ed.) *Agrarian History of England and Wales, Vol. 1, Part 2, AD 43–1042*, Cambridge

Jones, M. J. 1975 *Roman Fort-Defences to AD 117*, BAR British Series 21, Oxford

Jones, R. F. J. and Bird, D. G. 1972 Roman gold-mining in north-west Spain II: workings on the rio Duerna, *Journal of Roman Studies* 62, 59–74

Knight, H. H. 1856 On the methods employed by the ancients in working gold mines, *Archaeologia Cambrensis* 3, ii, 199–210, 264

Knorr, R. 1919 *Töpfer und Fabriken verzierter Terra-Sigillata des ersten Jahrhunderts*, Stuttgart

Knorr, R. 1952 *Terra-Sigillata-Gefässe des ersten Jahrhunderts mit Töpfernamen*, Stuttgart

Lamb, H. H. 1977 *Climate: present, past and future*, Volume 2, London

Lamb, H. H. 1982 *Climate, history and the modern world*, London

Landmark Surveys and Cambria Archaeology 1999 *Dolaucothi Gold Mines, Carmarthenshire*, unpublished report by Cambria Archaeology for The National Trust, March-October 1999

Lewis, M. J. T. 1997 *Millstone and Hammer: the Origins of Water Power*, University of Hull

Lewis, N. and Reinhold, M. 1966 *Roman Civilisation Sourcebook II: the Empire*, New York

Lewis, P. R. 1977 The Ogofau Roman Gold Mines at Dolaucothi, *National Trust Yearbook 1976–77*, 16pp

Lewis, P. R. and Jones, G. D. B. 1969 The Dolaucothi Gold Mines 1: The surface evidence, *Archaeological Journal* 49, 244–272

Lewis, P. R. and Jones, G. D. B. 1970 Roman gold-mining in north-west Spain, *Journal of Roman Studies* 60, 169–185

Locock, M. 1996 Bethel Square, Brecon: excavations in the Medieval Town, *Brycheiniog* 28 (1995–6), 35–79

Lowe, J. J. 1982 Three Flandrian pollen profiles from the Teith Valley, Perthshire, Scotland. II. Analysis of deteriorated pollen, *New Phytologist* 90, 371–385

Lowe, J. J. and Walker, M. J. C. 1986 Lateglacial and early Flandrian environmental history of the Isle of Mull, Inner Hebrides, Scotland, *Transactions of the Royal Society of Edinburgh: Earth Science* 77, 1–20

Lucas, J. 1981 Tripontium, third interim report, *Transactions of the Birmingham Archaeological Society* 91, 25–54

Maaskant-Kleibrink, M. 1978 *Catalogue of the Engraved Gems in the Royal Coin Cabinet, The Hague. The Greek, Etruscan and Roman Collections*, The Hague

McCarthy. M. R. 1991 *Roman Waterlogged Remains at Castle Street, Carlisle*, Cumberland and Westmorland Archaeological Society Research Series no. 5, Stroud

Manley, G. 1974 Central England temperatures: monthly means 1659–1973, *Quarterly Journal of the Royal Meteorological Society* 100, 389–406

Manning, W. H. 1968 The Dolaucothi Gold Mines, *Antiquity* 42, 299–302

Manning, W. H. 1975 Roman military timber granaries, *Sonderdruck aus dem Saalburg Jahrbuch* 22, 105–129

Manning, W. H. 1985 *Catalogue of the Romano-British Iron Tools, Fittings and Weapons in the British Museum*, London

Manning, W. H. 1993 *Report on the Excavations at Usk 1965–75. The Roman Pottery*, Cardiff

Manning, W. H. 2001 *Roman Wales: a Pocket Guide*, Cardiff

Manning, W. H. and Scott, I. R. 1979 Roman timber military gateways in Britan and on the German frontier, *Britannia* 10, 19–62

Margary, I. D. 1973 *Roman Roads in Britain* (3rd edition), London

Marvell, A. G. and Owen-John, H. S. 1997 *Leucarum: Excavations at the Roman Auxiliary Fort at Loughor, West Glamorgan 1982–84 and 1987–88*, Britannia Monograph 12, London

Mees, A. 1995 *Modelsignierte Dekorationen auf Sudgallischer Terra Sigillata*, Forschungen und Berichte zur Vor- und Fruhgeschicthte in Baden-Wurttenberg, band 54

Milne, G. 1985 *The Roman Port of London*

Moore, P. D. and Chater, E. H. 1969 The changing vegetation of west-central Wales in the light of human history, *Journal of Ecology* 57, 361–379

Moore, P. D., Webb, J. A. and Collinson, M. E. 1991 *Pollen Analysis*, (2nd edition), London

Murchison, R. I. 1839 *The Silurian System*, London

Murphy, K. 1995 *Water Mains Renewal at Pumpsaint, Carmarthenshire*, Dyfed Archaeological Trust Report, PRN 30673

Mytum, H. 1999 Castell Henllys, *Current Archaeology* 161, 164–172

Nash-Williams, V. E. 1932 The Roman legionary fortress at Caerleon in Monmouthshire. Report on the excavations carried out in Prysg Field, 1927–29. Part III, *Archaeologia Cambrensis* 87, 237–315

Nash-Williams, V. E. 1952 The Roman gold-mines at Dolaucothi (Carm), *Bulletin of the Board of Celtic Studies* 14, 79–84

Nelson, T. R. H. 1944 Gold mining in south Wales, *Mine and Quarry Engineering* 9, Jan, Feb, March issues, 3–10, 33–38, 55–60

Newman, R. and Wilkinson, P. 1997 Excavations at Llanmaes, near Llantwit Major, South Glamorgan, *Post-Medieval Archaeology* 30 (1996), 187–234

Niblett, R. 1999 *The Excavation of a Ceremonial Site at Folly Lane, Verulamium*, Britannia Monograph 14, London

Noël Hume, I. 1961 The Glass Bottle in Colonial Virginia, *Journal of Glass Studies* 3, 91–118

Nuber, H.-U. 1983 *Die Römischen Kastella bei Hofheim am Taunus, Main-Taunus-Freis. Neue Ausgrabungen 1969–1981 im römischen Steinkastell Hofheim*, Wiesbaden

O'Mahoney, C. 1985a West Wales fabrics, an interim note, *Medieval and Later Pottery in Wales* 8, 20–24

O'Mahoney, C 1985b Part II, the Pottery, in Murphy, K. and O'Mahoney, C. Pottery from the Tudor Merchant's House, Tenby, *Medieval and Later Pottery in Wales* 8, 26–39

O'Mahoney, C. 1995 *Pottery, Ridge Tile and Ceramic Waterpipe: Excavations at Carmarthen Greyfriars 1983–1990*, Dyfed Archaeological Trust Occasional Papers 2 (Electronic Version 1998, www.acadat.com/projects/cgfpot.pdf)

Opie, I. and Tatem, M. 1989 *A Dictionary of Superstitions*, Oxford

Page, N. 2003 Pumsaint, Coronation Hall, SN 6565 4055, *Archaeology in Wales* 42 (2002), 115

Pannuti, U. 1983 *Museo Archeologico Nazionale de Napoli. Catologo della Collezione Glittica I*, Rome

Papazian C. and Campbell E. 1992 Medieval Pottery and Roof Tiles in Wales AD 1100–1600, *Medieval and Later Pottery in Wales* 13 (whole volume)

Peacey, A. A. 1985 The Clay Pipes, in Shoesmith, R. (ed.) *Hereford City Excavations vol 3: The Finds*, CBA Reseach Report 56, London

Peacock, D. P. S. 1967 Romano-British pottery production in the Malvern district of Worcestershire, *Transactions of the Worcestershire Archaeological Society* 1 (1965–7), 15–28

Peacock, D. P. S. and Williams, D. F. 1986 *Amphorae and the Roman Economy*, London

Perring, D. 1989 Cellars and cults in Roman Britain, *Archaeological Journal* 146, 279–301

Petit, J-P., Mangin, M. and Brunella, P. (eds) 1994 *Les Agglomérations Secondaires: La Gaule Belgique, les Germanies et l'Occident romain*, Paris

Petit, J-P and Schaub, J. 1995 *Bliesbruck Reinheim: Parc*

Archéologique Européen, Guides Archéologiques de la France

Price, J. 1977 The Roman Glass, in Gentry, A. *et al.* Excavations at Lincoln Road, London Borough of Enfield, November 1974–March 1976, *Transactions of the London and Middlesex Archaeological Society* 28, 154–161

Price, J. 1978 Trade in Glass, in du Plat Taylor, J. and Cleere, H. (eds) *Roman Shipping and Trade: Britain and the Rhine Provinces*, CBA Research Report 24, 70–78, London

Proudfoot, B. 1976 The analysis and interpretation of soil phosphorus in archaeological contexts, in Davidson, D. A. and Shackley, M. L. (eds) *Geoarchaeology*, London, 93–113

RCAHMW 1917 *An Inventory of the Ancient Monuments of Carmarthenshire*, London

Rees, W. 1873 Loventium, *Archaeologia Cambrensis* 4, iv, 113–138

RIB = Collingwood, R.G. and Wright, R.P. 1965 *The Roman Inscriptions of Britain*, Oxford

RIC = Mattingly, H. and Sydenham, E. A. 1926, *The Roman Imperial Coinage, Vol. II* London

Richter, G. M. A. 1970 *The Sculpture and Sculptors of the Greeks* (4th edition), New Haven and London

Rivet, A. L. F. and Smith, C. 1979 *The Place-Names of Roman Britain*, London

Rogers, G. B. 1974 *Poteries Sigillés de la Gaule Centrale, Volume 1: Les Motifs Non Figurés*, 28th Supplement of *Gallia*, Paris

Round, A. 1992 Excavations on the Mansio site at Wall (Staffordshire), 1972–78, *South Staffordshire Archaeological and Historical Soc*iety 32 (1990–91), 1–78

Roussel, L. 1994 Mâlain – Mediolanum, in Bénard *et al.* 1994, 63–78

Sanchez-Palencia, F.-J. 1984–5 Los 'morteros' de Fresnedo (Allande) y Cecos (Ibias) y los lavaderos de oro romanos en el noroeste de la Península Ibérica, *Zephyrus* 37/8, 349–59

Sanchez-Palencia, F.-J. 1989 La explotacion del oro en la Hispania romana: sus inicios y precedentes, in Domergue, C. (ed.) *Mineria y Metallurgia en las Antiquas Civilizaciones Mediterraneas y Europeas* (Madrid), 35–49

Scollar, I., Tabbagh, A., Hesse, A. and Herzog, I. 1990 *Archaeological Prospecting and Remote Sensing*, Cambridge

Sena Chiesa, G. 1966 *Gemme del Museo Nazionale di Aqileia*, Associazione Nazionale per Aquileia

Sim, D. and Ridge, I. 2002 *Iron for the Eagles. The Iron Industry of Roman Britain*, Stroud

Smith, A. G. and Cloutman, E. W. 1988 Reconstruction of Holocene vegetation history in three dimensions at Waun-Fignen-Felen, an upland site in South Wales, *Philosophical Transactions of the Royal Society of London* B322, 159–219

Smith, A. G. and Green, C. A. 1995 Topogenous peat development and the late-Flandrian vegetation history at a site in upland South Wales, *The Holocene* 5, 172–183

Smith, R. F. 1987 *Roadside Settlements in Lowland Roman Britain*, BAR British Series 157, Oxford

Smyth, W. W. 1846 Note on the Gogofau or Ogofau mine near Pumsaint, Carmarthenshire, *Memoirs of the Geological Survey* (1846), 480–484

Sommer, C. S. 1984 *The Military Vici in Roman Britain: aspects of their origins, their location and layout, administration, function and end*, BAR British Series 129, Oxford

Stace, C. 1991 *New Flora of the British Isles*, Cambridge

Stansfield, J. A. and Simpson, G. 1958 *Central Gaulish Potters*, London

Stephens, G. R. 1985 Military aqueducts in Roman Britain, *Archaeological Journal* 142, 216–236

Stuiver, M. and Becker, B. 1993 High-precision decadal calibration of the radiocarbon time scale AD 1950–6000 BC, *Radiocarbon* 35, 35–65

Stuiver, M. and Reimer, P. J. 1993 Extended ^{14}C data base and revised CALIB 3.0 ^{14}C age calibration program, *Radiocarbon* 35, 215–230

Symonds, R. P. 1990 The problem of roughcast beakers and related fine wares, *Journal of Roman Pottery Studies 3*, 1–17

Thompson, F. H. 1976 The excavation of the Roman Amphitheatre at Chester, *Archaeologia* 105, 127–239

Timberlake, S. 2003 *Excavations on Copa Hill, Cwmystwyth (1986–1999). An Early Bronze Age Copper Mine within the uplands of Central Wales*, BAR British Series 348, Oxford

Tite, M. S. 1972 The influence of geology on the magnetic susceptibility of soils on archaeological sites, *Archaeometry* 14, 229–236

Tite, M. S. and Mullins, C. 1971 Enhancement of magnetic susceptibility of soils on archaeological sites, *Archaeometry* 13, 209–219

Toller, H. 1993 A Roman Road between Llandovery and Pumpsaint, unpublished report in Dyfed Archaeological Trust's *Sites and Monuments Record* and in the National Monuments Record, Aberystwyth

Turner, J. 1964 The anthropogenic factor in vegetational history 1: Tregaron and Whixall mosses, *New Phytologist* 63, 73–89

Turner, J. 1965 A contribution to the history of forest clearance, *Proceedings of the Royal Society* B161, 343–354

Vitali, D. and Zvald, V. 1999 Évolution du bâti: techniques locales de construction et influences méditerranéennes, in Gruel, K. and Vitali, D. (eds) *L'Oppidum de Bibracte: un bilan de onze années de recherche (1984–1995)*, *Gallia* 55 (1998), 34–43

Wade Martins, S. 1991 *Historic Farm Buildings*, London

Wahl, J. 1993 Três Minas: Vorbericht über die archäologischen Ausgrabungen im Bereich des römischen Goldbergwerks 1986–87, in Steuer, H. and Zimmerman, U. (eds) *Montanarchäologie in Europa*, Sigmaringen, 123–52

Walker, M. J. C. 1989 Pollen analyses and radiocarbon dates, in Austin *et al.* 1989, 179–189

Walker, M. J. C. 1993 Flandrian vegetation change and human activity in the Carneddau region of upland mid-Wales, in Chambers, F. M. (ed.) *Climatic Change and Human Impact on the Landscape*, London, 171–185

Wallace, C. and Webster, P. V. 1989 Jugs and lids in Black-burnished ware, *Journal of Roman Pottery Studies* 2, 88–91

Webster, J. and Webster, P. V. 1998 Second-century pottery from Caerleon derived from metal and samian prototypes, in Bird, J. (ed.) *Form and Fabric: Studies in Rome's Material Past in Honour of B. R. Hartley*, Oxbow Monograph 80, Oxford, 249–262

Webster, P. V. 1976 Severn Valley Ware: a preliminary study, *Transactions of the Bristol and Gloucestershire Archaeological Society* 94, 18–46

Webster, P. V. 1981 Cardiff Castle Excavations, *Morgannwg* 25, 201–211

Webster, P. V. 1991 Pottery and trade in Roman Wales, in Burnham and Davies 1991, 138–149

Webster, P. V. 1992 Roman Pottery in South East Wales, *Journal of Roman Pottery Studies* 5, 111–121

Westell, W. P. 1931 A Romano-British Cemetery at Baldock, Herts, *Archaeological Journal* 88, 247–301

White, K. D. 1984 *Greek and Roman Technology*, London

White, R. B. 1978 Excavations at Brithdir, near Dolgellau, 1974, in Boon, G. C. (ed.) *Monographs and Collections 1: Roman Sites*, Cambrian Archaeological Association

Wiliam, E. 1982 *Traditional Farm Buildings in North-East Wales, 1550–1900*, Cardiff

Wiliam, E. 1986 *The Historical Farm Buildings of Wales*, Cardiff

Wiliam, E. 1988 *Home-made Homes: Dwellings of the Rural Poor in Wales*, Cardiff

Williams, E. 1818 Historical account, or an inquiry into the situation of the Gold Mines of the Ancient Britons, *Cambrian Register* 3, 31–58

Williams, G. 1988 Recent work on rural settlement in later prehistoric and early historic Dyfed, *Antiquaries Journal* 68, 30–54

Williams, G. 2002 Excavations at Bryn Maen Caerau, Cellan, Ceredigion, *Archaeology in Wales* 41 (2001), 10–20

Williams, G. and Mytum, H. (Blockley, K. ed.) 1998 *Llawhaden, Dyfed: Excavation on a Group of Small Defended Enclosures, 1980–84*, BAR British Series 275, Oxford

Woodforde, J. 1983 *Farm Buildings*, London

Woodiwiss, S. (ed.) 1992 *Iron Age and Roman Salt Production and the Medieval Town of Droitwich*, CBA Research Report 81, London

Worth, R. H. 1967 *Worth's Dartmoor* (compiled from his published works and edited by G. M. Spooner and F. S. Russell), Newton Abbot

Young, C. J. 1977 *Oxfordshire Roman Pottery*, BAR British Series 43, Oxford

Zienkiewicz, J. D. 1992 Pottery from the excavations on the site of the Roman Legionary Museum, Caerleon, 1983–5, *Journal of Roman Pottery Studies* 5, 81–109